The Daily Telegraph
DICTIONARY
OF
MILITARY
QUOTATIONS

The Daily Telegraph
DICTIONARY
OF
MILITARY
QUOTATIONS

Edited by
Peter G. Tsouras

GREENHILL BOOKS, LONDON

Greenhill Books

The Daily Telegraph Dictionary of Military Quotations

First published 2005 by Greenhill Books/Lionel Leventhal Ltd
Park House, 1 Russell Gardens, London NW11 9NN

Copyright © Peter G. Tsouras 1992, 2000, 2005

The right of Peter G. Tsouras to be identified as the author of this work has been asserted by him in accordance with the Copyright, Designs and Patents Act 1988.

All rights reserved. No part of this publication may be reproduced, stored in or introduced into a retrieval system, or transmitted, in any form, or by any means (electronic, mechanical, photocopying, recording or otherwise), without the prior written permission of the publisher. Any person who does any unauthorised act in relation to this publication may be liable to criminal prosecution and civil claims for damages.

British Library Cataloguing-in Publication Data
The Daily Telegraph dictionary of military quotations
1. War – Quotations, maxims, etc.
2. Military art and science – Quotations, maxims, etc.
I. Tsouras, Peter
II. Daily Telegraph
III. Dictionary of military quotations
355'.02

ISBN 1-85367-666-7

Publishing History

The Daily Telegraph Dictionary of Military Quotations is a substantially revised and updated version of *The Greenhill Dictionary of Military Quotations* (2000), which in turn was a revised, updated and expanded version of *Warriors' Words: A Quotation Book* (Arms & Armour Press, 1992).

For more information on our books, please visit www.greenhillbooks.com, email sales@greenhillbooks.com, or telephone us within the UK on 020 8458 3614. You can also write to us at the above London address.

Typeset by Palindrome

Printed and bound in the USA

Contents

Preface	11	Boldness	53
		Bravery	55
The Quotations		Bureaucracy	56
A		**C**	
Ability	21	Calculation	58
Action	21	Camp Followers	59
Activity	22	The Captain	59
Adaptability/Adaptation	23	Casualties	59
Administration	24	Cavalry	62
Admirals	25	Censorship	65
Adversity	25	Centre of Gravity	65
Advice	26	Chance	66
Aggression	27	Change	67
Aggressiveness	27	Chaplains	68
Airborne	28	Character	69
Air Power	28	Chief of Staff	71
Alliances/Allies/Coalition Warfare	29	Chivalry	72
Amateurs	31	Citizen Soldiers	74
Ambition	33	Civilian Control of the Military	75
Analysis	34	Civil-Military Relations	76
Appeasement	34	Cohesion	77
Army		The Colours	78
The Nature of Armies	35	Combined Arms	80
Peacetime Army	37	Command	81
Standing Army	38	The Commander	83
Artillery	38	Commanding General/Commander-in-Chief	85
The Art of War	39		
Attack	42	Common Sense	87
Attrition	43	Communication	87
Audacity	44	Competence	87
Authority	45	Comradeship and The Band of Brothers	88
B			
Battle		Concentration	90
The Cost of Battle	46	Confidence	92
The Dynamics of Battle	46	Conscience	94
The Human Factor	48	Conscription	95
The Purpose of Battle	49	Contact	96
Battlefield	50	Contractors	97
The Bayonet	51	A Cool Head	97
Blitzkrieg	52	Council of War	98
Bloodshed	53	*Coup d'Oeil*	99

CONTENTS

Courage		Example	158
General	100	Exercises/Manoeuvres	163
Moral Courage	103	Exhortations	163
Physical Courage	105	Experience	
Court Martial	106	General	166
Cowardice/Cowards	106	Lessons Learned	167
Criticism	107	Exploitation	169

D

Danger	109
Daring	110
Death	110
Deception	112
Decision/Decisiveness	113
Declaration of War	115
Defeat	116
Defence	119
Defiance	119
Delay	121
Delegation of Authority	121
Demoralisation	122
Desertion/Deserters	123
Destiny	123
Destruction of the Enemy	124
Determination	125
Diplomacy/Negotiations/Treaties	126
Discipline	
The Effects of Discipline	127
Instilling Discipline	128
The Lack of Discipline	129
The Nature of Discipline	130
Divine Favour	134
Doctrine	135
Dogma	135
Do or Die	136
Drill and Ceremonies	140
Dulce et Decorum Pro Patria Mori	141
Duty	142

E

Economy of Force (Principle of War)	146
Endurance	147
The Enemy	
General	148
Know Your Enemy	150
Respect Your Enemy	152
Energy	154
Engineers/Sappers	154
Envelopment	155
Epitaphs and Memorials	156
Esprit de Corps	158

F

Failure	170
Fallen Comrades	170
Fallen Sons	171
Fame	172
Familiarity	174
'Famous Last Words'	174
Fate	175
Fatigue and Rest	176
Fear	178
Fighting	180
Firepower	182
First Battle	183
Flexibility	185
Fog of War	186
Follow-up	187
Folly	188
Force	188
Foresight/Anticipation	190
Fortifications	191
Fortune	191
Freedom	192
Friction	193
Friendly Fire/Fratricide	193
Funerals	194

G

Generals/Generalship	195
General Staff	199
Genius	200
The Gentleman	203
Glory	203
The Golden Bridge	206
Great Captains/Greatness	206

H

Health	208
Hero/Heroism	208
History	
General	209
Studying Military History	210
Writing Military History	213
Honour	214
Hope	216
Horses	217

CONTENTS

Humanity	217	Meeting Engagement	269
Humour	218	Memoirs/Diaries	269
Hunger	218	Mentor/Preferment	270

I

		Mercenaries	271
Idealism/Ideals	220	Militarism	272
Ignorance	220	Military Education	272
Imagination	221	Military Intelligence	275
Impedimenta	222	Military Justice	276
Incompetence	223	Military Life	277
Independence of Command	224	Military Music	278
The Indirect Approach	225	Military Science	279
Infantry	226	Military Service	279
Initiative		Mistakes	280
The Operational Dynamic	228	Mobilisation	282
The Personal Quality	229	Mobility/Movement	283
Innovation	232	Momentum	284
Inspiration	233	Moral Ascendancy/Moral Force	284
Instinct/Intuition	234	Morale	286
Integrity	234	Mothers	288
Intellect/Intelligence	236	Mutiny	290
Intrepidity	237		

N

Invention	237	Navy	291

J

		Necessity	291
Judgement	238	Neutrality	292
Junior Officers	239	Non-combatants	292
Justice	240	Non-commissioned Officers	293

K

		Numbers	294
Killing	240		

O

L

		Obedience	295
Last Words	242	The Objective/Maintenance of the	
The Leader	246	Aim (Principle of War)	296
Leadership	248	Obstacles	300
Lines of Communication	251	Offence and Defence	
Logisticians	251	The Interrelationship	300
Logistics	252	The Offensive (Principle of War)	302
Logistics and Generalship	253	The Officer	303
Looting/Pillage/Plunder/Rapine	254	Officers and Men	304
Loyalty	257	The Old School	308
Luck	259	Old Soldiers	308

M

		Operations	309
		Opinions	309
Man	260	Opportunity	310
Manoeuvre (Principle of War)	262	Order	312
Manhood/Manliness	263	Orders	312
March to the Sound of the Guns	263	Organisation	315
Marching	264	Originality and the Creative Mind	316
Marines	264		

P

Marksmanship	265		
Mass (Principle of War)	266	Pacification	317
Medals and Decorations	267	Pacifism/Anti-war Activism	318
Medical Corps	268	Panic	319

CONTENTS

Patience	320	Resistance to Reform	378
Patriotism	321	Resolution/Resolve	379
Pay	321	Responsibility/Accountability	380
Peace	323	Retirement	381
People and Army	324	Retreat	382
Perseverance	325	Revenge	384
Personal Presence of Commander	325	Revolution	385
Personal Risk of the Commander	329	Revolution in Military Affairs (RMA)	385
Physical Fitness	330	Risk	386
Planning/Plans	330	Rout	388
Politics and the Military	333	Ruse	388
Power	334		
Praise	334	**S**	
Prayers and Hymns	336	Sacrifice	389
Preparedness	338	Safety	391
The Press	340	Sailors/Seamen	392
The Principles of War	342	Salute	392
Prisoners of War (PoWs)	346	*Sang Froid*	393
The Profession of Arms	348	The School Solution	393
Promotion	352	Scorched Earth	394
Promptness	354	Sea Power	395
Propaganda	354	Secrecy	397
Public Opinion	356	Security (Principle of War)	398
Punishment and Rewards	356	Self-confidence	399
Pursuit	359	Self-control	400
Q		Sex	401
Quagmire	360	*Sic Transit Gloria Mundi*	401
Quality	360	Simplicity (Principle of War)	401
Quarter	361	Skill	402
Quotations and Maxims	361	Soldiers	402
R		Soldiers and Politicians	406
Rank	362	Speeches and the Spoken Word	407
Rashness	362	Speed	408
Readiness	362	Spirit	409
Reconciliation	363	Spit and Polish	410
Reconnaissance	364	Sport	411
Recruits/Recruitment	364	The Staff	411
Reflection/Meditation	365	Strategy	414
The Regiment	366	Strength	416
Regulars	367	Strike Weakness	416
Regulations	367	Subordinates	417
Reinforce Success	368	Success	418
Relief	368	Suicide	419
Religion	369	Superiority	420
Reminiscences	371	Surprise (Principle of War)	421
Reorganisation	373	Surrender	424
Replacements	374	The Sword and the Pen	427
Reports	374	**T**	
Reputation	375	Tactics	428
Reserves	377	Tactics and Strategy	430

CONTENTS

Talent for War	431
Tanks	432
Team/Teamwork	435
Tempo	435
Tenacity	436
Terrain	437
Terrorism	438
Theory and Practice	438
Those Left Behind	439
Time/Timing	441
Timidity	443
Tommy Atkins	444
Tradition	445
Training	
General	447
The Conduct of Training	448
Individual Training	450
Lack of Training	450
Physical Training	451
Small Unit Training	452
Train as You Fight	452
Treachery/Treason	453
The Troops	
Take Care of the Troops	453
Talk and Listen to the Troops	455
Truce	458
Trust	458
Truth	459

U

Uncertainty/The Unexpected/ The Unknown	460
Uniforms	460
Unity of Command (Principle of War)	461

V

Valour	463
Veterans	463
Victory	
General	464
The Consequences of Victory	465
The Elements of Victory	465
Generalship and Victory	466
The Necessity of Victory	466
Pyrrhic Victory	466
Reports of Victory	467
The Uses of Victory	468
Violence	468
Volunteers	469

W

War	
The Causes of War	470
Civil War	472
The Economy of War	472
Guerrilla/Partisan War	473
Just War	474
The Love of War	475
The Nature of War	477
The Objectives of War	479
Politics/Policy and War	480
Protracted War	481
The Results of War	482
War Correspondents	483
War Office/Pentagon	485
War on Terrorism	486
Weapons	487
Weather	488
Whitehall	489
Will	489
Wives	490
Women in War	492
Work	494
The Wounded	494
Wounds	495

Y

Youth and Age	496
Acknowledgements	499
Index	501

Preface
'I had it from Xenophon...'

In June 1758, the British Army was laying siege for the second time in that century to the great French fortress of Louisburg, the iron gate of Canada. Then Brigadier General James Wolfe hurried the French defenders back into their works with his skilled use of light infantry to turn one flank and then another. The rest of the British force was no less impressed with the manoeuvre than the dazed French now shut up in Louisburg. A brother officer, more perceptive than the rest, remarked that Wolfe's tactics reminded him of an account by Xenophon in the *Anabasis*, the March of the Ten Thousand Greeks. It is not recorded if Wolfe smiled. If he did, it was probably out of exasperation, as he replied, 'I had it from Xenophon, but our friends here are astonished at what I have done because they have read nothing.' (See p. 211.)

A century and a half later, another British soldier exemplified not the art of war, as Wolfe had at Louisburg, but the art of leadership and the nature of man in war. Private Richard Chuck was serving in Mesopotamia in 1916, in the company of the future Field Marshal Viscount Slim of Burma, then a young officer in one of the war-raised reserve battalions of his regiment. The soldier was a former Royal Naval seaman who had been discharged as an 'Incorrigible Rogue' just before the war and then had been conscripted into the Army; he was the bane of every NCO in the company, an affront to the neat and orderly side of Army life and seemingly indifferent to the reputation of his regiment, the Warwicks. In this he appeared little different from most of the soldiers in Slim's company, middle-aged family men who had not volunteered, but who had been conscripted and had not been well thought of for it. The company was assaulting Turkish positions along a dyke when machine-gun fire scythed through the ranks and mortars rained down on them. Half of the men went down, while the rest tried hopelessly to worm their way into the salt-pan desert. Slim saw doom stretching out its hand to his shrinking company. Then, despite the machine-gun fire and mortars, up stood the 'Rogue'. He bellowed, 'Heads up the Warwicks! Show the _____ yer cap-badges.' Slim then recounted:

> Even in that pandemonium a few men heard him as his great form came bullocking through them. They had no cap-badges for we wore Wolseley helmets, but they heard the only appeal that could have reached them – to the Regiment, the last hold of the British soldier when all else has gone. The half dozen around Chuck heard, their heads came up too; in a moment the whole line was running forward. We cursed the wire as it tore at us, but we were through, looking down on brown faces, grey uniforms, and raised hands and dirty white rags held aloft. [See p. 366.]

It is my hope that this book will help the aspiring soldier as well as the military professional and enthusiast, also to learn 'from Xenophon', the 'Incorrigible Rogue' and the almost 800 other individuals quoted. Theirs is much of the distilled wisdom

PREFACE

of 4,000 years of the profession of arms and experience of war, from the sack of Sumerian Ur c. 2000 BC to the present day. As conflicts from the end of the second millennium continue to smoulder, we dare not forget the eternal seriousness of war. This lament of Ur is not so different from more recent scenes in present-day Iraq, Rwanda and Kosovo.

> Dead men, not potsherds
> Covered the approaches,
> The walls were gaping,
> the high gates, the roads,
> were piled with dead.
> In the side streets, where feasting crows would gather,
> Scattered they lay.
> In all the streets and roadways bodies lay.
> In open fields that used to fill with dancers,
> they lay in heaps.
> The country's blood now filled its holes,
> like metal in a mould;
> Bodies dissolved – like fat left in the sun.*

There is a numbing sameness to such wails of horror throughout the history of war, but if war holds such constants in its left hand, it holds the new in its right. For war is ever the realm of the new and change; not to change passes you to the blood-soaked left hand of war, a grim reality check.

The great British military reformer, Major-General J.F.C. Fuller, saw in the slaughter of the Western Front the consequences of the refusal to recognize the need for change. He put his finger on an institutional cause.

> What for, indeed what for? Not to win a battle, for the impossibility of this is obvious to a rhinoceros. No; but to maintain the luxury of mental indolence in the head of some military alchemist. Thinking to some people is like washing to others. A tramp cannot tolerate a hot bath, and the average general cannot tolerate any change in preconceived ideas; prejudice sticks to his brain like tar to a blanket. [See p. 67]

The professionalisation of the armies of the English-speaking democracies have come a long way since a bitter Fuller penned this lament. However, one thing has remained a vital constant – the hard-fighting infantry, and in this the British are unsurpassed, though enthusiastically rivalled by their American, Australian, New Zealand, and Canadian cousins. Their hereditary enemies, the French, have grudgingly admitted it. To William of Poitiers in the Hundred Years War, 'These people, descended from the ancient Saxons (the fiercest of men) are always by nature eager for battle, and they could only be brought down by the greatest valour.' Or Louis XIV to one of his generals, 'Pay special heed to that part of the line that shall sustain the first shock of the English troops.' The Duke of Wellington gave the infantry the laurels for Waterloo. 'Our loss is immense particularly in the best of All Instruments, British Infantry. I never saw the Infantry behave so well.' (See pp 226–7.) Another great British captain, Field Marshal Montgomery, echoed the Iron

---

* Joan Oates, *Babylon* (Thames and Hudson, London, 1979), p. 52.

PREFACE

Duke, 'The least spectacular arm of the Army, yet without them you cannot win a battle. Indeed, without them, you can do nothing. Nothing at all, nothing.' (See p. 228.) A less exalted but just as accurate view was held by an anonymous British officer during the Iraq War, 'Quite frankly, the average British infantryman is far better... They're a tribe of feral monsters, but they're highly disciplined monsters. You don't want to get in their way.' (See p. 228.) And it was the British infantry that George Orwell had in mind when he reminded fat civilians of the source of their safety. 'We sleep safely in our beds because rough men stand ready in the night to visit violence on those who would harm us.' (See p. 340.)

This book of over 3,500 quotations, however, despite its reach across four millennia, is not the easy one-stop answer to the art of war. It is a place to start.

I have chosen over 400 subject areas for quotation. Basic terms, such as helmet and rifle, I have left to elementary experience and a general dictionary. The subject headings address concepts that are the essence of the profession of arms, the realm of ideas rather than objects, for example 'Change', 'The Indirect Approach' and 'War' itself, divided into eleven sub-categories. The human side of war is equally represented under headings such as 'The Regiment', 'Morale', 'Fear' and, let us never forget, 'Killing', 'Mothers', 'Those Left Behind', 'Wives' and more add the forgotten dimension of those who have suffered so much without even the warrior's passions as slight reward.

Included also are the matters of the soldier's heart, the living, spiritual dimensions of human beings in war – 'Cohesion', 'Comradeship', 'Fallen Comrades' and 'Religion'. King David's grief over fallen Jonathan lives on:

> How are the mighty fallen in the midst of the battle! O Jonathan, thou wast slain in thine high places.
> I am distressed for thee, my brother Jonathan: very pleasant has thou been unto me: thy love to me was wonderful, passing the love of women.
> How are the mighty fallen, and the weapons of war perished. [See p. 170.]

As does the desperate plea of Agamemnon rallying the Greeks, as Man-Killing Hector stormed their camp.

> Now be men, my friends! Courage, come, take heart!
> Dread what comrades say of you here in bloody combat!
> When men dread that, more men come through alive –
> when soldiers break and run, good-bye glory,
> good-bye all defenses. [See p. 88.]

And Shakespeare's rendition of Henry V's speech to his army before the Battle of Agincourt in 1415:

> We few, we happy few, we band of brothers;
> For he to-day that sheds his blood with me
> Shall be my brother. [See p. 88.]

Still bright also is the clear beauty of simple and utter faith, an inextricable part of 'The Nelson Touch':

PREFACE

When I lay me down to sleep, I recommend myself to the care of almighty God; when I awake I give myself up to his direction. Amidst all the evils that threaten me, I will look up to Him for help, and question not but that He will either avert them or turn them to my advantage. Though I know neither the time nor the place of my death, I am not at all solicitous about it, because I am sure that He knows them both, and that He will not fail to support and comfort me. [See p. 369.]

Here is a fraternity of soldiers, first and foremost. Most of them have long since turned to dust, but their deeds and experiences remain as vital and as insightful as when they trod the earth with thunder and acclaim. Follow them to their original sources – the histories, memoirs, textbooks and works of theory. Open an old book that smells, feels and looks of another time. Be your own Odysseus when he summoned the ghosts of Achilles and Agamemnon from Hades to speak to him. If you let them, they will speak across time to you of problems of leadership and command, of logistics, of tactics and strategy, the sinews of their art. They will speak to you of the human heart, comradeship, example, sacrifice, glory and death. They are not dead as long as they are remembered. And they are exhilarating company. If you are a serving soldier, the reward is not merely the satisfaction of a good read. It can also be, as Wolfe so aptly stated, the road to success, and at the price of a small butcher's bill.

Wherever possible, I have included the full quotation to provide the greatest understanding of the source's intent. All too often, the most glib part of a quotation has been lifted out of its context and turned around in its meaning. A good example is Admiral Farragut's admonition to one of his captains in battle, which is often cited as being a demand for the strict obedience to orders: 'I want none of this Nelson business in my squadron.' Here the reference is to Nelson at the Battle of Copenhagen, where he put his spyglass over the eyepatch covering his blind eye when looking at the signal to withdraw from action, saying, 'I have a right to be blind sometimes.' However, a more complete citation reveals Farragut to be very much in the Nelson tradition:

> On Farragut's quarterdeck: 'Captain, you begin early in your life to disobey orders. Did you not see the signal flying for near an hour to withdraw from action? ...I want none of this Nelson business in my squadron.'
> Later In Farragut's cabin: 'I have censured you, sir, on my quarterdeck, for what appeared to be a disregard of my orders. I desire now to commend you and your officers and men for doing what you believe right under the circumstances. Do it again whenever in your judgment it is necessary to carry out your conception of duty.' [See pp 313–14.]

The exercise of the profession of arms is the basic requirement for inclusion in this book. Into this fraternity, however, enter some of Athena's daughters – Joan of Arc with banner of fleurs-de-lis before the English forts surrounding Orleans:

> I used to say to them, 'go boldly in among the English,' and then I used to go boldly in myself. [See p. 160.]

But it would not be as warriors that women would first seriously partake of the man's business of war but as reformers of its most appalling failure to care properly

PREFACE

for the soldier's health in peace and war. Through sheer force of personality, sense of mission, and organisational efficiency Florence Nightingale was the driving force in the reformation of the health-care system of the British Army. As importantly, she taught the British officers corps to treat their soldiers like 'Christian men'. When society and their officers despised them, she observed,

> I have never seen so teachable & helpful a class as the Army generally.
> Give them opportunity promptly & securely to send money home and they will use it.
> Give them a schools & lectures & they will come to it.
> Give them books & a game & a Magic Lanthorn & they will leave off drinking.
> Give them suffering & they will bear it.
> Give them work & they will do it.
> I had rather to do with the Army generally than with any other class I have attempted to serve...
> If Officers would but think thus of their men, how much might be done for them. [See p. 404.]

The entire army mourned her passing as six sergeants carried her coffin to its grave.

Of necessity, only those who had the talent, time and inclination to write are represented here. Fate also plays a part: the fact that no first-hand account of Alexander survives is simply a great misfortune. While Alexander is quoted as far as the sources allow, it is nothing to what could have been included if his Royal Journal and Ptolemy's history or any other first-hand source had survived. Only the glittering fragments remain, such as when he faced down the mutiny of the Macedonians at Ophis in 323 BC:

> But does any man among you honestly feel that he has suffered more for me than I have suffered for him? Come now – if you are wounded, strip and show your wounds, and I will show mine. There is no part of my body but my back which has not a scar; not a weapon a man may grasp or fling the mark of which I do not carry upon me. I have sword-cuts from close fight; arrows have pierced me, missiles from catapults bruised my flesh; again and again I have been struck by stones or clubs – and all for your sakes: for your glory and gain. [See pp 158–9.]

We are left only with the consolation of Arrian's devoted and first-class history, written almost 500 years after the death of Alexander.

Unfortunately, accident and scholars more interested in literary merit and moral lessons, and not students of war, preserved what little survived the destruction of the classical world, the destruction of the Library of Alexandria and the later burning of the libraries of Constantinople by French barbarians in 1204. We can only shudder when we realise how close Homer and Thucydides came to extinction, yet somehow these giants survived to enrich our study of the art of war.

Some of the history of the ancient Middle East has come to life through archaeology. The deeds of Thutmose III, Rameses II and Sargon II are known again, but of their thoughts there is very little. Still, the deeds evoke powerful images, such as Rameses II's valour at Kadesh (1284 BC), where the Hittites surprised his camp and

PREFACE

cut him off from even his escort; he rallied his men by charging single-handed among the mass of Hittite chariots and cut a way through:

> I charged all countries, while I was alone, my infantry and my chariotry having forsaken me. Not one among them stood to turn about. I swear, as Re loves me, as my father, Atum, favors me, that as for every matter which his majesty has stated, I did it in truth, in the presence of my infantry and my chariotry. [See pp 158–9.]

Recent modern scholarship has brought to English significant works by Byzantine soldiers, chief among whom is the Emperor Maurice. His military classic, *The Strategikon*, together with his *Three Byzantine Military Treatises* were translated by George Dennis in 1984 and 1985. The Byzantine blend of the utterly practical and the religious comes through clearly:

> Before getting into danger, the general should worship God. When he does get into danger, then, he can with confidence pray to God as friend. [See p. 369.]

Similarly, Thomas Cleary has brought to English-speaking readers the works of the great Chinese captains, Zhuge Liang and Liu Ji in *Mastering the Art of War* (1989). The world of ancient China, where the scholar and general were often the same, was where the most subtle thoughts on war were set down and eagerly read and copied in a strong and unbroken tradition from the T'ai Kung (c. 1000 BC) to Mao Tse-tung. William Scott Wilson has done the same for the art of war in medieval Japan in *Ideals of the Samurai* (1982).

Caesar was the first great captain to immortalise the work of his sword with his pen. His fashion was not repeated on any scale in the West until the modern era, which has been increasingly rich in its memoirs, beginning in the eighteenth century with De Saxe's *Mes Reveries* and Suvorov's *The Science of Victory*, and accelerating so much that Churchill was heard to remark after World War II that his generals were selling themselves dearly.

Since the American Civil War, it has almost become *de rigueur* for senior officers to write their war memoirs. So we have almost all of them, from Alexander of Tunis to Zhukov. Others were lucky enough to have a friend or military secretary who was inspired to record events and words – Scipio Africanus had Polybius, and Belisarius had Procopius. To our great loss, the words of other great captains were never written down or have been lost. Lee the Incomparable and Nelson of Immortal Memory, on the other hand, were so beloved that every word they uttered was seized upon and preserved like gold nuggets. Wellington and Napoleon were survived by their stupendous collections of official business and correspondence, and by the numerous memoirs of their colleagues.

Of course, the soldier-scholar, by his very nature, is well-represented – Sun Tzu, Xenophon, Machiavelli, Clausewitz, Liddell Hart and Fuller. For Sun Tzu, the first of the great soldier-scholars to leave a legacy, which in many ways has never been surpassed, I have relied on one of the first translations, that by Lionel Giles in 1911. This continues to hold its own, despite a deluge of more recent efforts. We know more of the Prussian Carl von Clausewitz. No library soldier he – his writings were based on the many opportunities provided by Napoleon to amass a great deal of military experience in the field – his students laughed at the broken veins in his nose

PREFACE

and cheeks, making sly asides about his affection for the bottle. Silly boys, those rosy marks were caused by frostbite during Napoleon's invasion of Russia in 1812. Despite the intellectual and prose structure of German philosophy, one can still hear the whistle of the wind off the Russian snows. For this, we appreciate the splendid modern translation, which has become the indisputable standard, the Princeton University's edition of his *On War*, translated by Michael Howard and Peter Paret.

While frequently of value, the words of priest, scholar and diplomat are no more appropriate here than a lawyer's comments in a textbook on surgery. There is no substitute for wielding either scalpel or sword in medicine and war. I also include in this company the great national war leaders, such as Lincoln, Clemenceau, Churchill, Franklin Roosevelt, and they are joined by Thatcher, Blair and Bush. There is also something to be gained here from the occasional incompetent; as Wellington observed, 'I learnt what one ought not to do, and that is always something.' (See p. 168.)

Of course, I have exercised an editor's prerogative of being inconsistent here and there to please myself and, it is hoped, the reader. Some opportunities are too good to miss. For example, Ralph Waldo Emerson: 'I hate quotations; tell me what you know.' Two pillars of the Western canon that I have included, simply because of their manifest majesty and authority, are the Bible and Homer. What greater reason could there be for the Bible's inclusion than Exodus 15.3: 'The LORD is a man of war: The LORD is his name.'

Homer finds an especially honoured place. In *The Iliad*, I am convinced you can read many of the lines composed by the Achaean bards who accompanied their royal masters to Troy to record their glories, much as the Norse bards followed their Viking kings on campaign. The purpose was both informative and entertaining no matter how much they composed to royal taste.* In the hands of Homer, the story displays a profound and unsurpassed understanding of men in war:

> War – I know it well, and the butchery of men. Well I know, shift to the left, shift to the right my tough tanned shield. That's what the real drill, defensive fighting means to me. I know it all, how to charge in the rush of plunging horses – I know how to stand and fight to the finish, twist and lunge in the War-god's deadly dance.
> On guard! [See p. 180.]

And women, too, to whom war gives the deepest grief:

> 'That cry – that was Hector's honored mother I heard! My heart's pounding, leaping up in my throat, the knees beneath me paralyzed – Oh I know it... something terrible's coming down on Priam's children. Pray god the news will never reach my ears! Yes but I dread it so – what if great Achilles has cut my Hector off from the city, daring Hector, and driven him out across the plain, and all alone? – He may have put an end to that fatal headstrong pride that

---

* Michael Wood, *In Search of the Trojan War* (Facts on File Publications, New York and Oxford, 1985), pp. 10–11, 20–21, 91. *The Iliad* contains many references to specific features of the city and the local geography that only an eye-witness at the time could have made, for example, a trained observer – a bard – gathering details for his royal compositions. Some of these, I believe, have survived the endless retellings of the poems that eventually were compiled into *The Iliad*.

PREFACE

always seized my Hector – never hanging back with the main force of men, always charging ahead, giving ground to no man in his fury!'

So she cried, dashing out of the royal halls like a madwoman, her heart racing hard, her women close behind her. But once she reached the tower where soldiers massed she stopped on the rampart, looked down and saw it all – saw him dragged before the city, stallions galloping, dragging Hector back to Achaea's beaked warships – ruthless work. The world went black as night before her eyes, she fainted, falling backward, gasping away her life breath... she flung to the winds her glittering headdress, the cap and the coronet, braided band and veil, all the regalia golden Aphrodite gave her once, the day that Hector, helmet aflash in sunlight, led her home to Troy from her father's house with countless wedding gifts to win her heart. [See pp 490–1.]

But along with the grief's overrunning cup, women are the bulwark of civilisation's values for they teach their sons and remind their husbands of the value of what they must protect. Lord Mountbatten's mother sent him off to war with the words any Spartan mother would have been proud of:

The motto Papa and I chose when he got his arms as peer, will guide me as I know it will guide you: "In Honour Bound"... We who come from an old stock of privileged family, that has not had to worry over material existence, have inherited that sense of duty towards our fellow men, those especially whose nation we belong to, and who look, to us instinctively for example and guidance.

I know that you feel this too, more than ever in times like these. Let us live or die honourably. I am proud with the old feelings of our ancestors, that you my son once more are called to such high service. My love for you and my pride in you are too deep for selfish worries or repining. [See p. 290.]

The very best of the war correspondents, though engaged in what most soldiers consider an odious profession, somewhere between pest and spy, sometimes come close to Homer's standard. These exceptions are, like Homer – who put an unerring finger on the pulse of man in war – Kipling, the young Churchill, Constantin Simonov, Ernie Pyle and Max Hastings who were ever welcome in the mess.

Most soldiers quoted herein wrote at the end of their careers. Others were more prolific throughout their military service, and attempting to determine rank at the moment of literary conceptions presents an endless number of 'Gordian Knots'. Therefore, I have resorted to the highest rank held. Soldiers in antiquity are not referred to by rank, which would be difficult, if not impossible, to determine. Some great commanders are referred to by their titles alone, such as the great English dukes, Marlborough and Wellington, whose reputations outshone any insignia of military rank. Of course, Napoleon, although officially Napoleon I, Emperor of the French, is simply Napoleon. The name alone kept the whole of Europe in terror for almost 20 years. In the end, he was stripped of everything but that enduring reputation:

Many faults, no doubt, will be found in my career; but Arcole, Rivoli, the Pyramids, Marengo, Austerlitz, Jena, Friedland – these are granite: the tooth of envy is powerless here. [See p. 376.]

PREFACE

Titles of nobility are also included. Sometimes they were a delight – in those cases where they were earned – such as Marshal of France Michel Ney, Duc d'Elchingen and Prince de la Moskova! If anything, Ney was under-titled. Field Marshal Viscount Allenby of Meggido falls into that category as well. The British field marshal viscounts embody in their titles the far-flung campaigns of the British Army – Khartoum, Meggido, Alamein, Burma.

When possible, I have given the source and date of the quotation. Unfortunately, this has not always been viable; in such cases, I have cited the birth and death dates of the source.

Where there are several translations, mostly those from antiquity, I have selected those that best express the particular theme. The range of translation will come as a great surprise to the reader used to a single standard in more modern works.

The translations of Sun Tzu's *The Art of War* and Homer's *The Iliad* are infinitely varied, something readily apparent to anyone who has noticed whole shelves now devoted to different translations of both masterpieces.

Although I have chosen freely from among the rich range of translations, I have preferred to stick to the clarity of Lionel Giles' *The Art of War*. Homer yields endless insight, enriched by the nuances of his many translators. So rich, indeed, is this resource that the choice is daunting, but I have found the first and the last to be the best translations of *The Iliad*. They are John Dryden's great effort* of almost 300 years ago and Robert Fagles' breathtaking translation of 1990.

The delight of Fagles' translation is its fidelity to Homer's medium – the spoken not the written word, although that, too, is unsurpassed. Listening to Derek Jacobi's narration of this translation brings one closer to the golden time when the blind bard himself sang of the Anger of Achilles. I cannot recommend the book and the tape too highly.†

One imagines them all listening to the Bard still. At one table brilliant fire-eaters like Suvorov and Patton. At another are the scholar-soldiers – Sun Tzu, von Clausewitz, Fuller and Liddell Hart. By themselves, stiff and dour, are Heaven's captains – Tlacaelel and Cromwell. Over there are the mighty men of war, flinty commanders of great hosts – Scipio, Marlborough, Montgomery and Zhukov. By them, the knightly warriors *'sans peur et sans reproche'* – Belisarius, Saladin, Bayard and Lee. In the corner huddled together are the artists of the subtle strategem, the clever ones – Philip of Macedon, Wu Ch'i, Hannibal. And there, where the light pools around them from great windows, are those in love with adventure and glory – Rameses II, young again and boasting of his courage at Kadesh to a smiling Alexander, shining in the midst of everlasting renown.

Peter G. Tsouras

---

* When reading Dryden, the modern reader is struck by the odd sound of the gods' names in their Latin form. Until the middle of the last century, the Greek names were so unfamiliar that this Latinate convention was followed. Today the opposite applies.

† I strongly recommend the audio tape of Robert Fagles' translation of *The Iliad*, read by Derek Jacobi, who was certainly possessed by the Muse: Homer, *The Iliad*, Classics on Cassette, HBP 19495 (Penguin-HighBridge Audio, 1992).

ABILITY

The winds and waves are always on the side of the ablest navigators.
  Edward Gibbon, Decline and Fall of the Roman Empire, 1776.

Win with ability, not with numbers.
  Field Marshal Prince Aleksandr V. Suvorov, quoted in Danchenko and Vydrin, Military Pedagogy, 1973.

Great events hang by a thread. The able man turns everything to profit, neglects nothing that may give him one chance more; the man of less ability, by overlooking just one thing, spoils the whole.
  Napoleon, 25 September 1797, quoted in R.M. Johnston, ed., The Corsican, 1910.

This is a long tough road we have to travel. The men that can do things are going to be sought out just as surely as the sun rises in the morning. Fake reputations, habits of glib and clever speech, and glittering surface performance are going to be discovered…
  General of the Army Dwight D. Eisenhower, At Ease: Stories I Tell My Friends, 1967.

ACTION

The proof of battle is action, proof of words, debate.
No time for speeches now, it's time to fight.
  Homer, The Iliad, 16.730–732, c. 800 BC, tr. Robert Fagles, 1990; Patrocles chiding Meriones for engaging in taunts with Aeneas rather than in combat.

It is a common mistake in going to war to begin at the wrong end, to act first, and wait for disaster to discuss the matter.
  Thucydides, History of the Peloponnesian War, 1.78, c. 404 BC, tr. Richard Crawley, 1910; the Athenian delegation's argument to the Spartans to avoid war, 432 BC.

[In] the valiant man [are] invincibility, robustness, unconquerability. [He is] powerful, rugged, strong.
  The good valiant man [is] one who excels others – a victor, a conqueror, a taker of captives. He is reckless; he destroys, he charges the foe; he takes captives; he besieges, he sweeps away [the foe]. He glorifies himself, he glorifies [his exploits].
  The bad valiant man [is] vainglorious, a boaster that he is an eagle warrior, an ocelot warrior, a brave warrior. He pretends to be a brave warrior; he brags of himself, he boasts that he is a brave warrior.
  An Aztec account of the valiant man, quoted in Fray Bernardino de Sahagún, General History of the Things of New Spain, Book 10 – The People, 1961.

My business is to succeed, and I'm good at it. I create my Iliad by my actions, create it day by day.
  Napoleon, 1804, conversation with Pope Pius VII at Fountainbleu, quoted in Alfred Comte de Vigny, Servitude and Grandeur of Arms, 1857.

Action in war is like movement in a resistant element. Just as the simplest and most natural of movements,

ACTIVITY

walking, cannot easily be performed in water, so in war it is difficult for normal efforts to achieve even moderate results.
> Major-general Carl von Clausewitz, *On War*, 1.7, 1832, tr. Michael Howard and Peter Paret, 1976.

You ask me why I do not write something... [I] think one's feelings waste themselves in words. They ought to be distilled into actions and into actions which bring results.
> Florence Nightingale (1820–1910), quoted in Cecil Woodham-Smith, *Florence Nightingale*, 1951.

There is always hazard in military movements, but we must decide between the possible loss of inaction and the risk of action.
> General Robert E. Lee, letter to James A. Seddon, 8 June 1863, *The War of the Rebellion: A Compilation of the Official Records of the Union and Confederate Armies*, Vol. XLV, p. 868, US Government Printing Office, 1880–1901.

In tactics, action is the governing rule of war.
> Marshal of France Ferdinand Foch, *Precepts and Judgments*, 1919.

The essential thing is action. Action has three stages: the decision born of thought, the order or preparation for execution, and the execution itself. All three stages are governed by the will. The will is rooted in character, and for the man of action character is of more critical importance than intellect. Intellect without will is worthless, will without intellect is dangerous.
> Colonel-General Hans von Seekt, *Thoughts of a Soldier*, 1930.

When faced with the challenge of events, the man of character has recourse to himself. His instinctive response is to leave his mark on action, to take responsibility for it, to make it his own business... he embraces action with the pride of a master; for if he takes a hand in it, it will become his, and he is ready to enjoy success on condition that it is really his own, and that he derives no profit from it. He is equally prepared to bear the weight of failure, though not without a bitter sense of satisfaction. In short, a fighter who finds within himself all the zest and support he needs, a gambler more intent on success than profits, a man who pays his debts with his own money lends nobility to action. Without him there is but the dreary task of the slave; thanks to him, it becomes the divine sport of the hero.
> General Charles de Gaulle, *The Edge of the Sword*, 1932.

The relief which normally follows upon action after a long period of tension – that vast human sigh of relief – is one of the most recurrent phenomena of history, marking the onset of every great conflict. The urge to gain release from tension by action is a precipitating cause of war. (February 1929.)
> Captain Sir Basil Liddell Hart, *Thoughts on War*, 1944.

In critical and baffling situations, it is always best to return to first principle and simple action.
> Sir Winston S. Churchill (1874–1965).

Whatever the dangers of the action we take, the dangers of inaction are far, far greater.
> Prime Minister Tony Blair, speech to the Labour Party Conference, 2 Oct 2001.

ACTIVITY

If the enemy is taking his ease, he can harass him; if well supplied he can starve him; if quietly encamped, he can force him to move.

Appear at points which the enemy must hasten to defend; march swiftly to places where you are not expected.

ADAPTABILITY/ADAPTATION

Sun Tzu, *The Art of War*, 6, c. 500 BC, tr. Giles, 1910.

What do we do with a man who refuses to accept either good fortune or bad? This is the only general who gives his enemy no rest when he is victorious, nor takes any himself when he is defeated. We shall never have done with fighting him, it seems, because he attacks out of confidence when he is winning, and out of shame when he is beaten.

Hannibal, 210 BC, of the Roman Consul Marcellus upon whom he had just inflicted a reverse at Numistro, but who immediately offered battle again, quoted in Plutarch, *The Lives*, c. AD 100, (*Makers of Rome*, tr. Ian Scott-Kilvert, 1965).

I am up and about when I am ill, and in the most appalling weather. I am on horseback when other men would be flat out on their beds, complaining. We are made for action, and activity is the sovereign remedy for all physical ills.

Frederick the Great (1712–1786), quoted in Christopher Duffy, *Military Life of Frederick the Great*, 1986.

Accustom yourself to tireless activity...

Field Marshal Prince Aleksandr V. Suvorov (1729–1800), quoted in W. Lyon Blease, *Suvorof*, 1920.

The history of warfare so often shows us the very opposite of unceasing progress toward the goal, that it becomes apparent that immobility and inactivity are the normal state of armies in war, and action is the exception.

Major-General Carl von Clausewitz, *On War*, 3.16, 1832, tr. Michael Howard and Peter Paret, 1976.

We must make this campaign an exceedingly active one. Only thus can a weaker country cope with a stronger; it must make up in activity what it lacks in strength. A defensive campaign can only be made successful by taking the aggressive at the proper time. Napoleon never waited for his adversary to become fully prepared, but struck him the first blow.

Lieutenant-General Thomas J. 'Stonewall' Jackson, letter before the Battle of Chancellorsville, April 1863, quoted in Mary Anne Jackson, *Life and Letters of General Thomas J. Jackson*, 1892.

ADAPTABILITY/ADAPTATION

Do not repeat the tactics which have gained you one victory, but let your methods be regulated by the infinite variety of circumstances.

Sun Tzu, *The Art of War*, 6, c. 500 BC, tr. Giles, 1910.

As soon as they made these discoveries the Romans began to copy Greek arms, for this is one of their strong points: no people are more willing to adopt new customs and to emulate what they see is better done by others.

Polybius, *Histories*, 6.25, c. 125 BC (*The Rise of the Roman Empire*, tr. Ian Scott-Kilvert, 1987).

After the war, when I presented my order of recall, I met General Fuji, the Chief of the General Staff of the 1st Army, and told him I was anxious to know what changes in the Japanese Regulations would be introduced owing to their experiences in the war; he answered: 'So am I. We will wait to see what new Regulations for the Service Germany will issue on the basis of the reports that the officers who have been sent here will make, and we will translate these Regulations as we did the former ones.'

General Max Hoffman, *War Diaries and Other Papers*, II, 1929; observation while attached as an observer to the Japanese Army during the Russo-Japanese War, 1904–1905.

I have only one merit: I have forgotten what I taught and what I learned.

ADMINISTRATION

Marshal of France Ferdinand Foch, quoted in Monteilhet, *Les Institutions Militaires de la France*, 1932.

It is also greatly in the commander's own interest to have a personal picture of the front and a clear idea of the problems his subordinates are having to face. It is the only way in which he can keep his ideas permanently up to date and adapted to changing conditions. If he fights his battles as a game of chess, he will become rigidly fixed in academic theory and admiration of his own ideas. Success comes most readily to the commander whose ideas have not been canalised into any one fixed channel, but can develop freely from the conditions around him.

Field Marshal Erwin Rommel, *The Rommel Papers*, 1953.

In any problem where an opposing force exists, and cannot be regulated, one must foresee and provide for alternative courses. Adaptability is the law which governs survival in war as in life – war being but a concentrated form of the human struggle against environment.

Captain Sir Basil Liddell Hart, *Strategy*, 1954.

ADMINISTRATION

The good condition of my armies comes from the fact that I devote an hour or two every day to them, and when I am sent the returns of my troops and my ships each month, which fills twenty large volumes, I set every other occupation aside to read them in detail in order to discern the difference that exists from one month to another. I take greater pleasure in this reading than a young lady would get from reading a novel.

Napoleon to Joseph Bonaparte, 20 August 1806, *Correspondance de Napoléon Ier, publié par ordre de l'Empereur Napoléon III*, No. 10672, Vol. XIII, 1858–1870.

My Lord,

If I attempted to answer the mass of futile correspondence that surrounds me, I should be debarred from all serious business of campaigning. I must remind your Lordship – for the last time – that so long as I retain an independent position, I shall see that no officer under my Command is debarred, by attending to the futile drivelling of mere quill-driving in your Lordship's office, from attending to his first duty – which is, and always has been, so to train the private men under his command that they may, without question, beat any force opposed to them in the field.

Duke of Wellington, 1810, letter (attributed) to the Secretary of State for War.

My rule was to do the business of the day in that day.

The Duke of Wellington, quoted in Stanhope, *Conversations with the Duke of Wellington*, 1888.

I never knew what to do with a paper except to put it in a side pocket or pass it to a clerk who understood it better than I did.

General of the Army Ulysses S. Grant, quoted in S.L.A. Marshall, *The Armed Forces Officer*, 1951.

Although revolutionary government, none was ever so much under the domination of red tape as the one at Richmond. The martinets who controlled it were a good deal like the hero of Molière's comedy, who complained that his antagonist had wounded him by thrusting in carte, when according to the rule, it should have been in tierce. I cared nothing for the form of a thrust if it brought blood. I did not play with foils.

Colonel John S. Mosby, *Mosby's War Reminiscences*, 1887.

There has been a constant struggle on

ADVERSITY

the part of the military element to keep the end – fighting, or readiness to fight – superior to mere administrative considerations... The military man, having to do the fighting, considers that the chief necessity; the administrator equally naturally tends to think the smooth running of the machine the most admirable quality.
> Rear Admiral Alfred Thayer Mahan, *Naval Administration and Warfare*, 1908.

The more I see of war, the more I realize how it all depends on administration and transportation... It takes little skill or imagination to see where you would like your forces to be and when; it takes much knowledge and hard work to know where you can place your forces and whether you can maintain them there.
> Field Marshal Viscount Wavell of Cyrenaica (1883–1950).

...There must be a clear-cut, long-term relationship established between operational intentions and administrative resources. Successful administrative planning is dependent on anticipation of requirements.
> Field Marshal Viscount Montgomery of Alamein, *Memoirs of Field Marshal Montgomery*, 1958.

Success cannot be administered.
> Admiral Arleigh Burke, speech, 1962.

Paper-work will ruin any military force.
> Lieutenant-General Lewis B. 'Chesty' Puller, quoted in Burke Davis, *Marine!*, 1962.

ADMIRALS

There is in the naval profession a specialized technical mentality which blocked all my plans. No sooner had I proposed a new idea than I had [Admiral] Ganteaume and the whole navy on my neck: 'Sire, this is impossible.' – 'Why?' – 'Sire, the winds don't allow it, and then the doldrums, the currents' – and with that they stopped me short. How can a man argue with people who speak a different language?
> Napoleon, 1816, quoted in Christopher Herold, *The Mind of Napoleon*, 1955.

It is dangerous to meddle with admirals when they say they can't do things. They have always got the weather or fuel or something to argue about.
> Sir Winston Churchill, December 1941, to the American Secretary of War, Frank Knox.

The admirals are really something to cope with – and I should know. To change something in the na-a-vy is like punching a feather bed, you punch it with your right and you punch it with your left until you are finally exhausted and then you find the damn bed just as it was before you started punching.
> President Franklin D. Roosevelt (1882–1945), quoted in the *San Diego Union*, 20 November 1987.

[Admiral Halsey] was of the same aggressive type as John Paul Jones, David Farragut, and George Dewey. His one thought was to close with the enemy and fight him to the death. The bugaboo of many sailors, the fear of losing ships, was completely alien to his conception of fighting.
> General of the Army Douglas MacArthur, *Reminiscences*, 1964.

ADVERSITY

The noble and courageous man is known by his patience in adversity.
> The Inca Emperor Pachacutec (d. 1470), quoted in Garcilaso de la Vega, *Royal Commentaries of the Incas*, 1989.

The officers should feel the conviction that resignation, bravery, and faithful attention to duty are virtues without which no glory is possible and no army

ADVICE

is respectable, and that firmness amid reverses is more honorable than enthusiasm in success – since courage alone is necessary to storm a position whereas it requires heroism to make a difficult retreat before a victorious and enterprising enemy, always opposing to him a firm and unbroken front.
> Lieutenant-General Antoine-Henri Baron de Jomini, *Summary of the Art of War*, 1838, tr. Mendell and Craighill, 1862.

You're all right as long as you're winners; I'm a hell of a general when I'm winning, anybody is, but it's when you're not winning – and I have not always been winning, if you had been a British general at the start of a war you'd know that – it is then that the real test of leadership is made.
> Field Marshal Viscount Slim of Burma, 8 April 1952, speech to the US Army Command and General Staff College.

ADVICE

When men counsel reasonably, reasonable success ensues; but when in their counsels they reject reason, God does not choose to follow the wanderings of human fancies.
> Themistocles, 480 BC, urging the Peloponnesians not to abandon the defence of Salamis during the Persian invasion of Greece, quoted in Herodotus, *The Persian Wars*, 8, tr. George Rawlinson.

Generals ought to be given advice, in the first place from men of foresight, the experts who have specialized in military affairs, and who have learned from experience; secondly, from those who are on the spot, who see the terrain, who know the enemy, who can judge the right moment for action, who are, as it were, shipmates sharing the same danger. Therefore, if there is anyone who is confident that he can advise me about the best interest of the nation in the campaign which I am now about to conduct, let him not deny the state his services – let him come with me to Macedonia. I will assist him by providing his passage, his horse, his tent, yes, and his travelling money. If anyone finds this prospect too irksome, and prefers the ease of the city to the hardships of campaign, let him not steer the ship from his place on shore.
> Lucius Aemilius Paulus, 168 BC, upon his departure for the campaign in which he would defeat Perseus, King of Macedonia, at the Battle of Pydna, quoted in Livy, *The History of Rome*, 44.22, c. AD 17 (*Rome and the Mediterranean*, tr. Henry Bettenson, 1976).

I believe that a general who receives good advice from a subordinate officer should profit by it. Any patriotic servant of the state should forget himself when in that service, and look only to the interests of the state. In particular, he must not let the source of an idea influence him. Ideas of others can be as valuable as his own and should be judged only by the results they are likely to produce.
> Frederick the Great, *General Principles of War*, 1748.

There are generals who need no advice; who judge and decide for themselves, and their staffs merely execute orders. Such generals are stars of the first magnitude, appearing only once in a century.
> Field Marshal Helmuth Graf von Moltke, *Italian Campaigns of 1859*, 1904.

One must never be drawn off the job in hand by gratuitous advice from those who are not fully in the operational picture and who have no responsibility.
> Field Marshall Viscount Montgomery of Alamein, *The Memoirs of Field Marshal Montgomery*, 10, 1958.

'First Sea Lord, if this invasion happens, precisely what can we do?' I enquired. I shall never forget the quiet,

AGGRESSIVENESS

calm, confident answer.'

'I can put together a Task Force of destroyers, frigates, landing craft, support vessels,' he said. 'It will be led by the aircraft carriers HMS *Hermes* and HMS *Invincible*. It can be ready to leave in forty-eight hours.'

Once again, the hour had produced the man.

> Prime Minister Margaret Thatcher, 30 March 1982, of the First Sea Lord, Admiral Sir Henry Leach's decisive advice two days before the Argentine invasion of the Falkland Islands, quoted in the foreword of Sandy Woodward, *One Hundred Days: The Memoirs of the Falklands Battle Group Commander*, 1992, pp. xi–xii.

AGGRESSION

Let us engage in anger, convinced that nothing is more legitimate between adversaries than to claim to satisfy the whole wrath of one's soul in punishing the aggressor, and nothing more sweet, as the proverb has it, than the vengeance upon an enemy which it will now be ours to take. That enemies they are and mortal enemies you all know, since they came here to enslave our country, and if successful had in reserve for our men all that is most dreadful, and for our children and wives all that is most dishonorable, and for the whole city the name which conveys the greatest reproach.

> Gylippus, summer 413 BC, to the Syracusans before the last great naval battle with the Athenians in the harbour of Syracuse, quoted in Thucydides, *History of the Peloponnesian War*, 7.68, c. 404 BC, tr. Richard Crawley, 1910.

Men rise from one ambition to another: First, they seek to secure themselves against attack, and then they attack others.

> Niccolò Machiavelli, *Discourses*, 1517.

If aggressors are wrong above, they are right below.

> Napoleon, *Maxims*, 1804–1815.

There is never a convenient place to fight a war when the other man starts it.

> Admiral Arleigh A. '31-Knot' Burke (1901–1996)

We rule the skies and seas, and possess the power to rule the land when we are sufficiently aroused. But we have not learned to understand, much less to rule, minds and hearts and souls. The only moral we need to cull from the *Iliad* is that it is foolish to underestimate the complexity and determination of the killers from the other shore.

> Lieutenant-Colonel Ralph Peters, 'New Old Enemies', *Parameters*, Summer 1999.

AGGRESSIVENESS

Our Country will, I believe, sooner forgive an officer for attacking an enemy than for letting it [sic] alone.

> Admiral Viscount Nelson, 3 May 1794, letter during the attack on Bastin.

My rule is: if you meet the weakest vessel, attack; if it is a vessel equal to yours, attack; and if it is stronger than yours, also attack...

> Admiral Stepan O. Makarov (1849–1904).

There is one great maxim which throughout the history of war has more often than not proved successful, and this maxim is: 'When in doubt, hit out.' When the soldier, whether private or general, does not know what to do, he must strike; he must not stand still, for normally it is better to strike and fail than it is to sit still and be thrashed.

> Major-General J. F. C. Fuller, *Foundations of the Science of War*, 1926.

Hold what you've got and hit them where you can.

> Admiral of the Fleet Ernest J. King, December 1941, admonishment to the

AIRBORNE

hard-pressed commanders in the Pacific in the face of the Japanese onslaught after Pearl Harbor.

What counts is not necessarily the size of the dog in the fight – it's the size of the fight in the dog.
General of the Armies Dwight D. Eisenhower, 31 January 1958, Republican National Convention.

An army which thinks only in defensive terms is doomed. It yields initiative and advantage in time and space to the enemy – even an enemy inferior in numbers. It loses the sense of the hunter, the opportunist.
General Sir David Fraser, *And We Shall Shock Them*, 1983.

AIRBORNE

A standard question for a new man was why he had volunteered for parachuting and whether he enjoyed it. On one occasion, a bright-eyed recruit startled me by replying to the latter question with a resounding, 'No, sir.' 'Why then, if you don't like jumping did you volunteer to be a parachutist?' I asked. 'Sir, I like to be with people who do like to jump,' was the reply. I shook his hand vigorously and assured him that there were at least two of us of the same mind in the Division.
General Maxwell D. Taylor, *Swords and Ploughshares*, 1972.

I was confident that we were reasonably well prepared. There is enormous residual self-confidence in a parachute battalion. People don't wonder whether they can do things.
Lieutenant-Colonel Hugh Pike, 1982, commanding 3 Para during the Falklands War.

AIR POWER

Find the enemy and shoot him down, anything else is nonsense.
Captain Manfred Baron von Richthofen, 1917, opinion of air combat, quoted in Toliver and Constable, *Fighter Aces of the Luftwaffe*, 1977.

I have a mathematical certainty that the future will confirm my assertion that aerial warfare will be the most important element in future wars, and that in consequence not only will the importance of the Independent Air Force rapidly increase, but the importance of the army and navy will decrease in proportion.
General Giulio Douhet, *Command of the Air*, 1921.

It is probable that future war will be conducted by a special class, the air force, as it was by the armored knights of the Middle Ages.
Brigadier-General William 'Billy' Mitchell, *Winged Defense*, 1924.

Tom, you've never believed in air. Never get out from under the air umbrella; if you do, you'll be for it. And as you flutter up to heaven all you'll say is – 'My gosh, some sailor laid a hell of a mine for me!'
Marshal of the Royal Air Force Sir Arthur Harris, to Admiral Tom Phillips, commanding the squadron that included HMS *Prince of Wales* and *Repulse*, which were later sunk by Japanese carrier aircraft in the Far East in 1942, an action in which Phillips also perished, *Bomber Offensive*, 1947.

Victory, speedy and complete, awaits the side that employs air power as it should be employed.
Marshal of the Royal Air Force Sir Arthur Harris (1892–1984).

If we lose the war in the air, we lose the war, and we lose it quickly.
Field Marshal Viscount Montgomery of Alamein (1887–1976).

Tell him you will need more Air. And when you have told him that, tell him again from me that he will need *more*

ALLIANCES/ALLIES/COALITION WARFARE

Air! And when you have told him that for the second time, tell him from me for the third time that he will need MORE AIR!
> General of the Army Douglas MacArthur, advice to be delivered to Field Marshal Lord Mountbatten, quoted in Geoffrey Parret, *Old Soldiers Never Die: The Life of Douglas MacArthur*, 1996, p. 342.

Whereas to shift the weight of effort on the ground from one point to another takes time, the flexibility inherent in Air Forces permits them without change of base to be switched from one objective to another in the theatre of operations.
> Field Marshal Viscount Montgomery of Alamein (1887–1976).

Air warfare cannot be separated into little packets; it knows no boundaries on land and sea other than those imposed by the radius of action of the aircraft; it is a unity and demands unity of command.
> Air Chief Marshal Lord Tedder, *With Prejudice*, 1948.

Anyone who has to fight, even with the most modern weapons, against an enemy in complete command of the air, fights like a savage against modern European troops, under the same handicaps and with the same chances of success.
> Field Marshal Erwin Rommel, *Rommel Papers*, 1953.

The idea that superior air power can in some way be a substitute for hard slogging and professional skill on the ground in this sort of war is beguiling but illusory. Air support can be of immense value to an army; it may sometimes be its salvation. But we must have a care not to misread the lessons of... the closing years of World War II. The truth that we should take to heart is that armies can fight – and not only defensively – in the face of almost total air superiority... All this is cold comfort for anyone who hopes that air power will provide some cheap short cut to victory.
> Air Marshal Sir John Slessor, 'Air Power and World Strategy', *Foreign Affairs*, October 1954.

ALLIANCES/ALLIES/COALITION WARFARE

Alliances, to be sure, are good, but forces of one's own are still better.
> Frederick William of Brandenburg (the Great Elector), *Political Testament*, 1667.

'Tis our true policy to steer clear of permanent alliances with any portion of the foreign world.
> President George Washington, 17 September 1796, farewell address.

It is better to have a known enemy than a forced ally.
> Napoleon, *Political Aphorisms*, 1848.

The allies we gain by victory will turn against us upon the bare whisper of our defeat.
> Napoleon, *Political Aphorisms*, 1848.

Well, at any rate, he ought to be heartily grateful to old Blücher: had it not been for him, I know not where his Grace might have been today; but I know that I, at least, would not have been at St. Helena.
> Napoleon, on St Helena, 16 November 1816, speaking of the debt the Duke of Wellington owed to his Prussian ally at Waterloo, quoted in Emmanuel Las Cases, *The Life, Exile, and Conversations of the Emperor Napoleon by the Count de Las Cases*, IV/7, 1835, p. 222.

Granting the same aggregate of force, it is never as great in two hands as in one, because it is not perfectly concentrated.
> Rear Admiral Alfred Thayer Mahan, *Naval Strategy*, 1911.

ALLIANCES/ALLIES/COALITION WARFARE

It matters little what you send; we ask for one corporal and four men; but they must arrive at the very beginning. You will give them to me, I promise to do my best to get them killed; from that moment I shall be happy for I know that after that the whole of England will come as one man.
> Marshal of France Ferdinand Foch, 1909, conversation with Field Marshal Henry Wilson, quoted in C.R. Ballard, *Kitchener*, 1930, p. 211

I cannot believe that the British Army will refuse to do its share in this supreme crisis... history would severely judge your absence... Monsieur le Maréchal, the honor of England is at stake!
> Marshal of France Joseph Joffre, 5 September 1914; appeal to Field Marshal French, commanding the British Expeditionary Force (BEF), to participate in the counter-attack that would be the First Battle of the Marne, quoted in Barbara Tuchman, *The Guns of August*, 1962.

America has joined forces with the Allied Powers, and what we have of blood and treasure are yours. Therefore, it is that with loving pride we drape the colors in tribute of respect to this citizen of your great republic. And here and now, in the presence of the illustrious dead, we pledge our hearts and our honor in carrying this war to successful issue. Lafayette, we are here.
> Colonel Charles E. Stanton, 4 July 1917, Paris, as General John J. Pershing placed a wreath at the tomb of the Marquis de Lafayette, quoted in Henry F. Woods, *American Sayings*, 1945.

I was no more than conductor of an orchestra... A vast orchestra, of course... Say, if you like, that I beat time well! Has the music stopped? Are we tired of the tune? We must start a new one. Never stop... The true meaning of the unified command is not to give orders, but to make suggestions... One talks, one discusses, one persuades... One says, 'That is what should be done; it is simple; it is only necessary to will it.'
> Marshal of France Ferdinand Foch, commander of allied forces during World War I, quoted in Liddell Hart, *Foch: The Man of Orleans*, 1931.

Any alliance whose purpose is not the intention to wage war is senseless and useless.
> Adolf Hitler, *Mein Kampf*, 1925.

These two great organizations of the English-speaking democracies, the British Empire and the United States, will have to be somewhat mixed up together in some of their affairs for mutual and general advantage... I could not stop [this process] if I wished; no one can stop it. Like the Mississippi, it just keeps rolling along. Let it roll. Let it roll on full flood, inexorable, irresistible, benignant, to broader lands and better days.
> Sir Winston Churchill, 20 August 1940, as he was negotiating for the transfer of 50 US destroyers to the Royal Navy before American entry into the war, quoted in *H.C. Deb*. Vol. 354, Col. 1171.

In war I would deal with the Devil and his grandmother.
> Joseph Stalin (1879–1953).

If Hitler invaded hell, I would make at least a favourable reference to the devil in the House of Commons.
> Sir Winston S. Churchill, 1941, upon supporting the Soviet Union after the German invasion of 22 June.

There is only one thing worse than fighting with allies – and that is fighting without them.
> Sir Winston S. Churchill (1874–1965).

Allied effectiveness in World War II

AMATEURS

established for all time the feasibility of developing and employing joint control machinery that can meet the sternest tests of war. The key to the matter is a readiness, on highest levels, to adjust all nationalistic differences that affect the strategic employment of combined resources, and, in the war theater, to designate a single commander who is supported to the limit. With these two things done, success rests in the vision, the leadership, the skill, and the judgment of the professionals making up command and staff groups; if these two things are not done, only failure can result.
   General of the Army Dwight D. Eisenhower, *Crusade in Europe*, 1948.

In war it is not always possible to have everything go exactly as one likes. In working with Allies it sometimes happens that they develop opinions of their own.
   Sir Winston Churchill, *The Hinge of Fate*, 1950.

A too complete victory inevitably complicates the problem of making a just and wise peace settlement. Where there is no longer the counter-balance of an opposing force to control the appetites of the victors, there is no check on the conflict of views and interests between the parties to the alliance. The divergence is then apt to become so acute as to turn the comradeship of common danger into the hostility of mutual dissatisfaction – so that the ally of one war becomes the enemy in the next.
   Captain Sir Basil Liddell Hart, *Strategy*, 1954.

War without allies is bad enough – with allies it is hell!
   Air Marshal Sir John Slessor, *Strategy for the West*, 1954.

No other allies could have done it …

[Y]ou Americans and the British. You learn from each other. There is a special chemistry there.
   General Hans Speidel (Rommel's chief of staff in France), to New York Times correspondent, Drew Middleton, *Crossroads of Modern War*, 1983, p. 170.

I was operating with the starting assumption that there was no single target that was more important, if struck, than the principle of Alliance consensus and cohesion.
   General Wesley Clark, on holding the NATO alliance together during the bombing campaign against Serbia, quoted in 'The Commander's War', *Washington Post*, 21 September 1999, p. A16.

The mission needs to define the coalition, and we ought not to think that a coalition should define the mission.
   Secretary of Defense Donald Rumsfeld, 25 September 2001, *The Washington Post*, 26 September 2001, p. A7.

We were with you at the first. We will stay with you to the last.
   Prime Minister Tony Blair, to Britain's American ally in a speech to the Labour Party Conference, 2 October 2001.

I assure the Hon. friend that I speak frankly to our American allies, just as I do to our other allies. However, it is in the nature of that frankness that that is best done privately rather than publicly.
   Secretary of State for Defence Geoff Hoon, answering questions in the House of Commons, 10 January 2005

AMATEURS

There are three ways in which a ruler can bring misfortune upon his army:
(1) By commanding the army to advance or to retreat, being ignorant of the fact that it cannot obey. This is called hobbling the army.
(2) By attempting to govern the army in the same way as he administers his

kingdom, being ignorant of the conditions which obtain in an army. This causes restlessness in the soldier's [sic] minds.
(3) By employing the officers of his army without discrimination, through ignorance of the military principle of adaptation to the circumstances. This shakes the confidence of the soldiers.
> Sun Tzu, *The Art of War*, 3, c. 500 BC, tr. Giles, 1910.

I well know the character of that senseless monster the people, unable either to support the present or to foresee the future, always desirous of attempting the impossible, and of rushing headlong to its ruin. Yet your unthinking folly shall not induce me to permit your own destruction, nor to betray the trust committed to me by my sovereign and yours. Success in war depends less on intrepidity than on prudence to await, to distinguish, and to seize the decisive moment of fortune. You appear to regard the present contest as a game of hazard, which you might determine by a single throw of the dice; but I, at least, have learnt from experience to prefer security to speed. But it seems that you offer to reinforce my troops, and to march with them against the enemy. Where then have you acquired your knowledge of war? And what true soldier is not aware that the result of a battle must chiefly rest on the skill and discipline of the combatants? Ours is a real enemy in the field; we march to a battle, and not to a review.
> Count Belisarius, AD 538, when the population of Rome, in desperation, urged a battle during the siege by the Goths, in Mahon, *Life of Belisarius*, 1829.

The kind of person who could not lead a patrol of nine men is happy to arrange armies in his imagination, criticise the conduct of a general, and say to his misguided self: 'My God, I know I could do better if I was in his place!'
> Frederick the Great, 14 July 1745.

This is a very suggestive age. Some people seem to think that an army can be whipped by waiting for rivers to freeze over, exploding powder at a distance, drowning out troops, or setting them to sneezing; but it will always be found in the end that the only way to whip an army is to go out and fight it.
> General of the Army Ulysses S. Grant, January 1865, quoted in Horace Porter, *Campaigning with Grant*, 1906.

When, in complete security, after dinner in full physical and moral contentment, men consider war and battle, they are animated by a noble ardor which has nothing in common with reality. How many of them, however, at that moment would be ready to risk their lives? But oblige them to march for days and weeks to arrive at the battle front, and on the day of battle oblige them to wait a few minutes or hours to deliver it. If they were honest, they would testify how much the physical fatigue and the mental anguish that preceded action have lowered their morale, how much less eager to fight they are than a month before, when they arose from the table in a generous mood.
> Colonel Charles Ardant du Picq, *Battle Studies*, 1880, tr. Greely, 1957.

All those who criticize the dispositions of a general ought first to study military history, unless they have themselves taken part in a war in a position of command. I should like to see such people compelled to conduct a battle themselves. They would be overwhelmed by the greatness of their task, and when they realized the obscurity of the position, the exacting nature of enormous demands made on them, they would doubtless be more modest.
> General Erich Ludendorff, *My War Memories, 1914–1918*, 1919.

AMBITION

All the more honour to those gifted professional soldiers who have shown creative brains, but also all the more reason why we should preserve an open and receptive mind to the far larger stream of ideas which have come from 'amateurs,' from Roger Bacon and Leonardo da Vinci down to the Volunteers who simplified our cumbrous drill and developed our musketry and tactical instructions. Nor can we forget that the decisive new weapon of the World War, the tank, owed its introduction principally to 'amateurs' in the face of stubborn opposition from the supreme professional opinion. It may also be significant that the one soldier who contributed notably to its causation was an engineer and an imaginative writer. (March 1923.)

    Captain Sir Basil Liddell Hart, *Thoughts on War*, 1944.

The atmosphere in Washington in that period [1964] is difficult, indeed impossible, to re-create. There were conferences almost weekly, at government expense, to gather in judgments from all possible quarters on what was to be done to carry the war to a successful conclusion. These huddles would bring together professors and self-appointed experts on guerrilla warfare from all parts of the country. The collective contribution to the national cause may have been two degrees above zero. Government was proceeding according to the notion that, if enough persons are collected, some inspired thoughts of value will inevitably flow outward. It reminds me of the Army thesis that if two half-wits are assigned to a task, you get a whole wit, whereas the mathematical prospect is, more accurately, that what comes out will be a quarter-wit.

    Brigadier-General S.L.A. Marshall, 'Thoughts on Vietnam', in Thompson, *Lessons of Vietnam*, 1977.

Out there among you are the cynics, the people who scoff at what you're learning here. The people that scoff at character, the people that scoff at hard work. But they don't know what they're talking about, let me tell you. I can assure you that when the going gets tough and your country needs them, they're not going to be there. They will not be there, but you will... After Vietnam, we had a whole cottage industry develop basically in Washington, D.C., that consisted of a bunch of military fairies that had never been shot at in anger, who felt fully qualified to comment on the leadership ability of all the leaders of the United States Army. They were not Monday morning quarterbacks, they were the worst of all possible kind, they were Friday afternoon quarterbacks. They felt qualified to criticize us even before the game was played... And they are the same people who are saying, my goodness, we have terrible problems in the armed forces because there are no more leaders out there, there are no more combat leaders. Where are the Eisenhowers? Where are the Bradleys? Where are the MacArthurs? Where are the Audie Murphys?... Coming from a guy who's never been shot at in his entire life, that's a pretty bold statement.

    General H. Norman Schwarzkopf, 15 May 1991, address to the Corps of Cadets at West Point.

AMBITION

Don't you think that I have something worth being sorry about, when I reflect that at my age Alexander was already king over so many peoples, while I have never yet achieved anything really remarkable?

    Julius Caesar (100–44 BC), after reading an account of Alexander's triumphs, quoted in Plutarch, *The Lives*, c. AD 100 (*The Fall of the Roman Republic*, tr. Rex Warner, 1972).

ANALYSIS

A man's worth is no greater than the worth of his ambitions.
> Marcus Aurelius, *Meditations*, 7.3, c. 170, tr. Maxwell Staniforth.

I cannot, if I am in the field of glory, be kept out of sight: wherever there is anything to be done, there Providence is sure to direct my steps.
> Admiral Viscount Nelson, early 1797, letter to his wife, while stationed in Genoa, quoted in Robert Southey, *Life of Nelson*, 1813.

Ambition is the main driving power of men. A man expends his abilities as long as he hopes to rise; but when he has reached the highest round, he only asks for rest. I have created senatorial appointments and princely titles, in order to promote ambition, and, in this way, to make the senators and marshals dependent on me.
> Napoleon, quoted in F.M. Kircheisen, *Memoirs of Napoleon I*, 1929.

But all that... he will learn will be of little use to him if he does not have the sacred fire in the depths of his heart, this driving ambition which alone can enable one to perform great deeds.
> Napoleon, *Correspondance de Napoléon Ier, publié par ordre de l'Empereur Napoléon III*, XXXII, p. 379.

A different habit, with worse effect, was the way that ambitious officers, when they came in sight of promotion to the general's list, would decide that they would bottle up their thoughts and ideas, as a safety precaution, until they reached the top and could put these ideas into practice. Unfortunately, the usual result, after years of such self-repression for the sake of ambition, was that when the bottle was eventually uncorked the contents had evaporated.
> Captain Sir Basil Liddell Hart, quoted by Josiah Bunting, 'The Conscience of a Soldier', *Worldview*, 10/1973.

Ambition, not so much for vulgar ends, but for fame, glints in every mind.
> Sir Winston S. Churchill (1874–1965).

ANALYSIS

Napoleon once said: 'I have always liked analysis, if I were to be seriously in love, I should analyse my love bit by bit. Why? and How? are questions so useful that they cannot be too often asked. I conquered rather than studied history; that is to say, I did not care to retain and did not retain anything that could not give me a new idea; I disdained all that was useless, but took possession of certain results which pleased me.' In other words, he taught himself 'How to think' which is the harvest of all true education, civil or military.
> Major-General J.F.C. Fuller, *Lectures on F. S. R. II*, 1931, quoting Napoleon, 25 January 1803.

The analysts write about war as if it's a ballet... like it's choreographed ahead of time, and when the orchestra strikes up and starts playing, everyone goes out and plays a set piece.
What I always say to those folks is, 'Yes, it's choreographed, and what happens is the orchestra starts playing and some son of a bitch climbs out of the orchestra pit with a bayonet and starts chasing you around the stage.' And the choreography goes right out the window.
> General H. Norman Schwarzkopf, interview in *The Washington Post*, 5 February 1991.

APPEASEMENT

And that is called paying the Dane-geld;
But we've proved it again and again,
That if once you have paid him the
    Dane-geld
You never get rid of the Dane.
> Rudyard Kipling, *Dane-Geld*.

This is the second time there has come back from Germany to Downing Street

peace with honour. I believe it is peace in our time.
>   Prime Minister Neville Chamberlain, 30 September 1938, from 10 Downing Street upon his return from meeting Hitler at Munich, where Britain and France abandoned Czechoslovakia to its fate.

England has been offered a choice between war and shame. She has chosen shame and will get war.
>   Sir Winston S. Churchill, September 1938, speech in the House of Commons on Munich Agreement.

An appeaser is one who feeds the crocodile – hoping it will eat him last.
>   Sir Winston S. Churchill, *Reader's Digest*, December 1954.

Looking back over 12 years, we have been victims of our own desire to placate the implacable, to persuade towards reason the utterly unreasonable, to hope that there was some genuine intent to do good in a regime whose mind is in fact evil. Now the very length of time counts against us. You've waited 12 years. Why not wait a little longer?
>   Tony Blair, Speech in the House of Commons, 18 Mar 2003, final justification for war.

ARMY
The Nature of Armies

The ideal army would be the one in which every officer would know what he ought to do in every contingency; the best possible army is the one that comes closest to this. I give myself only half the credit for the battles I have won, and a general gets enough credit when he is named at all, for the fact is that a battle is won by the army.
>   Napoleon, 1817, conversation. Gaspard Gourgaud, *Sainte Hélène: Journal inédit de 1815 à 1818*, II, quoted in Christopher Herold, ed., *The Mind of Napoleon*, 1955.

War is a special activity, different and separate from any other pursued by man. This would still be true no matter how wide its scope, and though every able-bodied man in the nation were under arms. An army's military qualities are based on the individual who is steeped in the spirit and essence of this activity; who trains the capacities it demands, rouses them, and makes them his own; who applies his intelligence to every detail; who gains ease and confidence through practice, and who completely immerses his personality in the appointed task.
>   Major-General Carl von Clausewitz, *On War*, 3.5, 1832, tr. Michael Howard and Peter Paret, 1976.

When an army starts upon a campaign, it resolves itself speedily into two parts, one that means to keep out of harm's way if possible, and the other that always keeps with the colors.
>   Major-General George B. McClellan (1826–1885), quoted in *Tenting Tonight*, 1984.

The Army has its common law as well as its statute law; each officer is weighed in the balance by his fellows, and these rarely err. In the barrack, in the mess, on the scout, and especially in battle, a man cannot – successfully – enact the part of a hypocrite or flatterer, and his fellows will measure him pretty fairly for what he is.
>   General of the Army William T. Sherman (1820–1891).

An army is still a crowd, though a highly organized one. It is governed by the same laws… and under the stress of war is ever tending to revert to its crowd form. Our object in peace is so to train it that the reversion will become extremely slow.
>   Major-General J.F.C. Fuller, *Training Soldiers for War*, 1914.

ARMY

An army is to a commander what a sword is to a soldier, it is only worth anything in so far as it receives from him a certain imulsion.
> Marshal of France Ferdinand Foch, *Precepts and Judgments*, 1919.

All through history there have been recurring cycles of two schools of war; the one tending towards the adoption of small professional forces relatively well trained. The other towards the utilization of masses and mob psychology. Both schools have usually foundered on the rock of compromise, when the first sought to increase its numbers and the second its discipline...
> General George S. Patton, Jr., January 1928, *The Patton Papers*, I, 1972.

An army is an institution not merely conservative but retrogressive by nature. It has such natural resistance to progress that it is always insured against the danger of being pushed ahead too fast. Far worse and more certain, as history abundantly testifies, is the danger of it slipping backward. Like a man pushing a barrow up a hill, if the soldier ceases to push, the military machine will run back and crush him. To be deemed a revolutionary in the army is merely an indication of vitality, the pulse-beat which shows that the mind is still alive. When a soldier ceases to be a revolutionary it is a sure sign that he has become a mummy. (Jan. 1931.)
> Captain Sir Basil Liddell Hart, *Thoughts on War*, 1944.

An armed force differs from an armed mob, or crowd of men, not so much in its armament but in that it is disciplined, is organised, and is under control. A crowd may at times be courageous beyond belief, but it possesses little staying power, and is as quickly depressed as elated by outward circumstances. Courage in itself is insufficient, and to it must be added what is called morale, a complex quality depending upon honour to the cause, loyalty to the leader, confidence in one's skill and the skill of one's comrades, and in an innate feeling that everyone is doing his utmost to win and that no one will leave a comrade in the lurch. Morale endows a force of men with a feeling of superiority and invincibility, it is cultivated by common-sense, sound organisation, efficient arms and equipment, and the realisation that those in command are skilful, brave and just.
> Major-General J.F.C. Fuller, *Lectures on F. S. R. II*, 1931.

Their task is to destroy. The balance sheet of their activities shows a hideous total of broken lives, of wealth destroyed, of nations ground to powder, of work brought to nothing, of efforts frustrated, and happiness maimed and killed. The wreckage of war is beyond counting. Land left fallow, fire, and famine, such are its material consequences. But, in brooding on the evil, do we not tend to forget the children born into safety, the lives lived out in security as the result of the toil and sacrifice of our soldiers? But for the strength of armies what tribe, what city, and what states could ever have been established? Protected by that human shield, fields have been sown and reaped and men have been able to work on undisturbed. The clash of arms has made possible all material progress. History cannot measure the debt owed by the wealth of nations, their network of communications, their ships upon the seas, to the lust for conquest.
> General Charles de Gaulle, *The Edge of the Sword*, 1932.

The nature of armies is determined by the nature of the civilization in which they exist.
> Captain Sir Basil Liddell Hart, *The Ghost of Napoleon*, 1933.

ARMY

I have always held the view that an army is not merely a collection of individuals, with so many tanks, guns, machine-guns, etc., and that the strength of the army is not just the total of all these things added together. The real strength of an army is, and must be, far greater than the sum total of its parts; that extra strength is provided by morale, fighting spirit, mutual confidence between the leaders and the led and especially with the high command, the quality of comradeship, and many other intangible spiritual qualities.

> Field Marshal Viscount Montgomery of Alamein, *The Memoirs of Field Marshal Montgomery*, 1958.

A man can be selfish, cowardly, disloyal, false, fleeting, perjured, and morally corrupt in a wide variety of other ways and still be outstandingly good in pursuits in which other imperatives bear than those upon the fighting man. He can be a superb creative artist, for example, or a scientist in the very top flight, and still be a very bad man. What the bad man cannot be is a good sailor, or soldier, or airman. Military institutions thus form a repository of moral resource that should always be a source of strength within the state.

> General Sir John Hackett, 'The Military in the Service of the State', *War, Morality, and the Military Profession*, 1986.

The Army, despite its long-standing reputation for enforced homogeneity, is not an ant colony. Within the Army's vast hierarchy – which determines everything from career patterns to day-to-day chains of command – there is great diversity of experience. Orders are issued and obeyed, and yet highly individualistic people grow and flourish.

It took me many years to appreciate the subtle benefits that grow out of these paradoxical, sometimes contradictory elements. When the Army is functioning at its best, it can be stable without being stultifying. It can take risks without posing undue risks to people's lives. It can create individual 'laboratories,' within which offices and noncommissioned officers can experiment with the tools and techniques of leadership.

> Lieutenant-General William G. Pagonis, *Moving Mountains*, 1992.

Peacetime Army

Our long garrison life spoiled us, and effeminacy and desire for and love of pleasure, have weakened our military virtues. The entire nation must pass through the School of Misfortune, and we shall either die in the crisis, or a better condition will be created, after we have suffered bitter misery, and after our bones have decayed.

> Field Marshal August Graf von Gneisenau (1760–1831), quoted in Balck, *Development of Tactics*, 1922.

One should be careful not to compare this expanded and refined solidarity of a brotherhood of tempered, battle-scarred veterans with the self-esteem and vanity of regular armies which are patched together only by service-regulations and drill. Grim severity and iron discipline may be able to preserve the military virtues of a unit, but it cannot create them. These factors are valuable, but they should not be overrated. Discipline, skill, good will, a certain pride, and high morale, are the attributes of an army trained in times of peace. They command respect, but they have no strength of their own. They stand or fall together. One crack, and the whole thing goes, like a glass too quickly cooled.

> Major-General Carl von Clausewitz, *On War*, 3.5, 1832, tr. Michael Howard and Peter Paret, 1976.

It must be obvious, therefore, that periods of tranquility are rich in

ns of friction between soldiers and statesmen, since the latter are for ever trying to find ways of saving money, while the former are constantly urging increased expenditure. It does, of course, occasionally happen that a lesson recently learned, or an immediate threat, compels them to agree.
General Charles de Gaulle, *The Edge of the Sword*, 1932.

It is in peace that regulations and routine become important and that qualities of boldness and originality are cramped. It is interesting to note how little of normal peace soldiering many of our best generals had – Cromwell, Marlborough, Wellington, and his lieutenants, Graham, Hill, Craufurd.
Field Marshal Viscount Wavell of Cyrenaica, *Soldiers and Soldiering*, 1953.

Our traditional fault – the one against which I coped hopelessly in earlier days of the Eighth Air Force – went all the way back to Langley Field. There they had the usual Base commander and Group commander. The Base commander wanted to mow the grass; the Group commander wanted to fly his airplanes... Answer? They mowed the grass.
  Because why? Because the Base commander made out the efficiency report on the Group commander. He got rated whether his grass was cut or not, or whether his buildings were painted. By gad, that's what he was going to do: mow grass.
General Curtis LeMay, *Mission With LeMay*, 1965.

Standing Army

Fritz, pay close attention to what I am going to say to you. Always keep up a good and strong army – you won't have a better friend and you can't survive without it. Our Neighbors want nothing more than to bring about our ruin – I am aware of their intentions, and you will come to know them as well. Believe me, don't let wishful thinking run away with you – stick to what is real. Always put your trust in a good army and in hard cash – they are the things which keep rulers in peace and security.
Frederick William I, 1731, advice to his son, the future Frederick the Great.

The Jealousies of a standing Army, and the Evils to be apprehended from one, are remote; and in my judgment, situated and circumstanced as we are, not at all to be dreaded; but the consequence of wanting one, according to my Ideas, ...is certain, and inevitable Ruin.
General George Washington, *The Writings of George Washington*, VI, 1932, p. 112.

The country must have a large and efficient army, one capable of meeting the enemy abroad, or they must expect to meet him at home.
The Duke of Wellington, letter, 28 January 1811.

The terrible power of a standing army may usually be exercised by whomever can control its leaders, as a mighty engine is set in motion by the cranking of a handle.
Winston S. Churchill, *The River War*, 1899.

One of the main arguments against armies is their futility; but, if this be true, this argument can with equal force be directed against peaceful organizations; for surely it is just as futile to keep vast numbers of a nation on the brink of starvation and prostitution, as happens in nearly all civilized countries to-day, as it is to keep an insignificant minority of this same nation on the brink of war.
Major-General J.F.C. Fuller, *The Reformation of War*, 1923.

ARTILLERY

Blessed be those happy ages that were strangers to the dreadful fury of these devilish instruments of artillery, whose

THE ART OF WAR

inventor I am satisfied is now in hell, receiving the reward of his cursed invention, which is the cause that very often a cowardly base hand takes away the life of the bravest gentleman.
Miguel de Cervantes, *Don Quixote*, 1615.

Ultimo ratio regum (The final argument of kings).
Motto, c. 1660, inscribed on French artillery by order of Louis XIV.

The object of artillery should not consist of killing men on the whole of the enemy's front, but to overthrow it, to destroy parts of this front... then they obtain decisive effects; they make a gap.
Field Marshal Francois Comte de Guibert, *Essai général de tactique*, 1773.

It used to be our custom to form regiments from the largest men possible. This was done for a reason, for in the early wars it was men and not cannon that decided victory, and battalions of tall men advancing with the bayonet scattered the poorly assembled enemy troops – with the first attack. Now artillery has changed everything. A cannon ball knocks down a man six feet tall just as easily as one who is only five feet seven. Artillery decides everything, and infantry no longer do battle with naked steel.
Frederick the Great, *A History of My Own Times*, 1789, tr. Holcroft.

The artillery, like the other arms, must be collected in mass if one wishes to obtain decisive results.
Napoleon, *Correspondance de Napoléon Ier, publié par ordre de l'Empereur Napoléon III*, XIX, No. 15338, p. 116.

In siege warfare, as in the open field, it is the gun which plays the chief part; it has effected a complete revolution... It is with artillery that war is made.
Napoleon, 1809, after the Battle of Loebau, *Correspondance de Napoléon Ier, publié par ordre de l'Empereur Napoléon III*, XXX, p. 447.

Leave the artillerymen alone. They are an obstinate lot.
Napoleon, quoted in Bellemy, *Red God of War*, 1987.

Taylor: 'What are you using, Captain, grape or canister?'
Bragg: 'Canister, General.'
Taylor: 'Single or double?'
Bragg: 'Single.'
Taylor: 'Well, double-shot your guns and give 'em hell.'
General Zachary Taylor to Captain Braxton Bragg, 23 February 1847, at the Battle of Buena Vista, in K. Jack Bauer, *The Mexican War*, 1974.

You can't describe the moral lift,
When in the fight your spirit weary
Hears above the hostile fire
Your own artillery.
Shells score the air like wavy hair
From a forward battery,
As regimental cannon crack
While, from positions further back,
In bitter sweet song overhead
Crashing discordantly
Division's pounding joins the attack;
Mother like she belches shell;
Gloriously it flies, and well,
As, with a hissing, screaming squall,
A roaring furnace, giving all,
she sears a path for the infantry...
Aleksandr Tvardovskiy, 1943, from the narrative poem, *Vasily Tyorkin*, tr. Chris Bellemy, *Red God of War*, 1987.

THE ART OF WAR

In military affairs nothing is more important than certain victory. In employing the army nothing is more important than obscurity and silence. In movement nothing is more important than the unexpected. In planning nothing is more important than not being knowable.
The T'ai Kung, c. 1100 BC (*The Seven

THE ART OF WAR

Military Classics of Ancient China, tr. Ralph Sawyer, 1993).

The art of war is, in the last result, the art of keeping one's freedom of action.
 Xenophon (430–c. 355 BC).

It is bias to think that the art of war is just for killing people. It is not to kill people, it is to kill evil. It is a strategem to give life to many people by killing the evil of one person.
 Yagyu Munenori, late 16th century, quoted in Thomas Cleary, *The Japanese Art of War*, 1991.

The art of war is divided between art and strategem. What cannot be done by force, must be done by strategem.
 Frederick the Great, *Instructions to His Generals*, 1747, tr. Phillips, 1940.

The three military arts. First – Apprehension, how to arrange things in camp, how to march, how to attack, pursue, and strike; for taking up position, final judgement of the enemy's strength, for estimating his intentions.
  Second – Quickness... This quickness doesn't weary the men. The enemy doesn't expect us, reckons us 100 versts away, and if a long way off to begin with – 200, 300 or more – suddenly we're on him, like snow on the head; his head spins. Attack with what comes up, with what God sends; the cavalry to begin, smash, strike, cut off, don't let slip, hurra!
  Brothers do miracles!
  Third – Attack. Leg supports leg, arm strengthens arm; many men will die in the volley; the enemy has the same weapons, but he doesn't know the Russian bayonet. Extend the line – attack at once with cold steel; extend the line without stopping... the Cossacks to get through everywhere... In two lines is strength; in three, half as much again; the first breaks, the second drives into heaps, the third overthrows.
 Field Marshal Prince Aleksandr V. Suvorov, *The Science of Victory*, 1796.

One may teach tactics, military engineering, artillery work about as one teaches geometry. But knowledge of the higher branches of war is only acquired by experience and by a study of the history of the wars of great generals. It is not in a grammar that one learns to compose a great poem, to write a tragedy.
 Napoleon, quoted in Foch, *Principles of War*, 1913.

The art of war consists, with a numerically inferior army, in always having larger forces than the enemy at the point which is to be attacked or defended. But this art can be learned neither from books nor from practice: it is an intuitive way of acting which properly constitutes the genius of war.
 Napoleon, 1797, dictation, *Correspondance de Napoléon Ier, publié par ordre de l'Empereur Napoléon III*, Vol. III, p. 163, quoted in Christopher Herold, ed., *The Mind of Napoleon*, 1955.

The whole art of war consists in a well-reasoned and extremely circumspect defensive, followed by rapid and audacious attack.
 Napoleon, during the Prussian campaign of 1806, *Correspondance de Napoléon Ier, publié par ordre de l'Empereur Napoléon III*, Vol. XIII, No. 10558, p. 10.

The art of war is a simple art; everything is in the performance. There is nothing vague about it; everything in it is common sense; ideology does not enter into it.
 Napoleon, 1818, conversation, quoted in Gaspar Gougraud, *Sainte Hélène: Journal inédit de 1815 à 1818*, II, quoted in Christopher Herold, ed., *The Mind of Napoleon*, 1955.

The art of war is an immense study, which comprises all others.
 Napoleon, quoted in Christopher Herold, ed., *The Mind of Napoleon*, 1955.

THE ART OF WAR

The art of war is no more than the art of augmenting the chances which are in our favor.
Napoleon (1769–1821).

The whole art of war consists in getting at what lies on the other side of the hill, or, in other words, what we do not know from what we do know.
The Duke of Wellington (1769–1852), quoted in *The Croker Papers*, III, 1885, and David Chandler, *The Campaigns of Napoleon*, 1966.

The art of war deals with living and with moral forces. Consequently, it cannot attain the absolute, or certainty; it must always leave a margin for uncertainty, in the greatest things as much as in the smallest. With uncertainty in one scale, courage and self-confidence must be thrown into the other to correct the balance. The greater they are, the greater the margin that can be left for accidents.
Major-General Carl von Clausewitz, *On War*, 1.1, 1832, tr. Michael Howard and Peter Paret, 1976.

Of all theories on the art of war, the only reasonable one is that which, founded upon the study of military history, admits a certain number of regulating principles, but leaves to natural genius the greatest part in the general conduct of a war without trammeling it with exclusive rules.
  On the contrary, nothing is better calculated to kill natural genius and to cause error to triumph, than those pedantic theories, based upon the false idea that war is a positive science, all the operations of which can be reduced to infallible calculations.
Lieutenant-General Antoine-Henri Baron de Jomini, *Summary of the Art of War*, 1838, tr. Mendell and Craighill, 1862.

Always mystify, mislead and surprise the enemy if possible; and when you strike and overcome him, never let up in pursuit so long as your men have strength to follow, for an army routed, if hotly pursued, becomes panic-stricken, and can then be destroyed by half their number. The other rule is, never fight against heavy odds, if by any possible maneuvering you can hurl your own force on only a part, and that the weakest part, of your enemy and crush it. Such tactics will win every time, and a small army may thus destroy a large one in detail, and repeated victory will make it invincible.
Lieutenant-General Thomas J. 'Stonewall' Jackson, 1862, quoted by John D. Imboden in Underwood and Buel, ed., *Battles and Leaders of the Civil War*, II, 1887–1888.

The art of war is simple enough. Find out where your enemy is. Get at him as soon as you can. Strike at him as hard as you can and as often as you can, and keep moving on.
General of the Army Ulysses S. Grant (1822–1885) – attributed.

One does simply what one *can* in order to apply what one *knows*.
Marshal of France Ferdinand Foch, *The Principles of War*, 1913.

They forget that the whole art of war is to gain your objective with as little loss as possible.
Field Marshal Viscount Montgomery of Alamein, 8 November 1917, letter to his mother, quoted in Nigel Hamilton, *Monty*, 1981.

War is an art and as such is not susceptible of explanation by fixed formula.
General George S. Patton, Jr., 'Success in War', *The Infantry Journal Reader*, 1931.

Whereas the other arts are, at their height, individual, the art of war is essentially orchestrated.
Captain Sir Basil Liddell Hart, *Thoughts on War*, 1944.

ATTACK

I've been studying the art of war for forty-odd years. When a surgeon decides in the course of an operation to change its objective... he is not making a snap decision but one based on knowledge, experience and training. So am I.
> General George S. Patton, Jr., on being accused of making snap decisions.

The conduct of war, like the practice of medicine, is an art, and because the aim of the physician and surgeon is to prevent, cure, or alleviate the diseases of the human body, so should the aim of the statesman and soldier be to prevent, cure, or alleviate the wars which inflict the international body.
> Major-General J.F.C. Fuller, *The Conduct of War*, 1961.

ATTACK

The King also boasts of his wisdom in military matters. He knows when to strike and when to avoid combat... 'attacking him who attacks... since, if one is silent after attack, it strengthens the heart of the enemy. Valiance is eagerness, cowardice is to slink back,' he tells us, adding 'he is truly a coward who is repelled upon his border.'
> Sesostris III (1887–1849 BC), Pharaoh of Egypt, quoted in James Breasted, *Ancient Records of Egypt*, I, 1906.

Forward, even with only a spear.
> Samurai proverb, quoted in Inoguchi and Nakajima, *The Divine Wind*, 1958.

Decline the attack unless you can make it with advantage.
> Field Marshal Maurice Comte de Saxe, *My Reveries*, 1732, tr. Phillips, 1940.

Gentlemen, the enemy stands behind his entrenchments, armed to the teeth. We must attack him and win, or else perish. Nobody must think of getting through any other way. If you don't like this, you may resign and go home.
> Frederick the Great, 5 December 1757, to his officers before the Battle of Leuthen.

The impact of an army, like the total mechanical coefficients, is equal to the mass multiplied by the velocity.
> Napoleon, 30 July 1800, quoted in R.M. Johnston, *The Corsican*, 1910.

Attack inspires a soldier, it adds to his power, rouses his self-reliance, and confuses the enemy. The side attacked always overestimates the strength of the attacker.
> Field Marshal August Graf von Gneisenau (1760–1831), quoted in R.E. Dupuy, *The Military Heritage of America*, 1956.

The attack should be like a soap bubble, which distends itself until it bursts.
> Major-General Carl von Clausewitz, *On War*, 1832, tr. Michael Howard and Peter Paret, 1976.

I was too weak to defend, so I attacked.
> General Robert E. Lee (1807–1870) – attributed.

When we are fighting there is no need to think of defense. A positive attack is the best form of defense... the vital point in actual warfare is to apply to the enemy what we do not wish to be applied to ourselves and at the same time not to let the enemy apply it to us... we must always forestall them.
> Admiral Marquis Heihachiro Togo, 15 May 1905, quoted in Dennis and Peggy Warner, *The Tide at Sunrise*, 1974.

My centre is giving way, my right is pushed back, situation excellent, I am attacking.
> Marshal of France Ferdinand Foch, 8 September 1914, message to Marshal Joffre at the First Battle of the Marne, quoted in Liddell Hart, *Reputations: Ten Years After*, 1928.

The normal purpose of an attack is the infliction of death wounds and

ATTRITION

destruction on the enemy troops with a view to establish both physical and moral ascendancy over them. The gaining of ground in such a combat is simply an incident; not an object.
> General George S. Patton, Jr., 1 November 1926, letter to his wife, *The Patton Papers*, I, 1972.

Napoleon once said, 'I attack to be attacked.' What he meant was that he threw forward a small fraction of his forces for the enemy to bite on, and when his adversary's jaws were fixed he moved up his large reserves – the capital of his tactical bank – and struck his real blow...
> Major-General J.F.C. Fuller, 'Co-ordination of the Attack', *The Infantry Journal*, 1/1931.

Attack is the chief means of destroying the enemy, but defence cannot be dispensed with. In attack the immediate object is to destroy the enemy, but at the same time it is self-preservation, because if the enemy is not destroyed, you will be destroyed. In defence the immediate object is to preserve yourself, but at the same time defence is a means of supplementing attack or preparing to go over to the attack. Retreat is in the category of defence and is a continuation of defence, while pursuit is a continuation of attack. It should be pointed out that destruction of the enemy is the primary object of war and self-preservation the secondary, because only by destroying the enemy in large numbers can one effectively preserve oneself. Therefore attack, the chief means of destroying the enemy, is primary, while defence, a supplementary means of destroying the enemy and a means of self-preservation, is secondary. In actual warfare the chief role is played by defence much of the time and by attack for the rest of the time, but if war is taken as a whole, attack remains primary.
> Mao Tse-tung, *On Protracted War*, May 1938.

We are so outnumbered there's only one thing to do. We must attack.
> Admiral Andrew Cunningham, 11 November 1940, before attacking the Italian fleet at Taranto.

When the situation is obscure, attack.
> Colonel-General Heinz Guderian (1888–1954) – attributed.

ATTRITION

We lost fifty per cent more men than did the enemy, and yet there is sense in the awful arithmetic propounded by Mr. Lincoln. He says that if the same battle were to be fought over again, every day, through a week of days, with the same relative results, the army under Lee would be wiped out to its last man, the Army of the Potomac would still be a mighty host, the war would be over, the Confederacy gone, and peace would be won at a smaller cost of life than it will be if the week of lost battles must be dragged out through yet another year of camps and marches, and deaths in hospitals rather than upon the field. No general yet found can face the arithmetic, but the end of the war will be at hand when he shall be discovered.
> A summary of President Abraham Lincoln's comments to one of his secretaries after the Battle of Fredericksburg in late December 1862.

My object in war was to exhaust Lee's army. I was obliged to sacrifice men to do it. I have been called a butcher. Well, I never spared lives to gain an object; but then I gained it, and I knew it was the only way.
> General of the Army Ulysses S. Grant, quoted in Chancellor, *An Englishman in the American Civil War*, 1971.

The old wars were decided by their

AUDACITY

episodes rather than by their tendencies. In this (modern) war the tendencies are far more important than the episodes. Without winning any sensational victories, we may win... Germany may be defeated more fatally in the second or third year of the war than if the Allied armies had entered Berlin in the first.
 Sir Winston S. Churchill, November 1915, speech in the House of Commons.

In the stage of the wearing out struggle losses will necessarily be heavy on both sides, for in it the price of victory is paid.
 Field Marshal Earl Douglas Haig, *Dispatches*, 21 March 1919.

An enemy may be worn out by physical and moral action; this, though the usual method of defeating him, is also, frequently, the most uneconomical method, for the process of disintegration is mutually destructive.
 Major-General J.F.C. Fuller, *The Reformation of War*, 1923.

Of what use is decisive victory in battle if we bleed to death as a result of it?
 Sir Winston S. Churchill (1874–1965).

But attrition is a two-edged weapon and, even when skilfully wielded, puts a strain on the users. It is especially trying to the mass of the people, eager to see a quick finish – and always inclined to assume that this can only mean the enemy's finish.
 Captain Sir Basil Liddell Hart, *Strategy*, 1954.

Attrition is not a strategy. It is, in fact, irrefutable proof of the absence of any strategy. Any commander who resorts to attrition admits his failure to conceive of an alternative. He rejects warfare as an art... He uses blood in lieu of brains.
 General Dave R. Palmer, *Summons of the Trumpet*, 1978.

General Westmoreland thought he had found the answer to the question of how to win this war: He would trade one American life for ten or eleven or twelve North Vietnamese lives, day after day, until Ho Chi Minh cried uncle. Westmoreland would learn, too late, that he was wrong; that the American people didn't see a kill ratio of 10–1 or even 20–1 as any kind of bargain.
 Lieutenant-General Harold G. Moore, *We Were Soldiers Once... And Young*, 1992.

AUDACITY

Impetuosity and audacity often achieve what ordinary means fail to achieve.
 Niccolò Machiavelli, *Discourses*, 1517.

With audacity one can undertake anything, but not do everything.
 Napoleon, *Maxims of War*, 1831.

Never forget that no military leader has ever become great without audacity.
 Major-General Carl von Clausewitz, *Principles of War*, 1812.

They want war too methodical, too measured; I would make it brisk, bold, impetuous, perhaps sometimes even audacious.
 Lieutenant-General Antoine-Henri Baron de Jomini, *Summary of the Art of War*, 1838, tr. Mendell and Craighill, 1862.

In war nothing is impossible, provided you use audacity.
 General George S. Patton, Jr., *War As I Knew It*, 1947.

If you take a chance, it usually succeeds, presupposing good judgement.
 Lieutenant-General Sir Gifford Martel, quoted in S.L.A. Marshall, *The Armed Forces Officer*, 1951.

Audacity is nearly always right, gambling nearly always wrong.
 Captain Sir Basil Liddell Hart, quoted in Richard Simpkin, *Race to the Swift*, 1985.

AUTHORITY

Audace, audace, toujours audace.
[Audacity, audacity, always audacity.]
  Motto on plaque outside US Army Command and General Staff College, Fort Leavenworth, Kansas.

AUTHORITY

When soldiers forge ahead and do not dare to fall back, this means they fear their own leaders more than they fear the enemy. If they dare to fall back and dare not forge ahead, this means they fear the enemy more than they fear their own leaders. When a leader can get his troops to plunge right into the thick of raging combat, it is his authority and sternness that brings this about.

The rule is 'To be awesome and yet caring makes a good balance.'
  Liu Ji (1310–1375), *Lessons of War* (*Mastering the Art of War*, tr. Thomas Cleary, 1989).

I like to convince people rather than stand on mere authority.
  The Duke of Wellington (1769–1852), quoted in John Keegan, *The Mask of Command*, 1987.

'Where did you get these articles?' he [an officious officer] inquired...'

She paid no heed to his interrogations, and, indeed, did not hear them, so completely absorbed was she in her work of compassion. Watching her with admiration for her skill, administrative ability, and intelligence, – for she not only fed the wounded men, but temporarily dressed their wounds in some cases, – he approached her again: –

'Madam, you seem to combine in yourself a sick-diet kitchen and a medical staff. May I inquire under whose authority you are working?'

Without pausing in her work, she answered him, 'I have received my authority from the Lord God Almighty; have you anything that ranks higher than that?'
  Mary 'Mother' Bickerdyke, a volunteer caring Union wounded in the American Civil War, quoted in Mary A. Livermore, *My Story of the War*, 1887.

Discipline and morale influence the inarticulate vote that is constantly taken by masses of men when the order comes to move forward – a variant of the crowd psychology that inclines it to follow a leader. But the Army does not move forward until the motion has carried. 'Unanimous consent' only follows cooperation between the individual men in the ranks.
  Lieutenant-General James G. Harbord, *The American Army in France*, 1936.

But we can go one step beyond General Harbord's suggestion that the multiplied individual acceptance of a command alone gives that command authority. It is not less true that the multiplied rejection of a command nullifies it. In other words authority is the creature rather than the creator of discipline and obedience. In the more recent experiences of our arms, under the stresses of battle, there are many instances of troops being given orders, and refusing to obey. In every case, the root cause was lack of confidence in the wisdom and ability of those who led. When a determining number of men in ranks have lost the will to obey, their erstwhile leader has *ipso facto* lost the capacity to command. *In the final analysis, authority is contingent upon respect far more truly than respect is founded upon authority.*
  Brigadier-General S.L.A. Marshall, *The Armed Forces Officer*, 1951.

The commander must establish personal and comradely contact with his men but without giving away an inch of his authority.
  Field Marshal Erwin Rommel, *The Rommel Papers*, 1953.

B

BATTLE
The Cost of Battle

I hope to God that I have fought my last battle. It is a bad thing to be always fighting. While in the thick of it I am too much occupied to feel anything; but it is wretched just after. It is quite impossible to think of glory. Both mind and feelings are exhausted. I am wretched even at the moment of victory, and I always say that, next to a battle lost, the greatest misery is a battle gained. Not only do you lose those dear friends with whom you have been living, but you are forced to leave the wounded behind you. To be sure, one tries to do the best for them, but how little that is! At such moments every feeling in your breast is deadened. I am now just beginning to regain my natural spirits, but I never wish for any more fighting.
> The Duke of Wellington, July 1815, to Lady Shelley after Waterloo.

After the battle of Eylau, the Emperor spent several hours daily on the battlefield... To visualize the scene one must imagine, within the space of one square league, nine or ten thousand corpses; four or five thousand dead horses; rows upon rows of Russian field packs; the remnants of guns and swords; the ground covered with cannon balls, shells, and other ammunition; and twenty-four artillery pieces, near which could be seen the corpses of the drivers who were killed while trying to move them – all this sharply outlined on a background of snow. A sight such as this should inspire rulers with the love of peace and the hatred of war.
> Napoleon, 2 March 1807, 64th Bulletin (Napoleon composed most of the Bulletins personally), quoted in Christopher Herold, ed., *The Mind of Napoleon*, 1955.

Between a battle lost and a battle won, the distance is immense and there stand empires.
> Napoleon, 15 October 1813, on the eve of the Battle of Leipzig.

The Dynamics of Battle

Ares drove them [Trojans], fiery-eyed
   Athena drove the Argives,
and Terror and Rout and relentless
   Strife stormed too,
sister of manslaughterer Ares, Ares'
   comrade-in-arms –
Strife, only a slight thing when she first
   first rears her head
but her head soon hits the sky as she
   strides across the earth.
Now Strife hurled down the leveler
   Hate amidst both sides,
wading into the onslaught, flooding men
   with pain.
> Homer, *The Iliad*, 4.510–16, c. 800 BC, tr. Robert Fagles, 1990.

The result of a battle depends on the instantaneous flash of an idea. When you are about to give battle concentrate all your strength, neglect nothing; a battalion often decides the day.
> Napoleon, 30 July 1800, quoted in R.M. Johnston, ed., *The Corsican*, 1910.

The battle is completely lost; but it is only two o'clock, we have time to gain

BATTLE

another to-day.
> General Louis Desaix, 14 June 1800, to Napoleon during the Battle of Marengo; quoted in Louis de Bourrienne, *The Memoirs of Napoleon Bonaparte*, II, 1855.

There is a moment in engagements when the least manoeuvre is decisive and gives the victory; it is the one drop of water which makes the vessel run over.
> Napoleon (1769–1821), 'Precis des guerres de J. Cesar', *Correspondance de Napoléon Ier, publié par ordre de l'Empereur Napoléon III*, Vol. XXXII, 1858–1870.

The business of the English Commander-in-Chief being first to bring an Enemy's fleet to Battle on the most advantageous terms to himself, (I mean that of laying his Ships close on board the Enemy, as expeditiously as possible); and secondly to continue them there until the Business is decided.
> Admiral Viscount Nelson, 1805, from the Order to the Fleet before the Battle of Trafalgar.

The issue of a battle is the result of a single instant, a single thought. The adversaries come into each other's presence with various combinations; they mingle; they fight for a length of time; the decisive moment appears; a psychological spark makes the decision; and a few reserved troops are enough to carry it out.
> Napoleon, 4–5 December, 1815, conversation on Saint Helena, quoted in Las Casas, cited in Christopher Herold, ed., *The Mind of Napoleon*, 1955.

A battle sometimes decides everything; and sometimes the most trifling thing decides the fate of a battle.
> Napoleon, 9 November 1816, letter on Saint Helena.

A battle is a dramatic action which has its beginning, its middle, and its end. The battle order of the opposing armies and their preliminary maneuvers until they come to grips form the exposition. The countermaneuvers of the army which has been attacked constitute the dramatic complication. They lead in turn to new measures and bring about the crisis, and from this results the outcome or denouement.
> Napoleon, dictation on Saint Helena, quoted in Christopher Herold, ed., *The Mind of Napoleon*, 1955.

Battle is the bloodiest solution. While it should not simply be considered as mutual murder – its effect... is rather a killing of the enemy's spirit than of his men – it is always true that the character of a battle, like its name, is slaughter (*Schlacht*), and its price is blood.
> Major-General Carl von Clausewitz, *On War*, 4.11, 1832, tr Michael Howard and Peter Paret, 1976.

The major battle is therefore to be regarded as concentrated war, as the center of gravity of the entire conflict or campaign. Just as the focal point of a concave mirror causes the sun's rays to converge into a perfect image and heats them to maximum intensity, so all forces and circumstances of war are united and compressed to maximum effectiveness in battle.
> Major-General Carl von Clausewitz, *On War*, 4.11, 1832, tr. Michael Howard and Peter Paret, 1976.

Battles are won by slaughter and manoeuvre. The greater the general, the more he contributes to the manoeuvre, the less he demands in slaughter.
> Sir Winston S. Churchill, *The World Crisis*, II, 1923.

Battle is an orgy of disorder. No level lawns or marker flags exist to aid us strut ourselves in vain display, but rather groups of weary wandering men

BATTLE

seek gropingly for means to kill their foe. The sudden change from accustomed order to utter disorder – to chaos, but emphasize the folly of schooling to precision and obedience where only fierceness and habituated disorder are useful.
General George S. Patton, Jr., 27 October 1927, lecture 'Why Men Fight', *The Patton Papers*, I, 1972.

But battle must not be seen as some kind of smooth-running conveyor belt on which the various technical combat resources are merged. Battle is a complex and fickle thing. So command and control must be ready to deal with abrupt changes in the situation, and sometimes to reshape an earlier plan radically.
Marshal of the Soviet Union Mikhail N. Tukhachevskiy, 1931–32, quoted in Richard Simpkin, *Deep Battle*, 1987.

Battle should no longer resemble a bludgeon fight, but should be a test of skill, a manoeuvre combat, in which is fulfilled the great principle of surprise by striking 'from an unexpected direction against an unguarded spot.'
Captain Sir Basil Liddell Hart, *Thoughts on War*, 1944.

While the battles the British fight may differ in the widest possible ways, they have invariably two common characteristics – they are always fought uphill and always at the junction of two or more map sheets.
Field Marshal Viscount Slim of Burma, *Unofficial History*, 1959.

The Human Factor

It would be a service to humanity and to one's people to dispel this illusion and show what battles are. They are buffooneries, and none the less buffooneries because they are made terrible by the spilling of blood. The actors, heroes in the eyes of the crowd, are only poor folk torn between fear, discipline and pride. They play some hours at a game of advance and retreat, without ever meeting, closing with, even seeing closely, the other poor folk, the enemy, who are as fearful as they but who are caught in the same web of circumstance.
Colonel Charles Ardant du Picq, *Battle Studies*, 1880, tr. Greely, 1957.

It is always necessary in battle to do something which would be impossible for men in cold blood.
Colonel Louis Louzeau de Grandmaison, February 1911, lecture at the *École de Guerre*.

Battles are decided in favor of the troops whose bravery, fortitude, and especially, whose endurance, surpasses that of the enemy's; the army with the higher breaking point wins the decision.
General of the Army George C. Marshall, *Memoirs of My Service in the World War* (written 1919–1923), published 1976.

Battle is not a terrifying ordeal to be endured. It is a magnificent experience wherein all the elements that have made man superior to the beasts are present: courage, self-sacrifice, loyalty, help to others, devotion to duty. As you go in, you will perhaps be a little short of breath, and your knees may tremble... This breathlessness, this tremor, are not fear. It is simply the excitement every athlete feels just before the whistle blows – no, you will not fear for you will be borne up and exhalted by the proud instinct of our conquering race. You will be inspired by a magnificent hate.
General George S. Patton, Jr., 20 December 1941, address to the Second Armored Division, *The Patton Papers*, II, 1974.

When things are going badly in battle the best tonic is to take one's mind off one's own troubles by considering what a rotten time one's opponent must be having.

BATTLE

Battle is more than a combination of fire and movement. It is the integration of fire, movement, and consciousness. The commander, therefore, cannot rest content with guiding the fire and directing the movement; he must guide the soldier's mental reactions to battle. Hence the commander is responsible for the mental preparation of his men no less than for their physical and technical training and their being brought to battle. Paternal concern for the soldier and his welfare does not mean pampering him; far from it. Soldiers wrapped in cotton-wool will fall helpless victims to the terrors of war, being unprepared to meet the most terrible of all dangers. Sincere concern, on the other hand, will win the soldier's confidence and will prepare him to face the most trying circumstances.
General Yigal Allon, *The Making of Israel's Army*, 1970.

The Purpose of Battle

Battles concerning which one cannot say why and to what purpose they have been delivered are commonly the resource of ignorant men.
Field Marshal Maurice Comte de Saxe, *My Reveries*, 1732, tr. Phillips, 1940.

The man who does things without motive or in spite of himself is either insane or a fool. War is decided by battles, and it is not finished except by them. They have to be fought, but it should be done opportunely and with all the advantages on your side... Advantages are procured in battles every time that you determine to fight, or when a battle that you have meditated upon for a long time is a consequence of the maneuvers that you have made to bring it on.
Frederick the Great, *Instructions to his Generals*, 1747, tr. Phillips, 1940.

Field Marshal Viscount Wavell of Cyrenaica, *Other Men's Flowers*, 1944.

Battles decide the fate of a nation. In war it is absolutely necessary to come to decisive actions either to get out of the distress of war or to place the enemy in that position, or even to settle a quarrel which otherwise perhaps would never be finished. A wise man will make no movement without good reason, and a general of an army will never give battle if it does not serve some important purpose. When he is forced by his enemy into battle it is surely because he will have committed mistakes which force him to dance to the tune of his enemy.
Frederick the Great, *Instruction militaire*, 1761.

Maxim 15. The first consideration of a general who offers battle should be the glory and honour of his arms; the safety and preservation of his men is only the second; but it is in the enterprise and courage resulting from the former that the latter will most assuredly be found...
Napoleon, *The Military Maxims of Napoleon*, tr. George D'Aguilar, 1831 (David Chandler, ed., 1987).

War as we study it, positive in its nature, permits only of positive answers: there is no result without cause; if you seek the result, develop the cause, employ force.
 If you wish your opponent to withdraw, beat him; otherwise nothing is accomplished, and there is only one means to that end: the battle.
Marshal of France Ferdinand Foch, *Principles of War*, 1913.

Battles are the principal milestones in secular history. Modern opinion resents this uninspiring truth, and historians often treat the decisions of the field as incidents in the dramas of politics and diplomacy. But great battles, won or lost, change the entire course of events, create new standards of values, new

moods, new atmospheres, in armies and nations, to which all must conform.
Sir Winston S. Churchill, *Marlborough*, 1933.

Battles in which no one believes should not be fought.
Field Marshal Viscount Wavell of Cyrenaica, 'Recollections' (unpublished), 1946.

BATTLEFIELD

...A dancing floor of Ares.
Epaminondas (c. 418–362 BC) describing the Boeotian Plain, quoted in Plutarch, *The Lives, c.* AD 100 (*The Age of Alexander*, tr. Ian Scott-Kilvert, 1965).

The battlefield is the place:
Where one toasts the divine liquor in war,
Where are stained red the divine eagles,
Where the jaguars howl,
Where all kinds of precious stones rain from ornaments
Where wave headdresses rich with fine plumes,
Where princes are smashed to bits.
Aztec poem, quoted in Michael D. Coe, *Mexico: From the Olmec to the Aztecs*, 1994.

The village of Fuentes de Oñoro, having been the field of battle, has not been much improved by the circumstance.
The Duke of Wellington, 3–5 May 1811, *The Dispatches of Field Marshal the Duke of Wellington, During His Various Campaigns in India, Denmark, Portugal, Spain, the Low Countries and France*, VII, 1834–1838.

It was a beautiful, calm, moonlight night. Suddenly a dog, which had been hiding under the clothes of a dead man, came up to us with a mournful howl, and then disappeared again immediately into his hiding place. He would lick his master's face, then run up to us again, only to return once more to his master. It seemed as if he were asking both for help and revenge. Whether it was the mood of the moment, whether it was the place, the time, the weather, or the action itself, or whatever it was, it is certainly true that nothing on any battlefield ever made such an impression on me. I involuntarily remained still, to observe the spectacle. This dead man, I said to myself, has perhaps friends, and he is lying here abandoned by all but his dog! What a lesson nature teaches us by means of an animal.
Napoleon, 1–3 December 1815, conversation on St Helena about one of the battles of the Italian campaign of 1796–1797, reported by Las Casas, quoted in Kircheisen, ed., *The Memoirs of Napoleon I*, 1929.

Mountains, rivers, grass and trees: utter desolation
For ten miles the wind smells of blood from this new battlefield.
The steed advances not, men speak not.
General Maruseki Nogi, after the Battle of Nanshan, 1904, quoted in Dennis and Peggy Warner, *The Tide at Sunrise*, 1974.

The dominant feeling of the battlefield is loneliness.
Field Marshal Viscount Slim of Burma, June 1941, to the officers of the Tenth Indian Infantry Division.

The battlefield is cold. It is the loneliest place which men may share together.
To the infantry soldier new to combat, its most unnerving characteristic is not that it invites him to a death he does not seek. To the extent necessary, a normal man may steel himself against the chance of death.
The harshest thing about the field is that it is empty. No people stir about. There is little or no sign of action. Overall there is a great quiet which seems more ominous than the occasional tempest of fire.
It is the emptiness which chills a man's blood and makes the apple harden in his throat. It is the emptiness which grips him with paralysis. The small dangers

THE BAYONET

which he had faced in his earlier life had always paid their dividend of excitement. Now there is great danger, but there is not excitement about it.
> Brigadier-General S.L.A. Marshall, *Men Against Fire*, 1947.

I have always regarded the forward edge of the battlefield as the most exclusive club in the world.
> General Sir Brian Horrocks, *A Full Life*, 1960.

THE BAYONET

The onset of Bayonets in the hands of the Valiant is irresistable.
> Major-General John Burgoyne, extract from his orderly book, 1777.

The bullet is a mad thing; only the bayonet knows what it is about.
> Field Marshal Prince Aleksandr V. Suvorov, *The Science of Victory*, 1796.

When bayonets deliberate, power escapes from the hands of the government.
> Napoleon, *Political Aphorisms*, 1848.

The people never chafe themselves against naked bayonets.
> Napoleon, *Political Aphorisms*, 1848.

Rangers of Connaught! It is not my intention to expend any powder this evening. We'll do this business with cold steel.
> General Sir Thomas Picton, 6 April 1812, to the 88th Foot before the assault on Badajoz.

Have you not got your bayonets?
> Sir George Cathcart at Inkerman on 5 November 1854, when word came that his division was low on ammunition.

Col B.E. Bee: General, they are beating us back!
Stonewall Jackson: Then, sir, we will give them the bayonet!
> At the First Battle of Bull Run, 21 July 1861, *Charleston Mercury*, 25 July 1861.

Not a moment was about to be lost! Five minutes more of such a defensive, and the last roll-call would sound for us! Desperate as the chances were, there was nothing for it, but to take the offensive. I stepped to the colors. The men turned towards me. One word was enough, – 'BAYONET!' – It caught like fire, and swept along the ranks. The men took it up with a shout, one could not say, whether from the pit, or the song of the morning star! It was vain to order 'Forward.' No mortal could have heard it in the mighty hosanna that was winging the sky. Nor would he wait to hear. There are things still as the first creation, 'whose seed is in itself.' The grating clash of steel in fixing bayonets told its own story; the color rose in front; the whole line quivered for the start; the edge of the left-wing rippled, swung, tossed among the rocks, straightened, changed curve from scimitar to sickle-shaped; and the bristling archers swooped down upon the serried host – down into the face of half a thousand! Two hundred men!
> Major-General Joshua L. Chamberlain, of the bayonet charge of his 20th Maine at Little Round, Gettysburg on 2 July 1863, 'Through Blood and Fire at Gettysburg', *Hearst's Magazine*, 1913.

All too many of our men lost their lives in trying to 'close with the bayonet' and 'kill with cold steel' as prescribed, so that they were shot down at close range by cooler-headed opponents who realized that the bullet outreaches the bayonet until the range is closed to less than two yards. More wisely, most of the German troops did not even fix their bayonets, lest the drag should disturb their aim in firing.
> Captain Sir Basil Liddell Hart, *The Memoirs of Captain Liddell Hart*, I, 1965.

BLITZKRIEG

The laurels of victory are at the point of the enemy bayonets. They must be plucked there; they must be carried by a hand-to-hand fight if one really means to conquer.
   Marshal of France Ferdinand Foch, *Precepts and Judgments*, 1919.

I beg leave to remind the [Cavalry] Board that very few people have ever been killed with the bayonet or the saber, but the fear of having their guts explored with cold steel in the hands of battle maddened men has won many a fight.
   General George S. Patton, Jr., January 1941, reply to the Cavalry Board's solicitation of his views, *The Patton Papers*, II, 1974.

If you're close enough to stick 'em, you're close enough to shoot 'em.
   Bill Mauldin's 'Willie and Joe' to a recruit in the movie *Willie and Joe*.

That weapon ceased to have any major tactical value at about the time the inaccurate and short-range musket was displaced by the rifle. But we have stubbornly clung to it – partly because of tradition which makes it inevitable that all military habits die a slow death, but chiefly because of the superstition that the bayonet makes troops fierce and audacious, and therefore likely to close with the enemy.
   Brigadier-General S.L.A. Marshall, *The Soldier's Load and the Mobility of a Nation*, 1980.

With their bayonets on their rifles, these young soldiers, looking grim and determined, moving forward through the thick mist of dawn – it is a sight I shall never forget. A lot of the enemy were killed with those bayonets.
   Lieutenant-Colonel Hugh Pike, 1982, of the attack of his battalion, 3 Para, at the Battle of Mount Longdon on 12 June 1982 during the Falklands War.

BLITZKRIEG

Military tactics are like unto water; for water in its natural course runs away from high places and hastens downwards. So in war, the way to avoid what is strong is to strike what is weak.

Water shapes its course according to the ground over which it flows; the soldier works out his victory in relation to the foe whom he is facing.

Therefore, just as water retains no constant shape, so in warfare there are no constant conditions.

He who can modify his tactics in relation to his opponent and thereby succeed in winning, may be called a heaven-born captain.
   Sun Tzu, *The Art of War*, 6, c. 500 BC, tr. Giles, 1910.

Mobility, Velocity, Indirect Approach...
   Colonel-General Heinz Guderian (1888–1954).

We had seen a perfect specimen of the modern Blitzkrieg; the close interaction on the battlefield of army and air force; the violent bombardment of all communications and of any town that seemed an attractive target; the arming of an active Fifth Column; the free use of spies and parachutists; and above all, the irresistable forward thrust of great masses of armour.
   Sir Winston S. Churchill, *The Gathering Storm*, 1948.

In the series of swift German conquests, the air force combined with the mechanized elements of the land forces in producing the paralysis and moral disintegration of the opposing forces and of the nations behind. Its effect was terrific, and must be reckoned fully as important as that of the panzer forces. The two are inseparable in any valuation of the elements that created the new style of lightning war – the blitzkrieg.
   Captain Sir Basil Liddell Hart, *Strategy*, 1954.

BOLDNESS

Sixteen days to Baghdad ain't a bad record.
>Lieutenant-General William Scott Wallace, V Corps Commander, *Army News Service*, 2 September 2003.

BLOODSHED

I do not want them to get used to shedding blood so young; at their age they do not know what it means to be a Moslem or an infidel, and they will grow accustomed to trifling with the lives of others.
>Saladin to Beha-ed-Din, when Saladin's children asked if they might kill a prisoner, 1191.

If you wish to be loved by your soldiers, husband their blood and do not lead them to slaughter.
>Frederick the Great, *Instructions to His Generals*, 1747, tr. Phillips, 1940.

Nothing honors a general more than keeping his serenity in danger and facing it when there are chances of winning; but nothing so much eclipses his name as the useless shedding of the blood of his men.
>General Jose de San Martin, 1818, quoted in Rojas, *San Martin*, 1957.

We are not interested in generals who win victories without bloodshed. The fact that slaughter is a horrifying spectacle must make us take war more seriously, but not provide an excuse for gradually blunting our swords in the name of humanity. Sooner or later someone will come along with a sharp sword and hack off our arms.
>Major-General Carl von Clausewitz, *On War*, 4.11, 1832, tr. Michael Howard and Peter Paret, 1976.

Wounds which in 1861, would have sent a man to the hospital for months, in 1865 were regarded as mere scratches, rather the subject of a joke than of sorrow. To new soldiers the sight of blood and death always has a sickening effect, but soon men become accustomed to it, and I have heard them exclaim on seeing a dead comrade borne to the rear, 'Well, Bill has turned up his toes to the daisies.'
>General of the Army William T. Sherman, *Memoirs of General W.T. Sherman*, 1875.

War is only a means to results... These being achieved, no man has the right to cause another drop of blood to be shed.
>Marshal of France Ferdinand Foch, quoted in Liddell Hart, *Through the Fog of War*, 1938.

What the American people want to do is fight a war without getting hurt. You cannot do that any more than you can go into a barroom brawl without getting hurt.
>Lieutenant-General Lewis B. 'Chesty' Puller, 1951, quoted in Davis, *Marine!*, 1962.

BOLDNESS

Thrust in fearlessly: however foreign a man may be, in every crisis it is the high face which will carry him through.
>Homer, *The Odyssey*, Bk 7, c. 800 BC, tr. T.E. Lawrence, 1932, Athena's admonition to Odysseus to boldly enter the feasthall of the Phaeacians.

My conclusion is, then, that as fortune is variable and men fixed in their ways, men will prosper as long as they are in tune with the times and will fail when they are not. However, I will say that in my opinion it is better to be bold than cautious, for fortune is a woman and whoever wishes to win her must importune and beat her, and we may observe that she is more frequently won by this sort than by those who proceed more deliberately.
>Niccolò Machiavelli, *The Prince*, 1513.

I wish to have no connection with any ship that does not sail fast; for I intend to go *in harm's way*.

BOLDNESS

Desperate affairs require desperate measures.
> Admiral John Paul Jones (1747–1792).

The measure may be thought bold, but I am of the opinion the boldest are the safest.
> Admiral Viscount Nelson, 24 March 1801, to Sir Hyde Parker urging vigorous action against the Russians and Danes.

Perhaps I should not insist on this bold maneuver, but it is my style, my way of doing things.
> Napoleon, 1813, letter to Prince Eugene.

But this noble capacity to rise above the most menacing dangers should also be considered as a principle in itself, separate and active. Indeed, in what field of human activity is boldness more at home than in war?
  A soldier, whether drummer boy or general, possesses no nobler quality; it is the very metal that gives edge and luster to the sword.
> Major-General Carl von Clausewitz, On War, 3.10, 1832, tr. Michael Howard and Peter Paret, 1976.

Boldness governed by superior intellect is the mark of a hero. This kind of boldness does not consist in defying the natural order of things and in crudely offending the laws of probability; it is rather a matter of energetically supporting that higher form of analysis by which genius arrives at a decision: rapid, only partly conscious weighing of the possibilities. Boldness can lend wings to intellect and insight; the stronger the wings then, the greater the heights, the wider the view, and the better results; though a greater prize, of course, involves greater risks.
> Major-General Carl von Clausewitz, On War, 3.10, 1832, tr. Michael Howard and Peter Paret, 1976.

While rashness is a crime, boldness is not incompatable with caution, nay, is often the quintessence of prudence.
> Major-General J.E.B. Stuart, General Order No. 26, Cavalry Tactics, July 30, 1863, Letters of Major General James E.B. Stuart, 1990.

'Safety first' is the road to ruin in war.
> Sir Winston S. Churchill, 3 November 1940, telegram to Anthony Eden.

A bold general may be lucky, but no general can be lucky unless he is bold. The general who allows himself to be bound and hampered by regulations is unlikely to win a battle.
> Field Marshal Earl Wavell, Generals and Generalship, 1941.

The fight was carried to the enemy at all times and in all places and he was driven from every place he held by the resolute attack of men who were not afraid to die. God favors the bold and the strong of heart...
> General Alexander A. Vandegrift, message to the Marine First Division on Guadalcanal indicating that elements of the division had taken nearby Tulagi island, Navy Department Press Release, 29 August 1942.

I can always pick a fighting man and God knows there are few of them. I am happy they are sending you to the front at once. I like generals so bold that they are dangerous. I hope they give you a free hand.
> General John J. Pershing, 21 October 1942, to General George Patton before his departure for the invasion of North Africa, quoted in The Patton Papers, II, 1974.

It is my experience that bold decisions give the best promise of success. But one must differentiate between strategical or tactical boldness and military gamble. A bold operation is one in which success is not a certainty but which in case of failure leaves one

BRAVERY

with sufficient forces in hand to cope with whatever situation may arise. A gamble, on the other hand, is an operation which can lead either to victory or to the complete destruction of one's forces. Situations arise where even a gamble may be justified – as, for instance, when in the normal course of events defeat is merely a matter of time, when the gaining of time is therefore pointless and the only chance lies in an operation of great risk.
> Field Marshal Erwin Rommel, *The Rommel Papers*, 1953.

BRAVERY

Without a sign his sword the brave
  man, draws,
And asks no omen but his country's
  cause.
> Homer, *The Iliad*, 12.283, c. 800 BC, tr. Alexander Pope, 1743.

They are surely to be esteemed the bravest spirits who, having the clearest sense of both the pains and pleasures of life, do not on that account shrink from danger.
> Thucydides, *The History of the Peloponnesian War*, 2.40, 404 BC, tr. Benjamin Jowett, 1881.

Whoever wants to see his own people again must remember to be a brave soldier: that is the only way of doing it. Whoever wants to keep alive must aim at victory. It is the winners who do the killing and the losers who get killed.
> Xenophon, *Anabasis*, 3.2, c. 360 BC (*The Persian Expedition*, tr. Rex Warner, 1965).

Few men are born brave; many become so through training and force of discipline.
> Flavius Vegetius Renatus, *Military Institutions of the Romans*, AD 378, tr. Clark, 1776.

The brave man [is] tall, very tall, small, fat, thin, very fat, very thin, somewhat like a stone pillar, moderately capable, good of appearance. The brave man [is] an eagle [or] an ocelot warrior, scarred, painted, courageous, brave, resolute.

The good, the true brave man [is] one who stands as a man, who is firm of heart, who charges, who strikes out at [the foe]. He stands as a man, he rallies, he takes courage; he charges, he strikes out at the foe. He fears no one, none can meet his gaze.

The bad brave man [is] one who leads others to destruction by his deception, who secretly puts one in difficulty; who visits others' houses; who yells; who slays others viciously, who treacherously forsakes one, who swoons with terror. He becomes frightened, he swoons with terror, he secretly puts one in difficulty.
> An Aztec account of the brave man, quoted in Fray Bernardino de Sahagún, *General History of the Things of New Spain*, Book 10 – The People, 1961.

He supposes all men to be brave at all times and does not realize that the courage of the troops must be reborn daily, that nothing is so variable, and that the true skill of a general consists in knowing how to guarantee it by his dispositions, his positions, and those traits of genius that characterize great captains.

…it is of all the elements of war the one that is most necessary to learn.
> Field Marshal Maurice Comte de Saxe, *My Reveries*, 1732, tr. Phillips, 1940.

One cannot think that blind bravery gives victory over the enemy.
> Field Marshal Prince Aleksandr V. Suvorov (1729–1800).

When soldiers brave death, they drive him into the enemy's ranks.
> Napoleon, 14 November 1806, to a regiment of chasseurs before the Battle of Jena.

BUREAUCRACY

No, shoot them all, I do not wish them to be brave.
> Lieutenant-General Thomas J. 'Stonewall' Jackson, 2 June 1862, to Colonel John Patton who deplored the necessity of killing three especially brave Union cavalrymen, quoted in Patton, 'Reminiscences of Jackson', Virginia Historical Society.

I could not help admiring in their misfortunes those brave men who were about to sign a capitulation which would deprive them of the forts they had so confidently and gallantly defended, and who, when all was over and no more could be done, gave up without brag or bluster, and made no excuses for their failure.

I cannot help admiring bravery even in a bad cause, and when I went over the works and saw to what a dreadful hammering the Confederates had been subjected, I thought it not without honor for any one to have fought on the other side.
> Admiral David Dixon Porter, describing the bearing of the officers of Fort Jackson upon its surrender on 27 April 1862, *Incidents and Anecdotes*, 1885.

Each man must think not only of himself, but think of his buddy fighting beside him. We don't want yellow cowards in this army, to send back to the States after the war and breed more like them. The brave men will breed brave men. One of the bravest men I saw in the African campaign was one of the fellows I saw on top of a telegraph pole in the midst of furious fire while we were plowing towards Tunis. I stopped and asked him what in the hell he was doing there at a time like that. He answered, 'Fixing the wire, sir.' Isn't it a little unhealthy right now, I asked. 'Yes sir, but this God damn wire has got to be fixed.'
> General George S. Patton, Jr, 1944, speech.

I do not believe that there is any man who would not rather be called brave than have any other virtue attributed to him.
> Field Marshal Viscount Slim of Burma, *Courage and Other Broadcasts*, 1957.

BUREAUCRACY

The ancients had a great advantage over us in that their armies were not trailed by a second army of pen-pushers.
> Napoleon (1769–1852) quoted in Christopher Herold, ed., *The Mind of Napoleon*, 1955.

Gentlemen:

Whilst marching to Portugal to a position which commands the approach to Madrid and the French forces, my officers have been diligently complying with your request which has been sent to H.M. ship from London to Lisbon and then by dispatch rider to our headquarters.

We have ennumerated our saddles, bridles, tents and tent poles, and all manner of sundry items for which His Majesty's Government holds me accountable. I have dispatched reports on the character, wit, spleen of every officer. Each item and every farthing has been accounted for, with two regrettable exceptions for which I beg you your indulgence.

Unfortunately, the sum of one shilling and ninepence remains unaccounted for in one infantry battalion's petty cash and there has been a hideous confusion as to the number of jars of raspberry jam issued to one cavalry regiment during a sandstorm in western Spain. This reprehensive carelessness may be related to the pressure of circumstances since we are at war with France, a fact which may come as a bit of surprise to you gentlemen in Whitehall.

This brings me to my present purpose, which is to request elucidation of my instructions from His Majesty's

BUREAUCRACY

Government, so that I may better understand why I am dragging an army over these barren plains. I construe that perforce it must be one of two alternative duties, as given below. I shall pursue one with the best of my ability but I cannot do both.
    1. To train an army of uniformed British clerks in Spain for the benefit of the accountants and copy-boys in London, or perchance,
    2. To see to it that the forces of Napoleon are driven out of Spain.
    The Duke of Wellington (1769–1852), to the War Office during the Peninsular War (1808–1814) – attributed.

Although a revolutionary government, none was ever so much under the domination of red tape as the one at Richmond. The martinets who controlled it were a good deal like the hero of Moliere's comedy, who complained that his antagonist had wounded him by thrusting in carte, when according to the rule, it should have been in tierce. I cared nothing for the form of a thrust if it brought blood. I did not play with foils.
    Colonel John S. Mosby, *Mosby's War Reminiscences*, 1887.

It [the bureaucracy] ...tends to overvalue the orderly routine and observance of the system by which it receives information, transmits orders, checks expenditures, files returns, and in general, keeps with the Service the touch of paper; in short, the organization which has been created for facilitating its own labors.
    Rear Admiral Alfred Thayer Mahan, *Naval Administration and Warfare*, 1908.

This man, with all his charm, sincerity and self-sacrificing hard work, was one of the type who, through a profound sense of duty, coupled with a microscopic vision of the reality of things, spent hours every day blowing blue pencil dust into the military machine, and in consequence nearly brought our part of it to a standstill. To get a new establishment through his office was like playing golf on a course with a hundred bunkers between each green. Every item was scrutinised, criticised and discussed, and then re-scrutinised, re-criticised and re-discussed *ad infinitum*, with the result that delays were never ending. It was bureaucracy in a nightmare.
    Major-General J.F.C. Fuller, *Memoirs of an Unconventional Soldier*, 1936.

While giving lip service to the humanitarian values and while making occasional spectacular and extravagant gestures of sentimentality, those whose task it was to shape personnel policy have tended to deal with man power as if it were motor lubricants or sacks of potatoes. They have destroyed the name and tradition of old and honored regiments with the stroke of a pen, for convenience's sake. They have uprooted names and numbers which had identity with a certain soil and moved them willy-nilly to another soil. They have moved men around as if they were pegs and nothing counted but a specialist classification number. They have become fillers-of-holes rather than architects of the human spirit.
    Brigadier-General S.L.A. Marshall, *Men Against Fire*, 1947.

C

CALCULATION

Now the general who wins a battle makes many calculations in his temple ere the battle is fought. The general who loses a battle makes but few calculations beforehand. Thus do many calculations lead to victory, and few calculations to defeat: How much more no calculation at all! It is by attention to this point that I can see who is likely to win.
  Sun Tzu, *The Art of War*, 1, c. 500 BC, tr. Giles, 1910.

The reasonable course of action in any use of arms starts with calculation. Before fighting, first assess the relative sagacity of the military leadership, the relative strength of the enemy, the size of the armies, the lay of the land, and the adequacy of provisions. If you send troops out only after making these calculations, you will never fail to win.
  Liu Ji (1310–1375) *Lessons of War*, quoted in Thomas Cleary, *Mastering the Art of War*, 1989.

War is essentially a calculation of probabilities.
  Napoleon (1761–1821).

Military science consists in calculating all the chances accurately in the first place, and then in giving accident exactly, almost mathematically, its place in one's calculations. It is upon this point that one must not deceive oneself, and yet a decimal more or less may change all. Now this apportioning of accident and science cannot get into any head except that of a genius... Accident, hazard, chance, call it what you may, a mystery to ordinary minds, becomes a reality to superior men.
  Napoleon (1769–1821), quoted in Mme. de Rémusat, *Memoirs 1802–08*, 1895, cited in Chandler, *The Campaigns of Napoleon*, 1966.

In short, absolute, so-called mathematical factors never find a firm basis in military calculations. From the very start there is an interplay of possibilities, probabilities, good luck and bad that weaves its way throughout the length and breadth of the tapestry. In the whole range of human activities, war most closely resembles a game of cards.
  Major-General Carl von Clausewitz, *On War*, 1.1, 1832, tr. Michael Howard and Peter Paret, 1976.

In seeking to upset the enemy's balance, a commander must not lose his own balance. He needs to have the quality which Voltaire described as the keystone of Marlborough's success – 'that calm courage in the midst of tumult, that serenity of soul in danger, which the English call a cool head.' But to it he must add the quality for which the French have found the most aptly descriptive phrase – *'le sens du practicable'*. The sense of what is possible, and what is not possible – tactically and administratively. The combination of both these two 'guarding' qualities might be epitomised as the power of cool calculation. The sands of history are littered with the wrecks of finely conceived plans that capsized for want of ballast.
  Captain Sir Basil Liddell Hart,

CASUALTIES

introduction to Erwin Rommel, *The Rommel Papers*, 1953.

CAMP FOLLOWERS

The demi-brigades, are expressly forbidden to carry with them more women than the laundresses the law provides for. Every woman found with the army and not duly authorized shall be publicly whipped.
  Napoleon, 28 March 1797, at Goritz, quoted in R.M. Johnston, ed., *The Corsican*, 1910.

The Field Marshal desires that when the commanding officers... send non-commissioned officers or soldiers to England, from the corps under their command respectively, on any account whatever, they will take care that such men do not take with them from this country any woman who is not married to the person with whom she may cohabit... any individual taking a woman over to England, is to have a certificate from his commanding officer that she is his wife.
  The Duke of Wellington, 6 January 1818, at Cambrai, *Selections from the General Orders of Field Marshal the Duke of Wellington During His Various Commands 1799–1818*, Gurwood, ed., 1847.

THE CAPTAIN

The rest are nothing once the captain's down!
  Homer, *The Odyssey*, 22.266, c. 800 BC, tr. Robert Fagles, 1990.

A brave captain is a root, out which as branches the courage of his soldiers doth spring.
  Sir Philip Sidney (1554–1586).

The company is the true unit of discipline, and the captain is the company. A good captain makes a good company, and he should have the power to reward as well as punish. The fact that soldiers would naturally like to have a good fellow for their captain is the best reason why he should be appointed by the colonel, or by some superior authority, instead of being elected by the men.
  General of the Army William T. Sherman, *Memoirs of General W.T. Sherman*, II, 1875.

A loud-mouthed, profane captain who is careless of his personal appearance will have a loud-mouthed, profane, dirty company. Remember what I tell you. Your company will be the reflection of yourself. If you have a rotten company it will be because you are a rotten captain.
  Major C.A. Bach, US Army, 1917, farewell instructions to the graduating student officers, Fort Sheridan, Wyoming.

The company commander is a living example to every man in his organization. To be an officer means to set an example for the men. The officer must be his soldier's incarnation of soldiery, his model. If the German officer is inspired by this mission, the best and deepest qualities of his soul will be awakened; his life's aim will be fulfilled if he succeeds, through knowledge, demeanor, and conviction, in forcing his troops to follow him. This is the manly purpose for which it is worth-while to stake life in order to win life.
  German Army, *Company Commander*, quoted in US War Department, *German Military Training*, 17 September 1942.

CASUALTIES

Mourn not for houses and lands, but for men; men may gain these, but these will not gain men.
  Pericles, 432 BC, urging war against Sparta, Thucydides, *The Peloponnesian War*, 1.143, c. 404 BC, tr. Benjamin Jowett, 1881.

Quintus Fabius, upon being urged by his son to seize an advantageous position at the expense of a few men, asked: 'Do you want to be one of those few?'

CASUALTIES

Fabius Maximus (d. 203 BC), quoted in Frontinus, *Stratagems*, 4.6, c. AD 84–96, tr. Charles E. Bennett, 1925.

Quintilius Varus, Give me back my legions (*Quintili Vare, legiones redde*).
Augustus Caesar, AD 9, in despair after the loss of three irreplaceable legions in the Battle of the Teutoberger Wald, quoted in Suetonius, *On the Life of the Caesars*, 23, AD 121.

O my lord, I do not marvel at such deaths, nor do they fill me with fear. This is how wars are fought! Remember Huitzilihuitl the Elder, our forebear and king, who died in Colhuacan before we were born; behold how he left behind him eternal glory as a valiant man. The Aztec nation needs bold men such as those who lie before you. This is Mexico-Tenochtitlan, and men who are even more courageous than these will rise here. How long and how deeply must we mourn the deceased? If we stay here weeping we shall not be able to accomplish more important matters.
Tlacaelel, c. 1450, to his brother, Motecuhzoma I, on the death of two of their brothers in the last campaign of the war to subjugate the Chalcas, quoted in Diego Durán, *The History of the Indies of New Spain*, 1581, tr. Doris Heyden, 1994.

All great captains, chieftains, and men of charge have holden for a maxim to preserve by all means possible the lives of their soldiers, and not to employ and hazard them upon every light occasion, and therewithal to esteem the preservation of the lives of a very few of their soldiers before the killing of great numbers of the enemy.
John Smythe (c. 1580–1631), *Certain Discourses Military*, 1590.

It is very difficult to do one's duty. I was considered a barbarian because at the storming of the Praga 7,000 people were killed. Europe says that I am a monster. I myself have read this in the papers, but I would have liked to talk to people about this and ask them: is it not better to finish a war with the death of 7,000 people rather than to drag it on and kill 100,000.
Field Marshal Prince Aleksandr V. Suvorov (1729–1800), on the Polish deaths in the storming of the fortified Praga suburb in Warsaw, 15 October 1794, during the Polish revolt, quoted in Philip Longworth, *The Art of Victory*, 1966.

We had a great battle yesterday; victory is mine, but my losses are very heavy; the enemy's losses, which were heavier, do not console me. The great distance at which I find myself makes my losses even more accutely felt.
Napoleon, 9 February 1807, after the Battle of Eylau, quoted in R.M. Johnston, ed., *The Corsican*, 1910.

I arrived here about two this day, and received your letter of the 17th... Your loss, by all accounts, has been very large; but I hope that it will not prove so large as was first supposed. You could not be successful in such an action without a large loss, and we must make up our minds to affairs of this kind sometimes, or give up the game.
The Duke of Wellington, in reply to General Beresford's report on the great losses he had suffered at the Battle of Albuera on 16 May 1811, *The Dispatches of Field Marshal the Duke of Wellington, During His Various Campaigns in India, Denmark, Portugal, Spain, the Low Countries, and France*, VII, 1834–1838, p. 558.

Men who march, scout, and fight, and suffer all the hardships that fall to the lot of soldiers in the field, in order to do vigorous work must have the best bodily sustenance, and every comfort that can be provided. I knew from practical experience on the frontier that my efforts in this direction would not only be appreciated, but required by

CASUALTIES

personal affection and graditude; and, further, that such exertions would bring the best results to me. Whenever my authority would permit I saved my command from needless sacrifices and unnecessary toil; therefore, when hard or daring work was to be done I expected the heartiest response, and always got it. Soldiers are averse to seeing their comrades killed without compensating results, and none realize more quickly than they the blundering that often takes place on the field of battle. They want some tangible indemnity for the loss of life, and as victory is an offset the value of which is manifest, it not only makes them content to shed their blood, but also furnishes evidence of capacity in those who commanded them. My regiment had lost very few men since coming under my command, but it seemed, in my eyes of all who belonged to it, that casualties to the enemy and some slight successes for us had repaid every sacrifice, and in consequence I had gained not only their confidence as soldiers but also their esteem and love as men, and to a degree far beyond what I then realized.
  General of the Army Philip H. Sheridan, *Personal Memoirs*, I, 1888.

His majesty's millions conquer the
  strong foe.
Field battles and siege result in
  mountains of corpses.
How can I, in shame, face their fathers?
Songs of triumph today, but how many
  have returned?
  General Maresuki Nogi, 1905, at the surrender of Port Arthur, quoted in Dennis and Peggy Warner, *The Tide at Sunrise*, 1974.

Let there be dismissed at once, as preposterous, the hope that war can be carried on without some one or something being hurt.
  Rear Admiral Alfred Thayer Mahan, *Some Neglected Aspects of War*, 1907.

War consumes men; that is its nature... That the large masses which were led into battle would suffer heavy casualties, in spite of all tactical counter-measures, was unfortunately a matter of course.
  General Erich von Ludendorff, *My War Memories, 1914–1918*, 1919.

After all, the most distressing and the most expensive thing in war is – to get men killed.
  Major-General J.F.C. Fuller, *Memoirs of an Unconventional Soldier*, 1936.

A big butcher's bill is not necessarily evidence of good tactics.
  Field Marshal Viscount Wavell of Cyrenaica, 1940, reply to Churchill when reproached for evacuating British Somaliland with only 260 casualties.

I was very careful to send to Mr. Roosevelt every few days a statement of our casualties. I tried to keep before him all the time the casualty results because you get hardened to these things and you have to be very careful to keep them always in the forefront of your mind.
  General of the Army George C. Marshall (1880–1959).

In battle, casualties vary directly with the time you are exposed to effective fire. Your own fire reduces the effectiveness and volume of the enemy's fire, while rapidity of attack shortens the time of exposure. A pint of sweat will save a gallon of blood.
  General George S. Patton, Jr., *War As I Knew It*, 1947.

War is a nasty, dirty, rotten business. It's all right for the Navy to blockade a city, to starve the inhabitants to death. But there is something wrong, not nice, about bombing that city.

Marshal of the Royal Air Force Sir Arthur 'Bomber' Harris (1892–1984).

Far from being a handicap to command, compassion is the measure of it. For unless one values the lives of his soldiers and is tormented by their ordeals, he is unfit to command. He is unfit to appraise the cost of an objective in terms of human lives.

To spend lives, knowingly, deliberately – even cruelly – he must steel his mind with the knowledge that to do less would cost only more in the end. For if he becomes tormented by the casualties he must endure, he is in danger of losing sight of his strategic objectives. Where the objective is lost, the war is prolonged and the cost becomes infinitely worse.
General of the Army Omar Bradley, *A Soldier's Story*, 1951.

As in all battles the dead and wounded came chiefly from the best and the bravest.
Field Marshal Lord Carver, *El Alamein*, 1962.

Every waking and sleeping moment, my nightmare is the fact that I will give an order that will cause countless numbers of human beings to lose their lives. I don't want my troops to die. I don't want my troops to be maimed.

It is an intensely personal, emotional thing for me... Any decision you have to make that involves the loss of human life is nothing you do lightly. I agonize over it.

It's not purely a question of accomplishing the mission... But it's a question of accomplishing the mission with a minimum loss of human life and within an effective time period.
General H. Norman Schwarzkopf, interview in *The Washington Post*, 5 February 1991.

...The question of casualties. Norman Schwarzkopf and I were very close on this point. He, still haunted by hideous memories of Vietnam, was anxious to keep American casualties to a minimum. I felt the same. Strongly as I supported the international crusade against Saddam Hussein, I did not see that this war was worth a lot of British dead...
General Sir Peter de la Billière, *Storm Command*, 1992.

Desert Storm left one awful legacy: It imposed the idea that you must be able to fight the wars of the future without suffering losses. The idea of zero-kill as an outcome has been imposed on American generals. But there is no such thing as a clean or risk-free war. You condemn yourself to inactivity if you set that standard.
General Phillippe Morillon, French Commander in Bosnia 1993, in Jim Hoagland, 'Even America Gets the Blues', *The Washington Post*, 14 December 1993.

Sending Americans into battle is the most profound decision a president can make. The technologies of war have changed. The risks and suffering have not.

For the brave Americans who bear the risk, no victory is free from sorrow.
President George W. Bush, 28 January 2003, State of the Union Address, Washington Post, 29 January 2003, p. A11.

CAVALRY

I am also sure that there are young men who can be filled with enthusiasm for serving in the cavalry if one describes the splendour of a cavalryman's life...
Xenophon, *How To Be A Good Cavalry Commander*, c. 360 BC (*Hiero the Tyrant and Other Treatises*, tr. Robin Waterfield, 1997).

Those who face the dizzying heights and cross the dangerous defiles, who can shoot at a gallop as if in flight, who are in the vanguard when advancing and in the rearguard when withdrawing, are called cavalry generals.

CAVALRY

Zhuge Liang (AD 180–234), *The Way of the General* (*Mastering the Art of War*, tr. Thomas Cleary, 1989).

Why then should it seem wonderful that a firm and compact body of infantry should be able to sustain a cavalry attack, especially since horses are prudent animals and when they are apprehensive of danger cannot easily be brought to rush into it? You should also compare the force that impels them to advance with that which makes them retreat; you will then find that the latter is much more powerful than the former. In the one case, they feel nothing but the prick of a spur, but in the other they see a rank of pikes and other sharp weapons presented to them. So, you can see from both ancient and modern examples that good infantry will always be able not only to make headway against cavalry but generally to get the better of them. But if you argue that the fury with which the horses are driven to charge an enemy makes them consider a pike no more than a spur, I answer that even though a horse has begun to charge, he will slow down when he draws near the pikes and, when he begins to feel their points, will either stand stock still or wheel off to the right or left. To convince yourself of this, see if you can can ride a horse against a wall; I fancy you will find very few, if any – however spirited they may be – that can be made to do that.

Niccolò Machiavelli, *The Art of War*, 1521.

As for the cavalry attack, I have considered it necessary to make it so fast and in such close formation for more than one reason: (1) so that this large movement will carry the coward along with the brave man; (2) so that the cavalryman will not have time to reflect; (3) so that the power of our big horses and their speed will certainly overthrow whatever tries to resist them; and (4) to deprive the simple cavalryman of any influence in the decision of such a big affair.

Frederick the Great, *Instructions to His Generals*, 1747, tr. Phillips, 1940.

Without cavalry, battles are without result.

Napoleon, *Correspondance de Napoléon Ier, publiée par ordre de l'Empereur Napoléon III*, XXXI, 1858–1870, p. 427.

Maxim 51. Victor or vanquished, it is of the greatest importance to have a body of cavalry in reserve, either to take advantage of victory or to secure a retreat. The most decisive battles lose half their value to the conqueror when the want of cavalry prevents him from following up his success and depriving the enemy of the power of rallying.

Napoleon, *The Military Maxims of Napoleon*, tr. George D'Aguilar, 1831 (David Chandler, ed., 1987).

Maxim 86. A cavalry general should be a master of practical science, know the value of seconds, despise life and not trust to chance.

Napoleon, *The Military Maxims of Napoleon*, tr. Burnod, 1827.

Our cavalry is the most delicate instrument in our whole machine. Well managed, it can perform wonders, and will always be of use, but it is easily put out of order in the field… if any serious accident were to happen to a large body …while we should be forward in the plains, we are gone, and with us our political system, our allies, &c…The General Officer of a brigade or division of cavalry must have eyes and ears about him…

The Duke of Wellington, 16 March 1813, to E. Cooke, quoted in Julian Rathbone, ed., *Wellington's War: His Peninsular Dispatches*, 1984.

Useful as cavalry may be against simple

CAVALRY

bodies of broken, demoralized troops, still opposed to the whole, it becomes again only the auxiliary arm, because the troops in retreat can reemploy fresh reserves to cover their movement, and therefore, at the next trifling obstacle of ground, by combining all arms they can make a stand with success.

> Major-General Carl von Clausewitz, *On War*, 4.12, 1832, tr. Michael Howard and Peter Paret, 1976.

The principal value of cavalry is derived from its rapidity and mobility. To these characteristics may be added its impetuosity, but we must be careful lest a false application be made of the last.

> Lieutenant-General Antoine Henri Baron de Jomini, *Summary of the Art of War*, 1838, tr. Mendell and Craighill, 1862.

An attack of cavalry should be sudden, bold, and vigorous. The cavalry which arrives noiselessly but steadily near the enemy, and then, with one loud yell leaps upon him without a note of warning, and giving no time to form or consider anything but the immediate means of flight, pushing him vigorously every step with all the confidence of victory achieved, is the true cavalry; while a body of men equally brave and patriotic, who halt at every picket and reconnoiter until the precious surprise is over, is not cavalry.

> Major-General J.E.B. Stuart, General Order No. 26, Cavalry Tactics, July 30, 1863, *Letters of Major General James E.B. Stuart*, 1990.

I believe I was the first cavalry commander who discarded the sabre as useless and consigned it to museums for the preservation of antiquities. My men were as little impressed by a body of cavalry charging them with sabres as though they had been armed with cornstalks... I think my command reached the highest point of efficiency as cavalry because they were well armed with two six-shooters and their charges combined the effect of fire and shock.

> Colonel John S. Mosby, *Memoirs*, 1917.

The cavalry is always the aristocratic arm which loses very lightly, even if it risks at all. At least it has the air of risking all, which is something at any rate. It has to have daring and daring is not so common. But the merest infantry engagement in equal numbers costs more than the most brilliant cavalry raid.

> Colonel Charles Ardant du Picq, *Battle Studies*, 1880, tr. Greely, 1957.

In one respect a cavalry charge is very like ordinary life. So long as you are all right, firmly in your saddle, your horse in hand, and well armed, lots of enemies will give you a wide berth. But as soon as you have lost a stirrup, have a rein cut, have dropped your weapon, are wounded, or your horse is wounded, then is the moment when from all quarters enemies rush upon you.

> Sir Winston S. Churchill, *The River War*, 1899.

Stubborn and unshaken infantry hardly ever meet stubborn and unshaken cavalry. Either the infantry run away and are cut down in flight, or they keep their heads and destroy nearly all the horsemen by their musketry.

> Sir Winston S. Churchill, *The River War*, 1899.

'No longer like Frederick at the end of the day do we hurl our jingling squadrons upon the tottering foe.' No indeed; better not: they might still have one of those machine guns with them which refuses to totter.

> General Sir Ian Hamilton, comment in his introduction to General Hans von Seekt, *Thoughts of a Soldier*, 1930.

In war only what is simple can succeed. I visited the staff of the Cavalry Corps. What I saw there was not simple.

Field Marshal Paul von Hindenburg, at the

CENTRE OF GRAVITY

1932 German Army manoeuvres, quoted in Guderian, *Panzer Leader*, 1953.

As to the Army, heaven knows that we have suffered enough from what is known in the services and outside as the 'cavalry mind.' For the past century and more, even to the outbreak of war in 1939, nobody had much chance of promotion to high command in the army unless he was a cavalry soldier, and it is perfectly well known that in order to get cavalry soldiers as far as high command, it was often necessary to excuse them from the more difficult qualifications in the intervening stages of their career. Unfortunately, they were too apt to develop the mentality of the animals they were so enthusiastic about. If some such provision had not been made, the working infantry, sappers, and gunners would generally have reached the higher posts first, but, as it is, how many of the high commanders of the last 150 years, up to 1939, were cavalry officers? Because of the cavalry influence, and consequently, of this system of making appointments, the idea of using cavalry inevitably persisted for a quarter of a century after the presence of a horse on a battlefield could only be considered a symptom of certifiable lunacy.

Air Marshal Sir Arthur 'Bomber' Harris, *Bomber Offensive*, 1947.

CENSORSHIP

The publishers have to understand that we're never more than a miscalculation away from war and that there are things we're doing that we just can't talk about.

President John F. Kennedy (1917–1963).

Vietnam was the first war fought without censorship. Without censorship, things can get terribly confused in the public mind.

General William C. Westmoreland, *A Soldier Reports*, 1976.

I'm not allowed to say how many planes joined the raid but I counted them all out, and I counted them all back.

Brian Hanrahan, 1 May 1982, reporting on BBC TV the British attack on Port Stanley airport during the Falklands War.

It is our great pride that the British media are free. We ask them, when the lives of some of our people may be at stake through information or through discussion that can be of use to the enemy ...to take that into account on their programmes. It is our pride that we have no censorship. That is the essence of a free country. But we expect the case for freedom to be put by those who are responsible for doing so...

Prime Minister Margaret Thatcher, 11 May 1982, in the House of Commons during the Falklands War, quoted in *The Falklands Campaign: A Digest of Debates in the House of Commons*, 1982, p. 232.

CENTRE OF GRAVITY

One must keep the dominant characteristics of both belligerents in mind. Out of these characteristics a certain center of gravity develops, the hub of all power and movement, on which everything depends. That is the point against which all our energies should be directed.

Major-General Carl von Clausewitz, *On War*, 8.4, 1832, tr Michael Howard and Peter Paret, 1976.

Here's how an offense might look. But there's one more important thing: the Republican Guard is another center of gravity of Iraq's army. While you're running your phase-one campaign over Iraq... I want the Republican Guard bombed the very first day, and I want them bombed every day after that. They're the heart and soul of his army and therefore they will pay the price.

General H. Norman Schwarzkopf, on the planning for the air portion of Desert Storm, *It Doesn't Take a Hero*, 1992.

CHANCE

If it really was Achilles who reared besides the ships,
all the worse for him – if he wants his fill of war.
I for one, I'll never run from his grim assault.
I'll stand up to the man – see if he drags off glory
or I drag it off myself! The god of war is impartial:
he hands out death to the man who hands out death.

Homer, *The Iliad*, 18.359–362, c. 800 BC, tr. Robert Fagles, 1990; Hector, rallying the Trojan host after they had been driven back from their attempt at burning the Greek ships by Achilles.

Consider the vast influence of accident in war, before you are engaged in it. As it continues, it generally becomes an affair of chances, chances from which neither of us is exempt, and whose event we must risk in the dark.

Thucydides, *History of the Peloponnesian War*, 1.78, c. 404 BC, tr. Richard Crawley, 1910; speech of the Athenian delegation to the Spartans arguing against war, 432 BC.

They did not reflect on the common chances of war where such trifling causes as groundless suspicion or sudden panic or superstitious scruples frequently produce great disasters, or when a general's mismanagement or a tribune's mistake is a stumbling block to an army.

Julius Caesar, *The Civil War*, c. 45 BC, (*The Gallic War and Other Writings of Julius Caesar*, tr. Moses Hadas, 1957).

The French are what they were in Caesar's time, and as he described them, brave to excess but unstable... As it is impossible for me to make them what they ought to be, I get what I can out of them and try to leave nothing of importance to chance.

Field Marshal Maurice Comte de Saxe, September 1746, letter to Frederick the Great.

In war something must be allowed to chance and fortune seeing it is, in its nature, hazardous and an option of difficulties.

Major-General Sir James Wolfe, 5 November 1757, letter to Major Rickson.

The art of war lies in calculating odds very closely to begin with, and then in adding exactly, almost mathematically, the factor of chance. Chance will always remain a sealed mystery for average minds.

Napoleon, 28 December 1796, in Milan, quoted in R.M. Johnston, ed., *The Corsican*, 1910.

Something must be left to chance; nothing is sure in a sea fight above all.

Admiral Viscount Nelson, 9 October 1805, in a memorandum to the fleet off Cadiz before the Battle of Trafalgar.

War is the realm of chance. No other human activity gives it greater scope: no other has such incessant and varied dealings with this intruder. Chance makes everything more uncertain and interferes with the whole course of events.

Major-General Carl von Clausewitz, *On War*, 3.3, 1832, tr. Michael Howard and Peter Paret, 1976.

One must consider two factors, the first of which is known: one's own will, the other unknown: the will of the opponent. To these must be added factors of another kind, impossible to foresee, such as the weather, sickness, railroad accidents, misunderstandings, mistakes, in short all factors of which man is neither the creator nor master, let them be called luck, fate or providence. War must not, however, be waged arbitrarily or blindly. The laws of chance show that these factors are bound to be as often favorable as un-

CHANGE

favorable to one or the other opponent.
Field Marshal Helmuth Graf von Moltke, cited in Foch, *Principles of War*, 1913.

Always remember, however sure you are that you can easily win, there would not be a war if the other man did not think he also had a chance.
Sir Winston S. Churchill, *My Early Life*, 1930.

CHANGE

The essence of the principles of warriors is responding to changes; expertise is a matter of knowing the military. In any action it is imperative to assess the enemy first. If opponents show no change or movement, then wait for them. Take advantage of change to respond accordingly, and you will benefit.
  The rule is 'The ability to gain victory by changing and adapting according to opponents is called genius.'
Liu Ji (1311–1375), *Lessons of War*, quoted in Thomas Cleary, *Mastering the Art of War*, 1989.

There is nothing more difficult to take in hand, more perilous to conduct, or more uncertain in its success, than to take the lead in the introduction of a new order of things.
Niccolò Machiavelli, *The Prince*, 1513.

Gentlemen: I hear many say, 'What need so much ado and great charge in caliver, musket, pike, and corselet? Our ancestors won many battles with bows, black bills, and jacks.' But what think you of that?
Captain: Sir, then was then, and now is now. The wars are much altered since the fiery weapons first came up.
Robert Barret, *The Theory and Practice of Modern Wars*, 1598.

To say that the enemy will adopt the same measures is to admit the goodness of them; nevertheless they will probably persist in their errors for some time, and submit to be repeatedly defeated, before they will be reconciled to such a change – so reluctant are all nations to relinquish old customs.
Field Marshal Maurice Comte de Saxe, *My Reveries*, 1732, tr. Phillips, 1940.

The student will observe that changes in tactics have not only taken place after changes in weapons, which reasonably is the case, but that interval between such changes has been unduly long. An improvement of weapons is due to the energy of one or two men, while changes in tactics have to overcome the inertia of a conservative class.
Rear Admiral Alfred Thayer Mahan (1840–1914).

Victory always smiled on one who was able to renew traditional forms of warfare, and not on the one who hopelessly tied himself to those forms.
General Giulio Douhet (1869–1930).

Thus for a hundred years we find the French knights charging English archers; for another hundred years or so, cavalry charging musketeers and riflemen; and I suppose we shall see for yet another hundred years infantry charging tanks. What for, indeed what for? Not to win a battle, for the impossibility of this is obvious to a rhinoceros. No; but to maintain the luxury of mental indolence in the head of some military alchemist. Thinking to some people is like washing to others. A tramp cannot tolerate a hot bath, and the average general cannot tolerate any change in preconceived ideas; prejudice sticks to his brain like tar to a blanket.
Major-General J.F.C. Fuller, *The Foundations of the Science of War*, 1926.

Philosophers and scientists have shown that adaptation is the secret of existence. History, however, is a catalogue of failures to change in time with the need. And armies, which

because of their role should be the most adaptable of institutions, have been the most rigid – to the cost of the causes they upheld. Almost every great soldier of the past has borne witness to this truth. But it needs no such personal testimony, for the facts of history, unhappily, prove it in overwhelming array. No one can in honesty ignore them if he has once examined them. And to refrain from emphasizing them would be a crime against the country. For it amounts to complicity after the event, which is even more culpable when the life of a people, not merely of one person, is concerned. In the latter case there may be some excuse for discreet silence, as your testimony cannot restore the dead person to life. But in the former case there is no such excuse – because the life of a people will again be at stake in the future. (Dec. 1932.)
  Captain Sir Basil Liddell Hart, *Thoughts on War*, 1944.

The only thing harder than getting a new idea into the military mind is to get an old one out.
  Captain Sir Basil Liddell Hart, *Thoughts on War*, 1944.

In military affairs, especially in contemporary war, one cannot stand in one place; to remain, in military affairs, means to fall behind; and those who fall behind, as is well known, are killed.
  Joseph Stalin (1879–1930), quoted in Garthoff, *Soviet Military Doctrine*, 1953.

Through all this welter of change, your mission remains fixed, determined, inviolable – it is to win our wars.
  General of the Army Douglas MacArthur, Thayer Award Speech, US Military Academy at West Point, 12 May 1962.

The history of wars convincingly testifies... to the constant contradiction between the means of attack and defense. The appearance of new means of attack has always inevitably led to the creation of corresponding means of counteraction, and this in the final analysis has led to the development of new methods for conducting engagements, battles, and operations and the war in general.
  Marshal of the Soviet Union Nikolai V. Ogarkov, quoted in *Kommunist* (Moscow), 1978.

If you don't like change, you're going to like irrelevance a lot less.
  General Erick K. Shinseki, Chief of Staff, US Army, quoted in Peter J. Boyer, 'A Different War,' *The New Yorker*, July 1, 2002.

Even the greatest regiments cannot fight change.
  Secretary of State for Defence Geoff Hoon, on the decision to disband the Black Watch and several other ancient regiments, *The Scotsman*, 28 October 2004.

CHAPLAINS

[General Montgomery insisted that] everyone must be imbued with the burning desire to 'kill Germans'. 'Even the padres – one per weekday and two on Sundays!'
  Major-General Sir Francis de Guingand, *Operation Victory*, 1947.

A chaplain visits our company. In a tired voice, he prays for the strength of our arms and for the souls of the men who are to die. We do not consider his denomination. Helmets come off. Catholics, Jews, and Protestants bow their heads and finger their weapons. It is front-line religion: God and the Garand.
  First Lieutenant Audie Murphy, *To Hell and Back*, 1949.

Our chaplains were crucial for morale. I used them with our front line troops. Because of this, a senior chaplain criticized me for exposing our chaplain too often to combat. My answer was:

CHARACTER

He is where the men need him, where there are wounded and dying. When he's back in the chapel, in the rear, none of the soldiers who really need him can get to him. I went so far as to close the chapel to encourage the chaplain to spend most of his time with troops in difficult areas. One of our chaplains in the Brigade was awarded the Medal of Honor, the highest valor award in the military.
    Major-General Bernard Loeffke, *From Warrior to Healer*, 1998.

CHARACTER

A man should be upright, not kept upright.
    Emperor Marcus Aurelius Antoninus, *Meditations*, c. AD 161–180, tr. George Long.

The success of my whole project is founded on the firmness of the conduct of the officer who will command it.
    Frederick the Great, *Instructions to His Generals*, 1747, tr. Phillips, 1940.

War must be carried on systematically, and to do it you must have men of character activated by principles of honor.
    General George Washington (1732–1799).

My character and good name are in my own keeping, Life with disgrace is dreadful. A glorious death is to be envied.
    Admiral Viscount Nelson, 10 March 1795, journal entry.

True character always pierces through in moments of crisis... There are sleepers whose awakening is terrifying.
    Napoleon, 8 November 1816, St Helena, quoted in Emmanuel Las Casas, *Mémorial de St. Hélène*, 1840.

We repeat again: strength of character does not consist solely in having powerful feelings, but in maintaining one's balance in spite of them. Even with the violence of emotion, judgment and principle must still function like a ship's compass, which records the slightest variations however rough the sea.
    Major-General Carl von Clausewitz, *On War*, 1.3, 1832, tr. Michael Howard and Peter Paret, 1976.

You may be whatever you resolve to be.
    A maxim of Lieutenant-General Thomas J. 'Stonewall' Jackson, originally from Joel Hawes, *Letters to Young Men, on the Formation of Character &c*, 1851.

Private and public life are subject to the same rules; and truth and manliness are two qualities that will carry you through this world much better than policy, or tact, or expediency, or any other word that was ever devised to conceal or mystify a deviation from a straight line.
    General Robert E. Lee, from his post-war writings, quoted in William Jones, *Life and Letters of Robert Edward Lee*, 1906.

An insurance company offered him a salary of $10,000 to become their president, and sent a distinguished Confederate soldier to secure his acceptance. To his reply that he could not discharge its duties without giving up the presidency of the College, and that he could not do that, the reply was made, 'We do not wish you to give up your present position, General, or to discharge any duties in connection with our company. The truth is that we only want your name connected with the company. That would amply compensate us for the salary we offer you.' General Lee's face flushed, and his whole manner indicated his displeasure as he replied, 'I am sorry, sir, that you are so little acquainted with my character as to suppose that my name is for sale at any price.'
    General Robert E. Lee, quoted in William Jones, *Life and Letters of Robert Edward Lee*, 1906.

CHARACTER

We know not of the future, and cannot plan for it much. But we can hold our spirits and our bodies so pure and high, we may cherish such thoughts and such ideals, and dream such dreams of lofty purpose, that we can determine and know what manner of men we will be whenever and wherever the hour strikes, that calls to noble action... no man becomes suddenly different from his habit and cherished thought.
> Major-General Joshua L. Chamberlain, address at Gettysburg, 3 October 1893, *Dedication to the Twentieth Maine Monuments on the Battlefield of Gettysburg*, 1895.

By greatness of character a general gains command over himself, and by goodness of character he gains command over his men, and those two moods of command express the moral side of generalship.
> Major-General J.F.C. Fuller, *The Foundations of the Science of War*, 1926

Some virtue is required which will provide the army with a new ideal, which, through the military elite, will unite the army's divergent tendencies and fructify its talent. This virtue is called character which will constitute the new ferment – character the virtue of hard times.
> General Charles de Gaulle, *The Edge of the Sword*, 1932.

War is the severest test of spiritual and physical strength. In war, character outweighs intellect. Many stand forth on the field of battle, who in peace would remain unnoticed.
> Generals Werner von Fritsch and Ludwig Beck, *Troop Leadership*, 1933, quoted in US War Department, 'What is the test of human character?'

Happily no democracy can be prepared in the German sense for war – we are of course paying in full for this freedom of the mind – but it should and can be prepared to fight evil; a free people is only ready to resist aggression when the Christian virtues flourish, for a man of character in peace is a man of courage in war.
> Lord Moran, 12 May 1943, preface to *The Anatomy of Courage*, 1945.

He must have 'character', which simply means that he knows what he wants and has the courage and determination to get it. He should have a genuine interest in, and a real knowledge of, humanity, the raw material of his trade; and, most of all, he must have what we call the fighting spirit, the will to win. You all know and recognize it in sport, the man who plays his best when things are going badly, who has the power to come back at you when apparently beaten, and who refuses to acknowledge defeat. There is one other moral quality I would stress the mark of the really great commander as distinguished from the ordinary general. He must have a spirit of adventure, a touch of the gambler in him.
> Field Marshal Viscount Wavell of Cyrenaica, *Soldiers and Soldiering*, 1953.

Intelligence, knowledge, and experience are telling prerequisites [for a commander]. Lack of these may, if necessary, be compensated for by good general staff officers. Strength of character and inner fortitude, however, are decisive factors. The confidence of the men in the ranks rests upon a man's strength of character.
> Field Marshal Erich von Manstein, *Lost Victories*, 1958.

Character is the bedrock on which the whole edifice of leadership rests. It is the prime element for which every profession, every corporation, every industry searches in evaluating a member of its organization. With it, the full worth of an individual can be developed. Without it – particularly in the military profession – failure in

CHIEF OF STAFF

peace, disaster in war, or, at best, mediocrity in both will result.
  General Matthew B. Ridgway, 'Leadership', *Military Review*, 10/1966.

A man of character in peace is a man of courage in war. Character is a habit. The daily choice of right and wrong. It is a moral quality which grows to maturity in peace and is not suddenly developed in war. The conflict between morality and necessity is eternal. But at the end of the day the soldier's moral dilemma is only resolved if he remains true to himself.
  General Sir James Glover, 'A Soldier and His Conscience', *Parameters*, 9/1983.

Glib, cerebral, detached people can get by in positions of authority until the pressure is on. But when the crunch comes, people cling to those they know they can trust – those who are not detached, but involved – those who have consciences, those who can repent, those who do not dodge unpleasantness. Such people can mete out punishment and look their charges in the eye as they do it. In different situations, the leader with the heart, not the bleeding heart, not the soft heart, but the Old Testament heart, the hard heart, comes into his own.
  Admiral James Stockdale, 'Educating Leaders', *Washington Quarterly*, Winter 1983.

CHIEF OF STAFF

Maxim 74. To know the country thoroughly; to be able to conduct a reconnaissance with skill; to superintend the transmission of orders promptly; to lay down the most complicated movements intelligibly, but in a few words and with simplicity; these are the leading qualifications which should distinguish an officer selected for the head of the staff.
  Napoleon, *The Military Maxims of Napoleon*, tr. George D'Aguilar, 1831 (David Chandler, ed., 1987).

Well! If I am to become a doctor, you must at least make Gneisenau an apothecary, for we two belong together always... Gneisenau makes the pills which I administer.
  Field Marshal Prince Gebhard von Blücher (1742–1819), describing his chief of staff, August von Gneisenau, upon receiving an honorary degree at Oxford after the Napoleonic Wars.

The best means of organizing the command of an army, in default of a general approved by experience, is to: (1) Give the command to a man of tried bravery, bold in the fight and of unshaken firmness in danger. (2) Assign as his chief of staff a man of high ability, of open and faithful character, between whom and the commander there may be perfect harmony. The victor will gain so much glory that he can spare some to the friend who has contributed to his success.
  Lieutenant-General Antoine-Henri Baron de Jomini, *Summary of the Art of War*, 1838, tr. Mendell and Craighill, 1862.

As I knew from my own experience, the relations between the Chief of Staff and his General, who has the responsibility, are not theoretically laid down in the German Army. The way in which they work together and the degree to which their powers are complementary are much more a matter of personality. The boundaries of their respective powers are therefore not clearly demarcated. If the relations between the General and his Chief of Staff are what they ought to be, these boundaries are easily adjusted by soldierly and personal tact and the qualities of mind on both sides.
  I myself have often characterized my relations with General Ludendorff as those of a happy marriage. In such a relationship how can a third party clearly distinguish the merits of the individuals? They are one in thought

and action, and often what the one says is only the expression of the wishes and feelings of the other.
Field Marshal Paul von Hindenburg, *Out of My Life*, 1920.

The commander himself takes the responsibility for giving orders and must listen to the advice of one man alone: the Chief of Staff assigned to him. The decision is taken by these two together and when they emerge from their privacy there is one decision and one only. They have reached it together; they two are one. If there was any difference of opinion, then by evening the two parties to this happy matrimonial council no longer know which gave way. The outside world and the history of war learn nothing of any tiff. In this unification of two personalities lies the security of command. It is all the same whether the order is signed by the General Officer Commanding or, according to our old custom, by the Chief of Staff on his behalf...Thus the relation between the two is built up entirely on confidence; if there is no confidence they should part without delay. A relation of this kind cannot be reduced to rule; it will and must always vary according to situation and personality. Thus it is obvious that the correct combination of personalities is of decisive importance for success.
Colonel General Hans von Seekt, *Thoughts of a Soldier*, 1930.

Have a good chief of staff.
Field Marshal Viscount Montgomery of Alamein (1887–1976), advice on the requirements for a general.

The Chief of Staff must be anonymous and must never attempt to take unto himself the powers of his leader; on the other hand, he must be prepared to give decisions on all matters of detail. He must, therefore, be completely in the mind of his boss, nothing being hidden from him, and being trusted absolutely.
Field Marshal Montgomery Viscount of Alamein, *The Path to Leadership*, 1960.

Between the commander and his chief of staff in a division or larger unit there should be thorough mutual respect, understanding, and confidence with no offical secrets between them. Together they form a single dual personality, and the instructions issuing from the chief of staff must have the same weight and authority as those of the commander himself.

But this does not mean that a commander who delegates such authority to his chief of staff can allow his chief to isolate him from the rest of the staff. If that happens, the commander will soon find himself out of touch, and the chief of staff will be running the unit.
General Matthew B. Ridgway, 'Leadership', *Military Review*, 10/1966

A good Chief of Staff becomes the voice of his commander: he reflects his commander's views, carries out his orders and interprets them in detail. If I had suddenly said to Ian, 'We're going to move the Division to a new location next week,' it would have been his responsibility to interpret that decision in the form of specific orders and to set the whole move in motion. A Chief of Staff has a responsible and demanding job, but also one of considerable influence: he needs to be a man with experience and judgment.
General Sir Peter de la Billière, *Storm Command*, 1992.

CHIVALRY

His helmet flashed as Hector nodded:
    'Yes, Ajax,
Since god has given you power, build
    and sense
and you are the strongest spearman of
    Achaea,
let us break off this dueling to the
    death,
at least for today. We'll fight again

CHIVALRY

tomorrow,
until some fatal power decides between
　our armies,
handing victory down to one side or
　another...

Come, let us give each other gifts,
　unforgetable gifts,
so any man may say, Trojan soldier or
　Argive,
'First they fought with heart-devouring
　hatred,
then they parted, bound by pacts of
　friendship."
With that he gave him his silver-
　studded sword,
slung in its sheath on a supple, well-cut
　sword-strap,
and Ajax gave his war-belt, glistening
　purple.
So both men parted, Ajax back to
　Achaea's armies,
Hector back to his thronging Trojans –
　overjoyed
to see him still alive, unharmed,
　striding back,
free of the rage and hands of Ajax still
　unconquered.
They escorted him home to Troy –
　saved, past all their hopes –
While far across the field the Achaean
　men-at-arms
escorted Ajax, thrilled with victory,
　back to Agamemnon.
　　Homer, *The Iliad*, 7.300, c. 800 BC, tr.
　　Robert Fagles, 1990.

Alexander on the following day entered the tent accompanied by Hephaestion, and that Darius's mother, in doubt, owing to the similarity of their dress, which of the two was the King, pro-strated herself before Hephaestion, because he was taller than his com-panion. Hephaestion stepped back, and one of the Queen's attendants rectified her mistake by pointing to Alexander; the Queen withdrew in profound embarrassment, but Alexander merely remarked that her error was of no account, for Hephaestion, too, was an Alexander – a 'protector of men'.
　　Arrian, the meeting of Alexander and the mother of Darius whom the Persian had abandoned in his flight after the Battle of Issus, 1 October 333 BC, *The Campaigns of Alexander*, 2.13, c. AD 150, tr. Aubrey de Sélincourt, 1987; in Greek, the name Alexander means 'Protector of Men'.

The treatment, in general, that we have received from the enemy since our surrender, has been perfectly good and proper. But the kindness and attention that have been shown to us, by the French officers in particular – their delicate sensibility of our situation, their generous and pressing offer of money, both public and private, to any amount – have really gone beyond what I can possibly describe; and will, I hope, make an impression on the breast of every officer, whenever the fortune of war should put any [of] them into our power.
　　General Marquis Cornwallis, 20 October 1781, letter to Sir Henry Clinton, announcing his surrender at Yorktown, quoted in Henry Lee, *Memoirs of the War in the Southern Department*, 1869.

As a soldier, preferring loyal and chivalrous warfare to organized assassination if it be necessary to make a choice, I acknowledge that my prejudices are in favour of the good old times when the French and English Guards courteously invited each other to fire first – as at Fontenoy – preferring them to the frightful epoch when priests, women and children throughout Spain plotted the murder of isolated soldiers.
　　Lieutenant-General Antoine Henri Baron de Jomini, *Summary of the Art of War*, 1838, tr. Mendell and Craighill, 1862.

Our Southern ideals of patriotism provided us with the concepts of chivalry. I tried to excell in these virtues, but others provided a truer interpretation of gallant conduct. A devoted champion of the South was one who possessed a heart intrepid, a spirit invincible, a patriotism too lofty to admit a selfish thought and a conscience that scorned to do a mean act. His legacy would be to leave a shining example of heroism and patriotism to those who survive.

    Major-General James E.B. Stuart, 3 December 1862, letter to R.H. Chilton, Virginia Historical Society.

We want to make our nation entirely one of gentlemen, men who have a strong sense of honour, of chivalry towards others, of playing the game bravely and unselfishly for their side, and to play it with a sense of fair play and happiness for all.

    Lieutenant-General Sir Robert Baden-Powell, Headquarters Gazette, 1914.

Chivalry governed by reason is an asset both in war and in view of its sequel – peace. Sensible chivalry should not be confounded with the quixotism of declining to use a strategical or tactical advantage, of discarding the supreme moral weapon of surprise, of treating war as if it were a match on the tennis court – such quixotism as is typified by the burlesque of Fontenoy, 'Gentlemen for France, fire first.' This is merely stupid. So also is the traditional tendency to regard the use of a new weapon as 'hitting below the belt,' regardless of whether it is inhuman or not in comparison with existing weapons...

    But chivalry... is both rational and far-sighted, for it endows the side which shows it with a sense of superiority, and the side which falls short with a sense of inferiority. The advantage in the moral sphere reacts on the physical.

    Captain Sir Basil Liddell Hart, A Greater Than Napoleon: Scipio Africanus, 1926.

Generals think war should be waged like the tourneys of the Middle Ages. I have no use for knights. I need revolutionaries.

    Adolf Hitler, to Herman Rauschning, 1940.

Let it not be thought that the age of chivalry belongs to the past.

    Sir Winston S. Churchill (1874–1965).

Rommel gave me and those who served under my command in the Desert many anxious moments. There could never be any question of relaxing our efforts to destroy him, for if ever there was a general whose sole preoccupation was the destruction of the enemy, it was he. He showed no mercy and expected none. Yet I could never translate my deep detestation of the regime for which he fought into personal hatred of him as an opponent. If I say, now that he is gone, that I salute him as a soldier and a man and deplore the shameful manner of his death, I may be accused of belonging to what Mr. Bevin has called the 'trade union of generals.' So far as I know, should such a fellowship exist, membership in it implies no more than recognition in an enemy of the qualities one would wish to possess oneself, respect for a brave, able and scrupulous opponent and a desire to see him treated, when beaten, in the way one would have wished to be treated had he been the winner and oneself the loser. This used to be called chivalry: many will now call it nonsense and say that the days when such sentiments could survive a war are past. If they are, then I, for one, am sorry.

    Field Marshal Sir Claude Auchinleck, the foreword to Desmond Young, The Desert Fox, 1950.

CITIZEN SOLDIERS

Cease to hire your armies. Go yourselves, every man of you, and stand in the ranks and either a victory beyond all victories in its glory awaits you, or

CIVILIAN CONTROL OF THE MILITARY

falling you shall fall greatly, and worthy of your past.

Demosthenes, *Third Philippic*, 341 BC; calling upon the Athenians to form citizen armies again rather than rely upon mercenaries to face the great danger posed by Philip of Macedon.

After they have elected the consuls, they proceed to appoint military tribunes; fourteen are drawn from those who have seen five years' service and ten from those who have seen ten. As for the rest, a cavalryman is required to complete ten years' service and an infantryman sixteen before he reaches the age of forty-six, except for those rated at less than 400 drachmae worth of property who are assigned to naval service. In periods of national emergency the infantry are called upon to serve for twenty years and no one is permitted to hold any political office until he has completed ten years' service.

Polybius, *Histories*, 6.19, c. 125 BC (*The Rise of the Roman Empire*, tr. Ian Scott-Kilvert, 1987).

When we assumed the Soldier, we did not lay aside the citizen; and we shall most sincerely rejoice with you in that happy hour when the establishment of American liberty, upon the most firm and solid foundations, shall enable us to return to our Private Stations in the bosom of a free, peaceful and happy Country.

General George Washington, 26 June 1775, address to the New York legislature, *Writings*, III, 1931, p. 305.

Remember, soldiers, that first and foremost you are citizens. Let us not become a greater scourge to our country than the enemy themselves.

General Lazare Carnot, 1792, address to the *Armée du Nord*.

These men are not an army; they are citizens defending their country.

General Robert E. Lee, 15 May 1864, to Lieutenant-General A.P. Hill, referring to the Army of Northern Virginia.

The armies of Europe are machines: the men are brave and the officers capable; but the majority of the soldiers in most of the nations of Europe are taken from a class of people who are not very intelligent and who have very little interest in the contest in which they are called upon to take part. Our armies were composed of men who were able to read, men who knew what they were fighting for, and could not be induced to serve as soldiers, except in an emergency when the safety of the nation was involved, and so necessarily must have been more than equal to men who fought merely because they were brave and because they were thoroughly drilled and inured to hardships.

General Ulysses S. Grant, *Personal Memoirs*, I, 1885.

Montgomery understood this 'civilian army' as few before him. The rigid old type discipline was not enforced. Human weaknesses were fully appreciated, and the man's lot made as easy for him as possible. This is why he was so lenient as regards dress, and why a certain amount of 'personal commandeering' – technically I suppose it might be called 'looting' – was winked at. Even when on occasions a unit did not behave in battle as well as it might have done, Montgomery gave them a chance to put things right. All in all he realized that the Prussian type of discipline was not suited to the 'civilian' soldier of the Empire.

Major-General Sir Francis de Guingand, *Operation Victory*, 1947.

CIVILIAN CONTROL OF THE MILITARY

Presbyterian cleric: ''Tis against the will of the nation: there will be nine in ten against you.'

CIVIL–MILITARY RELATIONS

Cromwell: 'But what if I should disarm the nine, and put a sword in the tenth man's hand?'
> Oliver Cromwell, shortly after dissolving Parliament in 1649.

It may be proper constantly and strongly to impress upon the Army that they are mere agents of Civil power: that out of Camp, they have no other authority than other citizens, that offences against the laws are to be examined, not by a military officer, but by a Magistrate; that they are not exempt from arrests and indictments for violations of the law.
> General George Washington, 27 March 1795, letter to Daniel Morgan, *Writings*, XXXIV, 1940, p. 160.

It is melancholy to be obliged to act against one's countrymen. I hope sincerely it will never be our case; but, as soldiers, we engaged not only to fight the foreigner; but also to support the Government and laws, which have long been in use and framed by wiser men than we.
> General Sir John Moore, 1798, address to his Irish troops in Cork.

So long as I hold a commission in the Army I have no views of my own to carry out. Whatever may be the orders of my superiors, and law, I will execute. No man can be efficient as a commander who sets his own notions above the law and those whom he is sworn to obey. When Congress enacts anything too odious for me to execute I will resign.
> General Ulysses S. Grant, 22 March 1862, letter to Congressman Elihu B. Washburne, *Personal Memoirs and Selected Letters*, 1990.

I have heard in such a way as to believe it, of your recently saying that both the army and the government needed a dictator… Only those generals who gain successes can set up dictators. What I ask you now is military success, and I will risk the dictatorship.
> President Abraham Lincoln, 26 January 1863, letter to Major-General Joseph Hooker.

It is always dangerous for soldiers, sailors or armies to play at politics.
> Sir Winston S. Churchill (1874–1965).

CIVIL–MILITARY RELATIONS

'So now they are openly recalling me,' he said, 'for years they have tried covertly to force my return by refusing me any money and reinforcements. The Roman people did not conquer Hannibal; many times their armies were decimated and put to rout. It is the senate of Carthage which has defeated me, through hatred and jealousy.'
> Hannibal, 203 BC, upon his recall from Italy to defend Carthage from a Roman landing in Africa, quoted in Livy, *Annals of the Roman People*, c. 25 BC–AD 17 (*Livy: A History of Rome*, 30, tr. Moses Hadas and Joe Poe, 1962).

The advance and retirement of the army can be controlled by the general in accordance with prevailing circumstances. No evil is greater than commands of the sovereign from the court.
> Chia Lin, Tang Dynasty, AD 618–905.

…That a general during the campaign need not follow court orders is sound military law.
> Emperor Kublai Khan (1215–1294).

A general to whom the sovereign has entrusted his troops should act on his own initiative. The confidence which the sovereign reposes in the general's ability is authorization enough to conduct affairs in his own way.
> Frederick the Great, *General Principles of War*, 1748.

Where the state is weak, the army rules.
> Napoleon, 9 January 1808, remark.

COHESION

To leave a great military enterprise or the plan for one to purely military judgment and decision is a distinction which cannot be allowed, and is even prejudicial; indeed, it is an irrational proceeding to consult professional soldiers on the plan of a war, that they may give a purely military opinion upon what the cabinet ought to do...
   Major-General Carl von Clausewitz, On War, 1832, tr. Graham, 1918.

The action of a cabinet in reference to the control of armies influences the boldness of their operations. A general whose genius and hands are tied by an Aulic council five hundred miles distant cannot be a match for one who has liberty of action, other things being equal.
   Lieutenant-General Antoine-Henri Baron de Jomini, *Summary of the Art of War*, 1838, tr. Mendell and Craighill, 1862.

These men are at this moment fighting in the hardest battle of the war, fighting it with heroism which I can find no phrase worthy to express. And it is we who for a mistake made in such and such a place, or which may not even have been made, demand explanations, on the field of battle of a man worn with fatigue. It is of this man that we demand to know whether on such and such a day he did such and such a thing! Drive me from this place if that is what you ask, for I will not do it.
   Georges Clemenceau, speech in the Chamber of Deputies, Paris, 4 June 1918, at the height of the last great German offensive of WWI, quoted in Lewis Copeland, ed., *The World's Great Speeches*, 1952.

A matter of more immediate importance is that the nation should realize the necessity for having educated leaders – trained statesmen – to conduct its war business, if and when war should again come along. This is a direction in which much-needed preparation can be made without the expenditure of cash, and it may be the means of saving tens of thousands of lives and hundreds of millions of money. In all trades and professions the man who aims at taking the lead knows that he must first learn the business he proposes to follow: that he must be systematically trained in it. Only in the business of war – the most difficult of all – is no special training or study demanded from those charged with, and paid for, its management.
   Field Marshal Sir William Robertson, *From Private to Field-Marshal*, 1921.

All the reproaches that have been levelled against the leaders of the armed forces by their countrymen and by the international courts have failed to take into consideration one very simple fact: that policy is not laid down by soldiers, but by politicians. This has always been the case and is so today. When war starts, the soldiers can only act according to the political and military situation as it then exists. Unfortunately it is not the habit of politicians to appear in conspicuous places when the bullets begin to fly. They prefer to remain in some safe retreat and to let the soldiers carry out 'the continuation of policy by other means.'
   Colonel-General Heinz Guderian, *Panzer Leader*, 1953.

While political goals shape war and influence it from one moment to the next, it is the soldier's duty to insure that he will not be confronted with insoluble tasks by his political leaders (as happened in June 1941).
   General Reinhard Gehlen, *The Service*, 1972.

COHESION

Every man reported and one [the King] mustered an army of Pharaoh which was under command of this king's son.

THE COLOURS

He made troops, commanded by commanders, each man with his village...
   Pharaoh Ammenhotep III (1411–1375 BC), quoted in James Breasted, *Ancient Records of Egypt*, II, 1906.

The worst cowards, banded together, have their power.
   Homer, *The Iliad*, 13.281, *c.* 800 BC, tr. Robert Fagles, 1990.

Having formed his company... he (the captain) will then arrange comrades. Every Corporal, Private and Bugler will select a comrade of the rank differing from his own, i.e. front and rear rank, and is never to change him without permission of his captain. Comrades are always to have the same berth in quarters; and that they may be as little separated as possible, in either barracks or the field, will join the same file on parade and go on the same duties with arms when it is with baggage also.
   British Army *Regulations of the Rifle Corps*, 1800, prepared under the aegis of Sir John Moore.

Victory and disaster establish indestructable bonds between armies and their commanders.
   Napoleon, *Political Aphorisms*, 1848.

A wise organization ensures that the personnel of combat groups changes as little as possible, so that comrades in peacetime shall be comrades in war.
   Colonel Charles Ardant du Picq, *Battle Studies*, 1880, tr. Greely, 1957.

The thing in any organization is the creation of a soul.
   General George S. Patton, July 1941, *The Patton Papers*, II, 1974.

...The inherent unwillingness of the soldier to risk danger on behalf of men with whom he has no social identity. When a soldier is unknown to the men who are around him he has relatively little reason to fear losing the one thing that he is likely to value more highly than life – his reputation as a man among other men.

However much we may honor the 'Unknown Soldier' as the symbol of sacrifice in war, let us not mistake the fact that it is the 'Known Soldier' who wins battles. Sentiment aside, it is the man whose identity is well known to his fellows who has the main chance as a battle effective.
   Brigadier-General S.L.A. Marshall, *Men Against Fire*, 1947

Commitment is the thing. Soldiers in the ranks have rarely, over the ages, fought for king, country, freedom, or moral principle. More than anything else, men have fought and winners have won because of a commitment – to a leader and to a small brotherhood where the ties that bind are mutual respect and confidence, shared privation, shared hazard, shared triumph, a willingness to obey, and determination to follow... bravery, as we recognize it and reward it, is far more than anything else a matter of commitment to the hero's leader or peers.
   Lieutenant-General Victor H. 'The Brute' Krulak, *First to Fight*, 1984.

I was deluded by the conventional wisdom which maintains that it is the personal linkages that give a group its unity. I was slow to comprehend the truth; that comrades-in-arms unconsciously create from their particulate selves an imponderable entity which goes its own way and has its own existence, regardless of the comings and goings of the individuals who are its constituent parts. ...Once out of it, it ceases to exist for you – and you for it.
   Farley Mowat, *And No Birds Sang.*

THE COLOURS

Landing was difficult ...while our soldiers were hesitating, chiefly because

THE COLOURS

of the depth of the water, the standard-bearer of the Tenth, with a prayer that his act might rebound to the success of the legion, cried, 'Leap down, men, unless you want to betray your eagle to the enemy; I, at least shall have done my duty to my country and my general.' As he said this, in a loud voice, he threw himself overboard and began to advance against the enemy with the eagle. Then our men called upon one another not to suffer such a disgrace, and with one accord leaped down from the ships. Seeing this their comrades from the nearest ships followed them and advanced close to the enemy.
   Julius Caesar, of the first Roman landing in Britain, *The Gallic War*, c. 51 BC, tr. Moses Hadas, 1957.

Inquisitor: Did you ask your saints if, by virtue of your banner, you would win every battle you entered, and be victorious?
Joan: They told me to take it up bravely and God would help me.
Inquisitor: Which helped you more – you your banner, or your banner you?
Joan: As to whether victory was my banner's or mine, it was all our Lords's.
   Joan of Arc, 17 March 1431, *Joan of Arc In Her Own Words*, 1996.

The soldiers should make it an article of faith never to abandon their standard. It should be sacred to them; it should be respected; and every type of ceremony should be used to make it respected and precious.
   Field Marshal Maurice Comte de Saxe, *My Reveries*, 1732, tr. Phillips, 1940.

Soldiers, there are your colours. These eagles will serve you as points for rallying. They will be everywhere your Emperor judges necessary for the defence of his throne and his people. You will swear to give your life to defend him and constantly to uphold them by your courage on your way to victory.
   Napoleon, 5 December 1804, the presentation of the new regimental eagles to the French Army, quoted in Georges Blond, *The Grand Armée*, 1979, 1999, p.33.

When the proud and sensitive sons of Dixie came to a full realization of the truth that the Confederacy was overthrown and their leader had been compelled to surrender his once invincible army, they could no longer control their emotions, and tears ran like water down their shrunken faces. The flags which they still carried were objects of undisguised affection. These Southern banners had gone down before overwhelming numbers; and torn by shells, riddled by bullets, and laden with the powder and smoke of battle, they aroused intense emotion in the men who had so often followed them to victory. Yielding to overpowering sentiment, these high-mettled men began to tear the flags from the staffs and hide them in their bosoms, as they wet them with burning tears.
   The Confederate officers faithfully endeavored to check this exhibition of loyalty and love for the old flags. A great majority of them were duly surrendered; but many were secretly carried by devoted veterans to their homes, and will be cherished forever as honored heirlooms.
   There was nothing unnatural or censurable in all this. The Confederates who clung to these pieces of battered bunting knew they would never again wave as martial ensigns above embattled hosts; but they wanted to keep them, just as they wanted to keep the old canteen with a bullet-hole through it, or the rusty gray jacket that had been torn by canister. They loved those flags, and will love them forever, as mementoes of the unparalleled struggle. They cherish them because they represent the consecration and courage not only of Lee's army but of all the Southern armies, because they

symbolize the bloodshed and the glory of nearly a thousand battles.
> General John B. Gordon, describing the surrender at Appomattox, *Reminiscences of the Civil War*, 1903.

And hence it was that when one man in every two, or even two in every three, had fallen in Hoghton's Brigade the survivors were still in line by their colours, closing in towards the tattered silk which represented the ark of their covenant – the one thing supremely important in the World.
> Sir John Fortescue, of the Battle of Albuera, 1811, quoted in Richardson, *Fighting Spirit*, 1978.

A man is not a soldier until he is no longer homesick, until he considers his regiment's colours as he would his village steeple; until he loves his colours, and is ready to put hand to sword every time the honour of the regiment is attacked.
> Marshal of France Thomas R. Bugeaud (1778–1846), quoted in Thomas, *Les Transformations de l'Armée Française*, 1887.

In the small closed world of the military, great victories, great defeats, and great sacrifices are never forgotten. They are remembered with battle streamers attached to unit flags. Among the scores of streamers that billow and whirl around the flags of all the battalions of the 1st Cavalry Division there is one deep-blue Presidential Unit Citation streamer that says simply: PLEIKU PROVINCE.
  School children no longer memorize the names and dates of great battles, and perhaps that is good; perhaps that is the first step on the road to a world where wars are no longer necessary. Perhaps. But we remember those days and our comrades, and long after we're gone that long blue streamer will still caress proud flags.
> Lieutenant-General Harold G. Moore, of the Battle of the Ia Drang Valley, 14–17 November 1967, *We Were Soldiers Once... And Young*, 1992.

COMBINED ARMS

You hear that Phillip goes where he pleases not by marching his phalanx of infantry, but by bringing in his train light infantry, cavalry, archers, mercenaries, and other such troops.
> Demosthenes, quoted in Duncan Head, *Armies of the Macedonian and Punic Wars*, 1982.

Maxim 47. Infantry, cavalry, and artillery are nothing without each other. They should always be so disposed in cantonments as to assist each other in case of surprise.
> Napoleon, *The Military Maxims of Napoleon*, tr. George D'Aguilar, 1831 (David Chandler, ed., 1987).

We know well what happens when a single arm is opposed to two others.
> Major-General Carl von Clausewitz (1780–1831).

It is not so much the mode of formation as the proper combined use of the different arms which will ensure victory.
> Lieutenant-General Antoine-Henri Baron de Jomini, *Summary of the Art of War*, 1838, tr. Mendell and Craighill, 1862.

There is still a tendency in each separate unit... to be a one-handed puncher. By that I mean that the rifleman wants to shoot, the tanker to charge, the artilleryman to fire... That is not the way to win battles. If the band played a piece first with the piccolo, then with the brass horn, then with the clarinet, and then with the trumpet, there would be a hell of a lot of noise but no music. To get the harmony in music each instrument must support the others. To get harmony in battle, each weapon must support the other. Team play wins. You musicians of Mars must not wait for

COMMAND

the band leader to signal you... You must each of your own volition see to it that you come into this concert at the proper place and at the proper time.
>General George S. Patton, Jr., 8 July 1941, address to the men of the Second Armored Division, The Patton Papers, II, 1974.

COMMAND

As far as I am concerned, I would rather be the first man here than the second in Rome.
>Julius Caesar (100–44 BC), while 'crossing the Alps he came to a small native village with hardly any inhabitants and altogether a miserable-looking place,' quoted in Plutarch, The Lives, c. AD 125 (The Fall of the Roman Republic, tr. Rex Warner, 1972).

A prince should therefore have no other aim or thought, nor take up any other thing for his study, but war and its organization and discipline, for that is the only art that is necessary to one who commands, and it is of such virtue that it not only maintains those who are born princes, but often enables men of private fortune to attain to that rank.
>Niccolò Machiavelli, The Prince, 1513.

In all well-ordered militias the commendation and sufficiency of all generals, colonels, captains, and other officers hath consisted in knowing how to command, govern, and order their armies, regiments, bands, and companies, and to win the love of their soldiers by taking great care of their healths and safeties, as also by all examples of virtue and worthiness not only by instruction but also by action in their own persons, venturing their lives in all actions against the enemy amongst them, and therewithal accounting of them in sickness and health or wounds received as of their own children; and whereas, again, all colonels and captains of horsemen according to all discipline have used to serve amongst their horsemen on horseback, and all colonels and captains of footmen, yea, even the very lieutenants general and kings themselves, if their armies and forces of the field have consisted more of footmen than of horsemen, have always used by all discipline military upon the occasion of any battle to put their horses from them and to serve on foot, and to venture their lives in the former ranks.
>John Smythe (c. 1580–1631), Certain Discourses Military, 1966.

I cannot obey any longer. I have tasted the pleasure of command, and I cannot renounce it. My decision is taken. If I cannot be master, I shall quit France.
>Napoleon, 1798, before the Egyptian expedition, to Miot de Melito, quoted in Louis de Bourrienne, The Memoirs of Napoleon Bonaparte, I, 1855.

I like to walk alone...
>The Duke of Wellington, 7 July 1801, to his brother, Henry Wellesley, after the news that he had been superseded in command after his brilliant victories in India by a senior officer, Wellington MSS, quoted in Elizabeth Longford, Wellington: Years of the Sword, 1970.

We want Men Capable Of Command – who will fight and reduce their soldiers to strict obedience.
>Major-General Andrew Jackson, 15 September 1812, letter to Brigadier-General John Coffee, Library Of Congress.

How can I trust a man to command others who cannot command himself?
>General Robert E. Lee (1807–1870), quoted in Douglas Southall Freeman, Douglas Southall Freeman on Leadership, 1993.

The only prize much cared for by the powerful is power. The prize of the general is not a bigger tent, but command.

COMMAND

Justice Oliver Wendell Holmes, February 15, 1913, speech at Harvard Law School, *The Occasional Speeches of Oliver Wendell Holmes*, 1962.

It is often argued that the possession of original ideas is different from, or even incompatible with, the power of executive command. The gulf between these two faculties is narrower than is commonly supposed. The conception of ideas implies mental initiative, while the moral courage required to express them publically and with vigour in the face of tradition-bound seniors indicates a strong personality, for the role of pioneer or reformer is rarely a popular one. These two qualities of mental initiative and strong personality, or determination, go a long way towards the power of command in war – they are, indeed, the hallmark of the Great Captains. (March 1923.)

Captain Sir Basil Liddell Hart, *Thoughts on War*, 1944.

Your greatness does not depend upon the size of your command, but on the manner in which you exercise it.

Marshal of France Ferdinand Foch, quoted in Aston, *Biography of Foch*, 1929.

You are probably busier than I am. As a matter of fact, commanding an army is not such a very absorbing task except that one has to be ready at all hours of the day and night... to make some rather momentous decision, which frequently consists of telling somebody who thinks he is beaten that he is not beaten.

General George S. Patton, Jr., letter, 11 October 1944, *The Patton Papers*, II, 1974.

At the top there are great simplifications. An accepted leader has only to be sure of what it is best to do, or at least have his mind made up about it. The loyalties which center upon number one are enormous. If he trips, he must be sustained. If he makes mistakes, they must be covered. If he sleeps, he must not be wantonly disturbed. If he is no good, he must be pole-axed.

Sir Winston S. Churchill, *Their Finest Hour*, 1949.

From General Marshall I learned the rudiments of effective command. Throughout the war I deliberately avoided intervening in a subordinate's duties. When an officer performed as I expected him to, I gave him a free hand. When he hesitated, I tried to help him. And when he failed, I relieved him.

General of the Army Omar Bradley, *A Soldier's Story*, 1951.

I believe in 'personal' command, i.e. that a commander should never attempt to control an operation or a battle by remaining at his H.Q. or be content to keep touch with his subordinates by cable, W/T or other means of communication. He must as far as possible see the ground for himself to confirm or correct his impressions of the map; his subordinate commanders to discuss their plans and ideas with them; and the troops to judge of their needs and their morale. All these as often as possible. The same of course applies to periods of preparation and periods between operations. In fact, generally the less time a commander spends in his office and the more he is with the troops the better.

Field Marshal Viscount Wavell of Cyrenaica, *Soldiers and Soldiering*, 1953.

As to the moral factors in command, it is always worth while to bear in mind the following:
(a) Two-thirds of the reports which are received in war are inaccurate; never accept a single report of success or disaster as necessarily true without confirmation.
(b) Always try to devise means to deceive and outwit the enemy and

THE COMMANDER

throw him off balance; the British in war are usually very lacking in this low cunning.
(c) Attack is not only the most effective but easiest form of warfare and the moral difference between advance and retreat is incalculable. Even when inferior in numbers, it pays to be as aggressive as possible.
(d) Finally, when things look bad and one's difficulties appear great, the best tonic is to consider those of the enemy.
Field Marshal Viscount Wavell of Cyrenaica, *Soldiers and Soldiering*, 1953

It was good fun commanding a division in the Iraq desert. It is good fun commanding a division anywhere. It is one of the four best commands in the service – platoon, battalion, a division, and an army. A platoon, because it is your first command, because you are young, and because, if you are any good, you know the men in it better than their mothers do and love them as much. A battalion, because it is a unit with a life of its own; whether it is good or bad depends on you alone; you have at last a real command. A division, because it is the smallest formation that is a complete orchestra of war and the largest in which every man can know you. An army, because the creation of its spirit and its leadership in battle give you the greatest unit of emotional and intellectual experience that can befall a man.
Field Marshal Viscount Slim of Burma, *Defeat Into Victory*, 1956.

Command must be direct and personal.
Field Marshal Viscount Montgomery of Alamein, *The Memoirs of Field Marshal Montgomery*, 1958.

It was my first command. It was the small unit, the group unit. I knew the four squadrons. I could never know any other four squadrons as well as I knew those four, because, when I had more rank and more command and more responsibility, I would be bound to lose out on the proportionate intimacies. I went on to command air wings and air divisions, and then to command even larger assemblages in the field. And finally to command all of SAC; and eventually to my job as Chief of Staff of the United States Air Force.
But the 305th – It was my 305th, our 305th.
General Curtis LeMay, *Mission With LeMay*, 1965.

THE COMMANDER

Master Sun said: 'The traits of the true commander are: wisdom, humanity, respect, integrity, courage, and dignity. With his wisdom he humbles the enemy, with his humanity he draws the people near to him, with his respect he recruits men of talent and character, with his integrity he makes good on his rewards, with his courage he raises the morale of the men, and with his dignity he unifies his command. Thus, if he humbles his enemy, he is able to take advantage of changing circumstances; if the people are close to him, they will be of a mind to go to battle in earnest; if he employs men of talent and wisdom, his secret plans will work; if his rewards and punishments are invariably honored, his men will give their all; if the morale and courage of his troops is heightened, they will of themselves be increasingly martial and intimidating; if his command is unified, the men will serve their commander alone.
Sun Tzu, c. 500 BC, a passage recovered from Wang Fu's (AD 76–157) *Advice to the Commander* and not found in *The Art of War*, quoted in *Sun Tzu: The Art of Warfare*, tr. and ed. Roger Ames, 1993.

An army of lions commanded by a deer will never be an army of lions.
Napoleon, 'Précis des guerres de Frederic II', *Correspondance de Napoléon Ier, publiée par ordre de l'Empereur Napoléon III*, Vol. XXX, 1858–1870.

THE COMMANDER

No victory is possible unless the commander be energetic, eager for responsibilities and bold undertakings; unless he possess and can impart to all the resolute will of seeing the thing through; unless he be capable of exerting a personal action composed of will, judgment, and freedom of mind in the midst of danger.
> Marshal of France Ferdinand Foch, *Precepts and Judgments*, 1919.

There is required for the composition of a great commander not only massive common sense and reasoning power, not only imagination, but also an element of legerdemain, an original and sinister touch, which leaves the enemy puzzled as well as beaten.
> Sir Winston S. Churchill, *The World Crisis*, 1923.

Initiative and the desire to crawl into any crack in the enemy combat formations must be the main qualities of every commander.
> Marshal of the Soviet Union Mikhail N. Tukhachevskiy (1893–1937), quoted in Savkin, *The Basic Principles of Operational Art and Tactics*, 1972.

Sorting out muddles is really the chief job of a commander.
> Field Marshal Viscount Wavell of Cyrenaica, 'Training of the Army for War', *Journal, Royal United Services Institution*, 2/1933.

In the profoundest sense, battles are lost and won in the mind of the commander, and the results are merely registered in his men.
> Captain Sir Basil Liddell Hart, *Colonel Lawrence: The Man Behind the Legend*, 1934.

A commander should have a profound understanding of human nature, the knack of smoothing out troubles, the power of winning affection while communicating energy, and the capacity for ruthless determination where required by circumstances. He needs to generate an electrifying current, and to keep a cool head in applying it.
> Captain Sir Basil Liddell Hart, *Thoughts on War*, 1944.

In my experience, all very successful commanders are prima donnas, and must be so treated. Some officers require urging, others require suggestions, very few have to be restrained.
> General George S. Patton, Jr., *War As I Knew It*, 1947.

The commander is the backbone of the military unit. The command personnel constitute the skeleton which holds the separate limbs of an army together and supports them as a single, closely-knit military organization. The capability, devotion and courage of the combatants – in other words the practical ability of the army – are related directly to the ability of their commanders, of each of their commanders individually within the limits of his particular responsibilities, of all their commanders together in so far as they are collectively responsible for the army as a whole and all its undertakings.
> General Yigal Allon, *The Making of Israel's Army*, 1970.

I have developed almost an obsession as to the certainty with which you can judge a division, or any other large unit, merely by knowing its commander intimately. Of course, we have had pounded into us all through our school courses that the exact level of a commander's personality and ability is always reflected in his unit – but I did not realize, until opportunity came for comparisons on a rather large scale, how infallibly the commander and unit are almost one and the same thing.
> General of the Army Dwight D.

COMMANDING GENERAL/COMMANDER-IN-CHIEF

Eisenhower, *At Ease: Stories I Tell My Friends*, 1967.

Men are of no importance. What counts is who commands.
General Charles de Gaulle, *New York Times Magazine*, 12 May 1968.

For confidence, spirit, purposefulness, aggressiveness flow down from the top and permeate a whole command. And in the same way do anxiety and lack of resolution on the part of a commander put their indelible stamp upon his men.
General Matthew B. Ridgeway, quoted in *Military Review*, 6/1987.

COMMANDING GENERAL/COMMANDER-IN-CHIEF

The responsibility for a martial host of a million lies in one man. His is the trigger of its spirit.
Wu Ch'i (430–381 BC).

A commander-in-chief therefore, whose power and dignity are so great and to whose fidelity and bravery the fortunes of his countrymen, the defense of their cities, the lives of the soldiers, and the glory of the state, are entrusted, should not only consult the good of the army in general, but extend his care to every private soldier in it. For when any misfortunes happen to those under his command, they are considered as public losses and imputed entirely to his misconduct.
Flavius Vegetius Renatus, *Military Institutions of the Romans*, c. AD 378, tr. Clark, 1776.

I have formed a picture of a general commanding which is not chimerical – I have seen such men.
 The first of all qualities is COURAGE. Without this the others are of little value, since they cannot be used. The second is INTELLIGENCE, which must be strong and fertile in expedients. The third is HEALTH.
 He should possess a talent for sudden and appropriate improvisation. He should be able to penetrate the minds of other men, while remaining impenetrable himself. He should be endowed with the capacity of being prepared for everything, with activity accompanied by judgment, with skill to make a proper decision on all occasions, and with exactness of discernment.
Field Marshal Maurice Comte de Saxe, *My Reveries*, 1732, tr. Phillips, 1940.

Let what will arrive, it is the part of the general-in-chief to remain firm and constant in his purposes: he must be equally superior to elation in prosperity and depression in adversity, for in war good and bad fortune succeed each other by turns, and form the ebb and flow of military operations.
Field Marshal Prince Raimondo Montecuccoli (1609–1680), *Commentarii Bellici*, 1740, quoted in Barker, *The Military Intellectual and Battle*, 1975.

Remember that you are a commander-in-chief and must not be beaten; therefore do not undertake anything with your troops unless you have some strong hope of success.
The Duke of Wellington, 11 May 1809, to General Beresford, *The Dispatches of Field Marshal the Duke of Wellington, During His Various Campaigns in India, Denmark, Portugal, Spain, the Low Countries, and France*, IV, 1834–1838, p. 296.

Maxim 73. The first qualification in a general-in-chief is a cool head – that is, a head which receives just impressions, and estimates things and objects at their real value. He must not allow himself to be elated by good news, or depressed by bad.
 The impressions he receives, either successively or simultaneously in the course of the day, should be so classed as to take up only the exact place in his mind which they deserve to occupy;

COMMANDING GENERAL/COMMANDER-IN-CHIEF

since it is upon a just comparison and consideration of the weight due to different impressions that the power of reasoning and of right judgment depends.

Some men are so physically and morally constituted as to see everything through a highly coloured medium. They raise up a picture in the mind on every slight occasion and give to every trivial occurrence a dramatic interest. But whatever knowledge, or talent, or courage, or other good qualities such men may possess, nature has not formed them for the command of armies, or the direction of great military operations.
> Napoleon, *The Military Maxims of Napoleon*, tr. George D'Aguilar, 1831 (David Chandler, ed., 1987).

The character of the man is above all other requisites in a commander-in-chief.
> Lieutenant-General Antoine-Henri Baron de Jomini, *Summary of the Art of War*, 1838, tr. Mendell and Craighill, 1862.

In my youthful days I used to read about commanders of armies and envied them what I supposed to be a great freedom in action and decision. What a notion! The demands upon me that must be met make me a slave rather than a master. Even my daily life is circumscribed with guards, aides, etc., etc., until sometimes I want nothing so much as complete seclusion.
> General of the Army Dwight D. Eisenhower, 27 May 1943, letter to his wife, Mamie, quoted in *Letters to Mamie*, 1978.

In my opinion, generals – or at least the Commanding General – should answer their own telephones in the daytime. This is not particularly wearisome because few people call a general, except in emergencies, and then they like to get him at once.
> General George S. Patton, *War As I Knew It*, 1947.

There are three types of commanders in the higher grades:
1. Those who have faith and inspiration, but lack the infinite capacity for taking pains and preparing for every foreseeable contingency – which is the foundation of all success in war. These fail.
2. Those who possess the last-named quality to a degree amounting to genius. Of this type I would cite Wellington as the perfect example.
3. Those who, possessing this quality, are inspired by a faith and conviction which enables them, when they have done everything possible in the way of preparation and when the situation favours boldness, to throw their bonnet over the moon. There are moments in war when, to win all, one has to do this. I believe such a moment occurred in August 1944 after the Battle of Normandy had been won, and it was missed. Nelson was the perfect example of this – when he broke the line at St. Vincent, when he went straight in to attack at the Nile under the fire of the shore batteries and with night falling, and at the crucial moment at Trafalgar.
> Field Marshal Viscount Montgomery of Alamein, *The Memoirs of Field Marshal Montgomery*, 1958.

The acid test of an officer who aspires to high command is his ability to be able to grasp quickly the essentials of a military problem, to decide rapidly what he will do, to make it quite clear to all concerned what he intends to achieve and how he will do it, and then to see that his subordinate commanders get on with the job. Above all, he has got to rid himself of all irrelevant detail, he must concentrate on essentials.
> Field Marshal Viscount Montgomery of Alamein, *Memoirs of Field Marshal Montgomery*, 1958.

A [commander] of great armies in the field must have an inner conviction

which, though founded closely on reason, transcends reason.
> Field Marshal Montgomery of Alamein, *The Path to Leadership*, 3, 1960.

COMMON SENSE

When one is intent on an object, common sense will usually direct one to the right means.
> The Duke of Wellington (1769–1852), quoted in Philip Henry Stanhope, *Notes of Conversations with the Duke of Wellington, 1831–1851*, 1888.

I attribute it entirely to the application of good sense to the circumstances of the moment.
> The Duke of Wellington (1769–1852), when asked to what characteristic of mind he attributed his invariable success, quoted in William Fraser, *Words of Wellington*, 1889.

The art of war does not require complicated maneuvers; the simplest are the best, and common sense is fundamental. From which one might wonder how it is generals make blunders; it is because they try to be clever. The most difficult thing is to guess the enemy's plan, to sift the truth from all the reports that come in. The rest merely requires common sense; it is like a boxing match, the more you punch the better it is. It is also necessary to read the map well.
> Napoleon, 29 January 1818, St Helena, quoted in R.M. Johnston, ed., *The Corsican*, 1910.

While exceptional ability is desirable in a general, common sense is essential. Yet there is a peculiar danger in the latter quality unless it is accompanied by the former faculty. For many men of strong common sense but limited education are apt to feel uneasy in dealing with those who are intellectually better equipped. And the too common result is that they become more than normally sensitive to the arguments, especially the negative arguments, of intellectual second-raters, clever but uncreative – the class which produces most of the mouthpieces of orthodoxy. And the purely practical man is too often led to yield, against his better judgment, because he feels incapable of combating their objections. (April 1929.)
> Captain Sir Basil Liddell Hart, *Thoughts On War*, 1944.

COMMUNICATION

If intercommunication between events in front and ideas behind are not maintained, then two battles will be fought – a mythical headquarters battle and an actual front-line one, in which case the real enemy is to be found in our own headquarters. Whatever doubt exists as regards the lessons of the 1st war, this is one which cannot be controverted.
> Major-General J.F.C. Fuller, quoted in George Marshall, ed., *Infantry in Battle*, 1939.

COMPETENCE

The officers know that I myself am not ashamed to work at this. ...Suvorov was Major, and Adjutant, and everything down to Corporal; I myself looked into everything and could teach everybody.
> Field Marshal Prince Aleksandr V. Suvorov (1729–1800), quoted in W. Lyon Blease, *Suvorof*, 1926.

One must understand the mechanism and power of the individual soldier, then that of a company, a battalion, a brigade and so on, before one can venture to group divisions and move an army. I believe I owe most of my success to the attention I always paid to the inferior part of tactics as a regimental officer. There are few men in the Army who knew these details better than I did; it is the foundation of all

COMRADESHIP AND THE BAND OF BROTHERS

military knowledge.
> The Duke of Wellington (1769–1852), quoted in John Croker, *The Croker Papers*, I, 1884.

The true way to be popular with the troops is not to be free and familiar with them, but to make them believe you know more than they do.
> General of the Army William T. Sherman, 11 November 1864, letter to Rev. Henry Lay.

A competent leader can get efficient service from poor troops, while on the contrary an incapable leader can demoralize the best of troops.
> General of the Armies John J. Pershing, *My Experiences in the World War*, 1931.

COMRADESHIP AND THE BAND OF BROTHERS

And it came to pass, when he made an end of speaking unto Saul, that the soul of Jonathan was knit with the soul of David, and Jonathan loved him as his own soul.

And Saul took him that day, and would let him go no more home to his father's house.

Then Jonathan and David made a covenant, because he loved him as his own soul.

And Jonathan stripped himself of the robe that was upon him, and gave it to David, and his garments, even to his sword, and to his bow, and to his girdle.
> I Samuel 18:1–4, after David slew Goliath.

Now be men, my friends! Courage, come, take heart!
Dread what comrades say of you here in bloody combat!
When men dread that, more men come through alive –
when soldiers break and run, good-bye glory,
good-bye all defenses.
> Homer, *The Iliad*, 5.610–14, c. 800 BC, tr. Robert Fagles, 1990; Agamemnon rallying the Achaeans.

It is further recorded in these documents that the soldiers were passionately eager to seem him; some hoped for a sight of him while he was still alive; others wished to see his body, for a report had gone round that he was already dead, and they suspected, I fancy, that his death was being concealed by his guards. But nothing could keep them from a sight of him, and the motive in almost every heart was grief and a sort of helpless bewilderment at the thought of losing their king. Lying speechless as the men filed by, he yet struggled to raise his head, and in his eyes there was a look of recognition for each individual as he passed...
> Arrian, of Alexander the Great's farewell to the Macedonians on his deathbed, 323 BC, *The Campaigns of Alexander*, 7.26 c. BC 150, tr. Aubrey de Sélincourt, 1987.

We few, we happy few, we band of brothers;
For he to-day that sheds his blood with me
Shall be my brother.
> William Shakespeare, *Henry V*, Act IV, 3.59; speech of Henry V at before the Battle of Agincourt, 25 October 1415.

My first wish would be that my military family, and the whole Army, should consider themselves as a band of brothers, willing and ready to die for each other.
> General George Washington, 21 October 1798, letter to Henry Knox, *Writings*, XXXVI, 1941, p. 508.

I had the happiness to command a band of brothers.
> Admiral Lord Horatio Nelson, August 1798, to Admiral Lord Howe after the battle of the Nile.

A mysterious fraternity born out of smoke and danger of death.
> Stephen Crane, *The Red Badge of Courage*, 1893.

COMRADESHIP AND THE BAND OF BROTHERS

These voices, these quiet words, these footsteps in the trench behind me recall me at a bound from the terrible loneliness and fear of death by which I had been almost destroyed. They are more to me than life, those voices, they are more than motherliness and more than fear, they are the strongest, most comforting thing there is anywhere, they are the voices of my comrades.

Erich Maria Remarque, *All Quiet On the Western Front*, 1929.

The soldier's field of activity is man, who controls science, technics and material. The army is a combination of many men with the same serious aim. This gives the soldier's profession a quite peculiar bond of unity, a corporate sense which we call comradeship. The term is extremely comprehensive. If we start out from the notion of responsibility we find that 'comradeship' means 'one for all,' for each man bears, in his own way and in his own place, a share of the responsibility for the welfare, the ability, the achievements, and the life of others. For the senior, the leader, the superior, this means the duty of correcting, of training, of supervising others; for the junior, the novice, the subordinate, it means the duty of conscious, voluntary subordination. Love and confidence are the two great components of comradeship.

Colonel-General Hans von Seekt, *Thoughts of a Soldier*, 1930.

I have always tried to crack a joke or two before, and you have been friendly and laughed at them. But today I am afraid I have run out of jokes, and I don't suppose any of us feel much like laughing. The Kelly has been in one scrap after another, but even after we have had men killed the majority survived and brought the old ship back. Now she lies in fifteen hundred fathoms and with her more than half our shipmates. If they had to die, what a grand way to go, for now they all lie together in the ship we loved… We have lost her, but they are still with her. There may be less than half the Kelly left, but I feel that each of us will take up the battle with even stronger heart… You will all be sent to replace men who have been killed in other ships, and the next time you are in action remember the Kelly. As you ram each shell home into the gun, shout 'Kelly!' and so her spirit will go on inspiring us until victory is won. I should like to add that there isn't one of you I wouldn't be proud and honoured to serve with again. Goodbye, good luck, and thank you all from the bottom of my heart.

Admiral Lord Louis Mountbatten of Burma (when he was a lowly destroyer captain), his farewell to the survivors of the crew of the *Kelly* in Alexandria after the loss of the gallant ship in the Battle of Crete, 1941.

Why does the soldier leave the protection of his trench or hole in the ground and go forward in the face of shot and shell? It is because of the leader who is in front of him and his comrades who are around him. Comradeship makes a man feel warm and courageous when all his instincts tend to make him cold and afraid.

Field Marshal Viscount Montgomery of Alamein, *A History of Warfare*, 1968.

In such a unit as the 101st, it was a constant task to impress on the officers the folly, indeed the unfairness, of unnecessarily exposing themselves to enemy fire. The strength of a fighting outfit is the mutual respect of all its members of whatever rank. Shared danger breeds admiration for the hard, utter intolerance for the weak, and a fierce loyalty to comrades. Students of military history have often tried to determine why some men fight well and others run away. It never seemed to me that ideological motives or political or

CONCENTRATION

moral concepts had much to do with it. If I could get any of my men to discuss a matter so personal as their honest reaction to combat, they would tell me that they fought, though admittedly scared, because 'I couldn't let the other boys down' or 'I couldn't look chicken before 'Dog Company'.' These are simple reasons for simple virtues in simple men whom it is an ennobling privilege to command. For their officers, that privilege carries with it the responsibility to stay alive and look after them. This was a sound precept but one hard to impress on the officers of the 101st who, like their men, didn't want to 'look chicken.'

General Maxwell Taylor, *Swords and Ploughshares*, 1972.

This day I am proud to be a Canadian, and I am prouder still to belong to the 'band of brothers' who volunteered for overseas service. Today I don't despise the [Canadians] who refuse overseas duty – I pity them, for they will never know what it is to be a man among men. Men! Not supermen, just men, who have learned to act together from no selfish motives but as comrades willing to risk death for one another. After the busting of the Hitler Line, the squirmings of the politicians and the war profiteers and the other gutless wonders at home won't bother us as much. By God, we *know* who *we* are, and what we amount to.

Captain Farley Mowat, Hastings and Prince Edward Regiment, *My Father's Son: Memories of War and Peace*, 1992.

Another and far more transcendent love came to us unbidden on the battlefields, as it does on every battlefield in every war man has ever fought. We discovered in that depressing, hellish place, where death was our constant companion, that we loved each other. We killed for each other, we died for each other, and we wept for each other. And in time we came to love each other as brothers. In battle our world shrank to the man on our left and the man on our right and the enemy all around. We held each other's lives in our hands and we learned to share our fears, our hopes, our dreams as readily as we shared what little else good came our way.

Lieutenant-General Harold G. Moore, *We Were Soldiers Once… And Young*, 1992.

CONCENTRATION

The few [are the ones] who prepare against others; the many [are the ones] who make others prepare against them.

Sun Tzu, *The Art of War*, 6, c. 500 BC, tr. Ralph Sawyer, 1985.

One should never risk one's whole fortune unless supported by one's entire forces.

Niccolò Machiavelli, *Discourses*, 1517.

Petty geniuses attempt to hold everything; wise men hold fast to the most important points. They parry great blows and scorn little accidents. There is an ancient apothegm: he who would preserve everything, preserves nothing. Therefore, always sacrifice the bagatelle and pursue the essential! The essential is to be found where big bodies of the enemy are. Stick to defeating them decisively, and the detachments will flee by themselves or you can hunt them down without difficulty.

Frederick the Great, *Instructions to His Generals*, 1747, tr. Phillips, 1940.

True military art consists of the ability to be stronger than the enemy at the given moment.

Military art is the art of separating for life and uniting for battle.

On the battlefield there is no such thing as a surplus battalion or squadron.

Commanders who save fresh forces for operations after the battle will almost always be defeated.

CONCENTRATION

Napoleon's sayings on concentration, quoted in V.Ye. Savkin, *Basic Principles of Operational Art and Tactics*, 1972.

The best strategy is always *to be very strong*; first in general, and then at the decisive point. Apart from the effort needed to create military strength, which does not always emanate from the general, there is no higher and simpler law of strategy than of *keeping one's forces concentrated*.
Major-General Carl von Clausewitz, *On War*, 3.11, 1832, tr. Michael Howard and Peter Paret, 1976.

Concentration sums up in itself all the other factors, the entire alphabet of military efficiency in war.
Lieutenant-General Antoine-Henri Baron de Jomini, quoted in Richard Simpkin, *Race to the Swift*, 1985.

Another rule – never fight against heavy odds, if by any possible maneuvering you can hurl your own forces on only a part, and then the weakest part, of your enemy and crush it. Such tactics will win every time, and a small army may thus destroy a large one in detail, and repeated victory will make it invincible.
Lieutenant-General Thomas J. 'Stonewall' Jackson, 1862, quoted in G.F.R. Henderson, *Stonewall Jackson*, 1898.

I always make it a rule to get there first with the most men.
Lieutenant-General Nathan Bedford Forrest (1821–1877). Note: This statement has often been rendered popularly as 'Git thar furst with the mostest.'

To remain separated as long as possible while operating and to be concentrated in good time for the decisive battle, that is the task of the leader of large masses of troops.
Field Marshal Helmuth Graf von Moltke, *Instructions for Superior Commanders of Troops*, 1869.

In any military scheme that comes before you, let your first question be, Is this consistent with the requirement of concentration?
Rear Admiral Alfred Thayer Mahan, *Naval Strategy*, 1911.

One must not rely on the heroism of the troops. Strategy must furnish tactics with tasks easy to accomplish. This is obtained in the first place by the concentration in the place of the main blow of forces many times superior to those of the enemy.
Marshal of the Soviet Union Mikhail N. Tukhachevskiy, 1920, quoted in Leites, *The Soviet Style in War*, 1982.

The Principle of Concentration. Concentration, or the bringing of things or ideas to a point of union, pre-supposes movement; movement of ideas, especially in an army, is a far more difficult operation than the movement of men. Nevertheless, unless ideas, strategical, tactical and administrative, be concentrated, cohesion of effort will not result; and in proportion as annuity of action is lacking, so will an army's strength, moral and physical, be squandered in detail until a period be arrived at in which the smallest result will be obtained from every effort.
Major-General J.F.C. Fuller, *The Reformation of War*, 1923.

All through history, from the days of the great phalanx of the Roman Legion, the master law of tactics remains unchanged; this law is that to achieve success you must be superior at the point where you intend to strike the decisive blow.
Field Marshal Viscount Montgomery of Alamein, 'The Growth of Modern Infantry Tactics', *Antelope*, 1/1925.

My strategy is one against ten, my tactic ten against one.

CONFIDENCE

Mao Tse-tung, quoted in Dennis and Ching Ping Bloodworth, *The Chinese Machiavelli*, 1976.

I don't care how many tanks you British have so long as you keep splitting them up the way you do. I shall just continue to destroy them piecemeal.
Field Marshal Erwin Rommel, 1941, to a captured British brigadier, quoted in Alan Moorehead, *The March to Tunis*, 1943.

The principles of war, not merely one principle, can be condensed into a single word – 'concentration.' But for truth this needs to be amplified as the 'concentration of strength against weakness.' And for any real value it needs to be explained that the concentration of strength against weakness depends on the dispersion of your opponent's strength, which in turn is produced by a distribution of your own that gives the appearance, and partial effect of dispersion. Your dispersion, his dispersion, your concentration – such is the sequence, and each is a sequel. True concentration is the fruit of calculated dispersion.
Captain Sir Basil Liddell Hart, *Strategy*, 1954.

CONFIDENCE

the troops must have confidence in the orders of their seniors. The orders of their superiors is the source whence discipline is born.
Wu Ch'i (430–381 BC), *Art of War*, quoted in Sun Tzu, *The Art of War*, c. 500 BC, tr. Samuel Griffith, 1969.

Remember, that already danger has often threatened you and you have looked it triumphantly in the face; this time the struggle will be between a victorious army and an enemy already once vanquished.
Alexander the Great, October 333 BC, address to his army before the Battle of Issus, quoted in Arrian, *The Campaigns of Alexander*, 2.7, c. ad 150, tr. Aubrey de Sélincourt, 1987.

The consul Valerius Laevinus, having caught a spy within his camp, and having entire confidence in his own forces, ordered the man to be led around, and observing that, for the sake of terrifying the enemy, his army was open to inspection by the spies of the enemy, as often as they wished.
Frontinus, *Stratagems*, 4.7, c. AD 84–96, tr. Charles E. Bennett, 1925.

Troops are not to be led into battle unless confident of success.
Flavius Vegetius Renatus, *Military Institutions of the Romans*, c. AD 378, tr. Clark, 1776.

When the general leads his men out to battle, he should present a cheerful appearance, avoiding any gloomy look. Soldiers usually estimate their prospects by the appearance of the general.
The Emperor Maurice, *The Strategikon*, c. AD 600 (*Maurice's Strategikon*, tr. George Dennis, 1984).

There is plenty of time to win this game, and to thrash the Spaniards too.
Sir Francis Drake, 20 July 1588, while playing bowls when the Spanish Armada was sighted.

Every man passed through my hands, and he was told that nothing more remained for him to know, if only he did not forget what he had learned. Thus he was given confidence in himself, the foundation of bravery.
Field Marshal Prince Aleksandr V. Suvorov (1729–1800), quoted in W. Lyon Blease, *Suvorof*, 1920.

A great and good general is... in himself an host; for his influence, insinuating itself into every member of the military body, connects and binds the whole together imperceptibly, but firmly and

CONFIDENCE

securely. Such confidence in a leader is the charm against a panic.
  Robert Jackson, *A Systematic View of the Formation, Discipline, and Economy of Armies*, 1804.

I am thinking of the French I am going to fight. I have not seen them since the campaigns in Flanders [1793–1794] when they were capital soldiers, and a dozen years of victory under Bonaparte must have made them better still. 'Tis enough to make one thoughtful. But though they may overwhelm me, I don't think they will outmanoeuvre me. First, because I am not afraid of them, as every one else seems to be, and secondly, because (if all I hear about their system is true) I think it a false one against steady troops. I suspect all the continental armies are half-beaten before the battle begins. I at least will not be frightened beforehand.
  The Duke of Wellington, early July 1808, conversation with John Croker on the eve of the Peninsular War, cited in Charles Oman, *Wellington's Army*, 1913.

The Spaniards make excellent soldiers. What spoils them is that they have no confidence in their officers – this would ruin any soldiers – and how should the Spaniards have confidence in such officers as theirs?
  The Duke of Wellington, during the Peninsular War (1808–1813), quoted in John Croker, *The Croker Papers*, III, 1884.

Depend upon it. It requires time for a general to inspire confidence or to feel it; for you will never have confidence in yourself until others have confidence in you.
  The Duke of Wellington to Earl Stanhope, *Notes on the Conversations with the Duke of Wellington*, 1888.

Confidence, the dawn of victory, inspired the whole line.
  Major-General Winfield Scott, May 1814, letter, quoted in Eisenhower, *Agent of Destiny*, 1997.

No matter what may be the ability of the officer, if he loses the confidence of his troops, disaster must sooner or later ensue.
  General Robert E. Lee, 8 August 1863, letter to Jefferson Davis.

Confident language by a military commander is not usually regarded as evidence of competence.
  General Joseph E. Johnston, 18 July 1864; his tart rejoinder to President Davis' relief of him, citing in part his lack of confidence that he could defeat Sherman who was at the gates of Atlanta, *Narrative*, 1874.

I never saw a more confident army. The soldiers think I know everything and that they can do anything.
  General William T. Sherman, letter to his wife after taking Savanna, 16 December 1864, *Home Letters*, 1909.

If confidence be indeed 'half the battle', then to undermine the enemy's confidence is more than the other half – because it gains the fruits without an all-out fight. (April 1929.)
  Captain Sir Basil Liddell Hart, *Thoughts on War*, 1944.

Late one night during the war, after a certain serious decision had been taken, my chief coadjutor came to me full of doubts as to whether our decision had been right. 'Let it be,' I answered. 'Only military academies fifty years hence will know for certain whether we did right or wrong.'
  Colonel General Hans von Seekt, *Thoughts of a Soldier*, 1930.

For two weeks Puller had commanded the rear of the First Marine Division, cut off in the Chosin Reservoir region by hundreds of thousands of Chinese Communist troops. The colonel was

CONSCIENCE

visiting a hospital tent where a priest administered last rites to Marine wounded when a messenger came:

'Sir, do you know they've cut us off? We're entirely surrounded.'

'Those poor bastards,' Puller said, 'They've got us right where we want 'em. We can shoot in every direction now.'

> Lieutenant-General Lewis B. 'Chesty' Puller, 1950, as narrated in Burke Davis, *Marine!*, 1962.

However desperate the situation, a senior commander must always exude confidence in the presence of his subordinates. For anxiety, topside, can spread like cancer down through the command.

> General of the Army Omar Bradley, *A Soldier's Story*, 1951.

The general who sees that the soldier is well fed and looked after, and who puts him into a good show and wins battles, will naturally have his confidence. Whether he will also have his affection is another story... But does it matter to a general whether he has his men's affection so long as he has their confidence? He must certainly never court popularity. If he has their appreciation and respect it is sufficient. Efficiency in a general his soldiers have a right to expect; geniality they are usually right to suspect. Marlborough was perhaps the only great general to whom geniality was always natural.

> Field Marshal Viscount Wavell of Cyrenaica, *Soldiers and Soldiering*, 1953.

Probably one of the greatest assets a commander can have is the ability to radiate confidence in the plan and operations even (perhaps especially) when inwardly he is not too sure about the outcome.

> Field Marshal Viscount Montgomery of Alamein, *The Memoirs of Field Marshal Montgomery*, 1958.

Allied commands depend on mutual confidence. How is mutual confidence developed? You don't command it... By development of common understanding of the problems, by approaching these things on the widest possible basis with respect to each other's opinions, and above all, through the development of friendships, this confidence is gained in families and in Allied Staffs.

> General of the Army Dwight D. Eisenhower, quoted in Joint Pub 1, *Joint Warfare of the US Armed Forces*, 1991.

CONSCIENCE

Labor to keep alive in your breast that little spark of celestial fire called conscience.

> General George Washington (1732–1799).

In this solemn hour it is a consolation to recall and to dwell upon our repeated efforts for peace. All have been ill-starred, but all have been faithful and sincere. This is of the highest moral value – and not only moral value, but practical value – at the present time, because of the wholehearted concurrence of scores of millions of men and women, whose cooperation is indispensable and whose comradeship and brotherhood are indispensable, is the only foundation upon which the trial and tribulation of modern war can be endured and surmounted. This moral conviction alone affords that ever-fresh resilience which renews the strength and energy of people in long, doubtful and dark days. Outside, the storms of war may blow and the lands may be lashed with the fury of its gales, but in our own hearts this Sunday morning there is peace. Our hands may be active, but our consciences are at rest.

> Sir Winston S. Churchill, 3 September 1939, upon the British declaration of war against Germany, *Blood, Sweat, and Tears*, 1941.

CONSCRIPTION

The only guide to a man is his conscience; the only shield to his memory is the rectitude and sincerity of his actions. It is very imprudent to walk through life without this shield, because we are so often mocked by the failure of our hopes and the upsetting of our calculations; but with this shield, however the fates may play, we march always in the ranks of honour.
  Sir Winston S. Churchill, 12 November 1940, tribute in the House of Commons to the late Neville Chamberlain.

Despite his doubts about the universal validity of the deliverances of conscience, our model officer recognizes that there is in himself an instinctive resistance to actions inconsistent with principles of behavior he learned to follow early in his career. Perhaps this is the voice of his professional conscience; if so, he is happy to have one to keep him straight and will give it due heed. However, he is most unsympathetic with officers who use conscience as an excuse for dereliction of duty or the avoidance of dangerous or unpleasant tasks. In his view, such conduct is worthy of the disdain accorded the soldier who does not discover until the eve of battle that he is a conscientious objector.
  General Maxwell D. Taylor, 'A Do-It-Yourself Professional Code for the Military', *Parameters*, 12/1980.

Once a soldier's conscience is aroused, it defines a line he dare not cross and deeds he dare not commit, regardless of orders, because those very deeds would destroy something in him which he values more than life itself. If the path of military operations and this line of a soldier's conscience collide, disobedience and mutiny erupt.
  Lieutenant-General Sir James Glover, 'A Soldier and His Conscience', *Parameters*, 9/1983.

CONSCRIPTION

You have the impudence to talk of the conscription in France; it wounds your pride because it fell upon all ranks. Oh, how shocking, that a gentleman's son should be obliged to defend his country, just as if he were one of the mob!
  The conscription did not crush a particular class like your press-gang, nor the rabble, because they were poor. My rabble would have become the best educated in the world. All my exertions were directed to illuminate the mass of the nation instead of brutalizing them by ignorance and superstition.
  Napoleon, 18 February 1818, on St Helena, quoted in R.M. Johnston, ed., *The Corsican*, 1910.

Conscription forms citizen armies. Voluntary enlistment forms armies of vagrants and good-for-nothings. The former are guided by honor; mere discipline controls the latter.
  Napoleon (1769–1821), quoted in Christopher Herold, ed., *The Mind of Napoleon*, 1955.

The South, as we all knew, were conscripting every able-bodied man between the ages of eighteen and forty-five; and now they had passed a law for the further conscription of boys from fourteen to eighteen, calling them the junior reserves, and men from forty-five to sixty to be called the senior reserves. The latter were to hold the necessary points not in immediate danger, and especially those in the rear. General Butler, in alluding to this conscription, remarked that they were thus 'robbing both the cradle and the grave,' an expression which I afterwards used in writing a letter to Mr. Washburn.
  General Ulysses S. Grant, of the last months of the American Civil War, *Personal Memoirs*, II, 1885.

Grant says the Confederates, in their endeavors to get men, have robbed the

CONTACT

cradle and the grave; if that is the case, I must say their ghosts and babies fight well!
  Major-General George G. Meade, comment to his staff, 12 December 1864.

The system of conscription has always tended to foster quantity at the expense of quality.
  Captain Sir Basil Liddell Hart, *The Untimeliness of a Conscript Army*, 1950.

Any recruiter will tell you that the incentive for enlistment is it [being drafted] is inevitable if you don't.
  Lieutenant-General Lewis B. Hershey, director of the Selective Service System, address at the American University, Washington, DC, 11 December 1966, quoted in *The New York Times*, 12 December 1966.

CONTACT

At last the armies clashed at one strategic point, they slammed their shields together, pike-scraped pike with the grappling strength of fighters armed in bronze and their round shields pounded, boss on welded boss, and the sound of struggle roared and rocked the earth. Screams of men and cries of triumph breaking in one breath, fighters killing, fighters killed, and the ground streamed blood. Wildly as two winter torrents raging down from the mountains, swirling into a valley, hurl their great waters together, flash floods from the wellsprings plunging down in a gorge and miles away in the hills a shepherd hears the thunder – so from the grinding armies broke the cries and crash of war.
  Homer, *The Iliad*, 4.517–528, c. 800 BC, tr. Robert Fagles, 1990.

Another woman, in reply to her son who declared that the sword he had was a small one, said: 'Then extend it by a stride.'
  An anonymous Spartan mother, quoted in Plutarch, *The Lives*, c. AD 100 (*Plutarch on Sparta*, tr. Richard Talbert, 1988).

At the Luxembourg, during the provisional consulate, he [Abbé Siéyès] often awakened his colleague Napoleon, and harassed him about the new plots which he heard of every moment from his private police. 'But have they corrupted our guard?' Napoleon used to say. 'No.' 'Then go to bed. – In war, as in love, my dear Sir, we must come to close quarters to conclude matters. It will be time enough to be alarmed when our 600 men are attacked.
  Napoleon, November 1799, quoted in Emmanuel Las Casas, *The Life, Exile, and Conversations of the Emperor Napoleon*, II, 1890.

Wherever the enemy goes, let our troops go also.
  General of the Army Ulysses S. Grant, 1 August 1864, dispatch to General Halleck about Sheridan's operations in the Shenandoah Valley.

Contact (a word which perhaps better than any other indicates the dividing line between tactics and strategy).
  Rear Admiral Alfred Thayer Mahan, *The Influence of Sea Power Upon History*, 1890.

If your sword is too short, take one step forward.
  Admiral Marquis Heihachiro Togo, 15 May 1905.

Contact is information of the most tangible kind, an enemy met with is an enemy at grips, and, as in a wrestling match, contact is likely to be followed by much foot-play. Time still remains the decisive factor, time wherein to modify a plan according to the information contact gains.
  Major-General J.F.C. Fuller, *Armoured Warfare*, 1943.

Boys, we've been seeking the enemy and now we have found them. They are in

A COOL HEAD

front of us, behind us, and on both our flanks.
> Lieutenant-General Lewis B. 'Chesty' Puller, December 1950, at the Chosin Reservoir when the Chinese entered the Korean War, Puller Collection, Marine Corps Historical Division.

CONTRACTORS

There is nothing performed by contractors which may not be much better executed by intelligent officers. They make immense fortunes at the expense of the state which ought to be saved. They destroy the army, horse and foot and even hospitals, by furnishing the worst of everything.
> Major-General Henry Lloyd (1720–1783), *History of the Late War in Germany, 1766–1782*, quoted in Liddell Hart, *The Sword and the Pen*, 1976.

[Civilian-controlled] bureaux, not to speak of the contractors, are the born enemies of all that tends to put the details of military administration into military hands.
> Field Marshal Francois Comte de Guibert (1744–1790), quoted in Alfred Vagts, *A History of Militarism*, 1937.

In the councils of government we must guard against the acquisition of unwarranted influence, whether sought or unsought, by the military-industrial complex. The potential for the disastrous rise of misplaced power exists and will persist.
> President Dwight D. Eisenhower, 17 January 1961, farewell address as President.

A COOL HEAD

The great thing about Grant, I take it, is his perfect coolness and persistency of purpose. I judge he is not easily excited – which is a great element in an officer – and he has the *grit* of a bulldog. Once let him get his 'teeth' *in*, and nothing can shake him loose.
> Abraham Lincoln, 1864, response when the author asked him what he thought of U.S. Grant, quoted in F.B. Carpenter, *Six Months in the White House with Abraham Lincoln*, 68, 1866

A man who cannot think clearly and act rationally in the bullet zone is more suited for a monastery than the battlefield.
> Major General J.F.C. Fuller, *Generalship: Its Diseases and Their Cure*, 1936

Those who matter don't worry, and those who worry don't matter.
> Field Marshal William Slim, favourite motto, quoted in Ronald Lewin, *Slim: The Standard Bearer*, 1976, p. 176.

Having survived the first emerging sign of panic in my Ops Room [aboard HMS *Hermes* upon the sinking of the HMS *Sheffield* by an Argentine Missile on 4 May 1982], I proceeded to divorce myself from the details of the rescue and salvage work. Like any military man, I am not allowed to throw an attack of the 'wobblies' on these sorts of occasion. Never to panic if all possible. And I was working hard to convey to my staff an atmosphere which I hoped was one of calm and confidence. It's amazing what you can get away with sometimes.
> Admiral Sir J.F. 'Sandy' Woodward, *One Hundred Days: The Memoirs of the Falklands Battle Group Commander*, 1992, p. 16.

You know what? We've got a good plan. Look, we're entering a difficult phase. The press will seek to find divisions among us. They will try and force on a strategy that is not consistent with victory. We've been at this only 19 days. Be steady. Don't let the press panic us. Resist the second-guessing. Be confident but patient. We are going to continue this thing through Ramadan. We've got to be cool and steady. It's all going to work.

COUNCIL OF WAR

President George W. Bush, 26 October 2001, quoted in Bob Woodward, 'Doubts and Debate Before Victory over the Taliban,' *Washington Post*, 18 November 2002, p. A01.

Mr. President I'm finer than the hair on a frog's back,'
General Tommy Franks, Commander, CENTCOM, when asked by President George Bush how he was doing just before the invasion of Iraq, quoted in *The Online NewsHour* with Jim Lehrer, n.d

COUNCIL OF WAR

Come with me instantly, sword in hand, if you wish to save our country. The enemy's camp is nowhere more truly than in the place where such thoughts can rise!
Scipio Africanus, on bursting into a council of panic-stricken Patricians who were planning to abandon Rome after the disaster of Cannae in 216 BC, quoted in Livy, *The History of Rome*, c. AD 17 (*The War With Hannibal*, tr. Aubrey de Sélincourt, 1972).

My saddle is my council chamber.
Saladin, c. 1180, quoted by Ibn al-Athir in *Recueil des Historiens des Croisades*, 1872–1906.

You have been with your council and I have been with mine. Believe me that the counsel of my Lord will be accomplished and will stand, and the counsel of yours will perish.
Joan of Arc, 6 May 1429, after the capture of the English fortress of the Augustines, *Joan of Arc In Her Own Words*, 1996.

Otherwise we should have wasted all our time in discussions, diplomatical, tactical, enigmatical; they would have smothered me, and the enemy would have settled our arguments by smashing up our tactics.
Field Marshal Prince Aleksandr V. Suvorov (1729–1800).

If a man consults whether he is to fight, when he has the power in his own hands, it is certain that his opinion is against fighting.
Admiral Viscount Nelson, August 1801, letter.

Hold no council of war, but accept the views of each, one by one... the secret is to make each alike... believe he has your confidence.
Napoleon, 12 January 1806, letter to Joseph Bonaparte.

Whenever I hear of Councils of War being called, I always called them as 'cloaks for cowardice' – so said the brave Boscawen, and from him I imbibed this sentiment.
Admiral Lord St Vincent, 1809, speech at the opening of Parliament.

Maxim 65. The same consequences which have uniformly attended long discussions and councils of war will follow at all times. They will terminate in the adoption of the worst course, which in war is always the most timid, or, if you will, the most prudent. The only true wisdom in a general is determined courage.
Napoleon, *The Military Maxims of Napoleon*, tr. George D'Aguilar, 1831 (David Chandler, ed., 1987).

I never held a council of war in my life.
General of the Army Ulysses S. Grant (1822–1885) – attributed.

If the commander... feels the need of asking others what he ought to do, the command is in weak hands.
Field Marshal Helmuth Graf von Moltke, 20 January 1890, letter.

Not every act is favoured with such happy conception or such easy birth. Meetings, discussion, committees, councils of war, etc., are the enemies of vigorous and prompt decision and their

COUP D'OEIL

danger increases with their size. They are mostly burdened with doubts and petty responsibilities, and the man who pleads for action ill endures the endless hours of discussion.
Colonel-General Hans von Seekt, *Thoughts of a Soldier*, 1930.

Why, you may take the most gallant sailor, the most intrepid airman, or the most audacious soldier, put them at a table together – what do you get? The sum total of their fears.
Sir Winston S. Churchill (1874–1965).

It is essential to understand the place of the 'conference' when engaged on active operations in the field. By previous thought, by discussion with his staff, and by keeping in close touch with his subordinates by means of visits, a commander should know what he wants to do and whether it is possible to do it. If a conference of his subordinates is then necessary, it will be for the purpose of giving orders. He should never bring them back to him for such a conference; he must go forward to them. Then nobody looks over his shoulder. A conference of subordinates to collect ideas is the resort of a weak commander.
Field Marshal Viscount Montgomery of Alamein, *The Memoirs of Field Marshal Montgomery*, 1958.

I could almost hear my father's voice telling me as he had so many years before, 'Doug, councils of war ever breed timidity and defeatism.'
General of the Army Douglas MacArthur, *Reminiscences*, 1964.

COUP D'OEIL

The ability of a commander to comprehend a situation and act promptly is the talent which great men have of conceiving in a moment all the advantages of the terrain and the use that they can make of it with their army. When you are accustomed to the size of your army you soon form your coup d'oeil with reference to it, and habit teaches you the ground that you can occupy with a certain number of troops.

Use of this talent is of great importance on two occasions. First, when you encounter the enemy on your march and are obliged instantly to choose ground on which to fight. As I have remarked, within a single square mile a hundred different orders of battle can be formed. The clever general perceives the advantages of the terrain instantly; he gains advantage from the slightest hillock, from a tiny marsh; he advances or withdraws a wing to gain superiority; he strengthens either his right or his left, moves ahead or to the rear, and profits from the merest bagatelles.
Frederick the Great, *Instructions to His Generals*, 1747, tr. Phillips, 1940.

...The ability to assess a situation at a glance, to know how to select the sight for a camp, when and how to march, and where to attack.
Field Marshal Prince Aleksandr V. Suvorov, *The Science of Victory*, 1796.

There is a gift of being able to see at a glance the possibilities offered by the terrain... One can call it the *coup d'oeil* and it is inborn in great generals.
Napoleon (1769–1821).

On the field of battle the happiest inspiration (*coup d'oeil*) is often only a recollection.
Napoleon, cited in Richard Simpkin, *Race to the Swift*, 1985.

Coup d'oeil therefore refers not alone to the physical but, more commonly, to the inward eye. The expression, like the quality itself, has certainly always been more applicable to tactics, but it must also have its place in strategy, since here as well quick decisions are often needed.

COURAGE

Stripped of metaphor and of the restrictions imposed by the phrase, the concept merely refers to the quick recognition of a truth that the mind would ordinarily miss or would perceive only after long study and reflection.
> Major-General Carl von Clausewitz, On War, 1.3, 1832, tr. Michael Howard and Peter Paret, 1976.

A general thoroughly instructed in the theory of war but not possessed of military *coup d'oeil*, coolness, and skill, may make an excellent strategic plan and be entirely unable to apply the rules of tactics in presence of an enemy. His projects will not be successfully carried out, his defeat will be probable. If he is a man of character he will be able to diminish the evil results of his failure, but if he loses his wits he will lose his army.
> Lieutenant-General Antoine-Henri Baron de Jomini, *Summary of the Art of War*, 1838, tr. Mendell and Craighill, 1862.

The problem is to grasp, in innumerable special cases, the actual situation which is covered by the mist of uncertainty, to appraise the facts correctly and to guess the unknown elements, to reach a decision quickly and then to carry it out forcefully and relentlessly.
> Field Marshal Helmuth Graf von Moltke (1800–1891).

Nine-tenths of tactics are certain, and taught in books: but the irrational tenth is like the kingfisher flashing across the pool and that is the test of generals. It can only be ensured by instinct, sharpened by thought practicing the strokes so often that at the crisis it is as natural as a reflex.
> Colonel T.E. Lawrence, 'The Science of Guerrilla Warfare', *Encyclopaedia Britannica*, 1929.

A vital faculty of generalship is the power of grasping instantly the picture of the ground and the situation, of relating the one to the other, and the local to the general. It is that flair which makes the great executant. (Oct. 1933.)
> Captain Sir Basil Liddell Hart, *Thoughts on War*, 1944.

The acid test of an officer who aspires to high command is his ability to be able to grasp quickly the essentials of a military problem, to decide rapidly what he will do, to make it quite clear to all concerned what he intends to achieve and how he will do it, and then to see that his subordinate commanders get on with the job. Above all, he has got to rid himself of all irrelevant detail; he must concentrate on the essentials, and on those details and only those details which are necessary to the proper carrying out of his plan – trusting his staff to effect all necessary co-ordination.
> Field Marshal Viscount Montgomery of Alamein, *The Memoirs of Field Marshal Montgomery*, 1958.

COURAGE
General

The principle on which to manage an army is to set up one standard of courage which all must reach.
> Sun Tzu, *The Art of War*, 11, c. 500 BC, tr. Giles, 1910.

Nature has set nothing so high that it cannot be surmounted by courage. It is by using methods of which others have despaired that we have Asia in our power.
> Alexander the Great, to the men who had volunteered to scale the Sogdian Rock, 328 BC, quoted in Curtius, *The History of Alexander*, VII, 11.10, 1st century AD, p. 173.

The courage of the soldier is heightened by the knowledge of his profession.
> Flavius Vegetius Renatus, *Military Institutions of the Romans*, c. AD 378, tr. Clark, 1776.

COURAGE

The courage of the troops must be reborn daily... nothing is so variable... the true skill of the general consists in knowing how to guarantee it.
Field Marshal Maurice Comte de Saxe, *My Reveries*, 1732, tr. Phillips, 1940.

The more comfort the less courage there is.
Field Marshal Prince Aleksandr V. Suvorov, quoted in *Soviet Military Review*, 11/1979.

Pay not attention to those who would keep you far from fire: you want to prove yourself a man of courage. If there are opportunities, expose yourself conspicuously. As for real danger, it is everywhere in war.
Napoleon, 2 February 1806, to Joseph, *Correspondance de Napoléon Ier, publiée par ordre de l'Empereur Napoléon III*, No. 9738, Vol. XI, 1858–1870, p. 573.

Courage is like love; it must have hope for nourishment.
Napoleon, *Maxims*, 1804–1815.

War is the realm of danger; therefore *courage* is the soldier's first requirement. Courage is of two kinds: courage in the face of personal danger, and courage to accept responsibility, either before the tribunal of some outside power or before the court of one's own conscience.
Major-General Carl von Clausewitz, *On War*, 1.3, 1832, tr. Michael Howard and Peter Paret, 1976.

The well-known Spanish proverb, 'He was a brave man on such a day,' may be applied to nations as to individuals. The French at Rossbach were not the same people as at Jena, nor the Prussians at Prenzlau as at Dennewitz.
Lieutenant-General Antoine-Henri Baron de Jomini, *Summary of the Art of War*, 1838, tr. Mendell and Craighill, 1862.

One man with courage makes a majority.
American saying attributed to Major-General Andrew Jackson (1767–1845).

There is, of course, such a thing as individual courage, which has a value in war, but familiarity with danger, experience in war and its common attendants, and personal habit, are equally valuable traits, and these are the qualities with which we usually have to deal in war. All men naturally shrink from pain and danger, and only incur their risk from some higher motive, or from habit; so that I would define true courage to be a perfect sensibility of the measure of danger, and a mental willingness to incur it, rather than that insensibility to danger of which I have heard far more than I have seen. The most courageous men are generally unconscious of possessing the quality; therefore, when one professes it too openly, by words or bearing, there is reason to mistrust it. I would further illustrate my meaning by describing a man of true courage to be one who possesses all his faculties and senses perfectly when serious danger is actually present.
General of the Army William T. Sherman, *Memoirs of General W.T. Sherman*, 1875.

In sport, in courage, and in the sight of Heaven, all men meet on equal terms.
Sir Winston S. Churchill, *The Malakand Field Force*, 1898.

Courage... is that firmness of spirit, that moral backbone, which, while fully appreciating the danger involved, nevertheless goes on with the undertaking. Bravery is physical; courage is mental and moral. You may be cold all over; your hands may tremble; your legs may quake; your knees may be ready to give way – that is fear. If nevertheless, you go forward; if in spite of this physical defection you continue to

COURAGE

lead your men against the enemy, you have courage. The physical manifestations of fear will pass away. You may never experience them but once.
> Major C.A. Bach, 1917, address to graduating new officers, Fort Sheridan, Wyoming.

I think the death penalty will cease pretty soon. The debates on it in the House make my blood boil. I wish I could talk to some of the old stagers for a few minutes, about funk & courage. They are the same quality, you know. A man who can run away is a potential V.C.
> Colonel T.E. Lawrence, letter to Ernest Thurtle, 1 April 1924, quoted in *The Letters of T.E. Lawrence*, 1988.

Courage is rightly esteemed the first of human qualities, because... it is the quality that guarantees all others.
> Sir Winston S. Churchill, *Great Contemporaries*, 1937.

...But courage which goes against military expediency is stupidity, or, if insisted upon by a commander, irresponsibility.
> Field Marshal Erwin Rommel (1881–1944).

Courage is fear holding on a minute longer.
> General George S. Patton, Jr., quoted in James Carlton, ed., *The Military Quotation Book*, 1990.

After a quarter of a century I am driven to ask again 'Is courage common?' I find it difficult now, living in soft security, to speak from the bench. Twice in my lifetime I have seen boys grow to men, only to be consumed by war, and I have come to think of this almost every day. War is only tolerable when one can take part in it, when one is a bit of the target and not a pensioned spectator. Yet when the death of husband or son or brother has grown distant and the world is free to think again without impiety that courage is not common, men will remember that all the fine things in war as in peace are the work of a few men; that the honour of the race is in the keeping of but a fraction of her people.
> Lord Moran, 12 May 1943, preface to *The Anatomy of Courage*, 1945.

There is no better ramrod for the back of a senior who is beginning to buckle than the sight of a junior who has kept his nerve. Land battles, as to the fighting part, are won by the intrepidity of men in grade from private to captain mainly. Fear is contagious, but courage is not less so. The courage of any one man reflects in some degree the courage of all those who are within his vision. To the man who is in terror and bordering on panic, no influence can be more steadying than that of seeing some other man near him who is retaining self-control and doing his duty.
> Brigadier-General S.L.A. Marshall, *The Armed Forces Officer*, 1950.

I don't believe there is any man, who in his heart of hearts, wouldn't rather be called brave than have any other virtue attributed to him. And this elemental, if you like unreasoning, male attitude is a sound one, because courage is not merely a virtue; it is *the* virtue. Without it there are no other virtues. Faith, hope, charity, all the rest don't become virtues until it takes courage to exercise them. Courage is not only the basis of all virtue; it is its expression. True, you may be bad and brave, but you can't be good without being brave.
> Field Marshal Viscount Slim of Burma, *Courage and Other Broadcasts*, 1957.

There is nothing like seeing the other fellow run to bring back your courage.
> Field Marshal Viscount Slim of Burma, *Unofficial History*, 1959.

Courage in the commander is a prior condition for courage in his men and

COURAGE

for their ability to carry out orders promptly and well. It takes courage and daring to overcome the fears of battle and the trials of armed combat. The commander must also have the courage of his convictions, defying when necessary entrenched, conventional ideas. He must dare to make his views heard in front of his superiors and among his colleagues when permitted to do so, notwithstanding the contrary opinions of those he is addressing.
General Yigal Allon, *The Making of Israel's Army*, 1970.

There are two kinds of courage, physical and moral, and he who would be a true leader must have both. Both are the products of the character-forming process, of the development of self-control, self-discipline, physical endurance, of knowledge of one's job and, therefore, of confidence. These qualities minimize fear and maximize sound judgment under pressure – with some of that indispensable stuff called luck – often bring success from seemingly hopeless situations.

Putting aside impulsive acts of bravery, both kinds of courage bespeak an untroubled conscience, a man at peace with God. An example is Colonel John H. Glenn who was asked after his first rocket flight if he had been worried, and who replied: 'I am trying to live the best I can. My peace had been made with my Maker for a of years, so I had no particular worries.'
General Matthew B. Ridgway, 'Leadership', *Military Review*, 10/1966.

There were… plenty of toughs in the Army, whose peace of mind came from a certain vacancy which had always passed for courage; in them freedom from fear was the outcome of the slow working of their minds, the torpor of their imagination.
Lord Moran, *Churchill: Taken from the Diaries of Lord Moran*, 19, 1966.

Moral Courage

As to moral courage, I have rarely met with two-o'clock-in-the-morning courage; I mean instantaneous courage.
Napoleon, 4–5 December 1815, quoted in Emmanuel Las Casas, *Mémorial de Sainte Hélène*, 1840.

I felt my duty to take up the cudgels. The country's safety was at stake, and I said so bluntly. The President turned the full vials of his sarcasm upon me. He was a scorcher when aroused. The tension began to boil over. For the third and last time in my life that paralyzing nausea began to creep over me. In my emotional exhaustion I spoke recklessly and said something to the general effect that when we lost the next war, and an American boy, lying in the mud with an enemy bayonet through his belly and an enemy foot on his dying throat, spat out his last curse, I wanted the name not to be MacArthur, but Roosevelt. The President grew livid. 'You must not talk that way to the President!' he roared. He was right, of course, right, and I knew it almost before the words had left my mouth. I said that I was sorry and apologized. But I felt my Army career was at an end. I told him he had my resignation as Chief of Staff. As I reached the door his voice came with that cool detachment which so reflected his extraordinary self-control, 'Don't be foolish, Douglas; you and the budget must get together on this.'

Dern had shortly reached my side and I could hear his gleeful tones, 'You've saved the Army.' But I just vomited on the steps of the White House.
General of the Army Douglas MacArthur, 1933, MacArthur's stand to save the National Guard from a mortal budget slash by Roosevelt, *Reminiscences*, 1964.

Determined leadership is vital, and nowhere is this more important than in the higher ranks. Other things being equal the battle will be a contest

COURAGE

between opposing wills.

Generals who become depressed when things are not going well, and who lack the drive to get things done, and the moral courage and resolution to see their plan through to the end, are useless in battle. They are, in fact, worse than useless – they are a menace – since any lack of moral courage, or any sign of wavering or hesitation, has very quick repercussion down below.

Field Marshal Viscount Montgomery of Alamein, 10 November 1942.

Victory is never final. Defeat is never fatal. It is courage that counts.

Sir Winston Churchill, quoted in *Rumsfeld's Rules*, revised edition, 2000.

Now these two types of courage, physical and moral, are very distinct. I have known many men who had marked physical courage, but lacked moral courage. Some of them were in high places, but they failed to be great in themselves because they lacked it. On the other hand, I have seen men who undoubtedly possessed moral courage very cautious about taking physical risks. But I have never met a man with moral courage who would not, when it was really necessary, face bodily danger. Moral courage is a higher and a rarer virtue than physical courage.

Field Marshal Viscount Slim of Burma, *Courage and Other Broadcasts*, 1957.

To teach moral courage is another matter – and it has to be taught because so few, if any, have it naturally. The young can learn it from their parents, in their homes, from school and university, from religion, from other early influences, but to inculcate it in a grown-up who lacks it requires not so much teaching as some striking emotional experience – something that suddenly bursts upon him, something in the nature of a vision. That happens rarely, and that is why you will find that most men with moral courage learnt it by precept and example in their youth.

Field Marshal Viscount Slim of Burma, *Courage and Other Broadcasts*, 1957.

Moral courage – not afraid to say or do what you believe to be right.

Field Marshal Viscount Montgomery of Alamein, *The Memoirs of Field Marshall Montgomery*, 1958.

It takes courage, especially for a young officer, to check a man met on the road for not saluting properly or for slovenly appearance, but, every time he does, it adds to his stock of moral courage, and whatever the soldier may say he has a respect for the officer who does put him up.

Field Marshal Viscount Slim of Burma, *Defeat Into Victory*, 1956.

It has long seemed to me that the hard decisions are not the ones you make in the heat of battle. Far harder to make are those involved in speaking your mind about some harebrained scheme which proposes to commit troops to action under conditions where failure seems almost certain, and the only results will be the needless sacrifice of priceless lives. When all is said and done, the most precious asset any nation has is its youth, and for a battle commander ever to condone the unnecessary sacrifice of his men is inexcusable. In any action you must balance the inevitable cost in lives against the objectives you seek to attain. Unless the results to be expected can reasonably justify the estimated loss of life the action involves, then for my part I want none of it.

General Matthew B. Ridgway, 'Leadership', *Military Review*, 9/1966.

I remember the day I was ready to go over to the Oval Office and give my four stars to the President and tell him, 'You have refused to tell the country they cannot fight a war without

COURAGE

mobilization; you have required me to send men into battle with little hope of their ultimate victory; and you have forced us in the military to violate almost every one of the principles of war in Vietnam. Therefore, I resign and will hold a press conference after I walk out of your door.'

I made the typical mistake of believing I could do more for the country and the Army if I stayed in than if I got out. I am now going to my grave with that lapse in moral courage on my back.

General Harold K. Johnson, quoted in Lewis Sorely, 'To Change a War', *Parameters*, Spring 1998.

Bravery is the quintessence of the soldier... But moral courage – the strength of character to do what one knows is right regardless of the personal consequences – is the true face of conscience. Sacking your best friend, facing up rather than turning the blind eye, accepting that the principle at stake is more important than your job... Such actions demand moral courage of a high order.

Lieutenant-General Sir James Glover, 'A Soldier and His Conscience', *Parameters*, 9/1983.

Physical Courage

As to physical courage, although sheer cowardice (i.e., a man thinking of his own miserable carcass when he ought to be thinking of his men) is fatal, yet, on the other hand, a reputation for not knowing fear does not help an officer in his war discipline: in getting his company to follow him as the Artillery of the Guard followed Drouot at Wagram. I noticed this first in Afghanistan in 1879 and have often since made the same observation. If a British officer wishes to make his men shy of taking a lead from him let him stand up under fire whilst they lie in their trenches as did the Russians on the 17th of July at the battle of Motienling. Our fellows are not in the least impressed by such bravado. All they say is, 'This fellow is a fool. If he cares so little for his own life, how much less will he care for ours.'

General Sir Ian Hamilton, *The Soul and Body of an Army*, 1921.

When it comes to combat, something new is added. Even if they have previously looked on him as a father and believed absolutely that being with him is their best assurance of successful survival, should he then show himself to be timid and too cautious about his own safety, he will lose hold of them no less absolutely. His lieutenant, who up till then under training conditions has been regarded as a mean creature or a sniveler, but on the field suddenly reveals himself as a man of high courage, can take moral leadership of the company away from him, and do it in one day.

On the field there is no substitute for courage, no other binding influence toward unity of action. Troops will excuse almost any stupidity; excessive timidity is simply unforgivable...

Brigadier-General S.L.A. Marshall, *The Armed Forces Officer*, 1950.

Complete cowards are almost non-existent. Another matter for astonishment is the large number of men and women in any group who will behave in an emergency with extreme gallantry. Who they will be you cannot tell until they are tested. I long ago gave up trying to spot potential VCs by their looks, but, from experience, I should say that those who perform individual acts of the highest physical courage are usually drawn from two categories. Either those with quick intelligence and vivid imagination or those without imagination and with minds fixed on the practical business of living. You might almost say, I suppose, those who

live on their nerves and those who have not got any nerves. The one suddenly sees the crisis, his imagination flashes the opportunity and he acts. The other meets the situation without finding it so very unusual and deals with it in a matter of fact way.
> Field Marshal Viscount Slim of Burma, *Courage and Other Broadcasts*, 1957.

COURT MARTIAL

Court! I have no time to order courts! I can't blame an officer who seeks to put his ship close to the enemy. Is there any other vessel you would like to have? Breese, make out Selfridge's orders to the Conestoga.
> Admiral David D. Porter, upon being asked when he would initiate a court of inquiry to try the commander of a gunboat who had just lost his vessal to a mine in the Yazoo river while manoeuvring to fire on the Rebel positions in December 1862, quoted in *Memoirs of Thomas O. Selfridge, Jr.*, 1928.

The popular conception of a court-martial is half a dozen blood-thirsty old Colonel Blimps, who take it for granted that anyone brought before them is guilty – damme, sir, would he be here if he hadn't done something? – and who at intervals chant in unison, 'Maximum penalty – death!' In reality courts-martial are almost invariably composed of nervous officers, feverishly consulting their manuals; so anxious to avoid a miscarriage of justice that they are, at times, ready to allow the accused any loophole of escape. Even if they do steel themselves to passing a sentence, they are quite prepared to find it quashed because they have forgotten to mark something 'A' and attach it to the proceedings.
> Field Marshal Viscount Slim of Burma, *Unofficial History*, 1959.

I wish to record the opinion that trial under the old Manual of Court Martial came closer to providing real justice than any other American system of jurisprudence: the accused was brought promptly to trial, while the memory of witnesses were fresh; court routines were simple and quickly accomplished, which resulted in trials themselves being short and to the point; the rules of evidence, while eminently fair to the accused, served to allow the truth to become known; and the court itself was composed of officers with a good understanding of the law. I agree with the statement, I believe by the eminent Elihu Root, that if I as the accused were innocent of a crime I would rather be tried by a military court than any other tribunal; but if guilty I would wish to be tried by any tribunal other than a military one.
> General Hamilton H. Howze, *A Cavalryman's Story*, 1996.

COWARDICE/COWARDS

Why have you not launched the attack? Are you standing in chariots of water? Did you also turn to water? …If you were to fall down before him, you …would either kill him, or at least, frighten him! But instead you acted like a 'fairy'!
> Hattusili I, 1650–1620 BC, letter to the commander of his army besieging the city of Ursa.

And tall Hector nodded, his helmet flashing:
'All this weighs on my mind too, dear woman.
But I would die of shame to face the men of Troy
and the Trojan women in their long robes
if I would shrink from battle now, a coward.'
> Homer, *The Iliad*, 6.5–522–525, c. 800 BC, tr. Robert Fagles, 1990; Hector telling his wife, Andromache, that he feared for what would become of her if he died in battle, but that he feared shame more.

CRITICISM

In other cities whenever a man shows himself to be a coward his only punishment is that he is called a coward... But in Sparta anyone would be ashamed to dine or to wrestle with a coward... In the streets he must get out of the way... he must support his unmarried sisters at home and explain to them why they are still spinsters, he must live without a wife at his fireside... he may not wander about comfortably acting like someone with a clean reputation or else he is beaten by his betters. I don't wonder that where such a load of dishonor burdens the coward death seems preferable instead of a dishonored and shameful life.
Xenophon, *Constitution of the Spartans*, 9.4–6, c. 360 BC.

Another woman, when her sons fled from a battle and reached her, said: 'In making your escape, vile slaves, where is it you've come to? Or do you plan to creep back in here where you emerged from?' At this she pulled up her clothes and exposed her belly to them.
Plutarch, quoting an anonymous Spartan woman, *The Lives*, c. AD 100, (*Plutarch on Sparta*, tr. Richard Talbert, 1988).

He was just a coward and that was the worst luck any man could have.
Ernest Hemingway, *For Whom the Bells Toll*, 30, 1940.

When I ask can war make any man a coward it is no answer to point to men who were cowards before they were soldiers. Such men went about wearing labels for all to read. From the first they were plainly unable to stand this test of men. They had about them the marks known to our calling of the incomplete man, the stamp of degeneracy. The whole miserable issue could have been foretold, the man was certain to crack when the strain came.
Lord Moran, *The Anatomy of Courage*, 1945.

All men are timid on entering any fight. Whether it is the first fight or the last fight, all of us are timid. Cowards are those who let their timidity get the better of their manhood.
General George S. Patton, Jr., *War As I Knew It*, 1947.

CRITICISM

I should be more faint-hearted than they make me, if, through fear of idle reproaches, I should abandon my own convictions. It is no inglorious thing to have fear for the safety of our country, but to be turned from one's course by men's opinions, by blame, and by misrepresentations, shows a man unfit to hold an office such as this, which, by such conduct, he makes the slaves of those whose errors it is his business to control.
Quintus Fabius Maximus, 'The Delayer', 218 BC, to those who criticised him for delaying, hence 'Fabian' tactics of harassment rather than open battle with Hannibal in the Second Punic War, quoted in Plutarch, *The Lives*, c. AD 100, tr. John Dryden.

...Room for a military criticism as well as a place for a little ridicule upon some famous transactions of that memorable day... But why this censure when the affair is so happily decided? To exercise one's ill-nature? No, to exercise the faculty of judging... The more a soldier thinks of the false steps of those that are gone before, the more likely he is to avoid them.
Major-General Sir James Wolfe (1727–1759), reflections on the Battle of Culloden, quoted in Liddell Hart, *Great Captains Unveiled*, 1927.

You see the dash which the Common Council of the city of London have made at me! I act with a sword hanging over me, which will fall upon me whatever may be the result; but they may do what they please. I shall not give

CRITICISM

up the game here as long as it can be played.

> The Duke of Wellington, on criticism of his 1809 campaign in Spain and retreat to Portugal, *The Dispatches of Field Marshal the Duke of Wellington, During His Various Campaigns in India, Denmark, Portugal, Spain, the Low Countries, and France*, V, p. 403, 1834–1838.

There's a man for you! He is forced to flee from an army that he dares not fight, but he puts eighty leagues of devastation between himself and his pursuers. He slows down the march of the pursuing army, he weakens it by all kinds of privation – he knows how to ruin it without fighting it. In all of Europe, only Wellington and I are capable of carrying out such measures. But there is a difference between him and myself: In France... I would be criticized, whereas England will praise him.

> Napoleon, c. 1810, conversation on Wellington's retreat to Portugal, quoted in Jean Antoine Chaptal, *Mes Souvenirs sur Napoléon*, 1893.

I always remember the Japanese soldier who outraged the senses of patriotism and duty in his superior officer by saying, 'In Osaka I would get five yen for digging this gun pit; here I only get criticism.'

> General Sir Ian Hamilton, *The Soul and Body of an Army*, 1921.

We must remember that man remains man and that his heart does not change. Secondly, we must remember that the means of war do not change and that the intelligence of man must keep pace with these changes. We must keep minds subtle and active, and never let ourselves be hypnotized by traditions; we must criticize ourselves, and criticize our criticisms; we must experiment and explore.

> Major-General J.F.C. Fuller, *Sir John Moore's System of Training*, 1925.

In an institution such as the Army the hierarchical system and the habit of subordination, most necessary in its right place, preclude criticism to a degree unknown in other professions, and that would be hardly comprehensible to a scientific mind. This reflection reminds me of the humorous, yet serious, warning of my beloved old chief: 'You'll learn that generals are as sensitive as prima-donnas.' I have long learned to appreciate his wisdom. (Feb. 1933.)

> Captain Sir Basil Liddell Hart, *Thoughts on War*, 1944.

I have benefited enormously from criticism and at no point did I suffer from any perceptible lack thereof.

> Sir Winston S. Churchill, quoted in Omar Bradley, 'On Leadership', *Parameters*, 9/1981.

I like commanders on land and sea and in air to feel that between them and all forms of public criticism the Government stands like a strong bulkhead. They ought to have a fair chance, and more than one chance.

> Sir Winston S. Churchill (1874–1965).

I feel that retired general officers should never miss an opportunity to remain silent concerning matters for which they are no longer responsible. Having said that, I believe a few general (no pun intended) comments are in order.

> General H. Norman Schwarzkopf, *It Doesn't Take a Hero*, 1992, his memoirs after retirement.

D

DANGER

The Romans also have an excellent method of encouraging young soldiers to face danger. Whenever any have especially distinguished themselves in a battle, the general assembles the troops and calls forward those he considers to have shown exceptional courage. He praises them first for their gallantry in action and for anything in their previous conduct which is particularly worthy of mention, and then he distributes gifts... These presentations are not made to men who have wounded or stripped an enemy in the course of a pitched battle, or at the storming of a city, but to those who during a skirmish or some similar situation in which there is no necessity to engage in single combat, have voluntarily and deliberately exposed themselves to danger.
Polybius, *Histories*, 6.39, c. 125 BC (*The Rise of the Roman Empire*, tr. Ian Scott-Kilvert, 1987).

Remember also that God does not afford the same protection in unprovoked as in necessary dangers.
Count Belisarius, April, AD 531.

If I had been censured every time I have run my ship, or fleets under my command, into great danger, I should long ago have been *out* of the Service, and never *in* the House of Peers.
Admiral Viscount Nelson, March 1805, letter to the Admiralty.

Danger is part of the friction of war. Without an accurate conception of danger we cannot understand war.
Major-General Carl von Clausewitz, *On War*, 1.4, 1832, tr. Michael Howard and Peter Paret, 1976.

Not till I see day light ahead do I want to lead, but when danger threatens and others slink away I am and will be at my post.
General of the Army William T. Sherman, August 1861, unaddressed letter.

Should the general consistently live outside the realm of danger, then, though he may show high moral courage in making decisions, by his never being called upon to breathe the atmosphere of danger his men are breathing, this lens will become blurred, and he will seldom experience the moral influences his men are experiencing.
Major-General J.F.C. Fuller, *Generalship: Its Diseases and Their Cure*, 1933.

The major battle is only a skirmish multiplied by one hundred. The frictions, confusions, and disappointments are the same. But there is this absolute difference – that the supreme trial of the commander in war lies in his ability to overcome the weaknesses of human nature in the danger, and those are matters which he cannot know in full unless he has served with men where danger lies.
Brigadier-General S.L.A. Marshall, *Men Against Fire*, 1947.

If anything, I'm a little concerned it may be too comfortable. If they are required to do a dangerous job in difficult circumstances, they must get used to those circumstances.

DARING

General Sir Michael Jackson, Chief of the General Staff, speaking of the jump-off conditions of British forces before the invasion of Iraq, *BBC News*, 17 March 2003.

DARING

In war... all action is aimed at probable rather than certain success. The degree of certainty that is lacking must in every cause be left to fate, chance, or whatever you like to call it. ...But we should not habitually prefer the course that involves the least uncertainty. That would be an enormous mistake... There are times when the utmost daring is the height of wisdom.
> Major-General Carl von Clausewitz, *On War*, 2.5, 1832, tr. Michael Howard and Peter Paret, 1976.

Who dares wins.
> Motto of the Special Air Service (SAS).

DEATH

Death cut him short. The end closed in around him.
Flying free of his limbs
his soul went winging down to the House of Death
wailing his fate, leaving his manhood far behind,
his young and supple strength.
> Homer, *The Iliad*, 16.1001–1005, c. 800 BC, tr. Robert Fagles, 1990; the death of Patroclus at the hands of Hector.

Both armies battled it out along the river banks –
they raked each other with hurtling bronze-tipped spears.
And Strife and Havoc plunged in the fight, and violent Death –
now seizing a man alive with fresh wounds, now one unhurt, now hauling a dead man through the slaughter by the heels,
the cloak on her back stained red with human blood.
So they clashed and fought like living, breathing men
grappling each other's corpses, dragging off the dead.
> Homer, *The Iliad*, 18.621–628, c. 800 BC, tr. Robert Fagles, 1990; one of the scenes on the shield fashioned by the smith god, Hephaestus, for brilliant Achilles.

I comprehend the secret, the hidden:
O my lords!
Thus we are,
We are mortal,
men through and through
we all will have to go away,
we all will have to die on earth.
Like a painting,
we will be erased.
Like a flower,
we will dry up
here on earth...
Think on this my lords,
eagles and ocelots,
though you be of jade,
though you be of gold
you also will go there
to the place of the fleshless.
> Nezahualcoyotl (1402–1472), King of Texcoco, quoted in Miguel Leon-Portilla, *Native Mesoamerican Spirituality*, 1980.

These three things you must always keep in mind: concentration of strength, activity, and a firm resolve to perish gloriously. These are the three principles of the military art which have disposed luck in all my military operations. Death is nothing, but to live defeated and without glory is to die every day.
> Napoleon, 12 December 1804, letter to General Lauriston, *Correspondance de Napoléon Ier, publié par ordre de l'Empereur Napoléon III*, 1858–1870.

In this *Templars*, there is only one well-drawn character: that of a man who wants to die! That is not true to life, that is useless. Gentlemen, one must wish to live and know how to die.
> Napoleon, 1 December 1805, philosophising to his staff on the play by Raynouard on the eve of the Battle of

DEATH

Austerlitz, quoted in Philippe Paul, *Un Aide de camp de Napoléon: Mémoires du comte de Ségur*, I, 1894–1895.

It is part of a sailor's life to die well.
Commodore Stephen Decatur, on the death of Captain James Lawrence of the USS *Chesapeake* after the engagement of 1 June 1813 against HMS *Shannon*.

My Father, we must all die sooner or later, and if my time has come nothing will hold it back. It is far, far, better to die with the joy of battle in the heart than to pine away with age, or like a sick ox in a kraal. I have lived by the spear and I shall die by it. That is a man's death. You would not deprive me of that, who are my friend as well as my father?
Mgobozi, October 1826, to his king, Shaka Zulu, who urged him not to throw away his life in battle the next day, quoted in E.A. Ritter, *Shaka Zulu*, 1955.

I would that all could look on death as a cheerful friend, who takes us from a world of trial to our true home. All our sorrows come from a forgetfulness of this great truth. I desire to look on the departure of my friends as a promotion to another and a higher sphere, as I do believe that to be the case with all.
General Charles 'Chinese' Gordon (1833–1885), quoted in Paul Charrier, *Gordon of Khartoum*, 1965.

I see the scene as clearly as on that December day of 1862. That noble, Christian soul passing away; the timid, stricken brother by his side, all of us repeating together the scriptures and his saying to me, pitifully, before consciousness left him, 'Please stay with me to the end, for Charlie never saw any one die.' When I rose from the side of the couch where I had knelt for hours, until the last breath had faded, I wrung the blood from the bottom of my clothing before I could step, for the weight about my feet. Dreadful days, dear dear sister...
Clara Barton, quoted in Percy Epler, *The Life of Clara Barton*, 1915.

Few are wholly dead:
Blow on a dead man's embers
And a live flame will start.
Robert Graves, 'To Bring the Dead to Life', *Collected Poems*, 1961.

And when he gets to Heaven,
To Saint Peter he will tell:
'One more Marine reporting, Sir –
I've served my time in Hell.'
Marine Grave inscription on Guadalcanal, 1942.

In blossom today, then scattered:
Life is so like a delicate flower.
How can one expect the fragrance
To last for ever?
Vice Admiral Takijiro Ohnishi, Commander of the Kamikaze Special Attack Force – a special calligraphic presentation to his staff after the organisation of the Kamikaze (1944), quoted in Inoguchi and Nakajimi, *The Divine Wind*, 1958.

Death stands at attention, obedient, expectant, ready to serve, ready to shear away the peoples en masse; ready, if called on, to pulverize, without hope of repair, what is left of civilization. He awaits only the word of command. He awaits it from a frail, bewildered being, long his victim, now – for one occasion – his Master.
Sir Winston S. Churchill, *The Gathering Storm*, 1948.

Death in combat is not the end of the fight but its peak, and since combat is a part, and at times the sum total of life, death which is the peak of combat, is not the destruction of life, but its fullest, most powerful expression.
General Moshe Dayan, address in honour of Natan Altermann, 1971.

DECEPTION

One thing he [General Patton] said always stuck with me, for it was contrary to what I had believed up to that moment, but when I had been in combat only a short while, I knew he was right. Speaking to all of us late one afternoon as we assembled in the North African sunset, he said, 'Now I want you to remember that no sonuvabitch ever won a war by dying for his country. He won it making the other poor dumb bastard die for his country.'
General James M. Gavin, *On to Berlin*, 1978.

DECEPTION

All warfare is based on deception. Hence, when able to attack, we must seem unable; when using our forces we must seem inactive; when we are near, we must make the enemy believe that we are away; when far away, we must make him believe we are near. Hold out baits to entice the enemy. Feign disorder, and crush him.
Sun Tzu, *The Art of War*, 1, c. 500 AD, tr. Giles, 1910.

A cavalry commander should also devise tricks of his own, suitable for his situation. The basic point is that deceit is your most valuable asset in war... If you think about it, you will find that the majority of important military successes have come about as a result of trickery. It follows, then, that if you are to take on the office of commander, you should ask the gods to allow you to count the ability to deceive among your qualifications, and should also work on it yourself.
Xenophon, *How To Be A Good Cavalry Commander*, c. 360 AD (*Hiero the Tyrant and Other Treatises*, tr. Robin Waterfield, 1997).

When he was censured for operating mostly by trickery and fraud in a way unworthy of Heracles, and for achieving no honest success, he used to laugh and say that fox-skin had to be stitched on wherever lion-skin would not stretch.
Plutarch, paraphrasing Lysander, (k. 395 AD), *The Lives*, c. AD 100, (*Plutarch on Sparta*, tr. Richard Talbert, 1988).

Warfare is deception.
Mohammed, quoted in Al-Muttaqi' Al-Hindi, 'Sayings Ascribed to the Prophet,' Girard Chaliand, ed., *The Art of War in History*, 1994, p. 390.

Though fraud in other activities be detestable, in the management of war it is laudable and glorious, and he who overcomes an enemy by fraud is as much to be praised as he who does so by force.
Niccolò Machiavelli, *Discourses*, III, 1517.

Napoleon has humbugged me, by God! He has gained twenty-four hours' march on me.
The Duke of Wellington, 15 June 1815, to the Duke of Richmond during the celebrated ball in Brussels as Napoleon began the Waterloo campaign, quoted in Andrew Roberts, *Napoleon and Wellington*, 2002, from The Earl of Malmsbury, *A Series of Letters*, 1870, Vol. II, p. 246.

Always mystify, mislead, and surprise the enemy, if possible; and when you strike and overcome him, never give up the pursuit as long as your men have strength to follow; for an army routed, if hotly pursued, becomes panic-stricken, and can then be destroyed by half their number.
Lieutenant-General Thomas J. 'Stonewall' Jackson, quoted in G.F.R. Henderson, *Stonewall Jackson*, 1898.

To achieve victory we must as far as possible make the enemy blind and deaf by sealing his eyes and ears, and drive his commanders to distraction by creating confusion in their minds.
Mao Tse-tung, *On Protracted War*, 1938.

'In wartime' I said, 'truth is so precious

DECISION/DECISIVENESS

that she should always be attended by a bodyguard of lies.' Stalin and his comrades greatly appreciated this remark when it was translated, and upon this note our formal conference ended gaily.
  Sir Winston S. Churchill, *The Second World War: Closing the Ring*, 1951.

DECISION/DECISIVENESS

Iacta alea est. (The die is cast.)
  Julius Caesar, upon crossing the Rubicon, the act which precipitated the Roman Civil War, 49 AD.

On the other hand, experience shows me that, in an affair depending on vigour and dispatch, the generals should settle their plan of operations so that no time may be lost in idle debate and consultations when the sword is drawn; that pushing on smartly is the road to success, and more particularly so in an affair of this sort; that nothing is to be reckoned an obstacle to your undertaking which is not found really so on trial; that in war something must be allowed to chance and fortune, seeing that it is in its nature hazardous, and an option of difficulties; that the greatness of an object should come under consideration, opposed to the impediments that lie in the way; that the honour of one's country is to have some weight; and that, in particular circumstances and times, the loss of a thousand men is rather an advantage to a nation than otherwise, seeing that gallant attempts raise its reputation... whereas the contrary appearances sink the credit of a country, ruin the troops, and create infinite uneasiness and discontent at home.
  Major-General Sir James Wolfe, quoted in Liddell Hart, *Great Captains Unveiled*, 1927.

Once you have made up your mind, stick to it; there is no longer any *if* or *but*...
  Napoleon, 18 February 1812, to Marshal Marmont, *Correspondance de Napoléon Ier, publié par ordre de l'Empereur Napoléon III*, No. 18503, Vol. XXIII, 1858–1870, p. 229.

The commander is compelled during the whole campaign to reach decisions on the basis of situations which cannot be predicted. All consecutive acts of war are, therefore, not executions of a premeditated plan, but spontaneous actions, directed by military tact. The problem is to grasp in innumerable special cases, the actual situation which is covered by the mist of uncertainty, to appraise the facts correctly and to guess the unknown elements, to reach a decision quickly and then to carry it out forcefully and relentlessly... It is obvious that theoretical knowledge will not suffice, but that here the qualities of mind and character come to a free, practical and artistic expression, although schooled by military training and led by experiences from military history or from life itself.
  Field Marshal Helmuth Graf von Moltke, quoted in Earle, ed., *The Makers of Modern Strategy*, 1943.

Once you have taken a decision, never look back on it.
  Field Marshal Viscount Allenby of Meggido, quoted in Archibald Wavell, *Allenby, Soldier and Statesman*, 1943.

Sometimes a commander who cannot make decisions will, by fussing and fuming, try to persuade himself and others that he is taking an active part in the operation. He will give his personal attention to details of secondary importance, and satisfy his desire to leave his impress on the action by interfering in an aimless sort of way.
  General Charles de Gaulle, *The Edge of the Sword*, 1932.

The first demand of war is decisive action. Everyone, the highest

DECISION/DECISIVENESS

commander and the most junior soldier, must be aware that omissions and neglect incriminate him more severely than mistake in the choice of means.

Generals Werner von Fritsch and Ludwig Beck, *Troop Leadership*, 1933, quoted in US War Department, *German Military Training*, 17 September 1942.

In forty hours I shall be in battle, with little information, and on the spur of the moment will have to make the most momentous decisions. But I believe that one's spirit enlarges with responsibility and that, with God's help, I shall make them and make them right.

General George S. Patton, Jr. (1885–1945).

After a battle is over people talk a lot about how decisions were methodically reached, but actually there's always a hell of a lot of groping around.

Admiral Frank J. Fletcher (1885–1973).

In making an attack, there can be no vacillation or indecision, even though a bayonet charge be involved. Any vacillation will result in greater casualties, loss of victory, and general discouragement of the whole force.

General Lin Piao, 1946, quoted in Ebon, *Lin Piao*, 1970.

The power of decision develops only out of practice. There is nothing mystic about it. It comes of a clear-eyed willingness to accept life's risks, recognizing that only the enfeebled are comforted by thoughts of an existence devoid of struggle.

Brigadier-General S.L.A. Marshall, *The Armed Forces Officer*, 1951.

Any man facing a major decision acts, consciously or otherwise, upon the training and beliefs of a lifetime. This is no less true of a military commander than of a surgeon who, while operating, suddenly encounters an unsuspected complication. In both instances, the men must act immediately, with little time for reflection, and if they are successful in dealing with the unexpected it is upon the basis of past experience and training.

Admiral of the Fleet Ernest J. King, *Fleet Admiral King, Naval Record*, 1952.

When all is said and done the greatest quality required in a commander is 'decision'; he must be able to issue clear orders and have the drive to get things done. Indecision and hesitation are fatal in any officer; in a C-in-C they are criminal.

Field Marshal Viscount Montgomery of Alamein, *The Memoirs of Field Marshal Montgomery*, 1958.

He (the leader) cannot afford to be ambiguous.

Ever since I was a boy and read about Gettysburg, I've thought that ambiguity was the reason for Lee's losing the battle. Lee said, 'General Ewell was instructed to carry the hill occupied by the enemy, if he found it practicable...' I call that leaning on a subordinate, most definitely. Ewell didn't find the attack practicable; so he didn't attack that late afternoon or early evening. During the night the Federals heavily reinforced, and were never driven from their ridges, but instead repelled every Confederate attack poised against them.

Lee left it up to Ewell to make a decision which I feel that great general should have made himself. And Lee was a great general. Figure it out if you can.

General Curtis LeMay, *Mission With LeMay*, 1965.

When one starts a war, one has to understand that this is a serious thing, that blood is not water. If, having measured seven times, you did decide to cut, then cut quickly and firmly, to the end. Hesitating and thinking are for the time before you start the war. By the way, thinking is useful at any time. But

DECLARATION OF WAR

once you start fighting for real, you have to understand: You fight until you win, not 'until elections.'
>General Aleksandr I. Lebed, *Nezavisimya Gazeta* (Moscow), 3 April 1996.

This house wanted this decision. Well it has it. Those are the choices. And in this dilemma, no choice is perfect, no. But on this decision hangs the fate of many things.
>Prime Minister Tony Blair, speech, 18 March 2003, to the House of Commons laying out the justification for war with Iraq.

So we came to the point of decision. Prime Ministers don't have the luxury of maintaining both sides of the argument. They can see both sides. But, ultimately, leadership is about deciding.
>Prime Minister Tony Blair, speech, 5 March 2004, on his decision to go to war with Iraq the year before.

DECLARATION OF WAR

The Roman ambassadors had been received at Carthage. After they had heard the Carthaginians' statement of their case, they spoke no word in reply, but the senior member of the delegation pointed to the bosom of his toga and declared to the Senate that in its folds he carried both peace and war, and that he would let fall from it whichever they instructed him to leave. The Carthagian Suffete answered that he should bring out whichever he thought best, and when the envoy replied that it would be war, many of the senators shouted at once, 'We accept it!' It was on these terms that the Senate and the Roman ambassadors parted.
>Polybius, on the Roman declaration of war against Carthage beginning the Second Punic War, 219 AD, *Histories (Rise of the Roman Empire*, tr. Ian Scott-Kilvert, 1987).

To such a task we can dedicate our lives and our fortunes, everything that we are and everything that we have, with the pride of those who know that the day has come when America is privileged to spend her blood and her might for the principles that gave her birth and happiness and the peace which she has treasured. God helping her, she can do no other.
>President Woodrow Wilson, 2 April 1917, asking Congress to declare war against Germany.

I am speaking to you from the cabinet room at 10 Downing Street. This morning the British Ambassador in Berlin handed the German Government a final note stating that, unless we heard from them by eleven o'clock that they were prepared at once to withdraw their troops from Poland, a state of war would exist between us. I have to tell you now that no such undertaking has been received, and that consequently this country is at war with Germany.
>Prime Minister Neville Chamberlain, 3 September 1940, announcing Great Britain's declaration of war against Germany.

Yesterday, December 7, 1941 – a date which will live in infamy – the United States was suddenly and deliberately attacked by the naval and air forces of the empire of Japan...

The facts of yesterday and today speak for themselves. The people of the United States have already formed their opinions and well understand the implications to the very life and safety of our nation.

As Commander-in-Chief of our armed forces, I have directed that all measures be taken for our defense.

Always remember the character of the onslaught against us.

No matter how long it may take us to overcome this premeditated invasion, the American people, in their righteous might, will win through to absolute victory...

With confidence in our armed forces,

with the unbounding determination of our people, we will gain the inevitable triumph. So help us God.

I ask that the Congress declare that since the unprovoked and dastardly attack by Japan, on December 7, 1941, a state of war has existed between the United States and the Japanese Empire.

> President Franklin D. Roosevelt, 8 December 1941, asking Congress to declare war on Japan after the attack on Pearl Harbor.

The ruling to kill the Americans and their allies – civilians and military – is an individual duty for every Muslim who can do it in any country in which it is possible to do it, in order to liberate the al-Aqsa Mosque and the holy mosque [Mecca] from their grip, and in order for their armies to move out of all the lands of Islam, defeated and unable to threaten any Muslim. This is in accordance with the words of Almighty God, 'and fight the pagans all together as they fight you all together'.

> Osama bin Ladin, *Fatwa Urging Jihad Against Americans*, 23 February 1998, London Al-Quds al-'Arabi, p. 3.

September 11th was for me a revelation. What had seemed inchoate came together. The point about September 11th was not its detailed planning; not its devilish execution; not even, simply, that it happened in America, on the streets of New York. All of this made it an astonishing, terrible and wicked tragedy, a barbaric murder of innocent people. But what galvanised me was that it was a declaration of war without limit. They killed 3,000. But if they could have killed 30,000 or 3000,000 they would have rejoiced in it. The purpose was to cause such hatred between Moslems and the West that a religious Jihad became reality; and the world engulfed by it.

> Prime Minister Tony Blair, 5 March 2004, speech, reviewing the link between terrorism and the decision to go to war with Iraq.

DEFEAT

Dead men, not potsherds
Covered the approaches,
The walls were gaping,
the high gates, the roads,
were piled with dead.
In the side streets, where feasting crows
    would gather,
Scattered they lay.
In all the streets and roadways bodies lay.
In open fields that used to fill with
    dancers,
they lay in heaps.
The country's blood now filled its holes,
like metal in a mould;
Bodies dissolved – like fat left in the sun.

> A Sumerian account of the fall of Third Dynasty Ur (c. 2000 BC), quoted in John Oates, *Babylon*, 1979.

Those who reached my boundary, their seed is not; their heart and their soul are finished forever and ever. As for those who had assembled before them on the sea, the full flame was in their front, before the harbor-mouths, and a wall of metal upon the shore surrounded them. They were dragged, overturned, and laid low upon the beach; slain and made heaps from stern to bow of their galleys. While all their things were cast upon the water. Thus I turned back the waters to remember Egypt; when they mention my name in their land, may it consume them, while I sit upon the throne of Harakhte, and the serpent-diadem is fixed upon my head, like Re.

> Rameses III, c. 1190 AD, record of the Northern War, quoted in James Breasted, *Ancient Records of Egypt*, 1906.

Once more may I remind you that you have beaten most of the enemy's fleet already; and, once defeated, men do not meet the same dangers with their old spirit.

> Phormio, address to the Athenian Navy,

DEFEAT

429 BC, quoted in Thucydides, *The Peloponnesian War*, c. 404 BC, tr. Benjamin Jowett, 1881.

Troops defeated in open battle should not be pampered or, even if it seems like a good idea, take refuge in a fortified camp or some other strong place, but while their fear is still fresh, they should attack again. By not indulging them they may with greater assurance renew the fighting.
  The Emperor Maurice, *The Strategikon*, c. AD 600 (*Maurice's Strategikon*, tr. George Dennis, 1984).

Broken spears lie in the road;
we have torn our hair in our grief.
The houses are roofless
and their walls are red with blood.

Worms are swarming
in the streets and plazas,
and walls are spattered with gore.
The water has turned red, as if it were
    dyed,
and when we drink it,
it has the taste of brine.

We have pounded our hands in despair
against the adobe walls,
for our inheritance, our city,
is lost and dead.
The shields of our warriors were its
    defense
but they could not save it.

We have chewed dry twigs and salt
    grasses;
We have filled our mouths with dust
    and bits of adobe;
We have eaten lizards, rats, and
    worms...
  An Aztec account of the fall of Tenochtitlan in 1521, about 1528, quoted in Miguel Leon-Portilla, *The Broken Spears*, 1962.

A general of great merit should be said to be a man who has met with at least one great defeat. A man like myself who has gone his whole life with victories alone and suffered no defeats cannot be called a man of merit, even though he gains in years.
  Asakura Norikage (1474–1555) *Soteki Waki*, c. 1550 (*Ideals of the Samurai*, tr. William Wilson, 1982).

Calm yourself, young man. One may be beaten by my army without dishonour!
  Napoleon, 2 December 1805, at Austerlitz to a captured young Russian officer who was demanding to be shot for having lost his artillery, quoted in Claude Manceron, *Austerlitz*, 1966.

Trophies apart, there is no accurate measure of loss of morale; hence in many cases the abandonment of the fight remains the only authentic proof of victory. In lowering one's colors one acknowledges that one has been at fault and concedes in this instance that both might and right lie with the opponent. This shame and humiliation, which must be distinguished from all other psychological consequences of the transformation of the balance, is an essential part of victory.
  Major-General Carl von Clausewitz, *On War*, 4.4, 1832, tr. Michael Howard and Peter Paret, 1976.

When this army is defeated and when I am driven from this line, it will be when I have so few men left they will not want for any trains.
  General of the Army Ulysses S. Grant, May 6, 1864, comment to a staff officer at the Battle of the Wilderness who said that the trains would be vulnerable if the army was defeated.

Men are easily elated or depressed by victory. As to being prepared for defeat, I certainly am not. Any man who is prepared for defeat would be half-defeated before he commenced. I hope for success; shall do all in my power to secure it and trust to God for the rest.

117

DEFEAT

Admiral David G. Farragut, letter to his wife, 11 April 1862, quoted in Loyall Farragut, *The Life of David Glasgow Farragut*, 1879.

He would sit and talk in the twilight…
Once, I remember he sat still for sometime by the window and his face looked so sad. He spoke of the Southern people, of their losses, privations, and sufferings, and also of our vain struggle. 'I cannot sleep,' he said, 'for thinking of it, and often I feel so weighted down with sorrow that I have to get up in the night and go out and walk till I thoroughly weary myself before I can sleep.' That was the only melancholy sentence I ever heard him utter, and the only time I ever saw that heartbroken look on his face.

An account of Robert E. Lee during his visit to a student from Washington College (who had broken his leg) at his home in Lexington by the student's mother, quoted in Charles Flood, *Lee: The Last Years*, 1981.

I can't understand it. I left a thousand men here.

Lieutenant-General Lord Chelmsford, overheard upon discovering the massacre of his forces at Isandlwana in Zululand, 22 January 1879, quoted in Ian Knight, *Brave Men's Blood*, 1990.

When I was a boy
the Sioux owned the world;
the sun rose and set on their land;
they sent ten thousand men to battle.
Where are the warriors today?
Who slew them? Where are their lands?
 Who owns them?
Sitting Bull (c. 1831–1890).

Errors and defeats are more obviously illustrative of principles than successes are… Defeat cries aloud for explanation; whereas success, like charity, covers a multitude of sins.

Rear Admiral Alfred Thayer Mahan, *Naval Strategy*, 1911.

A beaten general is disgraced forever.

Marshal of France Ferdinand Foch, *Precepts and Judgments*, 1919.

I have seen much war in my lifetime and I hate it profoundly. But there are worse things than war; and all of them come with defeat.

Ernest Hemingway, *Men at War*, 1942.

Bataan is like a child in a family who dies. It lives in our hearts.

General of the Army Douglas MacArthur, 9 April 1943, *New York Times*; on the first anniversary of the fall of Bataan.

Positions are seldom lost because they have been destroyed, but almost invariably because the leader has decided in his own mind that the position cannot be held.

General Alexander A. Vandegrift, 'Battle Doctrine for Front Line Leaders', Third Marine Division, 1944.

Man in war is not beaten, and cannot be beaten, until he owns himself beaten. Experience of all war proves this truth. So long as war persists as an instrument of policy, the objects of that policy can never be attained until the opponent admits his defeat. Total annihilation, even if it were possible, would recoil on the victor in the close-knit organization of the world's society. Moreover, it is not necessary to the military result. In all wars, and battles, decision is obtained at the moment when the survivors – normally the vast majority – realize that unless they yield their extinction has become inevitable.
 The history of war shows that man tends to give way directly he recognizes a stronger – when it dawns upon his mind that he has no further hope of victory by continuing the fight. That is why in battle, contrary to popular impression, actual shock is the rarest episode. When an assault takes place, the weaker side – weaker in either

DEFIANCE

numbers, morale, or momentum – have almost invariably surrendered or fled before the clash actually comes.
Captain Sir Basil Liddell Hart, *Thoughts on War*, 1944.

DEFENCE

Little minds try to defend everything at once, but sensible people look at the main point only; they parry the worst blows and stand a little hurt if thereby they avoid a greater one. If you try to hold everything, you hold nothing.
Frederick the Great, quoted in Hermann Foertsch, *The Art of Modern Warfare*, 1940.

To me death is better than the defensive.
Field Marshal Prince Aleksandr V. Suvorov (1729–1800), quoted in W. Lyon Blease, *Suvorof*, 1920.

What is the concept of defense? The parrying of a blow. What is its characteristic feature? awaiting the blow. It is this feature which turns any action into a defensive one; it is the only test by which defense can be distinguished from attack in war. Pure defense, however, would be completely contrary to the idea of war, since it would mean that only one side was waging it… But if we are really waging war, we must return the enemy's blows; and these offensive acts in a defensive war come under the heading of 'defense' – in other words, our offensive takes place within our own positions or theater of operations. Thus, a defensive campaign can be fought with offensive battles, and in a defensive battle, we can employ our divisions offensively. Even in a defensive position awaiting the enemy assault, our bullets take the offensive. So the defensive form of war is not a simple shield, but a shield made up of well-directed blows.
Major-General Carl von Clausewitz, *On War*, 6.1, 1832, tr. Michael Howard and Peter Paret, 1976.

He who stays on the defensive does not make war, he endures it.
Field Marshal Colmar Baron von der Goltz, *The Nation in Arms*, 1883.

DEFIANCE

When Xerxes wrote again: 'Deliver up your arms,' he wrote back: 'Come and take them.'
King Leonidas of Sparta, 480 AD, at the Battle of Thermopylae, quoted in Plutarch, *The Lives*, c. AD 100 (*Plutarch on Sparta*, tr. Richard Talbert, 1988).

Something about six feet of English earth – a bit more as he is such a big man.
King Harold Godwinson of England, 25 September 1066, before the Battle of Stamford Bridge when the Norwegian Viking King Harald Hardrada, who was reputed to be seven feet tall, asked what the English king would pay him to leave England, quoted in David Howarth, *1066: The Year of the Conquest*, 1978.

King of England, and you duke of Bedford… give up to the Maid sent here by the King of Heaven the key of all the noble cities of France you have taken and ravaged… I have come here… to drive you man for man from France… If you will not believe the Maid's message from God, wherever you happen to be we… shall make such a great hahaye as has not been made in France these thousand years.
Joan of Arc, 22 March 1429, letter to the English besieging Orléans.

Tell your master that this is the only territory that I will give him (pointing to the great ditch before the defenses of Malta). There lies the land which he may have for his own – provided only that he fills it with the bodies of his Janissaries.
Jean de la Valette, Grand Master of the Order of the Knights of St John of Jerusalem, 30 June 1565, his reply to the emissary of Mustapha Pasha, commander

DEFIANCE

of the Turkish Army besieging Malta, when offered terms of surrender, quoted in Ernle Bradford, *The Great Siege: Malta 1565*, 1961.

I have singed the beard of the king of Spain.
> Sir Francis Drake, after destroying a vast amount of shipping in the Harbour of Cadiz, 19 April 1587.

I know I have the body of a weak and feeble woman, but I have the heart and stomach of a king, and of a king of England too.
> Queen Elizabeth I, 29 July 1588, speech at Tilbury to her army as it awaited the Spanish Armada.

I will die in the last ditch.
> William of Orange, later William III of England (1650–1702), quoted in Hume, *History of Great Britain*, II, 1757.

Heads up, for God's sake! Those are bullets – not turds!
> General Lepic, 8 February 1807, to the Guard Horse-Grenadiers at Eylau as their losses to random fire lowered morale while they waited in reserve.

If it cost me my throne, I will bury the world under its ruins!
> Napoleon, 26 June 1813, reply to Prince Metternich's demands during the crisis of the 1813 campaign, quoted in R.M. Johnston, ed., *The Corsican*, 1910.

By the Eternal, they shall not sleep on our soil!
> Major-General Andrew Jackson, 23 December 1814, upon learning that the British Army had landed near New Orleans.

Merde! (Shit!)
> '*Le Mot de Cambronne*', the actual reply of General Cambronne at Waterloo when called upon to surrender the Old Guard, not the more poetic, 'The Guard dies but does not surrender!'

I shall never surrender nor retreat.
> Lieutenant-Colonel William Barrett Travis, 24 February 1836, reply to General Santa Anna when called upon to surrender the Alamo.

Tell him to go to hell.
> Major-General Zachary Taylor, 22 February 1847, reply to General Santa Anna's demand for his surrender at the Battle of Buena Vista.

Out of the question!
> The reply of Captain Jean Danjou, commanding the 3rd Company of the Foreign Legion Regiment at the battle of Camerone, 30 April 1863, in Mexico, in reply to the demand for surrender by Colonel Milan of the Mexican Army. The company's heroic last stand became the founding legend of the Foreign Legion. Afer Danjou was killed, the demand for surrender was made again and answered by the more earthy, '*Merde!*', from the enlisted ranks.

They shall not pass. (*Ils ne passeront pas.*)
> General Robert Nivelle, order of the day at at Verdun, 23 June 1916.

We shall not flag or fail. We shall fight in France, we shall fight on the seas and oceans, we shall fight with growing confidence and growing strength in the air, we shall defend our island, whatever the cost may be, we shall fight on the beaches, we shall fight on the landing grounds, we shall fight in the fields and in the streets, we shall fight in the hills; we shall never surrender.
> Sir Winston S. Churchill, speech in the House of Commons after the evacuation of Dunkirk, 4 June 1940.

We are waiting for the long-promised invasion. So are the fishes.
> Sir Winston S. Churchill, 21 October 1940, broadcast in French to the French people.

DELEGATION OF AUTHORITY

Nuts!
> Major-General Terry MacAullife, 22 December 1944, reply to the German demand that he surrender the 101st Airborne Division defending Bastogne.

DELAY

Every procrastinator is not a Fabius.
> King Gustavus II Adolphus, July 1630, to the colonel who could not make up his mind to surrender Stettin, quoted in Theodore Ayrault Dodge, *Gustavus Adolphus*, 1895.

This looks like a rerun of a bad movie, and I'm not interested in watching.
> George W. Bush, 21 January 2003, upon the attempts of the allies to delay war with Iraq by endless and fruitless negotiations.

DELEGATION OF AUTHORITY

According to my custom... I was present at the attack near the monastery of Svyanty Kryzh, but held my tongue, not wishing in the least to detract from the praiseworthy, skilful and brave commands of my subordinates.
> Field Marshal Prince Aleksandr V. Suvorov (1729–1800).

Be content to do what you can for the well-being of what properly belongs to you; commit the rest to those who are responsible.
> General Robert E. Lee, quoted in William Jones, *Life and Letters of Robert Edward Lee*, 1906.

I do not propose to lay down for you a plan of campaign... but simply lay down the work it is desirable to have done and leave you free to execute it in your own way. Submit to me, however, as early as you can, your plan of operation.
> General of the Army Ulysses S. Grant, instructions to General Sherman, 4 April 1864, *Personal Memoirs*, II, 1885.

When a detachment is made, the commander thereof should be informed of the object to be accomplished, and left as free as possible to execute it in his own way...
> General of the Army William T. Sherman, *Memoirs*, II, 1875.

You bring over the papers that are going to win or lose a war – and you sign the others.
> General of the Army Douglas MacArthur, c. 1919–1922, to Colonel Louis E. Hibbs, his adjutant while superintendent at West Point, when Hibbs asked MacArthur which papers he wanted to sign, quoted in Francis Miller, *General Douglas MacArthur*, 1942.

When I say that the gift of persuading others to do what is wanted is more than half the art of command, I mean by others the 'right men'. For the art of selection is the secret of leadership. Without this power of devolution staleness seizes upon the harassed leader. In general men given great responsibility work too hard.
> Lord Moran, *The Anatomy of Courage*, 1945.

A company commander can no more hope to supervise directly the acts of several hundred men in battle without reposing large faith in his lieutenants than the general can expect good results to come of by-passing his staff and his corps commanders and dealing directly with his divisions. But there are generals who have failed because they did not learn this lesson as captains.
> Brigadier-General S.L.A. Marshal, *Men Against Fire*, 1947.

They (our company commanders) are now bound hand and foot by a whole new list of mandatory subjects... You've delegated to the commander only the choice of what he is going to catch hell for.
> General Creighton Abrams (1914–1974), quoted in *Military Review*, 4/1985.

DEMORALISATION

[T]here were constant temptations to embroil myself in the details of various operations, a luxury that was not only hopelessly inefficient but actually downright dangerous. You cannot command any operation effectively if you involve yourself with any form of trivia whatsoever. You need every moment to think and to assimilate the broadest possible picture, in the effort to out-manoeuvre your opponent.

> Admiral Sir J.F. 'Sandy' Woodward, of his command philosophy in April, 1982, as the Falkland task force sailed towards battle, *One Hundred Days: The Memoirs of the Falklands Battle Group Commander*, 1992, p.117.

Each commander's plan had to be clear so that every other commander would understand what his colleagues were doing – what their objectives were, and how they intended to accomplish them. By the time the order presentations and reviews were completed, they each knew the overall plan in detail, they knew how their own roles fit into the overall concept, they knew what their neighbors were trying to do and where they would be. They also knew that I trusted them to carry out their missions with an absolute minimum of interference. The field assignments were theirs, and my philosophy (as each commander understood) was that they would handle them best.

> General Ariel Sharon, referring to the Battle of Agheila, 1967, *Warrior*, 1989.

Command is lonely, I said, and that was not just a romantic cliché. Sharing a problem with the boss, in this corps, would not be seen as weakness or failure, but as a sign of mutual confidence. On the other hand, they did not have to buck every decision up to me. 'I have a wide zone of indifference,' I said. 'I don't care if you hold reveille at five-thirty or five forty-five a.m. And don't ask me to decide.'

> General Colin Powell, *My American Journey*, 1995.

DEMORALISATION

When men are once checked in what they consider their special excellence, their whole opinion of themselves suffers more than if they had not at first believed in their superiority, the unexpected shock to their pride causing them to give way more than their real strength warrants...

> Gylippus, summer 413 AD, to the Syracusans before the last great naval battle with the Athenians in the harbour of Syracuse, quoted in Thucydides, *History of the Peloponnesian War*, 7.67, c. 404 AD, tr. Richard Crawley, 1910.

The men of both Lee's and Johnston's armies, like their brethren of the North, are as brave as men can be; but no man is so brave that he may not meet such defeats and disasters as to discourage and dampen his ardor for any cause, no matter how just he deems it.

> General of the Army Ulysses S. Grant, *Personal Memoirs of U.S. Grant*, II, 1885.

The Principle of Demoralization. As the principle of endurance has, as its primary object, the security of the minds of men by shielding their moral [morale] against the shock of battle, inversely the principle of demoralization has as its object the destruction of moral: first, in the moral attack against the spirit and nerves of the enemy's nation and government; secondly against this nation's policy; thirdly against the plan of its commander-in-chief, and fourthly against the moral of the soldiers commanded by him. Hitherto the fourth, the least important of these objectives, has been considered by the traditionally-minded soldier as the sole psychological objective of this great principle.

> Major-General J.F.C. Fuller, *The Reformation of War*, 1923.

DESERTION/DESERTERS

DESERTION/DESERTERS
The Poet's Shield
A perfect shield bedecks some Thracian now;
I had no choice: I left it in a wood.
Ah, well, I saved my skin, so let it go!
A new one's just as good.
> Archilochus, c. 648 AD, tr. Sir William Morris, quoted in *The Oxford Book of Greek Verse in Translation*, 1930.

To seduce the enemy's soldiers from their allegiance and encourage them to surrender is of especial service, for an adversary is more hurt by desertion than by slaughter.
> Flavius Vegetius Renatus, *Military Institutions of the Romans*, c. AD 378, tr. Clark, 1776.

An enemy soldier who deserts to us, apart from some plot, is of the greatest advantage, for the enemy is hurt by deserters more than if the same men were killed in action.
> The Emperor Maurice, *The Strategikon*, c. AD 600 (*Maurice's Strategikon*, tr. George Dennis, 1984).

If any of the enemy's troops desert him and come over to you, it is a great acquisition – provided they prove faithful; for their loss will be more than that of those killed in battle, although deserters will always be suspected by their new friends and odious to their old ones.
> Niccolò Machiavelli, *The Art of War*, 1521.

Come, come, let us fight another battle to-day: if I am beaten, we will desert together to-morrow.
> Frederick the Great (1712–1786), to a captured deserter, quoted in Campbell, *Frederick the Great*, 1843.

A soldier who offers to quit his rank, or offers to flag, is to be instantly put to death by the officer who commands that platoon, or by the officer or sergeant in the rear of that platoon; a soldier does not deserve to live who won't fight for his king and country.
> Major-General Sir James Wolfe, 1755, order to the 20th Foot at Canterbury.

Thomas Creevy: 'Do you calculate upon any desertion in Bonaparte's army?'
Wellington: 'Not upon a man from the colonel to the private in a regiment – both inclusive. We may pick up a marshal or two; but not worth a damn.'
> The Duke of Wellington, June 1815, before the Battle of Waterloo, quoted in Creevy, *The Creevy Papers*, 1934.

DESTINY
Ah, why did the bullets spare my life if it was only to lose it in this wretched way?
> Napoleon, 5 April 1821, near death on St. Helena, quoted in R.M. Johnston, ed., *The Corsican*, 1910.

I feel that my destiny is in the hands of the Almighty. This belief, more than any other facts or reason, makes me brave and fearless as a I am.
> Major-General George Armstrong Custer, 1863, letter to his wife.

This cannot be by accident; it must be by design. I was kept for this job.
> Winston S. Churchill, 1940, remark upon becoming Prime Minister to Lord Moran, *Churchill: Taken from the Diaries of Lord Moran*, 71, 1966.

I do not feel a person but an instrument of Destiny.
> General Charles de Gaulle, August 1944, upon entering Paris at its liberation.

I can almost hear the ticking of the second hand of destiny. We shall land at Inchon, and I shall crush them.
> General of the Army Douglas MacArthur, just before the amphibious landings at Inchon, 15 September 1950.

The glory of human nature lies in our seeming capacity to exercise conscious

DESTRUCTION OF THE ENEMY

control on our own destiny.
> Winston S. Churchill (1874-1965), quoted in C.E.M. Joad, 'Churchill the Philosopher,' in Charles Eade, ed., *Churchill by His Contemporaries*, 1953

Zhukov spun the wheel of history three times. First when he defended the Soviet Far East and smashed the Japanese forces. Then he conquered the Nazis in December 1941. His deeds were totally against military experience and doctrines, but he did them. And the third time was in the cold summer of 1953, when he executed Beriya.
> General Aleksandr I. Lebed, *Izvestiya* (Moscow), 20 November 1996.

I remind you that the road home leads through Baghdad. That's also where our next 'rendezvous with destiny' is.
> Major-General David H. Petraeus, Commander 101st AB, 21 March 2003, statement to his division as it shipped out for the Iraq war. The term was first used by Maj. Gen. William Lee in 1942, when the 101st was created at Fort Campbell, Ky, quoted by M.E. Sprengelmeyer, *Scripps Howard News Service*.

DESTRUCTION OF THE ENEMY

And furthermore it is my opinion that Carthage must be destroyed. (*Carthago delenda est*.)
> Marcus Porcius Cato, the Elder, *c.* 150 BC, quoted in Plutarch, *The Lives, c.* AD 100 (*The Makers of Rome*, tr. Ian Scott-Kilvert, 1965); the sentence with which he ended every speech in the Roman Senate based on his fear of a resurgent and vengeful Carthage.

When the enemy is driven back, we have failed, and when he is cut off, encircled and dispersed, we have succeeded.
> Field Marshal Prince Aleksandr V. Suvorov (1729–1800), quoted in Reznichenko, *Tactica*, 1987.

There is no way of dealing with the Frenchman but to knock him down – to be civil to them is to be laughed at, why they are enemies.
> Admiral Viscount Nelson, 11 January 1798, of the capitulation of the French garrison at Capua to Captain T. Louis, quoted in Robert Southey, *Life of Nelson*, 1813.

Had we taken ten sail and allowed the eleventh to escape, being able to get at her, I could never have called it well done.
> Admiral Viscount Nelson (1758–1805).

We may depend upon it that whenever we shall assemble an army, the French will consider its defeat and destruction their first object, particularly if Buonoparte should be at the head of the French troops himself...
> The Duke of Wellington, 1808, Portugal, advice to the British Cabinet on operations on the Iberian Peninsula, *The Dispatches of Field Marshal the Duke of Wellington, During His Various Campaigns in India, Denmark, Portugal, Spain, the Low Countries, and France*, IV, 1834–1838, p. 144.

Strike an enemy once and for all. Let him cease to exist as a tribe or he will live to fly at your throat again.
> Shaka Zulu, *c.* 1811, advice to King Dingiswayo on the treatment of the defeated Ndwandwes, quoted in E.A. Ritter, *Shaka Zulu*, 1955.

I will smash them, so help me god!
> Major-General Andrew Jackson, 1815, at the Battle of New Orleans.

The battles of Hannibal... have a particular character of stubbornness explained by the necessity for overcoming the Roman tenacity. It may be said that to Hannibal victory was not sufficient. He must destroy. Consequently he always tried to cut off all retreat for the enemy. He knew that with Rome, destruction was the only way of finishing the struggle.

DETERMINATION

He did not believe in the courage of despair in the masses; he believed in terror and he knew the value of surprise in inspiring it.
Colonel Charles Ardant du Picq, *Battle Studies*, 1880, tr. Greely, 1957.

Jomini's dictum that the organized forces of the enemy are the chief objective, pierces like a two-edged sword to the joints and marrow of many specious propositions...
Rear Admiral Alfred Thayer Mahan, *From Sail to Steam*, 1907.

Who wants a military leader incapable of understanding that the opponent is not going to surrender – that he must be crushed?
Joseph Stalin to H.G. Wells, quoted in M. Berchin and E. Ben-Horin, *The Red Army*, 1942.

In case opportunity for destruction of major portion of the enemy fleet is offered or can be created, *such destruction becomes the primary task*.
Admiral of the Fleet Chester W. Nimitz, operations order to Admiral Halsey in command of the Third Fleet in support of the invasion of the Philippines at Leyte Gulf, September 1944, quoted in Samuel Eliot Morison, *Leyte, June 1944–January 1945*, 1958.

We shall land at Inchon and I shall crush them.
General of the Army Douglas MacArthur, 23 August 1950, after being briefed on the difficulties of landing at Inchon, quoted in *The Congressional Record*, 15 January 1959, p. 669.

Did you ever hear the story of the *Bismarck*? I forget the British Admiral's name. He told the story himself. He wirelessed Churchill. 'The *Bismarck* is a wreck. The crew has left her but she won't sink. And I've got just enough oil in my ship to get home.' The Prime Minister (Churchill) sent him a cable, 'You stay there until the *Bismarck* is on the bottom; and we'll send out and tow you in.'
General of the Army Dwight D. Eisenhower, *The Military Churchill*, 1970.

We need to destroy not attack, not damage, not surround – I want you to destroy the Republican Guard.
General H. Norman Schwarzkopf, *It Doesn't Take a Hero*, 1992.

DETERMINATION

We will either find a way or make one.
Hannibal (247–183 AD).

True wisdom for a general is in vigorous determination.
Napoleon, '*Précis des guerres de Frederic II*', *Correspondance de Napoléon Ier, publié par ordre de l'Empereur Napoléon III*, No. 209, Vol. XXXII, 1858–1870.

Determination in a single instance is an expression of courage; if it becomes characteristic, a mental habit. But here we are referring not to physical courage but to the courage to accept responsibility, a courage in the face of moral danger. This has often been called *courage d'esprit*, because it is created by the intellect. That, however, does not make it an act of the intellect: it is an act of temperament. Intelligence alone is not courage; we often see that the most intelligent people are irresolute. Since in the rush of events a man is governed by feelings rather than by thought, the intellect needs to arouse the quality of courage, which then supports and sustains it in action.

Looked at in this way, the role of determination is to limit the agonies of doubt and the perils of hesitation when the motives for action are inadequate...
Major-General Carl von Clausewitz, *On War*, 1.3, 1832, tr. Michael Howard and Peter Paret, 1976.

DIPLOMACY/NEGOTIATIONS/TREATIES

And perhaps the most influential reflection of all: 'In war it is often less important what one does than how one does it. Strong determination and perseverance in carrying through a simple idea are the surest routes to one's objective.'
Field Marshal Helmuth Graf von Moltke (1800–1891), quoted in Kessel, *Moltke*, 1957.

It is true that in war determination by itself may achieve results, while flexibility without determination in reserve, cannot, but it is only the blending of the two that brings final success. The hardest test of generalship is to hold this balance between determination and flexibility.
Field Marshal Viscount Slim of Burma, *Defeat Into Victory*, 1956.

The proper response to difficulty is not to retreat. It is to prevail.
President George W. Bush, 21 September 2004, address to the United Nations

DIPLOMACY/NEGOTIATIONS/TREATIES

When the Argives were disputing land boundaries with the Spartans and were maintaining that theirs was the fairer claim, he [Lysander] drew his sword and said: 'The man who has this within his grasp argues best about land boundaries.'
Plutarch, quoting Lysander (k. 395 AD), *The Lives*, c. AD 100, (*Plutarch on Sparta*, tr. Richard Talbert, 1988).

In no way should a sworn agreement made with the enemy be broken.
The Emperor Maurice, *The Strategikon*, c. AD 600 (*Maurice's Strategikon*, tr. George Dennis, 1984).

All treaties are broken from considerations of interest; and in this respect republics are much more careful in the observance of treaties than princes... I speak of the breaking of treaties from some extraordinary cause; and here I believe, from what has been said, that the people are less frequently guilty of this than princes, and are therefore more to be trusted.
Niccolò Machiavelli, *Discourses*, 1517.

Diplomacy without arms is music without instruments.
Frederick the Great (1712–1786).

One can never foresee the consequences of political negotiations under the influence of military eventualities.
Napoleon (1769–1821).

Treaties are observed as long as they are in harmony with interests.
Napoleon, *The Military Maxims of Napoleon*, tr. Burnod, 1827.

May the diplomatists not again spoil with their pens, that which the armies have at so much cost won with their swords!
Field Marshal Gebhard von Blücher, 1815, at a dinner given by the Duke of Wellington in Paris, following the Allied victory over Napoleon at Waterloo.

Every negotiation which does not have power behind it... [is] ridiculous and fruitless.
Adolf Hitler, 1924.

To jaw-jaw is better than to war-war.
Sir Winston S. Churchill, 26 June 1954.

Diplomacy has rarely been able to gain at the conference table what cannot be gained or held on the battlefield.
General Walter Bedell Smith, 1954, after a conference on Indochina.

Treaties are like flowers and young girls – they last while they last.
General Charles de Gaulle, on a Franco-German treaty, *Time Magazine*, 12 July 1963.

DISCIPLINE

Our diplomacy is backed by strength, and we have the resolve to use that strength if necessary.
> Prime Minister Margaret Thatcher, 14 April 1982, in the House of Commons during the Falklands War, quoted in *The Falklands Campaign: A Digest of Debates in the House of Commons*, 1982, p. 75.

I accept the negotiations are more likely to succeed if military pressure is kept up. One must always consider the military options, and in doing so we must look after our soldiers and marines who have to undertake them...
> Prime Minister Margaret Thatcher, 26 April 1982, in the House of Commons during the Falklands War, quoted in *The Falklands Campaign: A Digest of Debates in the House of Commons*, 1982, p. 133.

There is no compromise possible with such people, no meeting of minds, no point of understanding with such terror. Just a choice: defeat it or be defeated by it. And defeat it we must.
> Prime Minister Tony Blair, speech to Labour Party Conference 2 October 2001.

Your wish to the crusaders should be as came in this verse of poetry: 'The only language between you and us is the sword that will strike your necks.'
> Osama bin Ladin, statement, 12 February 2003, *BBC News World Edition*.

There was never any doubt in my mind that the quality of people, command and control, the equipment and the depth of resolve of our country took this beyond the point of negotiation before the fight ever started. If we fight, we win.
> General Tommy Franks quoted by Joseph L. Galloway, 'Gen. Franks tells how Iraq war plan came together,' *Knight Ridder Newspapers*, 19 June 2003.

DISCIPLINE
The Effects of Discipline

Be men, my friends! Discipline fill your hearts!
Dread what comrades say of you here in bloody combat!
When men dread that, more men come through alive –
When soldiers break and run, good-bye glory,
good-bye all defenses!
> Homer, *The Iliad*, 15.651–655, c. 800 BC, tr. Robert Fagles, 1990; Telamonian Ajax spurring on the Achaeans as their army lay at bay before the Trojan onslaught.

For where there is no one in control nothing useful or distinguished can be done. This is roughly true of all departments of life, and entirely true where soldiering is concerned. Here it is discipline that makes one feel safe, while lack of discipline has destroyed many people before now.
> Xenophon, speech to the Greek officers after the defeat of Cyrus at Cunaxa, 401 BC, *Anabasis*, c. 360 BC (*The Persian Expedition*, 3.1, tr. Rex Warner).

Discipline is the soul of an army. It makes small numbers formidable; procures success to the weak, and esteem to all.
> General George Washington, 29 July 1759, Letter of Instructions to the Captains of the Virginia Regiments.

The more modern war becomes, the more essential appear the basic qualities that from the beginning of history have distinguished armies from mobs. The first of these is discipline. We very soon learned in Burma that strict discipline in battle and in bivouac was vital, not only for success, but for survival. Nothing is easier in jungle or dispersed fighting than for a man to shirk. If he has no stomach for advancing, all he has to do is flop into the undergrowth; in retreat, he can slink out of the rear guard, join up later, and swear he was the last to leave. A patrol leader can take his men a mile into the jungle, hide there, and

DISCIPLINE

return with any report he fancies. Only discipline – not punishment – can stop that sort of thing; the real discipline that a man holds to because it is a refusal to betray his comrades. The discipline that makes a sentry, whose whole body is tortured for sleep, rest his chin on the point of his bayonet because he knows, if he nods, he risks the lives of the men sleeping behind him. It is only discipline, too, that can enforce the precautions against disease, irksome as they are, without which an army would shrivel away. At some stage in all wars armies have let their discipline sag, but they have never won victory until they made it taut again; nor will they. We found it a great mistake to belittle the importance of smartness in turn out, alertness of carriage, cleanliness of person, saluting, or precision of movement, and to dismiss them as naive, unintelligent parade-ground stuff. I do not believe that troops can have unshakable battle discipline without showing those outward and formal signs, which mark the pride men take in themselves and their units and the mutual confidence and respect that exist between them and their officers. It was our experience in a tough school that the best fighting units, in the long run, were not necessarily those with the most advertised reputations, but those who when they came out of battle at once resumed a more formal discipline and appearance.

Field Marshal Viscount Slim of Burma, *Courage and Other Broadcasts*, 1957.

Instilling discipline

You pick out the big men! I'll make them brave.

Pyrrhus (319–272 BC) to his recruiters, quoted in Frontinus, *Stratagems*, 4.1, c. AD 84–96, tr. Charles E. Bennett, 1925.

After the organization of troops, military discipline is the first matter that presents itself. It is the soul of armies. If it is not established with wisdom and maintained with unshakable resolution you will have no soldiers. Regiments and armies will be only contemptible, armed mobs, more dangerous to their own country than to the enemy. Few orders are best, but they should be followed up with care; negligence should be punished without partiality and without distinction of rank or birth; otherwise, you will make yourself hated. One can be exact and just, and be loved at the same time as feared. Severity must be accompanied with kindness, but his should not have the appearance of pretense, but of goodness.

Field Marshal Maurice Comte de Saxe, *My Reveries*, 1732, tr. Phillips, 1940.

The commander should practice kindness and severity, should appear friendly to the soldiers, speak to them on the march, visit them while they are cooking, ask them if they are well cared for, and alleviate their needs if they have any. Officers without experience in war should be treated kindly. Their good actions should be praised. Small requests should be granted and they should not be treated in an overbearing manner, but severity is maintained about everything regarding the service. The negligent officer is punished; the man who answers back is made to feel your severity by being reprimanded with the authoritative air that superiority gives; pillaging or argumentative soldiers, or those whose obedience is not immediate should be punished.

Frederick the Great, *Instructions to His Generals*, 1747, tr. Phillips, 1940.

With severity, kindness is needed, or else severity is tyranny. I am strict in maintaining the health of the soldiers and a true sense of good conduct; kind soldierly strictness, and then general brotherhood. To me strictness by whim

DISCIPLINE

would be tyranny.
Field Marshal Prince Aleksandr V. Suvorov (1829–1800), *Dokumenty*, 1952.

My son, put a little order into your corps, it wants it badly. The Italians in particular commit atrocities, robbing and pillaging wherever they go. Shoot a few of them. Your affectionate father.
Napoleon, May 1813, letter to Prince Eugène who was to conduct a highly credible campaign against the Austrians.

Impress upon your officers that discipline cannot be attained without constant watchfulness on their part. They must attend to the smallest particulars of detail. Men must be habituated to obey or they cannot be controlled in battle, and the neglect of the least important order impairs the proper influence of the officer.

In recommending officers or men for promotion you will always, where other qualifications are equal, give preference to those who show the highest appreciation of the importance of discipline and evince the greatest attention to its requirements.
Robert E. Lee, 22 February 1865, Circular of the Army of Northern Virginia, quoted in William Jones, *Life and Letters of Robert Edward Lee*, 1906.

In camp, and especially in the presence of an active enemy, it is much easier to maintain discipline than in barracks in time of peace. Crime and breaches of discipline are much less frequent, and the necessity for courts-martial less. The captain can usually inflict all the punishment necessary, and the colonel should always.
General of the Army William T. Sherman, *Memoirs*, II, 1875.

Discipline is not made to order, cannot be created off-hand; it is a matter of the institution of tradition. The Commander must have absolute confidence in his right to command, must have the habit of command, pride in commanding. It is this which gives a strong discipline to Armies commanded by an aristocracy, whenever such a thing exists.
Colonel Charles Ardant du Picq, quoted in Ian Hamilton, *The Soul and Body of an Army*, 1921.

Discipline can only be obtained when all officers are so imbued with the sense of their awful obligation to their men and to their country that they cannot tolerate negligence. Officers who fail to correct errors or to praise excellence are valueless in peace and dangerous misfits in war.
General George S. Patton, Jr, *War As I Knew It*, 1947.

That's the crux of the matter. Discipline is something that is enforced, either by fear or by understanding. Even in an Army, it is not merely a question of giving orders; there is more to a soldier's discipline than blind obedience. To take men into your confidence is not a new technique invented in the last war. Good generals were doing that long before you and I got into khaki to save the world.
Field Marshal Viscount Slim of Burma, *Courage and Other Broadcasts*, 1957.

The Lack of Discipline
Popularity, however desirable it may be to individuals, will not form, or feed, or pay an army; will not enable it to march, and fight; will not keep it in a state of efficiency for long and arduous service.
The Duke of Wellington, letter from Portugal, 8 April 1811.

Without discipline and good order, not only is an army unfit to be opposed to an enemy in the field, but it becomes a positive injury to the country by which it is maintained… The Commander in Chief assures the General and other

DISCIPLINE

Officers... that at the same time that he will be happy to draw the notice of Government, and to extol their good conduct, he will not be backward in noticing any inattention on the part of the officers of the army to the duties required from them... or any breach of discipline and orders by the soldiers.

> The Duke of Wellington, 1 January 1813, address to the Spanish Army, pleased at its improvement over the previous four years, but still cautious as to the stability of that improvement, quoted in Julian Rathbone, ed., *Wellington's War: His Peninsular Dispatches*, 1984.

Of late we have been doing everything in our power, both by law and by publications, to relax the discipline by which alone such men can be kept in order. The officers of the lower ranks will not perform the duty required of them for the purpose of keeping their soldiers in order; and it is next to impossible to punish any officer for neglect of this description. As to the non-commissioned officers, as I have repeatedly stated, they are as bad as the men, and too near them, in point of pay and situation, by the regulations of late years, for us to expect them to do anything to keep the men in order. It is really a disgrace to have anything to say to such men as some of our soldiers are.

> The Duke of Wellington, 2 July 1813, to Lord Bathurst, prompted by the breakdown in discipline and looting after the Battle of Vittoria, *The Dispatches of Field Marshal the Duke of Wellington, During His Various Campaigns in India, Denmark, Portugal, Spain, the Low Countries, and France*, X, 1834–1838, p. 496.

The spirit which animates our soldiers and the natural courage with which they are so liberally endowed have led to a reliance upon these good qualities to the neglect of those measures which would increase their efficiency and contribute to their safety. Many opportunities have been lost and hundreds of valuable lives uselessly sacrificed for want of a strict observance of discipline.

> Robert E. Lee, 22 February 1865, Circular of the Army of Northern Virginia, quoted in William Jones, *Life and Letters of Robert Edward Lee*, 1906.

You cannot be disciplined in great things and undisciplined in small things... Brave, undisciplined men have no chance against the discipline and valor of other men. Have you ever seen a few policemen handle a crowd?

> General George S. Patton, Jr, 17 May 1941, address to officers and men of the Second Armored Division, *The Patton Papers*, II, 1974.

There was no tactical organization other than tribal group, and few preparations were made for a campaign. There was a total lack of planning; battles were headlong assaults in rough phalangial order in which the warriors rapidly exhausted themselves and became disorganized; courage shattered itself upon the rocks of discipline.

> Major-General J.F.C. Fuller, on the military organisation of the Gauls compared to the Romans, *Julius Caesar: Man, Soldier, Tyrant*, 1965.

The Nature of Discipline

It is not... by native gallantry, it is not by the exertion of bodily strength... that bodies... can contend effectually... bodies of men... must get into confusion unless regulated by discipline; unless accustomed to subordination, and obedient to command. I am afraid that panic is the usual attendance upon such confusion. It is then by the enforcement of rules of discipline, subordination, and good order, that such bodies can render efficient service to their King and Country; and can be otherwise than a terror to their friends, contemptible to their enemies, and burthen to the State.

DISCIPLINE

The Duke of Wellington, 7 October 1834, speech upon the presentation of Colours to the 93rd Highlanders, USJ, 1834, III, quoted in Philip Haythornthwaite, *The Armies of Wellington*, 1994.

It is the discipline which is the soul of armies, as indeed the soul of power in all intelligence. Other things – moral considerations, impulses of sentiment, and even natural excitement – may lead men to great deeds; but taken in the long run, and in all vicissitudes, an army is effective in proportion to its discipline.
Major-General Joshua L. Chamberlain, 1869, speech 'The Army of the Potomac'.

As the severity of military operations increases, so also must the sternness of the discipline. The zeal of the soldiers, their warlike instincts, and the interests and excitements of war may ensure obedience of orders and the cheerful endurance of perils and hardships during a short and prosperous campaign. But when fortune is dubious or adverse; when retreats as well as advances are necessary; when supplies fail, arrangements miscarry, and disasters impend, and when the struggle is protracted, men can only be persuaded to accept evil things by the lively realization of the fact that greater terrors await their refusal.
Sir Winston S. Churchill, *The River War*, 1899.

To be disciplined does not mean to keep silent, to do only what one thinks can be done without risk or being compromised, the art of avoiding responsibilities, but it means acting in the spirit of the orders received, and to that end assuring by thought and planning the possibility of carrying out such orders, assuring by strength of character the energy to assume the risks necessary in their execution. The laziness of the mind results in lack of discipline as much as does insubordination. Lack of ability and ignorance are not either excuses, for knowledge is within reach of all who seek it.
Marshal of France Ferdinand Foch, *Principles of War*, 1913.

It is a mistaken idea that precision of movement and smartness of appearance, which for the popular mind are often the whole meaning of the word 'discipline', can only be obtained by a Prussian discipline, where the individuality of man is ground out until only a robot-like body is left. A century ago, Sir John Moore... introduced a new discipline based on intelligence and comradeship instead of on sullenness and fear. His work was justified by its fruits, and not only did the results show that this discipline of the thinking man instead of automata produces a more efficient soldier for war, but also that the former can rival the latter in order and smartness. Intelligent men whose minds are disciplined in the best sense can acquire these qualities as well, and more quickly, than the barrack-square product. But the reverse order of progress from the Frederican discipline to initiative is not possible. (Sept. 1925.)
Captain Sir Basil Liddell Hart, *Thoughts on War*, 1944.

Discipline... is the basis of military efficiency... the major factor of true military discipline consists in securing the voluntary cooperation of subordinates, thereby reducing the number of infractions of the laws and regulations to a minimum; by laying down the doctrine that the true test of the existence of a high state of discipline in a military organization is found in its cheerful and satisfactory performance of duty under all service conditions; and by reminding officers that a happy and contented detachment is usually a well disciplined detachment.
Major-General John A. Lejeune, *Reminiscences of a Marine*, 1929.

DISCIPLINE

Commanding and obeying are the characteristics of the army, and the one is as hard as the other. Both are simplified in proportion as orders are given with prudence and intelligence and obedience is rendered with perception and confidence. Human nature demands compulsion when many are to be united for one purpose. Thus discipline becomes an inseparable feature of the army, and its nature and degree are the true measure of the army's efficiency. The more voluntary the nature of the discipline the better, but only a discipline that has become habit and matter of course can survive the test in the hour of danger.

Colonel-General Hans von Seekt, *Thoughts of a Soldier*, 1930.

We had no discipline in the sense in which it was restrictive, submergent of individuality, the Lowest Common Denominator of men. In peace-armies discipline meant the hunt, not of an average but of an absolute; the hundred per cent standard in which the ninety-nine were played down to the level of the weakest man on parade. The aim was to render the unit a unit, the man a type; in order that their effort might be calculable, and the collective output even in grain and bulk. The deeper the discipline, the lower was the individual excellence; also the more sure the performance.

By this substitution of a sure job for a possible masterpiece, military science made a deliberate sacrifice of capacity in order to reduce the uncertain element, the bionomic factor, in enlisted humanity.

Colonel T.E. Lawrence, *The Seven Pillars of Wisdom*, 1935.

It is the basic constituent of all armies. By virtue of discipline something resembling a contract comes into being between the leader and the subordinates. It is an understood thing that the latter owe obedience to the former, and that each single component of an organized body must, to the best of its ability, carry out the orders transmitted to him by a higher authority. In this way a fundamental attitude of good will is created which guarantees a minimum degree of cohesion. But it is not enough for a chief to bind his men into a whole through the medium of impersonal obedience. It is on the inner selves that he must leave the imprint of his personality. If he is to have a genuine and effective hold on his men, he must know how to make their wills part and parcel of his own, and so to inspire them as something of their own choosing. He must increase and multiply the effects of mere discipline and implant in those under him a sort of moral suggestion which goes far beyond reasoning, and crystallize round his own person all their potentialities of faith, hope, and devotion.

General Charles de Gaulle, *The Edge of the Sword*, 1932.

The idea that real discipline is best instilled by fear of punishment is a delusion – especially in these days of open-order fighting. For such discipline kills the highest military qualities – initiative, and intelligent fulfillment of the superior's intentions. Moreover, the common notion that discipline is produced by drill is a case of putting the cart before the horse. A well-drilled battalion has often proved a bad one in the field. By contrast a *good* battalion is often good at drill – because if the spirit is right, it likes to do all things well. Here is the real sequence of causation. (June 1935.)

Captain Sir Basil Liddell Hart, *Thoughts on War*, 1944.

Military discipline, in the narrowest sense of the term, is the enforcement of instant obedience to orders through

DISCIPLINE

threat of punishment, supplemented by the control reflexes established through drill. In the wider sense it is bound up with the soldierly spirit. So far as this is susceptible to analysis, its components appear to be pride of manhood, the pride of arms (of being an initiate in the martial cult), the confidence that comes from skill at arms, the sense of comradeship, the sense of duty, and the sense of loyalty – to comrades, commanders, corps, and country. (Jan. 1936.)
    Captain Sir Basil Liddell Hart, *Thoughts on War*, 1944.

In all armies, obedience of the subordinates to their superiors must be exacted... but the basis for soldier discipline must be the individual conscience. With soldiers, a discipline of coercion is ineffective, discipline must be self-imposed, because only when it is, is the soldier able to understand completely why he fights and how he must obey. This type of discipline becomes a tower of strength within the army, and it is the only type that can truly harmonize the relationship that exists between officers and soldiers.
    Mao Tse-tung, *On Protracted War*, 1938.

Within our system, that discipline is nearest perfect which assures to the individual the greatest freedom of thought and action while at all times promoting his feeling of responsibility toward the group. *These twin ends are convergent and interdependent for the exact converse of the reason that it is impossible for any man to feel happy and successful if he is in the middle of a failing institution.*
    Brigadier-General S.L.A. Marshall, *The Armed Forces Officer*, 1951.

It is only discipline that enables men to live in a community and yet retain individual liberty... You can have discipline without liberty but cannot have liberty without discipline... The self-discipline of the strong is the safeguard of the weak.
    Field Marshal Viscount Slim of Burma, *Courage and Other Broadcasts*, 1957.

When we speak of conscious discipline, it means that it is built up on the basis of political consciousness of the officers and men, and the most important method of maintaining discipline is education and persuasion, thus making the army men of their own accord, respect and remind each other to observe discipline. When we speak of strict discipline, it means that everyone in the army, regardless of rank or office must observe discipline and no infringements are allowed.
    General Vo Nguyen Giap, *People's War – People's Army*, 1961.

One of the primary purposes of discipline is to produce alertness. A man who is so lethargic that he fails to salute will fall an easy victim to the enemy.
    General George S. Patton, Jr., *War As I Knew It*, 1947.

A most potent factor in spreading this belief in efficiency of an organization is a sense of discipline. In effect, discipline means that every man, when things pass beyond his own authority or initiative, knows to whom to turn for further direction. If it is the right kind of discipline he turns in the confidence that he will get sensible and effective direction. Every step must be taken to build up this confidence of the soldier in his leaders. For instance, it is not enough to be efficient; the organization must look efficient. If you enter the lines of a regiment where the quarter guard is smart and alert, and the men you meet are well turned out and salute briskly, you cannot fail to get an impression of efficiency. You are right; ten to one that unit is efficient. If you go into a headquarters and find the clerks scruffy, the floor unswept, and dirty tea mugs staining flyblown papers

on the office tables, it may be efficient, but no visitor will think so.
Field Marshal Viscount Slim of Burma, *Defeat Into Victory*, 1956.

DIVINE FAVOUR

There shall not be any man be able to stand before thee all the days of thy life: as I was with Moses, so I will be with thee: I will not fail thee, nor forsake thee.
Joshua 1:1.

Therefore thus saith the LORD concerning the king of Assyria, He shall not come into this city, nor shoot an arrow there, nor come before it with shield, nor cast a bank against it.

By the way that he came, by the same shall he return, and shall not come into this city, saith the LORD.

For I will defend this city, to save it, for mine own sake, and for my servant David's sake.

And it came to pass that night, that the angel of the LORD sent out, and smote in the camp of the Assyrians an hundred fourscore and five thousand: and when they arose early in the morning, behold, they were all dead corpses.

So Sennacherib king of Assyria departed, and went and returned, and dwelt at Nineveh.
Samuel 19:32–35.

The Protectress of Athens
Our city, by the immortal gods' intent and Zeus' decree, shall never come to
  harm:
for our bold champion, of proud
  descent,
Pallas of Athens shields us with her
  arm.
Solon, c. 594 BC, tr. Gilbert Highet, quoted in *The Oxford Book of Greek Verse in Translation*, 1930.

Thou art invincible, my son!
The Prophetess of Delphi, when Alexander the Great dragged her to the temple of Apollo to consult the god on his invasion of the Persian Empire, after she had refused because the day was inauspicious – Alexander replied that he 'desired no further prophecy, but had from her the oracle which he wanted,' quoted in Plutarch, 'Alexander,' *Lives*, VII, p. 261.

I bring you news from God, that our Lord will give you back your kingdom, bringing you to be crowned at Rheims, and driving out your enemies. In this I am God's messenger. Do you set me bravely to work, and I will raise the siege of Orléans.
Joan of Arc, February–March 1429, to the Dauphin Charles, *Joan of Arc In Her Own Words*, 1996.

Come here, Tepanecs. Are you ignorant of the fact that Huitzilopotchtli, god of the Aztecs, fights for them and protects them with all his might? It seems to me folly to try to fight against the gods.
Nezahualcoyotal, King of Texcoco, c. 1430, to a delegation from the city of Coyohuacan pleading war against the Aztecs, quoted in Diego Durán, *The History of the Indies of New Spain*, 1581, tr. Doris Heyden, 1994.

Then you would have me disregard the protection of God?
Gustavus II Adolphus (1594–1632), when his staff urged him to keep a bodyguard, quoted in Theodore Ayrault Dodge, *Gustavus Adolphus*, 1895.

God made them as stubble to our swords.
Oliver Cromwell, 2 July 1644, of his victory at Marston Moor, quoted in Theodore Ayrault Dodge, *Gustavus Adolphus*, 1895.

...As it has been a kind of destiny, that has thrown me upon this service, I shall hope that my undertaking it is designed to answer some good purpose... I shall rely, therefore, confidently on that

DOGMA

Providence, which has heretofore preserved and been bountiful to me.
> General George Washington, 8 June 1775, to Martha Washington, *The Writings of George Washington*, III, p. 294, John C. Fitzgerald, ed., 1931.

The unerring hand of providence shielded my men.
> Major-General Andrew Jackson, 26 January 1815, letter to Robert Hays of his victory at the Battle of New Orleans, *Correspondence*, II, p. 156.

The finger of providence was upon me, and I escaped unhurt.
> The Duke of Wellington, 19 June 1815, of his victory at Waterloo, *Supplementary Despatches, Correspondence, and Memoranda of Field Marshal Arthur Duke of Wellington, KG, 1794–1818*, X, 1863, p. 531.

The angel of the Lord on the traditional white horse and clad all in white with flaming sword, faced the advancing Germans at Mons and forbade their further progress.
> Brigadier John Charteris, letter 5 September 1914, *At G.H.Q.*, 1931, of the legend of the Angel of Mons which British troops claimed to have seen in the heavens guarding the British Army.

DOCTRINE

A doctrine of war consists first in a common way of objectively approaching the subject; second, in a common way of handling it, by adapting without reserve the means to the goal aimed at, to the object.
> Marshal of France Ferdinand Foch, *Precepts and Judgments*, 1919.

The central idea of an army is known as its doctrine, which to be sound must be based on the principles of war, and which to be effective must be elastic enough to admit of mutation in accordance with change in circumstances. In its ultimate relationship to the human understanding this central idea or doctrine is nothing else than common sense – that is, action adapted to circumstances.
> Major-General J.F.C. Fuller, *The Foundations of the Science of War*, 1926.

Military doctrine refers to the point of view from which military history is understood and its experience and lessons understood. Doctrine is the daughter of history... Doctrine is needed so that in the realm of military thought an army not represent human dust, but a cohesive whole... should be predatory and stern, ruthless toward defeat and the defeated.
> General A.A. Svechin (1878–1938).

The crumbling of the whole system of doctrines and organization, to which our leaders had attached themselves, deprived them of their motive force. A sort of moral inhibition made them suddenly doubtful of everything, and especially of themselves. From then on the centrifugal forces were to show themselves rapidly.
> General Charles de Gaulle, *The Call to Honor*, 1955.

At the very heart of war lies doctrine. It represents the central beliefs for waging war in order to achieve victory... It is the building material for strategy. It is fundamental to sound judgment.
> General Curtis LeMay, USAF, Air Force Manual 1-1, *Basic Doctrine* (Washington, DC: Department of the Air Force, 1984) frontispiece.

DOGMA

To dogmatize upon that which you have not practiced is the prerogative of ignorance; it is like thinking that you can solve, by an equation of the second degree, a problem of transcendental geometry which would have daunted Legrange or Laplace.
> Napoleon (1769–1821).

DO OR DIE

The dogmas of the quiet past are inadequate to the stormy present. The occasion is piled high with difficulty, and we must rise with the occasion. As our case is new, so we must think anew, and act anew. We must disenthrall ourselves, and then we shall save our country.
>President Abraham Lincoln, 1 December 1862, 'Annual Message to Congress', *Collected Works of Abraham Lincoln*, V, 1953.

It will be better to offer certain considerations for reflection than to make sweeping dogmatic assertions.
>Rear Admiral Alfred Thayer Mahan (1840–1914).

In itself, the danger of a doctrine is that it is apt to ossify into a dogma, and to be seized upon by mental emasculates who lack virility of judgment, and who are only too grateful to rest assured that their actions, however inept, find justification in a book, which, if they think at all, is, in their opinion, written in order to exonerate them from doing so. In the past many armies have been destroyed by internal discord, and some have been destroyed by the weapons of their antagonists, but the majority have perished through adhering to dogmas springing from their past successes – that is, self-destruction or suicide through inertia of mind.
>Major-General J.F.C. Fuller, *The Reformation of War*, 1923.

Adherence to dogmas has destroyed more armies and lost more battles and lives than any other cause in war. No man of fixed opinions can make a good general; consequently, if this series of lectures, however hypothetical many of its contentions may seem, succeeds in unfixing dogmas, then certainly it will not have been written in vain.
>Major-General J.F.C. Fuller, *Lectures on F.S.R. III*, 1932.

Prejudice against novel methods is a phenomenon typical of an officer corps raised in a proven system. The Prussian Army was defeated for this reason by Napoleon. The same phenomenon showed up in this war, in German as well as British officer circles, where complicated theories obstructed the capability to see things in reality. A military dogma had been worked out in every last detail, and this was taken to represent the very peak of military wisdom. In their minds, the only military thought allowed acceptable was that which followed their own doctrine. Everything other than the rule was a game of chance; and it followed that success could only be the result of luck or accident. This attitude leads to fixed ideas, the consequences of which are incalculable.
>Field Marshal Erwin Rommel, letter of May/June 1942, quoted in John Pimlott, ed., *Rommel in His Own Words*, 1994.

DO OR DIE

Friends! Fighting Danaans! Aides-in-arms of Ares!
Fight like men, my comrades – call up your battle-fury!
You think we have reserves in the rear to back us up?
Some stronger wall to shield our men from disaster?
No, there's no great citadel standing near with towers
where we could defend ourselves and troops could turn the tide.
No – we're here on the plain of Troy – all Troy's in arms!
Dug in, backs to the sea, land of our fathers far away!
Fight – the light of safety lies in our fighting hands,
not spines gone soft in battle.
>Homer, *The Iliad*, 15.851–859, c. 800 BC, tr. Robert Fagles, 1990; Telamonian Ajax's cry for support as he single-handedly held off the Trojan attack on the Greek ships.

DO OR DIE

To fail to think fast when surrounded by the enemy is to have your back pressed to the wall;

And to fail to take the battle to the enemy when your back is to the wall is to perish.

> Sun Tzu, c. 500 BC, a passage recovered from later encyclopaedic works and commentaries and not found in *The Art of War*, quoted in *Sun Tzu: The Art of Warfare*, 1993, tr. Roger Ames, 1993.

As to the field of battle it is the abiding place of the dead. And he who decides to die will live, and he who wishes to live will die.

> Wu C'hi (431–381 BC), *Art of War*, 3.4 (*Three Military Classics of China*, tr. A.L. Sadler, 1944).

We have nothing left in the world but what we can win with our swords. Timidity and cowardice are for men who can see safety at their backs – who can retreat without molestation along some easy road and find refuge in the familiar fields of their native land; but they are not for you: you must be brave; for you there is no middle way between victory or death – put all hope of it from you, and either conquer, or, should fortune hesitate to favour you, meet death in battle rather than in flight.

> Hannibal, 218 BC, address to his troops after crossing the Alps into Italy as they faced their first Roman Army, quoted in Livy, *The History of Rome*, 21.44, c. AD 17 (*The War With Hannibal*, tr. Aubrey de Sélincourt, 1972).

Go, therefore, to meet the foe with two objects before you, either victory or death. For men animated by such a spirit must always overcome their adversaries, since they go into battle ready to throw away their lives.

> Scipio Africanus, 202 BC, address to the Roman Army before the Battle of Zama, quoted in Polybius, *Histories*, c. 125 BC,
quoted in Liddell Hart, *A Greater Than Napoleon*, 1926.

When men find they must inevitably perish, they willingly resolve to die with their comrades and with their arms in their hands.

> Flavius Vegetius Renatus, *Military Institutions of the Romans*, c. AD 378, tr. Clark, 1776.

Courage shall be firmer, heart all the keener, spirit the greater, as our might grows less. Here lies our leader, fatally cut down, the good man in the dust. Forever shall that man mourn who now thinks to leave the battle! I am far advanced in years and I shall not leave. I intend to lie by the side of my dear lord.

> Byrhtwold, one of Byrtnoth's huscarls, after the latter had been killed in battle against the Vikings and his household had gathered around to defend his body; from the fragmentary poem, *The Battle of Maldon*, tr. C.B. Hieatt, 1967; one of the most famous passages in Anglo-Saxon literature.

Sirs and fellows, as I am true knight and king, for me this day shall never England ransom pay.

> King Henry V of England, 25 October 1415, address to his army before the Battle of Agincourt.

Some commanders have forced their men to fight by depriving them of all means of saving themselves except victory; this is certainly the best method of making them fight desperately. This resolution is commonly heightened either by the confidence they put in themselves, their arms, armor, discipline, good order, and lately-won victories, or by the esteem they have for their general... or it is a result of the love for their country, which is natural to all men. There are various ways of forcing men to fight, but that is the strongest and most operative; it leaves

DO OR DIE

men no other alternative but to conquer or die.
> Niccolò Machiavelli, *The Art of War*, 1521.

I shall surely prevent him [the Turk]. And even if this siege, contrary to my expectation, should end in a victory for the enemy, I declare to you all that I have resolved that no-one in Constantinople shall ever see a Grand Master of our Holy Order there in chains. If, indeed, the very worst should happen and all be lost, then I intend to put on the uniform of a common soldier and throw myself, sword in hand, into the thick of the enemy and perish there with my children and my brothers.
> Jean de la Valette, Grand Master of the Order of the Knights of St John, 17 August 1565, his address to the council of the Order during the siege of Malta when informed that the Turks intended to drag him back to Constantinople in chains.

I thank you for doing your duty in warning me of the danger. And now face me towards the enemy.
> Admiral Lord Edward Hawke, 20 November 1759, at the Battle of Quiberon Bay, when he was warned by one of his officers of the extreme danger of the stormy shallow waters which protected the French fleet, quoted in Sandy Woodward, *One Hundred Days: The Memoirs of the Falklands Battle Group Commander*, 1982, p. 27.

We beat them tonight or Molly Stark's a widow.
> Brigadier-General John Stark, 16 August 1777, addressing his men before the Battle of Bennington.

If we had not driven them into hell… hell would have swallowed us.
> Field Marshal Prince Aleksandr V. Suvorov, 1787, of the Battle of Kinburn.

Sir Robert Calder: There are eight sail of the line, Sir John.
Jervis: Very well, Sir.
Calder: There are 20 sail of the line, Sir John.
Jervis: Very well, Sir.
Calder: There are 25 sail of the line, Sir John.
Jervis: Very well, sir.
Calder: There are 27 sail of the line, Sir John, near double our own.
Jervis: Enough of that, Sir. If there are 50 sail of the line, I will go through them. The die is cast.
> Admiral Lord St Vincent (Sir John Jervis), 14 February 1797, as the Spanish fleet came into sight before the Battle of Cape St Vincent, quoted in David Walder, *Nelson*, 1978.

Westminster Abbey or victory!
> Admiral Viscount Nelson, 14 February 1797, battle cry as he led a boarding party aboard a Spanish ship of the line during the Battle of Cape St Vincent; Westminster Abbey was the sepulchre of the greatest of the English, quoted in Robert Southey, *Life of Nelson*, 1813.

The troops behaved admirably; if they had not done so, not a man of us would have quitted the field.
> The Duke of Wellington, 10 September 1800, on the conduct of his men after his victory at Hyderabad, quoted in Arthur Bryant, *The Great Duke*, 1972.

Farewell, my dear Lady Hester. If I can extricate myself and those with me from our present difficulties, and if I can beat the French I shall return to you with satisfaction; but if not, it will be better that I should never quit Spain.
> Lieutenant-General Sir John Moore, 24 November 1808, letter to Lady Hester Stanhope, quoted in Carola Oman, *Sir John Moore*, 1953.

Tell the men to fire faster and not to give up the ship. Fight her til she sinks.
> Captain James Lawrence, 1813, order, as he lay dying, during the battle between his USS *Chesapeake* and HMS *Shannon*.

DO OR DIE

If your officer's dead and the sergeants look white,
Remember it's ruin to run from a fight;
So take open order, lie down, and sit tight,
An' wait for supports like a soldier.
> Rudyard Kipling, 'The Young British Soldier'.

It so often happens that, when men are convinced that they have to die, a desire to bear themselves well and to leave life's stage with dignity conquers all other sensations.
> Sir Winston S. Churchill, Savonarola, 1900.

We are about to engage in a battle on which the fate of our country depends and it is important to remind all ranks that the moment has passed for looking to the rear; all our efforts must be directed to attacking and driving back the enemy. Troops that can advance no farther must, at any price, hold on to the ground they have conquered and die on the spot rather than give way. Under the circumstances which face us, no act of weakness can be tolerated.
> Marshal of France Joseph Joffre, early September 1914, message to the French Army as he began the counter-attack of the Battle of the Marne.

Every position must be held to the last man: there must be no retirement. With our backs to the wall, and believing in the justice of our cause, each one of us must fight on to the end. The safety of our Homes and the Freedom of mankind alike depend on the conduct of each one of us at this critical moment.
> Field Marshal Earl Haig, Order of the Day, 12 April 1918, at the height of the German *Friedenstürm* offensive that nearly broke the British Army.

Today is Trinity Sunday. Centuries ago words were written to be a call and a spur to the faithful servants of truth and justice: 'Arm yourselves, and be ye men of valour, and be in readiness for the conflict; for it is better for us to perish in battle than to look upon the outrage of our nation and our altar. As the Will of God is in Heaven, even so let it be.'
> Sir Winston S. Churchill, first address as Prime Minister, 19 May 1940, *Winston S. Churchill: His Complete Speeches 1987–1963*, VI, 1974, p. 6223.

Of the 240 men on board this ship, 239 behaved as they ought to have and as I expected them to... One did not. I had him brought before me a couple of hours ago, and he himself informed me that he knew the punishment for desertion of his post could be death. You will therefore be surprised to know that I propose to let him off with a caution, one caution to him and a second one to myself, for having failed in four months to impress my personality and doctrine on each and all of you to prevent such an incident from occurring. From now on I will try to make it clear that I expect every one of you to behave in the way that the 239 did, and to stick to their post in action to the last. I will under no circumstances whatever again tolerate the slightest suspicion of cowardice or indiscipline, and I know from now on that none of you will present me with any such problems.

I want to make it clear to all of you that I shall never give the order 'abandon ship', the only way you can leave the ship is if she sinks beneath your feet.
> Earl Mountbatten of Burma, 1940 (as captain of HMS *Kelly*), quoted in Philip Ziegler, *Mountbatten*, 1985.

Here we will stand and fight; there will be no further withdrawal. I have ordered that all plans and instructions dealing with further withdrawal are to be burnt at once. We will stand and fight here. If we can't stay here alive, then let us stay here dead... Meanwhile, we ourselves will start to plan a great offensive; it will be the beginning of a campaign which will hit Rommel and his army for six, right out of Africa... The great point to remember is that we are going to finish with this chap Rommel once and for all. It will be quite easy. There is no about it. He is definitely a nuisance. Therefore we will hit him a crack and finish with him.

Field Marshal Viscount Montgomery of Alamein, 13 August 1942, order to the Eighth Army upon his assumption of command.

It is essential to impress on all officers that determined leadership will be very vital in this battle, as in any battle. There have been far too many unwounded prisoners taken in this war. We must impress on our officers, N.C.O.'s and men that when they are cut off or surrounded, and there appears to be no hope of survival, they must organize themselves into a defensive locality and hold out where they are. By so doing they will add enormously to the enemy's difficulties; they will greatly assist the development of our own operations; and they will save themselves from spending the rest of the war in a prison camp.

Nothing is ever hopeless so long as troops have stout hearts, and have weapons and ammunition.

Field Marshal Viscount Montgomery of Alamein, 6 October 1942, instructions for Operation Lightfoot, the Battle of El Alamein, quoted in Francis de Guingand, Operation Victory, 1947.

Bob, I'm putting you in command at Buna. Relieve Harding... [all] officers who won't fight... If necessary put sergeants in charge of battalions and corporals in charge of companies – anyone who will fight. Time is of the essence... I want you to take Buna, or not come back alive.

General of the army Douglas MacArthur, 29 November 1942, MacArthur's oral orders to General Robert L. Eichelberger, quoted in Jay Luvaas, ed., Dear Miss Em: General Eichelberger's War in the Pacific, 1972.

There will be no Dunkirk, there will be no Bataan, a retreat to Pusan would be one of the greatest butcheries in history. We must fight until the end. Capture by these people is worse than death itself. We will fight as a team. If some of us must die, we will all die fighting together. Any man who gives ground may be personally responsible for the death of thousands of his comrades.

Major-General Walton Walker, 29 July 1950, his 'stand or die' order as the North Koreans were driving the US Army back toward Pusan, quoted in Omar Bradley, A General's Life, p. 543.

We have always said that in our war with the Arabs we had a secret weapon – no alternative.

Prime Minister Golda Meir, Life, 3 October 1969.

DRILL AND CEREMONIES

Drill is necessary to make the soldier steady and skilful, although it does not warrant exclusive attention.

Field Marshal Maurice Comte de Saxe, My Reveries, 1732, tr. Phillips, 1940.

I had never looked at a copy of tactics from the time of my graduation. My standing in that branch of studies had been near the foot of the class. In the Mexican war in the summer of 1846, I had been appointed regimental quartermaster and commissary and had not been at a battalion drill since.

General of the Army Ulysses S. Grant, The Personal Memoirs of U.S. Grant, I, 1885.

Drill may be beautiful: but beauty is not perceptible when you are expecting a punishment every moment for not doing it well-enough. Dancing is beautiful: – because it's the same sort of

DULCE ET DECORUM EST PRO PATRIA MORI

thing, without the sergeant-major and the 'office'. Drill in the R.A.F. is always punitive: – it is always practice-drill, never exercise-drill or performance-drill. Airmen haven't the time to learn combined rhythm. If they did learn it, their (necessarily) individual work with screwdrivers & spanners would suffer. Rhythm takes months to acquire & years to lose.

> Colonel T.E. Lawrence, 6 November 1928, letter to Robert Graves; Lawrence's perspective at this time was that of an enlisted man, having resigned his wartime commission and enlisted in the RAF; quoted in Malcolm Brown, ed., *The Letters of T.E. Lawrence*, 1988.

The military manifestations of discipline are many and various. At one end of the scale may be placed the outward display, such as saluting and smartness of drill, the meaning and value of which are often misunderstood and misused both inside the Army and outside. Saluting should be in spirit the recognition of a comrade in arms, the respect of a junior for a senior – a gesture of brotherhood on both sides. Good drill should either be a ceremony for the uplifting of the spirit or a time-saver for some necessary purpose – never mere formalism or pedantry. No one who has participated in it or seen it well done should doubt the inspiration of ceremonial drill.

> Field Marshal Viscount Wavell of Cyrenaica, *Soldiers and Soldiering*, 1953.

If men are to give their best in war they must be united. Discipline seeks through drill to instill into all ranks this sense of unity, by requiring them to obey orders as one man. A Ceremonial parade, moreover, provides the occasion for men to express pride in their performance, pride in the Regiment or Corps and pride in the Profession of Arms.

> Field Marshal Viscount Alexander of Tunis, November 1968.

DULCE ET DECORUM EST PRO PATRIA MORI
It is sweet and proper to die for country

So fight by the ships, all together. And
    that comrade
who meets his death and destiny,
    speared or stabbed,
Let him die! He dies fighting for
    fatherland –
no dishonor there!
He'll leave behind wife and sons
    unscathed,
His house and estates unharmed – once
    these Argives
sail for home, the fatherland they love.

> Homer, *The Iliad*, 15.574–579, c. 800 BC, tr. Robert Fagles, 1990; Hector, rallying the Trojans to storm the Greek camp.

As for you, who now survive them, it is your business to pray for a better fate, but to think it your duty to preserve the same spirit and warmth of courage against your enemies; not judging of the expediency of this from a mere harangue – where any man indulging in a flow of words may tell you what you yourselves know as well as he, how many advantages there are in fighting valiantly against your enemies – but, rather making the daily-increasing grandeur of this community the object of your thoughts and growing quite enamored of it. And when it really appears great to your apprehensions, think again that this grandeur was acquired by brave and valiant men, by men who knew their duty, and in the moments of action were sensible of shame; who, whenever their attempts were unsuccessful, thought it no dishonor for their country to stand in need of anything their valor could do for it, and so made it the most glorious present. Bestowing thus their lives on the public, they have every one received a praise that will never decay, a sepulchre that will always be the most illustrious – not that in which their

DULCE ET DECORUM EST PRO PATRIA MORI

bones lie moldering, but that in which their fame is preserved, to be on every occasion, when honor is the employ of either word or act, eternally remembered. For the whole earth is the sepulchre of illustrious men...
> Pericles, 431 BC, funeral oration for the Athenian war dead at the beginning of the Peloponnesian War, quoted in Thucydides, *The History of the Peloponnesian War*, 2.43, c. 404 BC, tr. Richard Crowley, 1910.

And what of those who have died in battle? Their death was noble, their burial illustrious; almost all are commemorated at home by statues of bronze; their parents are held in honour, with all dues of money or service remitted, for under my leadership not a man among you has ever fallen with his back to the enemy.
> Alexander the Great, 323 BC, speech to his troops at Opis in Babylonia, quoted in Arrian, *The Campaigns of Alexander*, 7.10, c. AD 150, tr. Aubrey de Sélincourt, 1987.

Courageous Aztecs, Tezcocans, Tepances, and all the men from the provinces: there is nothing we do here but conquer or die. For this we have come. Our enemy shows valor and a brave heart and has decided to defend the city. I beg of you to demonstrate bravery in this endeavor, for to die is to live forever in perpetual glory and honor.
> Motecuhzoma II Xochoyotl, Aztec Emperor, 1506, exhorting his army before its assault on the city of Quetzaltepec, quoted in Durán, *The History of the Indies of New Spain*, 1581, tr. Doris Heyden, 1994.

I regret the brave men you have lost, but they are dead on the field of honour.
> Napoleon, 16 October 1806, at Weimar, to Marshal Davout after his victory at Auerstadt, quoted in R.M. Johnston, ed., *The Corsican*, 1910.

Easy, but willing to die if God and my country think I have fulfilled my destiny and done my duty.
> Major-General J.E.B. Stuart, when asked how he felt as he lay dying of the wounds suffered at Yellow Tavern the day before, 12 May 1864, quoted by Major-General Fitzhugh Lee, in Early, *Southern Historical Society Papers*, I, January–June 1876.

The warmest instincts of every man's soul declare the glory of the soldier's death. It is more appropriate to the Christian than to the Greek to sing: Glorious his fate, and envied is his lot, Who for his country fights and for it dies.
There is a true glory and a true honor; the glory of duty done – the honor of the integrity of principle.
> General Robert E. Lee, from his post-war writings, quoted in William Jones, *Life and Letters of Robert Edward Lee*, 1906.

If you could hear, at every jolt, the blood
Come gargling from the froth-corrupted lungs,
My friend, you would not tell with such high Zest
To children ardent for some desperate glory,
The old lie: *Dulce et decorum est Pro Patria Mori*
> Wilfred Owen, 'Dulce et decorum est,', 1917, written after his experiences with gas warfare on the Western Front.

It is foolish and wrong to mourn the men who died. Rather we should thank God that such men lived.
> General George S. Patton, Jr, 7 June 1945, attributed, speech at the Copely Plaza Hotel, Boston, Massachusetts.

DUTY

You tell me to put my trust in birds, flying off on their long wild wings?
Never.
I would never give them a glance, a second thought,
whether they fly on the right toward the dawn and sunrise

DUTY

or fly on the left toward the haze and coming dark!
No, no, put our trust in the will of mighty Zeus,
King of the deathless gods and men who die.
Bird signs!
Fight for your country – that is the best, the only omen!
> Homer, *The Iliad*, 12.274–82, c. 800 BC, tr. Robert Fagles, 1990; Hector, rebuking Polydamus for reliance on seers instead of on his sense of duty; a famous quotation from antiquity.

And therefore the general who in advancing does not seek personal fame, and in withdrawing is not concerned with avoiding punishment, but whose only purpose is to protect the people and promote the best interests of his sovereign, is the precious jewel of the state.
> Sun Tzu, *The Art of War*, c. 500 BC, tr. Samuel Griffith.

The sense of duty makes the victor.
> Flavius Vegetius Renatus, *Military Institutions of the Romans*, c. AD 378, tr. Clark, 1776.

I thank all for the advice which you have given me. I know that my going out of the city might be of some benefit to me, inasmuch as all that you foresee might really happen. But it is impossible for me to go away! How could I leave the churches of our Lord, and His servants the clergy, and the throne, and my people in such a plight? What would the world say about me? I pray you, my friends, in future do not say to me anything else but, 'Nay, sire, do not leave us!' Never, never will I leave you! I am resolved to die here with you!
> The Emperor Constantine XI, in response to advice to flee Constantinople before the Turkish siege of 1453, quoted in Mijatovich, *Constantine: The Last Emperor of the Greeks*, 1892.

If I go away to sea,
  I shall return a corpse awash;
If duty calls me to the mountain,
  A verdant sward will be my pall;
Thus for the sake of the Emperor
  I will not die peaceful at home.
> Ancient Japanese war song, quoted in Inoguchi and Nakajimi, *The Divine Wind*, 1958.

Lord Malinche, I have assuredly done my duty in defense of my city and my vassals, and I can do no more. I am brought by force as a prisoner into your presence and beneath your power. Take the dagger that you have in your belt, and strike me dead immediately.
> The Aztec Emperor Cuauhtémoc (1521) brought before Cortes upon his capture, which ended the siege of Tenochtitlan, quoted in Bernal Diaz, *The Conquest of New Spain*, c. 1565.

Duty is the great business of a sea officer; all private considerations must give way to it, however painful it may be.
> Admiral Viscount Nelson, 1786, letter to Frances Nisbet.

I have not a thought on any subject separated from the immediate object of my command.
> Admiral Viscount Nelson (1758-1805).

England expects every man will do his duty.
> Admiral Viscount Nelson, 21 October 1805, order to the fleet before the Battle of Trafalgar.

I am *nimmukwallah*, as we say in the east, that is, I have eaten of the King's salt and, therefore, conceive it to be my duty to serve with unhesitating zeal and cheerfulness where and whenever the King and his Government may think proper to employ me.
> The Duke of Wellington, 1805, upon accepting an obscure posting as brigade

DUTY

commander in a military backwater after his victories as an army commander in India, *The Dispatches of Field Marshal the Duke of Wellington, During His Various Campaigns in India, Denmark, Portugal, Spain, the Low Countries, and France*, IV, 1834–1838, p. 2.

The brave man inattentive to his duty, is worth little more to his country, than the coward who deserts her in the hour of danger.
Major-General Andrew Jackson, 8 January 1815, to troops who had abandoned their lines during the Battle of New Orleans.

And for myself, I have done my duty. I have identified my fate with that of the heroic dead, & whatever lies these sordid exploiteurs of human misery spread about us these officials, there is a right & a God to fight for & our fight has been worth fighting. I do not despair – nor complain – It has been a great cause.
Florence Nightingale, 8 March 1855, Scutari, The Barracks Hospital, letter to her sister, Parthenope, quoted in Goldie, ed., '*I have done my duty': Florence Nightingale in the Crimean War 1954–56*, 1987.

Neither let us be slandered from our duty by false accusations against us, nor frightened from it by menaces of destruction to the Government nor of dungeons to ourselves. LET US HAVE FAITH THAT RIGHT MAKES MIGHT, AND IN THAT FAITH, LET US TO THE END DARE TO DO OUR DUTY AS WE UNDERSTAND IT.
Abraham Lincoln, Cooper Institute Address, New York City, 27 February 1860, *Collected Works*, III, p. 547.

Don't flinch from that fire, boys; there's a hotter fire than that for those who don't do their duty.
Admiral David Glasgow Farragut, during the battle with the forts and flotilla guarding the city of New Orleans on 24 April 1862, quoted in Spears, *David G. Farragut*, 1905.

It is hard to say what would be the most wise policy to pursue toward this people, but for a soldier his duty is plain. He is to obey the orders of all those placed over him and whip the enemy wherever he meets him. 'If he can' should only be thought of after an unavoidable defeat.
General of the Army Ulysses S. Grant, letter to Congressman Elihu B. Washburne, 19 June 1862, after the Battle of Shiloh, quoted in Bruce Catton, *Grant Moves South*, 1960.

Shall I tell you when he was on the Rappahannock, and they telegraphed him his child was dying – his darling little Flora – that he replied that 'I shall have to leave my child in the hands of God; my duty to my country requires me here.'
Major-General J.E.B. Stuart, as recounted by Major-General Fitzhugh Lee, at a A.N.V. Banquet, 28 October 1875, quoted in Early, *Southern Historical Society Papers*, I, January–June 1876, p. 102.

Duty is the sublimest word in our language. Do your duty in all things. You cannot do more. You should never do less.
General Robert E. Lee, quoted in A.L. Long, *Memoirs of Robert E. Lee*, 1886.

Duty is ours; consequences are God's.
Lieutenant-General Thomas J. 'Stonewall' Jackson, one of his favourite maxims, quoted in Mary Anne Jackson, *Life and Letters of General Thomas J. Jackson*, 1892.

Defeat is a common fate of a soldier and there is nothing to be ashamed of in it. The great point is whether we have performed our duty. I cannot but express admiration for the brave manner in which the officers and men of your vessels fought in the late battle for two days continuously. For you, especially, who fearlessly performed your great task until you were seriously wounded, I beg to express my sincerest

DUTY

respect and also my deepest regrets. I hope you will take great care of yourself and recover as soon as possible.
> Admiral Marquis Heihachiro Togo, 3 June 1905, to the wounded Russian Admiral Rodzhesvensky at his bedside after the Battle of Tsushima.

We never fail when we
  try to do our duty –
We always fail when we
  neglect to do it.
> Lieutenant-General Lord Baden-Powell, *Rovering to Success*, 1908.

You are ordered abroad as a soldier of the King to help our French comrades against an invasion of a common enemy. You have to perform a task which will need your courage, your energy, your patience. Remember that the honour of the British Army depends on your individual conduct. It will be your duty not only to set an example of discipline and perfect steadiness under fire but also to maintain the most friendly relations with those whom you are helping in this struggle. In this new experience you may find temptations both in wine and women. You must entirely resist both temptations, and, while treating all women with perfect courtesy, you should avoid any intimacy. Do your duty bravely. Fear God. Honour the King.
> Field Marshal Earl Kitchener of Khartoum, a message to the soldiers of the British Expeditionary Force (BEF), 1914, to be kept by each soldier in his Active Service Pay Book, quoted in Sir G. Arthur, *Life of Kitchener*, III.

The higher the soldier rises on the military ladder, the graver becomes his duty; not in itself, for it only changes form – and no man can do more than his duty – but because to his own duty and his own honour is added the responsibility for the duty and the honour of his subordinates. Responsibility grows to immensity; at one time the lives and the honour of hundreds, of thousands, are at stake; at another, the security of the state itself.
> Colonel-General Hans von Seekt, *Thoughts of a Soldier*, 1930.

The destiny of mankind is not decided by material computation. When great causes are on the move in the world... we learn that we are spirits, not animals, and that something is going on in space and time and beyond space and time, which whether we like it or not, spells duty.
> Sir Winston S. Churchill, 1941, speech in Rochester, New York.

If I do my full duty, the rest will take care of itself.
> General George S. Patton, Jr, 8 November 1942, diary entry before the North African landings, *The Patton Papers*, II, 1974.

'Up until now'; I said, 'you have seen fit to take the Queen's shilling. Now you must stand by to front up and earn it the hard way.' I told them there was no possibility of anyone being allowed to opt out now, that this was actually what you joined the Navy for, whether you knew it or not, chaps. It's too late to change your mind, so best face up to it. The British sailor has a phrase for it, well known to all: 'You shouldn't have joined if you can't take a joke.'
> Admiral Sir J.F. 'Sandy' Woodward, 7 April 1982, address to the ships' crews of the Falklands task force, *One Hundred Days: The Memoirs of the Falklands Battle Group Commander*, 1992, p.80.

E

ECONOMY OF FORCE
Principle of War

One must know how to accept a loss when advisable, how to sacrifice a province (he who tries to defend everything saves nothing) and meanwhile march with *all* one's forces against the other forces of the enemy, compel them to battle, spare no effort for their destruction, and turn them against the others.
> Frederick the Great (1712–1786), quoted in Foch, *The Principles of War*, 1913.

I could lick those fellows any day, but it would cost me 10,000 men, and, as this is the last army England has, we must take care of it.
> The Duke of Wellington, 1810, during the Peninsular Campaign, quoted in Jac Weller, *On Wellington: The Duke and His Art of War*, 1998.

One of these simplified features, or aids to analysis, is always to make sure that all forces are involved – always to ensure that no part of the whole force is idle. If a segment of one's force is located where it is not sufficiently busy with the enemy, or if troops are on the march – that is, idle – while the enemy is fighting, then these forces are being managed uneconomically. In this sense they are being wasted, which is even worse than using them inappropriately. When the time for action comes, the first requirement should be that all parts must act: even the least appropriate task will occupy some of the enemy's forces and reduce his overall strength, while completely inactive troops are neutralized for the time being.
> Major-General Carl von Clausewitz, *On War*, 3.14, 1832, tr. Michael Howard and Peter Paret, 1976.

The principle of economy of forces consists in throwing all one's forces at a given time on one point, in using there all one's troops, and, in order to render such a thing possible, having them always in communication among themselves instead of splitting them and of giving to each a fixed and unchangeable purpose.
> Marshal of France Ferdinand Foch, *The Principles of War*, 1913.

To me, an unnecessary action, or shot, or casualty, was not only waste but sin.
> Colonel T.E. Lawrence (1888–1935).

The Principle of Economy of Force. Economy of force may be defined as the efficient use of all means: physical, moral and material, towards winning a war. Of all the principles of war it is the most difficult to apply, because of its close interdependence on the ever changing conditions of war. In order to economize the moral energy of his men, a commander must not only be in spirit one of them, but must ever have fingers on the pulse of the fighters. What they feel he must feel, and what they think he must think; but while they feel fear, experience discomfort and think in terms of easy victory or disaster, though he must understand what all these mean to the men themselves, he must in no way be obsessed by them. To him economy of

ENDURANCE

force first means planning a battle which his men *can* fight, and secondly, adjusting this plan according to the psychological changes which the enemy's resistence is producing on their endurance without forgoing his objective. This does not only entail his possessing judgment, but also foresight and imagination. His plan must never crystallize, for the energy of the battle front is always fluid. He must realize that a fog, or shower of rain, a cold night or unexpected resistance may force him to adjust his plan, and in order to enable him to do so, the grand tactical economy of force rests with his reserves, which form the staying power of the battle and the fuel of all tactical movement.
   Major-General J.F.C. Fuller, *The Reformation of War*, 1923.

Economy of force rightly means, not a mere husbanding of one's resources in man-power, but the employment of one's force, both weapons and men, in accordance with economic laws, so as to yield the highest possible dividend of success in proportion to the expenditure of strength. While husbanding the lives of the troops is a matter of common sense, full economy of force demands organized common sense – the habit of thinking scientifically, and weighing up tactical values and conditions. Economy of force is the supreme law of successful war.
   Surprise is the psychological and Concentration the physical means to thoroughly *economic* 'offensive action' – blows that will attain the maximum result at the minimum cost. Similarly, *information* and a correct *distribution* of our forces are the economic means to Security. *Flexibility* and *co-operation* increase Mobility – just as by choosing and steering the best course and by oiling and tuning-up the mechanism we get the best speed out of a motor-car. (Oct. 1924.)
   Captain Sir Basil Liddell Hart, *Thoughts on War*, 1944.

In cold truth our small band, which at the most comprised some 300 Europeans and about 11,000 Askari, had occupied a very superior position for the whole war. According to what English officers told me, 137 Generals had been in the field, and in all about 300,000 men had been employed against us. The enemy's losses in dead would not be put too high at 60,000 for the English Press notice stated that about 20,000 Europeans and Indians alone had died or been killed, and to that must be added the large number of black soldiers who fell. The enemy had left 140,000 horse and mules behind in the battle area. Yet in spite of the enormously superior numbers at the disposal of the enemy, our small force, the rifle strength of which was only about 1,400 at the time of the armistice, had remained in the field always ready for action and possessed of the highest determination.
   Major-General Paul von Lettow-Vorbeck, *East African Campaigns*, 1957. Note: The defence of German East Africa (1914–1918) is undoubtedly the greatest demonstration of economy of force in modern military history.

ENDURANCE

Now lead the way, wherever your
   fighting spirit bids you.
All of us right behind you, hearts intent
   on battle.
Nor do I think you'll find us short on
   courage,
long as our strength will last. Past his
   strength
no man can go, though he's set on
   mortal combat.
   Homer, *The Iliad*, 13.908–912, *c.* 800 BC, tr. Robert Fagles, 1990; Paris to his brother Hector during the battle for the Greek ships.

THE ENEMY

What can a soldier do who charges when out of breath?
  Flavius Vegetius Renatus, *Military Institutions of the Romans*, c. AD 378, tr. Clark, 1776.

The Principle of Endurance. Springing directly from the principle of determination is the principle of endurance. The will of the commander-in-chief and the will of his men must endure, that is they must continue in the same state. It is the local conditions, mental and material, which continually weaken this state and in war often threaten to submerge it. To the commander endurance consists, therefore, in power of overcoming conditions – by foresight, judgment and skill. These qualities cannot be cultivated at a moment's notice, and the worst place to seek their cultivation is on the battlefield itself. The commander-in-chief must be, therefore, a mental athlete, his dumbbells, clubs, and bars being the elements of war and his exercises the application of the principles of war to the conditions of innumerable problems.
  Major-General J.F.C. Fuller, *The Reformation of War*, 1923.

Collectively, in an army, endurance is intimately connected with numbers, and, paradoxical as it may seem, the greater the size of an army the less is its psychological endurance. The reason for this is a simple one: one man has one mind; two men have three minds – each his own and a crowd mind shared between them; a million men have millions and millions and millions of minds. If a task which normally requires a million men can be carried out by one man, this one man possesses pyschologically an all but infinitely higher endurance than any single man out of the million. Man, I will again repeat, is an encumbrance on the battlefield, psychologically as well as physically; consequently, endurance should not be sought in numbers, for one Achilles is worth a hundred hoplites.
  Major-General J.F.C. Fuller, *The Reformation of War*, 1923.

The civil comparison to war must be that of a game, a very rough and dirty game, for which a robust mind and body are essential.
  Field Marshal Viscount Wavell of Cyrenaica, quoted in S.L.A. Marshall, *The Armed Forces Officer*, 1950.

THE ENEMY
General

In practice we always base our preparations against an enemy on the assumption that his plans are good; indeed, it is right to rest our hopes not on the belief in his blunders, but on the soundness of our provisions. Nor ought we to believe that there is much difference between man and man, but to think that the superiority lies with him who is reared in the severest school.
  Archidamus II, (c. 469–427 BC), Eurypontid King of Sparta, quoted in *The History of the Peloponnesian War*, 1.84, c. 404 BC, tr. R. Crawley, 1910.

He remarked that the Spartans do not ask how many the enemy are, but where they are.
  Plutarch, paraphrasing Agis II (427–400 BC), son of Archidamus, Eurypontid King of Sparta, *The Lives*, c. AD 100 (*Plutarch on Sparta*, tr. Richard Talbert, 1988).

You need not imagine that victory will be as hard to win as the fame of our antagonists might suggest. Fortune is fickle: often a despised enemy has fought to the death, and a feather in the scale has brought defeat to famous nations and their kings. Take away the blinding brilliance of the name, and in what can the Romans be compared to you?
  Hannibal, 218 BC, address to his army before the Battle of the Ticinus, quoted in Livy, *Universal History*, c. AD 17 (*The War*

THE ENEMY

With Hannibal, tr. Aubrey de Sélincourt, 1972).

For every one must confess that there is no greater proof of the abilities of a general than to investigate, with the utmost care, into the character and natural abilities of his opponent.
   Polybius, *Histories*, 3.81, c. 125 BC (*Familiar Quotations from Greek Authors*, tr. Crauford Ramage, 1895).

It is better to avoid a tricky opponent than one who never lets up. The latter makes no secret of what he is doing, whereas it is difficult to find out what the other is up to.
   The Emperor Maurice, *The Strategikon*, c. AD 600 (*Maurice's Strategikon*, tr. George Dennis, 1984).

A general in all of his projects should not think so much about what he wishes to do as what his enemy will do; that he should never underestimate this enemy, but he should put himself in his place to appreciate difficulties and hindrances the enemy could not interpose; that his plans will be deranged at the slightest event if he has not foreseen everything and if he has not devised means with which to surmount the obstacles.
   Frederick the Great, *Instructions to His Generals*, 3, 1747, tr. Phillips, 1940.

What design would I be forming if I were the enemy?
   Frederick the Great, *General Principles of War*, 1748.

The maxim in war, that your enemy is ever to be dreaded until at your feet, ought to be held inviolate.
   Major-General Henry 'Light Horse Harry' Lee, *Memoirs of the War in the Southern Department*, 1869.

Fight the enemy with the weapons he lacks.
   Field Marshal Prince Aleksandr V. Suvorov, quoted in Ossipov, *Suvorov*, 1945.

In war one sees his own troubles and not those of the enemy.
   Napoleon, 30 April 1809, to Eugene, *Correspondance de Napoléon Ier, publiée par ordre de l'Empereur Napoléon III*, No. 15144, Vol. XVIII, 1858–1870, p. 545.

Napoleon (watching Smolensk burn): An eruption of Vesuvius! Isn't that a fine sight, *Monsieur le Grand Ecuyer*?
Caulaincourt: Horrible, Sire.
Napoleon: Bah! Remember, gentlemen, remember the words of a Roman emperor: 'A dead enemy always smells sweet.'
   Napoleon, the night of 17 August 1812, referring to the words of Vespasian, quoted in Armand de Caulaincourt, *With Napoleon in Russia*, 1935.

Maxim 8. A general-in-chief should ask himself frequently in the day, What should I do if the enemy's army appeared now in my front, or on my right, or on my left? If he has any difficulty in answering these questions he is ill posted, and should seek to remedy it.
   Napoleon, *Military Maxims of Napoleon*, tr. George D'Aguilar, 1831 (David Chandler, ed., 1987).

The President once dropped a few kind words about the enemy. They were human beings – were they not? One could not be completely remorseless, even in war. The line must be drawn somewhere. An elderly woman in the reception room flashed a question; how could he speak kindly of his enemies when he should rather destroy them. 'What, madam?' slowly as he gazed into her face, 'do I not destroy them when I make them my friends?'
   Abraham Lincoln, recounted by Carl Sandburg, *Abraham Lincoln*, III, 1939.

It is always proper to assume that the enemy will do what he should do.

THE ENEMY

While dispositions taken in peace times can be weighted at length, and infallibly lead to the result desired, such is not the case with the use of forces in war, with operations. In war, once hostilities are begun, our will soon encounters the independent will of the enemy. Our dispositions strike against the freely-made dispositions of the enemy.
Field Marshal Helmuth Graf von Moltke, quoted in Foch, *Principles of War*, 1913, p. 100.

I hope you have kept the enemy always in the picture. War-books so often leave them out.
Colonel T.E. Lawrence, 9 February 1928, letter to Colonel A.P. Wavell.

If we try to give an exact answer to the question who deserves the most credit for the victory at Tannenberg, we must also briefly consider the conduct of the enemy, without whose blunders the success would not have been possible.
General Max Hoffmann, *War Diaries and Other Papers*, II, 1929.

Never mistreat the enemy by halves.
Sir Winston S. Churchill (1874–1965).

An important difference between a military operation and a surgical operation is that the patient is not tied down. But it is a common fault of generalship to assume that he is. (May 1934.)
Captain Sir Basil Liddell Hart, *Thoughts on War*, 1944.

However absorbed a commander may be in the elaboration of his own thoughts, it is sometimes necessary to take the enemy into account.
Sir Winston S. Churchill (1874–1965) – attributed.

General Robert E. Lee, letter to Jefferson Davis, quoted in *Douglas Southall Freeman on Leadership*, 1993.

When speaking to a junior about the enemy confronting him, always understate their strength. You do this because the person in contact with the enemy invariably overestimates their strength to himself, so, if you understate it, you probably hit the approximate fact, and also enhance your junior's self-confidence.
General George S. Patton, Jr, *War As I Knew It*, 1947.

A battle plan is good only until enemy contact is made. From then on, your ability to execute the plan depends on what the enemy does. If your assessment of the enemy is good, then you probably have anticipated his actions or reactions and will be able to quickly develop or adjust your plan accordingly. If your assessment is bad, due to poor understanding of how the enemy fights or poor battlefield intelligence, then you will probably get surprised, react slowly, lose the initiative, and cause your unit to be defeated.
General H. Norman Schwarzkopf, 'Food for Thought', *How They Fight*, 1988.

Know Your Enemy
Hence the saying: If you know the enemy and know yourself, you need not fear the result of a hundred battles. If you know yourself but not the enemy, for every victory gained you will also suffer a defeat. If you know neither the enemy nor yourself, you will succumb in every battle.
Sun Tzu, *The Art of War*, 3, c. 500 BC, tr. Giles, 1910.

They are no match for us either in bodily strength or resolution, [and] will find their superiority in numbers of no avail. Our enemies are Medes and Persians, men who for centuries have lived soft and luxurious lives; we of Macedon for generations past have been trained in the hard school of danger and war. Above all, we are free men, and they are slaves.

THE ENEMY

There are Greek troops, to be sure, in Persian service – but how different is their cause from ours! They will be fighting for pay – and not much of it at that; we, on the contrary, shall fight for Greece, and our hearts will be in it.
>Alexander the Great, October 333 BC, exhortation to his men before the Battle of Issus, quoted in Arrian, *The Campaigns of Alexander*, 2.7, c. AD 150, tr. Aubrey de Sélincourt, 1987.

'To become the enemy' means to think yourself into the enemy's position. In the world people tend to think of a robber trapped in a house as a fortified enemy. However, if we think of 'becoming the enemy', we feel that the whole world is against us and that there is no escape. He who is shut inside is a pheasant. He who enters to arrest is a hawk. You must appreciate this.
>Miyamoto Musashi, *A Book of Five Rings*, 1645, tr. Victor Harris, 1984.

Never despise your enemy, whoever he is. Try to find out about his weapons and means, how he uses them and fights. Research into his strengths and weaknesses.
>Field Marshal Prince Aleksandr V. Suvorov, 1789 or 1790, letter, quoted in Philip Longworth, *The Art of Victory*, 1966.

It is very important to know the genius, character and talents of the enemy general; it is on this knowledge that one can develop plans...
>Napoleon (1769–1821), quoted in David Chandler, *The Campaigns of Napoleon*, 1966.

My die is cast... they may overwhelm me but I don't think they will outmanoeuvre me. First, because I am not afraid of them, as everyone else seems to be; and secondly, because if what I hear of their system of manoeuvres be true, I think it a false one, as against steady troops. I suspect all the continental armies are more than half beaten before the battle was begun. I, at least, will not be frightened beforehand.
>The Duke of Wellington, 1808, giving his opinion of the French, upon embarking on the Peninsular Campaign, quoted in J. Crocker, *The Crocker Papers*, I, 1884.

They do not know what they say. If it comes to a conflict of arms, the war will last at least four years. Northern politicians do not appreciate the determination and pluck of the South, and Southern politicians do not appreciate the numbers, resources, and patient perseverance of the North. Both sides forget that we are all Americans, and that it must be a terrible struggle if it comes to war. Tell General Scott that we must do all we can to avert war, and if it comes to the worst we must then do everything in our power to mitigate its evils.
>General Robert E. Lee, May 1861, interview with several Northern visitors after he had resigned his commission and moved to Richmond, quoted in William Jones, *Life and Letters of Robert Edward Lee, Soldier and Man*, 1906.

The natural disposition of most people is to clothe a commander of a large army whom they did not know with almost superhuman abilities. A large part of the National army, for instance, and most of the press of the country, clothed General Lee with just such qualities, but I had known him personally, and knew that he was mortal; and it was just as well that I felt this.
>General Ulysses S. Grant, *Personal Memoirs*, II, 1885.

Most people think Americans love luxury and that their culture is shallow and meaningless. It is a mistake to regard the Americans as luxury-loving and weak. I can tell you Americans are full of the spirit of Justice, fight and adventure. Also their thinking is very

advanced and scientific. Lindbergh's solo crossing of the Atlantic is the sort of valiant act which is normal for them. That is typically American adventure based on science.
> Admiral Isoroku Yamamoto, late 1940, an address to his old middle school in Nagaoaka, quoted in John Deane Potter, *Yamamoto*, 1965.

What kind of people do they [the Japanese] think we are?
> Sir Winston S. Churchill, speech to a joint session of the US Congress, 26 December 1941.

For more than a dozen years I have been frustrated by my inability in formal briefings, lectures, and documents, either in the mud of gunnery ranges or in the comfort of theaters, to adequately transmit anything sufficiently meaningful about the men behind the Soviet guns. When asked what the Soviet military is 'really like,' I have often joked that it's a lot like sex: Much of what you've heard about it isn't true; when it's good, it can be amazing; but when it's bad, it's inexpressibly embarrassing.
> Lieutenant-Colonel Ralph Peters, *Red Army*, 1980.

Respect Your Enemy

Buonaparte had never yet had to contend with an English officer, and he [Nelson] would endeavour to make him respect us.
> Admiral Viscount Nelson (writing in the third person), to the British authorities in Bombay notifying them of his victory at the Battle of the Nile, 1 August 1798, quoted in Robert Southey, *Life of Nelson*, 1813.

I hope that you will be introduced to Buonoparte. You will hear him plentifully abused here – but I own tho' he has done many acts which are unpardonable, yet I cannot help having admiration and respect for him.
> General Sir John Moore, early 1803, letter to Captain Gardiner, quoted in Carola Oman, *Sir John Moore*, 1953.

Moore is the only General now worthy to contend with me. I shall move against him in person.
> Napoleon, 19 December 1808, when he learned that Moore was audaciously moving on the French lines of communication in Spain, quoted in Carola Oman, *Sir John Moore*, 1953.

Nor blame them too much... nor us for not blaming them more. Although, as we believed, fatally wrong in striking at the old flag, misreading its deeper meaning and the innermost law of the people's life, blind to the signs of the times in the march of man, they fought as they were taught, true to such ideals as they saw, and put into their cause their best. For us they were fellow-soldiers as well, suffering the fate of arms. We could not look into those brave, bronze faces, and those battered flags we had met on so many fields where glorious manhood lent a glory to the earth that bore it, and think of personal hate and mean revenge. Whosoever had misled these men, we had not. We had led them back, home.
> Major-General Joshua L. Chamberlain, *The Passing of the Armies*, 1915.

As my command, in worn-out shoes and ragged uniforms, but with proud mien, moved to the designated point to stack their arms and surrender their cherished battle-flags, they challenged the admiration of the brave victors. One of the knightliest soldiers of the Federal army, General Joshua L. Chamberlain of Maine, who afterward served with distinction as governor of his State, called his troops into line, and as my men marched in front of them, the veterans in blue gave a soldierly salute to those vanquished heroes – a token of respect from Americans to Americans, a final and fitting tribute from Northern

THE ENEMY

to Southern chivalry.
General John B. Gordon, *Reminiscences of the Civil War*, 1903.

So 'ere's to you, Fuzzy-Wuzzy, at your 'ome in the Soudan;
You're a pore benighted 'eathen but a first-class fighting man;
An' ere's to you, Fuzzy-Wuzzy, with your 'ayrick 'ead of 'air –
You big black boundin' begger – for you broke a British Square!
Rudyard Kipling, 'Fuzzy-Wuzzy'.

Newspaper libels on Fritz's courage and efficiency were resented by all trench-soldiers of experience.
Robert Graves, *Goodbye to All That*, 1929.

Those heroes that shed their blood and lost their lives… You are now lying in the soil of a friendly country, therefore rest in peace. There is no difference between the Jonnies and the Mehmets to us where they lie side by side. Here in this country of ours… You, the mothers, who sent their sons from far away countries wipe away your tears. Your sons are now lying in our bosom and are in peace. After having lost their lives on this land they have become our sons as well.
Ataturk, a plaque erected after World War I to the fallen Australians and New Zealanders at ANZAC Cove, Gallipoli.

I have always said what I here expressly repeat in print, that the German people exhibited such astounding energy, doggedness, sturdy patriotism, courage, endurance, discipline, and readiness to die for their country, that, as a soldier, I cannot but salute them. They fought with lion-like bravery and an amazing tenacity against the whole world. The German soldier – and this means the German people – deserves unstinted respect.
General Aleksei A. Brusilov, *A Soldier's Notebook*, 1930.

Exceptions [to the panic among the Turkish troops] were the German detachments; and there, for the first time, I grew proud of the enemy who had killed my brothers. They were two thousand miles from home, without hope and without guides, in conditions mad enough to break the bravest nerves. Yet their sections held together, in firm rank, sheering through the wrack of Turk and Arab like armoured ships, high-faced and silent. When attacked they halted, took position, fired to order. There was no haste, no crying, no hesitation. They were glorious.
Colonel T.E. Lawrence, *The Seven Pillars of Wisdom*, 1935.

Remember that these enemies, whom we shall have the honor to destroy, are good soldiers and stark fighters. To beat such men, you must not despise their ability, but you must be confident in your own superiority.
General George S. Patton, Jr, December 1941, *The Patton Papers*, II, 1974.

The willingness of the Allied pilots to engage us in combat deserves special mention here, for, regardless of the odds, their fighters were always screaming in to attack. And it is important to point out that their fighter planes were clearly inferior in performance to our own Zeros. Furthermore, almost all of our pilots were skilled air veterans; coupled with the Zero's outstanding performance, this afforded us a distinct advantage. The men we fought then were among the bravest I have ever encountered, no less so than our own pilots who, three years later, went out willingly on missions from which there was no hope of return.
Saburo Sakai, *Samurai!*, 1957; describing the air battles around Lae, 180 miles north of Port Moresby in 1942.

I never underestimate the enemy… I look at his capabilities and assume he

ENERGY

has them until we find out differently.
> General H. Norman Schwarzkopf, 1991, quoted in Peter David, *Triumph in the Desert*, 1991.

If you are ferocious in battle, remember to be magnanimous in victory. We go to liberate, not to conquer. We are entering Iraq to free a people, and the only flag that will be flown in that ancient land is their own. Don't treat them as refugees, for they are in their own country. If there are casualties of war, then remember, when they woke up and got dressed in the morning they did not plan to die this day. Allow them dignity in death. Bury them properly and mark their graves. You will be shunned unless your conduct is of the highest order, for your deeds will follow you down history. Iraq is steeped in history. It is the site of the Garden of Eden, of the Great Flood and the birth of Abraham. Tread lightly there.'
> Lieutenant Colonel Tim Collins, British Army, appointed 18 March 2003, to his command before the beginning of Operation IRAQI FREEDOM.

ENERGY

Generals wage war through the armed forces, the armed forces fight by energy. Energy prevails when it is drummed up. If you can energize your troops, don't do it too frequently, otherwise their energy will easily wane. Don't do it at too great a distance either, otherwise their energy will be easily exhausted. You should drum up the energy of your soldiers when enemies are within a calculated critical distance, having your troops fight at close range. When enemies wane and you prevail, victory over them is assured.

The rule is 'Fight when full of energy, flee when drained of energy.'
> Liu Ji (1310–1375), *Lessons of War* (*Mastering the Art of War*, tr. Thomas Cleary, 1989).

If you wage war, do it energetically and with severity. This is the only way to make it shorter, and consequently less inhuman.
> Napoleon, 1799, letter to General Hedouville, quoted in Christopher Herold, ed, *The Mind of Napoleon*, 1955.

If we read history with an open mind, we cannot fail to conclude that, among all the military virtues, *the energetic conduct of war* has always contributed most to glory and success.
> Major-General Carl von Clausewitz, *On War*, 4.3, 1832, tr. Michael Howard and Peter Paret, 1976.

The true speed of war is not headlong precipitancy, but the unremitting energy which wastes no time.
> Rear Admiral Alfred Thayer Mahan (1840–1914).

The power to vivify an undertaking implies the energy sufficient to shoulder the burden of its consequence.
> General Charles de Gaulle, *The Edge of the Sword*, 1932.

Mental conception must be followed by immediate execution. This is a matter of energy and initiative. What the soldier needs is a combination of realist intellect and energy. Whatever is attempted must be carried through. The young officer must understand from the outset of his training that just as much energy is required of him as mental ability.
> Field Marshal Erwin Rommel, *The Rommel Papers*, 1953.

ENGINEERS/SAPPERS

Good engineers are so scarce, that one must bear with their humours and forgive them because we cannot be without them.
> Lord Galway, 1704, report from Spain to the Board of Ordnance, quoted in David Chandler, *The Art of War in the Age of Marlborough*, 1976.

ENVELOPMENT

Their science demands a great deal of courage and spirit, a solid genius, perpetual study and consumate experience in all the arts of war.
   Marshal of France Sebastien de Vauban, *A Manual of Siegecraft*, 1740.

The sappers raise permanent and field fortifications; they attack and defend fortresses. They accompany the army to war. On these three counts alone one can rest assured that no other military organization in the World possesses a higher degree of ability and patriotism than exists in our corps of engineers. Alexandria, Antwerp, Juliers – and five hundred more places constructed, restored or augmented – these achievements prove that the arts of Vauban have not fallen into decadence in the hands of the Marescots, the Chasseloups and the Haxos. All Europe has been covered by our redoubts and entrenchements.
   Napoleon (1769–1821), quoted in J. Ambert, *Equises d l'Armée Française*, 1840.

The siege of Badajoz was a most serious undertaking, and the weather did not favour us. The troops were up to their middles in mud in the trenches... The Assault was a terrible business, of which I saw the loss when I was ordering it. But we had brought matters to that state that we could do no more, and it was necessary to storm or raise the siege. I trust, however, that future armies will be equipped for sieges, with the people necessary to carry them on as they ought to be; and that our engineers will learn how to put their batteries on the crest of the glaçis, and to blow in the counterscarp, instead of placing them wherever the wall can be seen, leaving the poor officers and troops to get into and across the ditch as they can...
   The Duke of Wellington, 28 May 1812, to Major-General Murray on the absence of professional engineers, quoted in Julian Rathbone, *Wellington's War: His Peninsular Dispatches*, 1984.

Clearly, we would need great skill and even greater courage. But our first need was for men and equipment. It will be seen why the creation of engineer units was a top priority. In just over two years we succeeded in creating and training almost 40 engineer battalions, some of them highly specialized. It was our biggest coup and the foundation of our success. In his memoirs, the Israeli Chief of Staff, General David Elazaar, records that during a discussion in the Israeli High Command of the possibility of Egyptian forces attempting to cross the canal, General Dayan, the Israeli Defense Minister, said: 'To cross the canal the Egyptians would need the support of both the American and the Soviet engineer corps.' I do not hold Dayan's dismissive remark against him. I accept the tribute. I know how hard we worked.
   General Saad El Shazly, *The Crossing of the Suez*, 1980.

No Corps was more constantly in demand, so much master of so many tasks.
   General Sir David Fraser, of the Royal Engineers, *And We Shall Shock Them*, 1983.

Ubique (Everywhere); *Quo fas et gloria ducunt* (Wherever right and glory lead).
   Mottos of the Corps of Royal Engineers.

Essayons! (Let's try!)
   The motto of the US Army Corps of Engineers.

Domitius Corbulo used to say that the pick was the weapon with which to beat the enemy.
   Frontinus, *Stratagems*, 4.7, c. AD 84–96, tr. Charles E. Bennett, 1925; speaking of the distinguished Roman general who successfully campaigned against the Parthians in the reign of Nero.

ENVELOPMENT

It is an invariable axiom of war to secure your own flanks and rear and

endeavor to turn those of your enemy.
> Frederick the Great, *Instructions to His Generals*, 1747, tr. Phillips, 1940.

A general should show boldness, strike a decided blow, and maneuver upon the flank of his enemy. The victory is in his hands.
> Napoleon, *Maxims of War*, 1831.

The greater conditions of warfare have remained unchanged. The battle of extermination may be fought today according to the same plan as elaborated by Hannibal in long forgotten times. The hostile front is not the aim of the principle attack. It is not against that point that the troops should be massed and the reserves disposed; the essential thing is to crush the flanks.
> Field Marshal Alfred Graf von Schlieffen, *Cannae*, 1913.

Envelopment. An enemy may be enveloped and so placed at a severe disadvantage. Envelopment, whether accomplished by converging or overlapping, presupposes a flank, a flank which may be tactically rolled up, or, if turned, will expose the command and lines of communication behind it. The attack by envelopment is a very common action in war, which more often than not has led to victory.
> Major-General J.F.C. Fuller, *The Reformation of War*, 1923.

If I had worried about flanks, I could never have fought the war.
> General George S. Patton, Jr., *War As I Knew It*, 1947.

We just went right around the enemy and were behind him in no time at all.
> General H. Norman Schwarzkopf, 1991, of the great envelopment in the Gulf War, quoted in Peter David, *Triumph in the Desert*, 1991.

EPITAPHS AND MEMORIALS

O man, I am Cyrus son of Cambyses, who founded the empire of Persia and ruled over Asia. Do not grudge me my monument.
> The inscription over the tomb of Cyrus the Great (c. 600–529 BC), quoted in Arrian, *The Campaigns of Alexander*, 6.29, c. AD 150, tr. Aubrey de Sélincourt.

Go, tell the Spartans, thou who passeth by,
That here obedient to their will we lie.
> Simonides of Keos, the inscription on the monument at the Pass of Thermopylae for the 300 Spartans who fell in 470 AD during the invasion of Greece by the Persians, quoted in Mackail, ed., *Select Epigrams*.

Into the dark death cloud they passed,
  to set
Fame on their own dear land for fadeless wreath,
and dying died not. Valour lifts them yet
Into the splendour from the night beneath.
> Simonides, on the Spartan monument at Platea, 479 BC, tr. H. Macnaghten, quoted in *The Oxford Book of Greek Verse in Translation*, 1930.

If Valour's best be gallantly to die,
Fortune to us of all men grants it now.
We to set Freedom's crown on Hellas' brow
Laboured, and here in ageless honour lie.
> Simonides, on the Athenian monument at Platea, 480 BC, tr. W.C. Lawton, quoted in *The Oxford Book of Greek Verse in Translation*, 1930.

No Athenian, through my means, ever wore mourning.
> Pericles, 429 BC, his own epitaph on his deathbed, quoted in Plutarch, *Plutarch's Lives*, c. AD 100, tr. John Dryden.

For if I have accomplished any glorious feat, that will be my memorial. But if I have not, not even all the statues in the

EPITAPHS AND MEMORIALS

world – the products of vulgar, worthless men – would make any difference.
> Agesilaus, Eurypontid King of Sparta (400–360 BC), to his servants as he lay dying, quoted in Plutarch, *The Lives*, c. AD 100 (*Plutarch on Sparta*, tr. Richard Talbert, 1988).

Stranger, this man you behold was the guiding star of his country,
Claudius Marcellus by name, born of a glorious line;
Seven times consul, he led the armies of Rome into battle,
Death and destruction he dealt to all who invaded his land.
> Inscription on the statue of Claudius Marcellus, one of Hannibal's most dangerous foes, who fell in battle against the famed Carthaginian in 208 BC, quoted in Plutarch, *The Lives*, c. AD 100 (*Makers of Rome*, tr. Ian Scott-Kilvert, 1965).

From the sun rising above the marshes of Maeotia
There is no one who may equal me in deeds.
If it is right for anyone to rise into the region of the gods,
For me alone the greatest gate of heaven stands open.
> Epitaph of Scipio Africanus by Quintus Ennius.

A mighty name will remain behind me in the world.
> Genghis Khan, quoted in Jack Weatherford, *Genghis Khan and the Making of the Modern World*, 2004, p.129.

Here stood fewer than sixty men against an entire army. Its weight overwhelmed them. Life, sooner than courage, forsook these soldiers of France.
> The inscription to the last stand of the Third Company of the Foreign Legion Regiment at Cambrone, Mexico, on 30 April 1863, a desperate action so heroic that it became the Legion's founding legend.

In memory of the brave warriors who fell here in 1879 in defence of the old Zulu order.
> Inscription on the Ulundi battlefield monument, South Africa.

They shall grow not old, as we that are left grow old;
Age shall not weary them, nor the years condemn.
At the going down of the sun and in the morning
We will remember them.
> Laurence Binyon, 'For the Fallen', *The Winnowing Fan: Poems of the Great War in 1914*. This verse, which became the Ode for the Returned and Services League, has been used in association with commemoration services in Australia since 1921.

A Soldier of the Great War, known unto God.
> Inscription on the grave stones of the unidentified British dead of World War I, quoted in M.J. Cohen, *History in Quotations*, 2004

Here Rests In Honored Glory An American Soldier Known But To God.
> Inscription on the Tomb of the Unknown Soldier at Arlington Cemetery; the Unknown Soldier was interred on 11 November 1921, and the tomb itself built in 1931 and unveiled in April 1932.

When you go home
Tell them of us and say,
For their tomorrow
We gave our today.
> Inscription on the monument of the British Second Division on Garrison Hill at Kohima, India, commemorating the 17,587 British casualties at the Battle of Kohima, 5 April–30 May 1944.

This embattled shore, portal of freedom, is forever hallowed by the ideals, the valor and the sacrifices of our fellow countrymen.
> Inscription on the monument to the

ESPRIT DE CORPS

American dead on the Normandy battlefield, commemorating the Normandy Campaign, 6 June–1 August 1944.

'Not in vain' may be the pride of those who survived and the epitaph of those who fell.
> Sir Winston S. Churchill, 28 September 1944, speech in the House of Commons in tribute to the British First Airborne Division's sacrifice at the Battle of Arnhem.

ESPRIT DE CORPS

All that can be done with the soldier is to give him *esprit de corps* – i.e., a higher opinion of his own regiment than all the other troops in the country.
> Frederick the Great, *Military Testament*, 1768.

Personal bravery of a single individual does not decide on the day of battle, but the bravery of the unit, and the latter rests on the good opinion and the confidence that each individual places in the unit to which he belongs. The exterior splendour, the regularity of movements, the adroitness, and at the same time the firmness of the mass – all this gives the individual soldier the safe and calming conviction that nothing can withstand his particular regiment or battalion.
> Field Marshal Francois Comte de Guibert (1744–1790), quoted in Keegan, *Soldiers*, 1986.

...(*Esprit de Corps*) which does not forget itself in the heat of action and which alone makes true combatants. Then we have an army; and it is no longer difficult to explain how men carried away by passion, even men who know how to die without flinching, without turning pale, really strong in the presence of death, but without discipline, without solid organization, are vanquished by others individually less valiant, but firmly, jointly, and severally knit into a fighting unit.
> Colonel Charles Ardant du Picq, *Battle Studies*, 1880, tr. Greely, 1957.

I have alluded before to the *esprit de corps*, founded as it was upon the sentiment of saving life – sentiment to which appeal has never failed. Other factors went to strengthen it. It was braced by a high standard of results demanded, by determination to make good in spite of partial first success. But the strongest element in it was the faith in our weapon – the machine necessary to supplement the other machines of war, in order to break the stalemate produced by the great German weapon, the machine-gun – our mobile offensive answer to the immobile defensive man-killer.
> Major-General Hugh Elles, Commander of the British Tank Corps in WWI, quoted in the preface to William Ellis, *The Tank Corps*, 1919.

Well, the main thing – that I have remembered all my life – is the definition of *esprit de corps*. Now my definition – the definition I was taught, that I've always believed in – is that *esprit de corps* means love for one's military legion, in my case the United States Marine Corps. I also learned that this loyalty to one's Corps travels both ways, up and down.
> Lieutenant-General Lewis B. 'Chesty' Puller, 13 August 1956, quoted in Burke Davis, *Marine!*, 1962.

EXAMPLE

I charged all countries, while I was alone, my infantry and my chariotry having forsaken me. Not one among them stood to turn about. I swear, as Re Loves me, as my father, Atum, favors me, that as for every matter which his majesty has stated, I did it in truth, in the presence of my infantry and my chariotry.
> Rameses II, 1284 BC, in his own words of the Battle of Kadesh when the Hittites

EXAMPLE

surprised his camp and cut him off from even his escort; he rallied his men by charging single-handed into the mass of Hittite chariots to cut a way through, quoted in James Breasted, *Ancient Records of Egypt*, 1906.

The general shares heat and cold, labor and suffering, hunger and satiety with the officers and men. Therefore when the masses of the Three Armies hear the sound of the drum they are happy, and when they hear the sound of the gong they are angry. When attacking a high wall or crossing a deep lake, under a hail of arrows and stones, the officers will compete to be first to scale the wall. When the naked blades clash, the officers will compete to be the first to go forward. It is not because they like death and take pleasure in being wounded, but because the general knows their feelings of heat and cold, hunger and satiety, and clearly displays his knowledge of their labor and suffering.
  The T'ai Kung, c. 1100 BC (*The Seven Military Classics of Ancient China*, tr. Ralph Sawyer, 1993).

Before the army's watering hole has been reached, the commander does not speak of thirst; before the fires have food on them, the commander does not speak of hunger.
  Sun Tzu, c. 500 BC, a passage recovered from later encyclopaedic works and commentaries not found in *The Art of War*, quoted in *Sun Tzu: The Art of Warfare*, tr. Roger Ames, 1993.

All these soldiers have their eyes on you. If they see you are discouraged they will all be cowards; but if you show that you are making preparations against the enemy, and if you call on them, you may be sure they will follow you and try to imitate you. Perhaps it is fair to expect you to be a bit better than they are. You are captain, you see, and you are in command of troops and companies, and while there was peace, you had more wealth and honour; then now when war has come, we must ask you to be better than the mob, and to plan and labour for their behalf, if necessary.
  Xenophon, rallying the few surviving officers of the Ten Thousand to action, after most of the officers were murdered under truce by the Persians in 401 BC, *The March Up Country*, c. 360 BC, tr. W.H.D. Rouse.

Perhaps you will say, that in my position as your commander, I had none of the labours and distress which you had to endure to win for me what I have won. But does any man among you honestly feel that he has suffered more for me than I have suffered for him? Come now – if you are wounded, strip and show your wounds, and I will show mine. There is no part of my body but my back which has not a scar; not a weapon a man may grasp or fling the mark of which I do not carry upon me. I have sword-cuts from close fight; arrows have pierced me, missiles from catapults bruised my flesh; again and again I have been struck by stones or clubs – and all for your sakes: for your glory and your gain. Over every land and sea, across river, mountain, and plain I led you to the world's end, a victorious army.
  Alexander the Great, 323 BC, to his troops at Opis in Babylonia, quoted in Arrian, *The Campaigns of Alexander*, 7.10, c. AD 150, tr. Aubrey de Sélincourt, 1987.

In carrying out very critical operations the general ought not set himself apart as though such labor was beneath him, but he should begin the work and toil along with his troops as much as possible. Such behavior will lead the soldier to be more submissive to his officers, even if only out of shame, and he will accomplish more.
  The Emperor Maurice, *The Strategikon*, c.

EXAMPLE

AD 600 (*Maurice's Strategikon*, tr. George Dennis, 1984).

Essential to generalship is to share the pleasures and pains of the troops. If you encounter danger, do not abandon the troops to save yourself, do not seek personal escape from difficulties confronting you. Rather, make every effort to protect the troops, sharing their fate. If you do this, the soldiers will not forget you.

The rule is 'When you see danger and difficulty, do not forget the troops' ('Sima's Art of War').

Liu Ji (1310–1375), *Lessons of War* (*Mastering the Art of War*, tr. Thomas Cleary, 1989).

I used to say to them, 'go boldly in among the English,' and then I used to go boldly in myself.

Joan of Arc, 3 March 1430, to her inquisitors who intimated that her success was due to luck, quoted in Edward Creasy, *Decisive Battles of the World*, 1899.

I, your king and leader, will go ahead and show every one the path of honor.

Gustavus II Adolphus, 14 November 1632, to his commanders on the day before the Battle of Lützen, where he met his death, quoted in Theodore Ayrault Dodge, *Gustavus Adolphus*, 1895.

I will not send troops to dangers which I will not myself encounter.

The Duke of Marlborough (1650–1722), quoted in Sir Winston S. Churchill, *Marlborough*, 1933.

I am now at an age when I find no heat in my blood that gives me temptation to expose myself out of vanity; but as I would deserve and keep the kindness of this army, I must let them see that when I expose them, I would not exempt myself.

The Duke of Marlborough (1650–1722), quoted in Sir Winston S. Churchill, *Marlborough*, 1933.

Drill your soldiers well, and give them a pattern yourself.

Alexander V. Suvorov (1729–1800), quoted in W. Lyon Blease, *Suvorof*, 1920.

Well, I will let you see that I was a grenadier before I was a marshal, and still am one!

Marshal of France Jean Lannes, 23 April 1809, seizing a scaling ladder and advancing alone on the Austrian-held fortifications of Ratisbonne when his men hesitated, quoted in Baron Marbot, *The Memoirs of Baron Marbot*, I, 1892.

As if the forces in one individual after another become prostrated, and can no longer be excited and supported by an effort of his own will, the whole inertia of the mass gradually rests its weight on the will of the commander: by the spark in his breast, by the light of his spirit, the spark of purpose, the light of hope, must be kindled afresh in others.

Major-General Carl von Clausewitz, *On War*, 1832, tr. Michael Howard and Peter Paret, 1976.

I cannot consent to be feasting while my poor soldiers are nearly starving.

General Robert E. Lee, upon refusing all but the plainest fare when served an elegant dinner on a visit to a great house, quoted in William Jones, *Personal Reminiscences*, 1874.

Jackson is with you… Rally, brave men, and press forward. Your general will lead you. Jackson will lead you. Follow me!

Lieutenant-General Thomas J. 'Stonewall' Jackson, 9 August 1862, rallying his shattered troops during the Battle of Cedar Mountain, quoted in David Martin, *The Second Bull Run Campaign*, 1997.

As to my exposing myself unnecessarily, you need not be concerned. I know better than C—— where danger lies and where I should be. Soldiers have a right to see and know that the man who guides them is near enough to see

EXAMPLE

with his own eyes, and that he cannot see without being seen.
> General William T. Sherman, letter to his wife from a camp near Vicksburg, 28 January 1863, *Home Letters*, 1909.

There are times when a corps commander's life does not count.
> Major-General Winfield Scott Hancock, 3 July 1863, remark to an officer who cautioned him not to expose himself as he rode along the stone wall at the High Water Mark at Gettysburg during the Confederate bombardment, as an example to the men of his glorious Second Corps awaiting Pickett's Charge.

Regular discipline will bring men to any amount of endurance, but there is a natural fear of hidden dangers, particularly when so awfully destructive of human life as the torpedo, which requires more than discipline to overcome... After I saw the Tecumseh, struck by a torpedo, disappear almost instantly, beneath the waves... I determined at once, as I had originally intended to take the lead; and... I dashed ahead with the Hartford and the ships followed on, their officers believing they were going to a noble death with their commander-in-chief.
> Admiral David G. Farragut, report of the Battle of Mobile Bay, 5 August 1864, quoted in John Spears, *David G. Farragut*, 1905. A 'torpedo' was what is now known as a 'mine'.

I have never, on the field of battle, sent you where I was unwilling to go myself; nor would I now advise you to a course which I felt myself unwilling to pursue. You *have been* good soldiers; you *can be* good citizens. Obey the laws, preserve your honor, and the Government to which you have surrendered can afford to be, and will be magnanimous.
> Lieutenant-General Nathan Bedford Forrest, 9 May 1865, farewell to Forrest's Cavalry Corps at Gainesville, Georgia, quoted in *As They Saw Forrest*, 1956.

You can forget all your training; you have come out here to show your men how to die.
> Anonymous commanding officer of a British Guards battalion to a newly joined subaltern in 1917, quoted in Major General E.K.G. Sixsmith, *British Generalship in the Twentieth Century*, 1970, p. 99.

You should never forget the power of example. The young men serving as enlisted men take their cue from you. If you conduct yourselves at all times as officers and gentlemen should conduct themselves, the moral tone of the whole Corps will be raised, its reputation, which is most precious to all of us, will be enhanced, and the esteem and affection in which the Corps is held by the American people will be increased.
> Major-General John A. Lejeune, 19 September 1922, Letter No. 1. to the Officer of the Marine Corps from the Commandant.

Thus we see how surely the physical is the foundation of the moral, and how these physical defects, for defects they are in war, react upon a general's moral sense by subordinating it to intellectual achievements. More and more do strategical, administrative and tactical details occupy his mind and pinch out the moral side of his nature. Should he be a man of ability, he becomes a thinker rather than a doer, a planner rather than a leader, until morally he is as far removed from his men as a chess player is from the chessmen on his board. The more he is thrown back upon the intellectual side of war, the more sedentary he becomes, until a kind of military scholasticism enwarps his whole life.

The repercussion of such generalship on subordinate command has always been lamentable, because whatever a general may be, he is always the example

EXAMPLE

which the bulk of his subordinate commanders will follow. If he becomes an office soldier, they become office soldiers; not only because his work makes their work, but because his morale makes their morale: how can he order them into danger if he remains in safety? If the general-in-chief does not face discomfort and danger neither will they; if they do not, neither will their subordinates, until the repercussion exhausts itself in a devitalized firing line.

   Major-General J.F.C. Fuller, *Generalship: Its Diseases and Their Cure*, 1933.

Be an example to your men, both in your duty and in private life. Never spare yourself, and let the troops see that you don't, in your endurance of fatigue and privation. Always be tactful and well mannered and teach your subordinates to be the same. Avoid excessive sharpness or harshness of voice, which usually indicates the man who has shortcomings of his own to hide.

   Field Marshal Erwin Rommel, 1938, remarks to graduating cadets of the Wiener Neustadt Military School.

Be a model of valor by example and precept.

   Lieutenant-General Lewis B. 'Chesty' Puller, 1942, to his officers and NCOs before the landing of the First Marine Division on Guadalcanal, quoted in the Puller Collection, Marine Corps Historical Division.

Fear is always worse when man is isolated; it is least prominent when following an example. This consideration points to the true cellule of combat on which tactics should be built up. The little compact group formed of men who have long shared their training and recreation, following the leader they know and who has trained them, is the ideal formation for battle. Example is better, as an antidote to fear, than the companionship of equals.

   Captain Sir Basil Liddell Hart, *Thoughts on War*, 1944.

In cold weather, General Officers must be careful not to appear to dress more warmly than the men.

   General George S. Patton, Jr, *War As I Knew It*, 1947.

Only those who have disciplined themselves can exact disciplined performance from others. When the chips are down, when privation mounts and the casualty rates rise, when the crisis is at hand, which commander, I ask, receives the better response? Is it the one who has failed to share the rough going with his troops, who is rarely seen in the zone of aimed fire, and who expects much and gives little? Or is it the one whose every thought is for the welfare of his men, consistent with the accomplishment of the mission; who does not ask them to do what he has not already done and stands ready to do again when necessary; who with his men has shared short rations, the physical discomforts and rigors of campaign, and will be found at the crisis of action where the issues are to be decided?

   General Matthew B. Ridgway, 'Leadership', *Military Review*, 10/1966.

The power of example is very important to people under stress. For one thing it affords an outlet for hero worship, to which there seems to be an important and deep-rooted inclination in men. The person under stress is aware of inadequacies. He sees someone else apparently less burdened in this way. To some extent he identifies with that other person. This gives him some release. He is then likely to be grateful and become even more biddable. He will be even more open to the influence of suggestion and example than he was

EXHORTATIONS

before.
> General Sir John Hackett, *The Profession of Arms*, 1983.

I can't command these soldiers unless I'm willing to take the same risks that they do.
> Lieutenant-General William Scott Wallace, V Corps Commander, while walking about the streets of Baghdad, quoted on CBS Evening News, 29 May 2003.

EXERCISES/MANOEUVRES

So sensible were the Romans of the imperfections of valour without skill and practice that, in their language, the name of an Army, was borrowed from the word which signified exercise.
> (*Exercitus* in Latin means Army) Edward Gibbon, *Decline and Fall of the Roman Empire*, 1776.

[Exercises] cultivated self-reliance – the foundation of courage.
> Field Marshal Prince Aleksandr V. Suvorov (1729–1800), quoted in K. Ossipov, *Suvorov*, 1945.

If the exercise is subsequently discussed in the officers' mess, it is probably worth while; if there is argument over it in the sergeants' mess, it is a good excercise; while if it should be mentioned in the corporals' room, it is an undoubted success.
> Field Marshal Viscount Wavell of Cyrenaica, *Journal of the Royal United Services Institution*, May 1933.

To me, as to Clausewitz, 'war lies in the realm of chance'; then, surely, should peace exercises lie there also. Every problem, I held, is a new picture, yet set in an old frame – the principles of war. Peace exercises should, therefore, test out wits, and not merely knowledge of the regulations. The unexpected is, consequently, their essence; and to make certain of the unexpected, all that is necessary is to arrange a clash, and then leave it to Commanders to fight it out; not to win, but to prove which of the two is the livest man can apply the principles in the livest way.
> Major-General J.F.C. Fuller, *Memoirs of an Unconventional Soldier*, 1936.

War makes extremely heavy demands on the soldier's strength and nerves. For this reason make heavy demands on your men in peacetime exercises.
> Field Marshal Erwin Rommel, *Infantry Attacks*, 1944.

There are no bullets in maneuvers, and things sometimes get a little dull. But play the game; don't lie in the shade… Try, above all things, to use your imagination. Think this is war. 'What would I do if that man were really shooting at me?' That is the only chance, men, that you are going to have to practice. The next time, maybe, there will be no umpires, and the bullets will be very real, both yours and the enemy's.
> General George S. Patton, Jr, 17 May 1941, address to the officers and men of the Second Armored Division, *The Patton Papers*, II, 1974.

EXHORTATIONS

Be men, my friends! Discipline fill your
    hearts,
maintain your pride in the eyes of other
    men!
Remember, each of you, sons, wives,
    wealth, parents –
are mother and father dead or alive? No
    matter,
I beg you for their sakes, loved ones far
    away –
now stand and fight, no turning back,
    no panic.
> Homer, *The Iliad*, 15.769–774, tr. Robert Fagles, 1990; King Nestor of Pylos exhorting the Achaean Army at bay to stand and fight.

He… said that there was no need for

EXHORTATION

him to inspire them to the fight; they had long ago been inspired by their own bravery and by their many splendid exploits in the past; but he required each of them to encourage his own men; the infantry captains their companies, the squadron commanders their own squadrons, the battalion commanders their battalions, and the infantry commanders the phalanx of which each was placed in charge. In this battle, he pointed out, they were to fight, not as before, for Hollow Syria or Phoenicia or Egypt: it was the sovereignty of all Asia that was there and then to be decided. There was then no need for them to employ long speeches to make their men act with that sense of honour which was born in them, but they were to urge each man in the moment of danger to attend in his own place in the line to the require‑ments of order, to keep perfect silence when that was necessary in the advance, and by contrast to give a ringing shout when it was right to shout, and howl to inspire greatest terror when the moment came to howl; they themselves were to obey orders sharply and to pass them on sharply to their regiments, and every man should recall that neglect of his own duty brought the whole cause into common danger, while energetic attention to it contributed to their common success.

Arrian, describing Alexander's instructions to his officers before the Battle of Gaugamela (Arbela), 1 October 331 BC, *Anabasis of Alexander*, 3.9, c. AD 150, tr. P.A. Brunt, 1976.

Now is gode tyme, for alle Engelond prayeth for us; and therefore be of gode chere, and lette us go to our iorney... In the name of Almyghti God and Saynt George, avant banerer! And Saynte George, this day thyn help.

Henry V, 14 October 1415, at the Battle of Agincourt.

In God's name! Let us go bravely!

Joan of Arc, 6 May 1429, in the assault on the English fortress of the Augustines at Orleans, *Joan of Arc In Her Own Words*, 1996.

It is the great battle of the Cross and Koran, which is now to be fought. A formidable army of infidels are on the point of investing our island. We, for our part, are the chosen soldiers of the Cross, and if Heaven requires the sacrifice of our lives, there can be no better occasion than this. Let us hasten then, my brothers, to the sacred altar. There we will renew our vows and obtain, by our Faith in the Sacred Sacraments, that contempt for death which alone can render us invincible.

Jean de la Valette, Grand Master of the Order of the Knights of St John of Jerusalem, mid-May 1565, address to the Brethren of the Order just before the epic siege of Malta by the fleet and army of Suleyman the Magnificent, quoted in Ernle Bradford, *The Great Siege, Malta 1565*, 1961.

My loving people! My loving people! We have been persuaded by some that are careful of our safety, to take heed how we commit ourselves to armed multitudes from fear of treachery. But I do assure you, I do not desire to live to distrust the faithful and loving people.

Let tyrants fear! I have always so behaved myself that, under God, I have placed my chiefest strength and safe‑guard in the loyal hearts and goodwill of my subjects. Therefore, I am come amongst you as you see at this time, not for my recreation and disport, but being resolved, in the midst and heat of the battle, to live or die amongst you all – to lay down for my God and for my kingdoms, and for my people, my honor and my blood even in the dust!

I know I have the body of a weak and feeble woman, but I have the heart and stomach of a King, and of a King of

EXHORTATION

England too, and think foul scorn that Parma or Spain, or any Prince of Europe should dare to invade the borders of my realm, to which, rather than any dishonour should grow by me, I myself will take up arms.
    Elizabeth I, 19 August 1588, the Tilbury Speech, to the English Army assembled at Tilbury camp ready to repel the invasion of the Spanish army aboard the Armada.

Hark! I hear the drum. Forward in God's name! Jesu! Jesu!
    Gustavus II Adolphus, 15 November 1632, to his army as he led it into the Battle of Lützen, where he met his death, quoted in Theodore Ayrault Dodge, *Gustavus Adolphus*, 1895.

We have beaten this same enemy, you and I. To-day you fight not for Poland but for Christianity; not for your king, but for your God! I have but one order for you: Wherever you see your king, follow and fear not!
    Jan III Sobieski, 12 September 1683, as he led the Polish attack on the Ottoman Army besieging Vienna, quoted in Theodore Ayrault Dodge, *Gustavus Adolphus*, 1895.

Rascals, would you live forever? (*Ihr Racker, wollt ihr ewig leben?*)
    Frederick the Great, 18 June 1757, when the Guards hesitated at the Battle of Kolin.

Soldiers! Forty centuries behold you!
    Napoleon, 21 July 1798, to his army before the Battle of the Pyramids, quoted in R.M. Johnston, ed., *The Corsican*, 1910.

When, in the heat of the battle, passing along the line, I used to exclaim, 'Soldiers, unfurl your banners, the moment is come,' our Frenchmen absolutely leapt for joy. I saw them multiply a hundred fold. I then thought nothing impossible.
    Napoleon, at St Helena in 1816, quoted in Emmanuel Las Cases, *The Life, Exile, and Conversations of the Emperor Napoleon by the Count de Las Cases*, IV/7, 1835, p.245.

Up, Guards and at 'em!
    The Duke of Wellington, 18 June 1815, to the Brigade of Guards at Waterloo.

Up! Children of Zulu, your day has come. Up! and destroy them all!
    Shaka Zulu, 1819, in the Second Ndwandwe War, quoted in E.A. Ritter, *Shaka Zulu*, 1955.

Up, men, and to your posts! Don't forget today that you are from Old Virginia.
    Major-General George Pickett, 3 July 1863, to his division just before the charge at Getttysburg.

Come on, Lakotas! It's a good day to die!
    Crazy Horse, 25 June 1876, as he rallied the Lakota warriors to defend their women and children camped on the Little Big Horn River against Custer's attack.

Come on you sons of bitches! Do you want to live forever?
    Sergeant Dan Daly, 4 June 1918, leading marines at Belleau Wood.

Come, then, let us to the task, to the battle, to the toil – each to our part, each to our station. Fill the armies, rule the air, pour out the munitions, strangle the U-boats, sweep the mines, plough the land, build the ships, guard the streets, succour the wounded, uplift the downcast, and honour the brave. Let us go forward together in all parts of the Empire, in all parts of the Island. There is not a week, nor a day, nor an hour to lose.
    Sir Winston S. Churchill, 27 January 1940, speech as First Lord of the Admiralty at the Free Trade Hall in Manchester.

Sons of Empire, forget it not! There are such things as love, honour and the soul of man, which cannot be bought with a price, nor die with death.

EXPERIENCE

A General Instructional Background for the Young Soldier, Canadian Army, 1942

EXPERIENCE
General

I have not lived so long, Spartans, without having had the experience of many wars, and I see those among you of the same age as myself, who will not fall into the common misfortune of longing for war from inexperience or from the belief in its advantage and its safety.

Archidamus II, King of Sparta, 432 BC, speech on the prospects of war with Athens, quoted in Thucydides, *History of the Peloponnesian War*, c. 404 BC, tr. Richard Crawley, 1910.

The more your troops have been accustomed to camp duties on frontier stations and the more carefully they have been disciplined, the less danger they will be exposed to in the field.

Men must be sufficiently tried before they are led against the enemy.

Flavius Vegetius Renatus, *Military Institutions of the Romans*, c. AD 378, tr. Clark, 1776.

Never lead soldiers into combat before having made sufficient trial of their courage.

The Emperor Maurice, *The Strategikon*, c. AD 600 (*Maurice's Strategikon*, tr. George Dennis, 1989).

We have acquired this knowledge not simply from hearing about it but also from having been taught by a certain amount of experience. For one thing, the men who instructed and trained us in this method were the very ones... who invented it. Then, on our own, we have put it into practice and, as best we could, almost made it a part of us.

The Emperor Nikephoros II Phokas, *Skirmishing*, c. AD 969 (*Three Byzantine Military Treatises*, tr. George Dennis, 1985).

What is the good of experience if you do not reflect?

Frederick the Great, quoted in Fuller, *Decisive Battles of the U.S.A.*, 1953.

No battle can be won in the study, and theory without practice is dead.

Field Marshal Prince Aleksandr V. Suvorov (1729–1800), quoted in K. Ossipov, *Suvorov*, 1945.

Habits of practice give, to the soldier, such skill and management in the use of arms in the day of battle, as might be expected to be acquired by experience, in working, in unison, the separate parts of a machine of compound movement. The knowledge and ability, acquired by such experience, aided by a correct direction of powers in general movement, ensure the application of united impulse, at the proper time and in the proper circumstances of action, producing a powerful effect, and a calculable one, as depending upon a uniform rule. It is thus that experience of actual war imprints, upon the soldier, the character of a veteran – a courage; arising from knowledge of things, and a consciousness of superiority in the art of applying powers. Such courage is cool and tempered: that of unexperienced troops is impetuous, blind, and headlong – liable to mistake its purpose unless plain and prominent in all its aspects.

Robert Jackson, *A Systematic View of the Formation, Discipline, and Economy of Armies*, 1804.

[I] understood as much of military matters as I have ever done since.

The Duke of Wellington, of his experience in India 1796–1805; quoted in Lord Stanhope, *Notes of Conversations with the Duke of Wellington*, 1888.

Our business, like any other, is to be learned by constant practice and experience; and our experience is to be had in war, not at reviews.

General Sir John Moore (1761–1809).

EXPERIENCE

Maxim 77. Generals-in-chief must be guided by their own experience or their genius. Tactics, evolutions, the duties and knowledge of an engineer or an artillery officer may be learned in treatises, but the science of strategy is only to be acquired by experience, and by studying the campaigns of all the great captains.
    Napoleon, *Military Maxims of Napoleon*, tr. George D'Aguilar, 1831 (David Chandler, ed., 1987).

Battle experience overcomes friction, from troopers and riflemen up to the divisional commander.
    Major-General Carl von Clausewitz, *On War*, 1832, tr. Michael Howard and Peter Paret, 1976.

It is a frequent contention of the 'practical soldier' that length of experience in the field counts for more than depth of study as a means to a mastery of war. The inevitable corollary of this claim is that the man who has seen the most fighting must be the greatest authority on tactics.
    In this connection one may recall the story of Marshal Saxe and the aged general who urged the acceptance of his opinion on the ground that he had seen more campaigns than anyone else, whereupon Saxe replied that he had a mule which had been through twenty campaigns 'and was still a mule.' (March 1923.)
    Captain Sir Basil Liddell Hart, *Thoughts on War*, 1944.

Unlike most other professions, the military offers but rare opportunities for actual practice. Most of its operating experience is necessarily in the realm of make-believe. It is as if a surgeon was confined for his practice to the dissection of frogs and dead paupers. The rest of a soldier's training for command lies in the realm of theory.
    Captain Sir Basil Liddell Hart, *The Ghost of Napoleon*, 1933.

There is no better teacher of war than war.
    Mao Tse-tung, *On the Study of War*, 1936.

A division represents not only the lives of 15,000 men and millions of dollars' worth of equipment, but it also represents a priceless investment in months and years of training. In the 1st Division that investment had been multiplied beyond measure by its long experience in battle. Thus in quality the 1st was worth the equal of several inexperienced divisions. It had become an almost irreplaceable weapon in the Normandy invasion.
    General of the Army Omar Bradley, *A Soldier's Story*, 1951.

In Tunisia the Americans had to pay a stiff price for their experience, but it brought rich dividends. Even at that time, the American generals showed themselves to be very advanced in the tactical handling of their forces, although we had to wait until the Patton Army in France to see the most astonishing achievements in mobile warfare. The Americans, it is fair to say, profited far more than the British from their experience in Africa, thus confirming the axiom that education is easier than reeducation.
    Field Marshal Erwin Rommel, *The Rommel Papers*, 1953.

Direct experience is inherently too limited to form an adequate foundation either for theory or for application. At best it produces an atmosphere that is of value in drying and hardening the structure of thought. The great value of indirect experience lies in its greater variety and extent.
    Captain Sir Basil Liddell Hart, *Strategy: The Indirect Approach*, 1954.

Lessons Learned

I am not sorry that I went, notwithstanding what has happened. One may

EXPERIENCE

pick up something useful from among the most fatal errors.
> Major-General Sir James Wolfe, 1757, of the Rochefort Expedition.

Oh, well. It's all the same. The students are a credit to their teachers. The French have taught you some terrible lessons, and you understand, at length, the art of making war as it should be.
> A captured French veteran to his British captors, in Spain during the Peninsular War, Costello, *Adventures of a Soldier written by Himself*, 1852, quoted in Arthur Bryant, *The Great Duke*, 1972.

I learnt what one ought not to do, and that is always something.
> The Duke of Wellington, after the disastrous British campaign in Holland, 1794–1795, in which he served as the colonel of the 33rd Foot; quoted in Earl Stanhope, *Conversations with the Duke of Wellington*, 1888.

These beasts have learned something.
> Napoleon, 1813, comment during the Battle of Lützen on the improved performance of the Prussian Army compared to the 1806 campaign.

On 12 November, 1918, twenty-four hours after the Armistice, General Pershing visited the I Corps headquarters, and found the corps commander pouring over his maps. 'Don't you know the war's over?' asked a bemused Pershing. Liggett replied, 'I'm trying to see where we might have done better.'
> Major-General Hunter Liggett, 12 November 1918, quoted in Adolph Carlson, 'A Chapter Not Yet Written', in Robert Neilson, ed., *Sun Tzu and the Information Warfare*, 1997.

It is absolutely necessary to put the experience of the war in a broad light and collect this experience while the impressions won on the battlefield are still fresh and a major proportion of the experienced officers are still in leading positions... [committees were to write] short, concise studies on the newly-gained experiences of the war and consider the following points: a). What new situations arose in the war that had not been considered before the war? b). How effective were our pre-war views in dealing with the above situations? c). What new guidelines have been developed from the use of new weaponry in the war? d). Which new problems put forward by the war have not yet found a solution?
> Colonel-General Hans von Seekt, 1 December 1919, letter to the Truppenamt on the duties of the 57 committees he established to examine the lessons learned from WWI, quoted in James S. Corum, *Roots of Blitzkrieg*, 1992.

It is easy to be wise after the event; but I cannot help wondering why none of us realized what the modern rifle, the machine-gun, the aeroplane and the wireless telegraphy will bring about. It seems so simple when judged by actual results. The modern rifle and machine-gun add tenfold to the relative power of the defence as against the attack... I fell sure in my own mind that had we realized the true effect of modern appliances of war in August 1914 there would have been no retreat from Mons.
> General Sir John French, quoted in C.R. Ballad, *Kitchener*, 1930, pp. 224–25.

I claim we got a hell of beating. We got run out of Burma and it is as humiliating as hell. I think we ought to find what caused it, and go back and retake it.
> General Joseph Stillwell after the Burma Campaign of 1942, *New York Times*, 26 May 1942.

Lesson No. 1 from Fuller: 'To anticipate strategy, imagine.'

EXPLOITATION

Lesson No. 2 from Martel: 'Men, not weapons, will shape the future, so stick with fundamentals.'

Brigadier-General S.L.A. Marshall, *Bringing Up the Rear*, 1979, referring to Major-General J.F.C. Fuller and Lieutenant-General Sir Giffard Martel.

EXPLOITATION

The gods do not grant all their gifts to one man, Hannibal. You know how to conquer but not how to use your victory.

Maharbal Barca, the commander of the Carthaginian cavalry after the battle of Cannae, when his brother, Hannibal, preferred to rest the army rather than pursue, 2 August 216 BC, quoted in Livy, *Annals of the Roman People*, 22, c. 25 BC – AD 17 (Livy: *A History of Rome*, tr. Moses Hadas and Joe Poe, 1962).

Those who know how to win are much more numerous than those who know how to make use of their victories.

Polybius, *Histories*, c. 125 BC (*The Rise of the Roman Empire*, tr. Ian Scott-Kilvert, 1987).

To-day the victory had been the enemy's had there been any one among them to gain it.

Julius Caesar, 48 BC, of Pompey's success in the Battle of Dyrrhachium, which he failed to exploit, quoted in Plutarch, *The Lives*, c. AD 100, tr. John Dryden.

A strong pursuit, give no time for the enemy to think, take advantage of victory, uproot him, cut off his escape route.

Field Marshal Prince Aleksandr V. Suvorov (1729–1800), quoted in V.Ye. Savkin, *Basic Principles of Operational Art and Tactics*, 1972.

First gain the victory, and then make the best use of it you can.

Admiral Viscount Nelson, to his captains before the Battle of the Nile, 1 August 1797, quoted in Robert Southey, *Life of Nelson*, 1813.

My Brother: I shall start at one in the morning so as to reach Burgos before dawn; there I will make my arrangements for the day, for a victory is nothing, it must be turned to account.

Napoleon, 10 November 1808, to his brother Joseph, quoted in R.M. Johnston, ed., *The Corsican*, 1910.

It is a familiar experience that the winner's casualties in the course of an engagement show little difference from the loser's. Frequently there is no difference at all, and sometimes even an inverse one. The really crippling losses, those the vanquished does not share with the victor, only start with his retreat. The feeble remnants of badly shaken battalions are cut down by cavalry; exhausted men fall by the wayside; damaged guns and caissons are abandoned, while others are unable to get away quickly enough on poor roads and taken by the enemy's cavalry; small detachments get lost in the night and fall defenseless into the enemy's hands. Thus a victory usually only starts to gather weight after the issue has already been decided...

Major-General Carl von Clausewitz, *On War*, 4.4, 1832, tr. Michael Howard and Peter Paret, 1976.

For success in the attack, two major problems must be solved – *dislocation* and *exploitation*. One precedes and one follows the actual blow, which in comparison is a simple act. You cannot hit the enemy with effect unless you have first created the opportunity; you cannot make that effect decisive unless you exploit the second opportunity that comes before he can recover. (May 1930.)

Captain Sir Basil Liddell Hart, *Thoughts on War*, 1944.

F

FAILURE

He who in war fails to do what he undertakes, may always plead the accidents which invariably attend military affairs: but he who declares a thing to be impossible, which is subsequently accomplished, registers his own incapacity.
> The Duke of Wellington (1769–1852), quoted in William Fraser, *Words on Wellington*, 1889.

The service cannot afford to keep a man who does not succeed.
> Lieutenant-General Thomas J. 'Stonewall' Jackson (1824–1863), quoted in G.F.R. Henderson, *Stonewall Jackson*, 1898.

On the heels of the South African War came the sleuth-hounds pursuing the criminals, I mean the customary Royal Commissions. Ten thousand words of mine stand embedded in their Blue Books, cold and dead as so many mammoths in glaciers. But my long spun-out intercourse with the Royal Commissioners did have living issue – my Manchurian and Gallipoli notes. Only constant witnesses could have shown me how fallible is the unaided military memory or have led me by three steps to a War Diary: –
(1) There is nothing certain about war except that one side won't win.
(2) the winner is asked no questions – the loser has to answer for everything.
(3) Soldiers think of nothing so little as failure and yet, to the extent of fixing intentions, orders, facts, dates firmly in their own minds, they ought to be prepared.
  Conclusion: – in war, keep your own counsel, preferably in a note-book.
> General Sir Ian Hamilton, *Gallipoli Diary*, 1918.

A man's life is interesting primarily when he has failed – I well know. For it's a sign that he tried to surpass himself.
> Georges Clemenceau, conversation with Jean Martet, 1 June 1928, quoted in *Clemenceau, the Events of His Life as Told by Himself to His Former Secretary, Jean Martet*, tr. Milton Waldman, 1930.

Encouraging initiative presupposes an understanding by seniors and juniors alike, that innovation is imprecise, that error and false starts must be expected, that to try and fail is much to be preferred to never trying at all.
> Lieutenant-General Victor H. 'The Brute' Krulak, 'A Soldier's Dilemma', *Marine Corps Gazette*, 11/1986.

FALLEN COMRADES

How are the mighty fallen in the midst of the battle! O Jonathan, thou wast slain in thine high places.
  I am distressed for thee, my brother Johnathan: very pleasant has thou been unto me: thy love to me was wonderful, passing the love of women.
  How are the mighty fallen, and the weapons of war perished.
> King David of Israel, lament for his friend, the warrior prince, Jonathan, son of King Saul, II Samuel, 1.25–27.

Fast friends, forest-companions,
we made one bed and slept one sleep
in foreign lands after the fray.
Scathach's pupils, two together

FALLEN SONS

we'd set forth to comb the forest.
...
There is no man that ever ate,
no man that was ever born,
no joyous son of king or queen,
for whose sake I would do you harm.
...
When we were away with Scathach
learning victory overseas,
it seemed our friend would remain
unbroken till the day of doom.

I loved the noble way you blushed,
and loved your fine, perfect form.
I loved your blue clear eye,
your way of speech, your skillfulness [,]
...
your curled yellow hair
like a great lovely jewel,
the soft leaf-shaped belt
that you wore at your waist.

You have fallen to the Hound,
I cry for it, little calf.
The shield didn't save you
that brought you to the fray.

> The lament of the high hero Cuchulain for his fallen warrior foster brother, Ferdia, in the Irish epic, *Tain*, 1970, quoted in Cahill, *How the Irish Saved Civilization* tr. Thomas Kinsella, 1995.

FALLEN SONS

And, behold, Cushi came; and Cushi said, Tidings, my lord the king: for the LORD hath avenged thee this day of all them that rose up against thee.

And the king said unto Cushi, Is the young man Abasalom safe? And Cushi answered, The enemies of my lord the king, and all that rise against thee to do thee hurt, be as that young man is.

And the king was much moved, and went up to the chamber over the gate, and wept: and as he went, thus he said, O my son Absalom, my son, my son Absalom! Would God I had died for thee, O Absalom, my son, my son!

> King David, II Samuel, 18.31–33, *c.* 1000 BC, of the death of his rebellious son Absalom at the hands of his general, Joab, after the King had expressly ordered him to spare the wayward prince so beloved of his father.

But I, dear god, my life so cursed by fate . . .
I fathered hero sons in the wide realm of Troy
and now not a single one is left, I tell you.
Fifty sons I had when the sons of Achaea came,
nineteen born to me from a single mother's womb
and the rest by other women in the palace. Many,
most them violent Ares cut the knees from under.
But one, one was left me, to guard my walls, my people –
the one you killed the other day, defending his fatherland,
my Hector! It's all for him I've come to the ships now,
to win him back from you – I bring a priceless ransom.
Revere the gods, Achilles! Pity me in my own right,
remember your own father! I deserve more pity . . .
I have endured what no one on earth has ever done before –
I put to my lips the hands of the man who killed my son.

> Homer, *The Iliad*, 24.577–591, *c.* 800 BC, tr. Robert Fagles, 1990; Priam begging Achilles for the body of his son, Hector, breaker of horses.

No one is so foolish as to prefer peace to war, in which, instead of sons burying their fathers, fathers bury their sons.

> King Croesus of Lydia, to Cyrus the Great who had just rescued him from his own funeral pyre following his defeat at the hands of the founder of the Persian Empire in 546 BC, quoted in Herodotus, *The Persian Wars*, 1.87, *c*, 430 BC, tr. George Rawlinson, 1942.

FALLEN SONS

But neither should you to feel so much concern about death, knowing that we must regard birth as the beginning of man's course, and death as the end. He had died, as he who was even ever so reluctant would have died; but to die honourably is the part of one who is willing to die, and who had been taught what he ought to know. Happy therefore is Gryllus, and whoever chooses, not the greatest prolongation of life, but life distinguished by virtue; though the gods granted him, indeed, but a short life.

Xenophon, letter to Soteira of the death of his son, Gryllus, who 'fell with honour at the battle of Mantinea', 362 BC, *Xenophon's Minor Works*, c. 360 BC, tr. J.S. Watson, 1891.

It gives me great concern to undertake a task, which is not only a bitter renewal of my own grief, but must be a violent shock to an affectionate parent.

You have for your support, the assistance of religion, good sense, and the experience of the uncertainty of human happiness. You have for your satisfaction, that your son fell nobly in the cause of his country, honored and lamented by all his fellow-soldiers; that he led a life of honor and virtue, which must secure to him everlasting happiness.

When the keen sensibility of the passions begins to subside, these considerations will give you real comfort. That the Almighty may give you fortitude to bear this severest of strokes, is the earnest wish of your companion in affliction.

General Marquis Cornwallis, letter to the father of Lieutenant-Colonel Webster of the 33rd Foot, killed at the Battle of Guilford Courthouse, 15 March 1781, quoted in Henry Lee, *Memoirs of the War in the Southern Department*, 1869.

March 13.—The bright, beautiful Sabbath. Another day of solemn suspense. My son, my son! When I left Washington it seemed as if dear Ully's death was the greatest possible affliction. Now, I hardly know if I can recover his body. How thoughts of him flit by me. the last evening he was with me I gave him my last memorandum in case I fell. Dear boy, he goes before me! Here, too, I have a whole life to contemplate, from the baby in the cradle to the distinguished Colonel falling in battle; and yet I am in the vigor of life. My son!

Rear Admiral John A. Dahlgren, 13 March 1864, diary entry while attempting to recover the body of his son, Colonel Ulric Dahlgren, from the Confederates, quoted in Madeleine Vinton Dahlgren, *Memoir of John A. Dahlgren*, 1891.

Little Flora is broken hearted, but very brave, and Mother suffers as much and is even braver; for Mother has the true heroism of heart. Well, it is very dreadful; it is the old who ought to die, and not fine and gallant youth with the golden morning of life still ahead; but after all he died as the heroes of old died, as brave and fearless men must die when a great cause calls. If our country did not contain such men it would not be our country. I bitterly mourn that he was not married and does not leave his own children behind him; but the children's children of his brothers and of Ethel will speak of him with pride as long as our blood flows in the veins of man or woman.

President Theodore Roosevelt, 21 July 1918, letter to Archibald Roosevelt, of the death of his son Quentin a few days before in an aerial battle in France, Theodore Roosevelt Collection, quoted in Edward Reenan, Jr., *The Lion's Pride*, 1998.

FAME

The general who advances without coveting fame and retreats without fearing disgrace, whose only thought is to protect his country and do good service for his sovereign, is the jewel of the kingdom.

Sun Tzu, *The Art of War*, 10, c. 500 BC, tr. Giles, 1910.

FAME

Fame is the perfume of heroic deeds.
  Socrates (469–399 BC).

Which would you rather be – a victor at the Olympic games or the announcer of the victor?
  Themistocles, upon being asked whether he would rather be Achilles or Homer, from Plutarch, *Apothegms, Themistocles*, c. AD 125.

O fortunate youth, to have found Homer as the herald of your glory!
  Alexander the Great at the tomb of Achilles at Troy, from Cicero, *Pro Archia*, 62 BC (*Selected Works of Cicero*, tr. Harry Hubbell, 1948).

There were some who thought he was too eager for fame, and indeed the desire of glory is the last infirmity cast off even by the wise.
  Cornelius Tacitus, *History*, 4.5, AD 98 (*The Complete Works of Tacitus*, tr. Alfred Church and William Brodribb, 1942); of Helvidius Priscus.

But with me immortality is the recollection one leaves in the memory of man. That idea prompts to great actions. It would be better for a man never to have lived than to leave behind him no traces of his existence.
  Napoleon, 1802, conversation with his private secretary, Louis de Bourrienne, *Memoirs of Napoleon Bonaparte*, II, 1855.

You have miscalculated the heights to which misfortune, the injustice and persecution of your government, and your own conduct have raised the Emperor. His head wears more than an imperial crown – it wears a crown of thorns.
  It is not in your power, or in that of the like of you, to obscure the radiance of that crown.
  Napoleon, autograph letter to Sir Hudson Lowe, governor of St Helena, 25 July 1817, *Correspondance de Napoléon Ier, publié par ordre de l'Empereur Napoléon III*, XXXII, quoted in Christopher Herold, ed., *The Mind of Napoleon*, 1955.

Of all the passions that inspire man in battle, none, we have to admit, is so powerful and so constant as the longing for honor and renown. The abuse of these noble ambitions has certainly inflicted the most disgusting outrages on the human race; nevertheless their origins entitle them to be ranked among the most elevated in human nature. In war they act as the essential breath of life that animates the inert mass. Other emotions may be more common and more venerated – patriotism, idealism, vengeance, enthusiasm of every kind – but there are no substitute for a thirst for fame and honor.
  Major-General Carl von Clausewitz, *On War*, 1.3, 1832, tr. Michael Howard and Peter Paret, 1976.

The road to failure is the road to fame – such apparently must be the verdict on posterity's estimate of the world's greatest figures. The flash of the meteor impresses the human imagination more than the remoter splendour of the star, fixed immutably in the high heavens. Is it that final swoop earthwards, the unearthly radiance ending in the common dust, that, by its evidence of the tangible or the finite, gives to the meteor a more human appeal? So with luminaries of the human system, provided that the ultimate fall has a dramatic note, the memory of spectacular failure eclipses that of enduring success. Again, it may be that the completeness of his course lends individual emphasis to the great failure, throwing his work into clearer relief, whereas the man whose efforts are crowned with permanent success builds a stepping-stone by which others may advance still farther, and so merges his own fame in that of his successors.
  Captain Sir Basil Liddell Hart, *A Greater Than Napoleon: Scipio Africanus*, 1926.

FAMILIARITY

I'm apparently not famous in Baghdad.
> General H. Norman Schwarzkopf, 1991, quoted in Peter David, *Triumph in the Desert*, 1991.

FAMILIARITY

The leader, mingling with the vulgar host,
Is in the common mass of matter lost.
> Homer, *The Odyssey*, 4.397, c. 800 BC, tr. Alexander Pope, 1725–1726.

Be easy and condescending in your deportment to your officers, but not too familiar, lest you subject yourself to a want of that respect, which is necessary to support a proper command.
> General George Washington, *The Writings of George Washington*, IV, p. 81, John C. Fitzgerald, ed., 1931.

Soldiers must never be witnesses to the discussions of the commanders.
> Napoleon, 31 March 1805, to Marshal de Moncey, *Correspondance de Napoléon Ier, publié par ordre de l'Empereur Napoléon III*, No. 8507, Vol. X, p. 279, 1858–1870.

Efficiency in a general, his soldiers have a right to expect; geniality they are usually right to suspect.
> Field Marshal Viscount Wavell of Cyrenaica, *Generals and Generalship*, 1941.

'FAMOUS LAST WORDS'

The Lord God is my armour!
> Gustavus Adolphus, refusing the steel body armour offered by his aides, which would have saved his life from the bullets and swords of Wallenstein's Croatian cavalry at the Battle of Lützen, 16 September 1632.

I did not mean to be killed today.
> Vicomte de Turenne, on a reconnaissance just before being struck by a cannonball at the Battle of Sasbach, 27 July 1675.

An alliance with Russia would be very valuable. Only women and children are capable of supposing that (I) would lose myself in the deserts of Russia.
> Napoleon, 14 March 1807, quoted in R.M. Johnston, ed, *The Corsican*, 1910.

I tell you Wellington is a bad general, the English are bad soldiers; we will settle the matter by lunch time.
> Napoleon, 18 June 1815, on the battlefield of Waterloo at 8:30 am; the battle began in the early afternoon, quoted in R.M. Johnston, ed., *The Corsican*, 1910.

The Rebel army is now the legitimate property of the Army of the Potomac... The enemy is now in my power, and God Almighty cannot deprive me of them.
> Major-General Joseph Hooker, 1 May 1863, *The War of the Rebellion: A Compilation of the Official Records of the Union and Confederate Armies*, XXV, Part 2, p. 328, US Government Printing Office, 1880–1901; just as his army was about to engage Lee at the Battle of Chancellorsville.

They couldn't hit an elephant at this dist...
> Major-General John Sedgwick, 9 May 1864, killed instantly by a sniper while carelessly peering over the parapet during the Battle of Spotsylvania.

Hurrah, boys, we've got them!
> General George Armstrong Custer, as the 7th Cavalry closed on the great Indian encampment along the Little Big Horn River, 25 June 1876, quoted by Giovanni Martini in Benteen, *The Custer Fight*, 1933.

My only fear is that the *Zulu* will not fight.
> Lieutenant-General Lord Chelmsford, commanding British forces, just before the invasion of Zululand, 1879.

You will be home before the leaves have fallen from the trees.
> Kaiser Wilhelm II, to his troops at the beginning of World War I, quoted in Barbara Tuchman, *The Guns of August*, 1962.

FATE

Japan should never be so foolish as to make enemies of Great Britain and the United States.
> Admiral Isoroku Yamamoto, c. 1937, quoted in Masanori Ito, *The End of the Imperial Japanese Navy*, 1956.

The Ruhr will not be subjected to a single bomb. If an enemy bomber reaches the Ruhr, my name is not Hermann Goering: you can call me Meier!
> Reichsmarshal Hermann Goering, 9 August 1939, as Germany prepared for war, quoted in William L. Shirer, *The Rise and Fall of the Third Reich*, 1960.

It is significant that despite the claims of air enthusiasts no battleship has yet been sunk by bombs.
> Caption to photograph of the USS *Arizona* in the Army-Navy football game programme, 29 November 1941.

There is no immediate prospect of an invasion.
> Field Marshal Gerd von Rundstedt, evening 5 June 1944, message to his forces in the field, quoted in F.H. Hinsley, *British Intelligence in the Second World War*, 1984, III, pt. 2, p. 64.

The atom bomb will never go off, and I speak as an expert in explosives.
> Admiral William Leahy to President Harry Truman before the first atomic bomb was dropped on Hiroshima, 1945.

If you want to accept his proposal I will not prostrate myself on the road and I will not resign; but you might as well know that it is superfluous.
> Moshe Dayan, at 8.00 am, 6 October 1973, during an emergency meeting of the Israeli Cabinet, dismissing the proposal of the Chief of Staff of the Israeli Defence Force that mobilisation be ordered in the face of the Egyptian and Syrian preparations to attack Israel, quoted in Haim Herzog, *The War of Atonement*, 1975.

The Iraqi people are capable of fighting to the victorious end which God wants... the blood of our martyrs will burn you.
> Saddam Hussein, August 1990, quoted in Pimlott and Badsey, *The Gulf War Assessed*, 1992.

[General Pavel] Grachev promised to the president to solve the Chechen crisis in two hours with an airborne regiment. The president believed him.
> Lieutenant-General Aleksander I. Lebed, of the Russian Defence Minister's rash promise in early December 1994 to easily suppress the Chechens in Groznyy, but which resulted in months of brutal and bloody city fighting, quoted in *Moshovshiya Novosti* (Moscow) No. 7, 29 January–5 February1995.

FATE

Fate is the same for the man who holds back, the same if he fights hard. We are all held in a single honour, the brave with the weaklings. A man dies still if he has done nothing, as one who has done much.
> Homer, *The Iliad*, 9.318–320, c. 800 BC, tr. Richmond Lattimore, 1951; Achilles refusing Agamemnon's gifts to settle their dispute.

Fate often spares the undoomed man when his courage lasts.
> *Beowulf*, c. AD 1000 (*My Name is Beowulf*, tr. Frederick Rebsamen, 1971).

What better fate could overtake me than to die doing my duty as king, in which place it has pleased heaven to set me?
> Gustavus II Adolphus (1594–1632), citing the example of Alexander the Great when taken to task by his senior general for unnecessarily risking his life, quoted in Theodore Ayrault Dodge, *Gustavus Adolphus*, 1895.

I could see clearly enough the fatal hour

coming! My star was growing paler; I felt the reins slipping from my fingers; and I could do nothing. Only a thunderstroke could save us. I had, therefore, to fight it out; and day by day, by this or that fatality, our chances were becoming more slender!
> Napoleon, 19 October 1813, after the defeat at Leipzig, quoted in R.M. Johnston, *The Corsican*, 1910.

A comet! It was the omen foretold the death of Caesar!
> Napoleon, 2 April 1821, on St Helena, quoted in R.M. Johnston, *The Corsican*, 1910.

FATIGUE AND REST

And grant me to buckle upon my shoulders that armour of thine, in hope that the Trojans may take me for thee, and so desist from war, and the warlike sons of the Achaeans may take breath, wearied as they are; for scant is the breathing-space in battle. And lightly might we that are unwearied drive men that are wearied with the battle back to the city from the ships and the huts.
> Homer, *The Iliad*, 16, c. 800 BC, tr. A.T. Murray, 1928; Patroclus to Achilles as the Trojans stormed the Greek camp.

...And this I know, that I wake earlier than you – and watch, that you may sleep.
> Alexander the Great, 323 BC, to his officers and men at Opis, quoted in Arrian, *The Campaigns of Alexander*, 7.9, c. AD 150, tr. Aubrey de Sélincourt, 1987.

I have decided to engage successively and without halt one body of troops after the other, until harassed and worn out the enemy will be unable to further resist.
> Mehmet II the Conquerer, 1453, Ottoman Sultan, at the Siege of Constantinople, quoted in J.F.C. Fuller, *A Military History of the Western World*, 1967.

Gentlemen, the dispositions for tomorrow, or rather for today since midnight has gone, cannot be altered. You have heard them read out. We shall do our duty. But before a battle, there is nothing more important than to sleep well. Gentlemen, let us take some rest.
> Field Marshal Prince Mikhail I. Kutuzov, 1 December 1805, the evening before the Battle of Austerlitz, quoted in Parkinson, *Fox of the North*, 1976.

In military, public or administrative affairs there is a need for deep thought as well as deep analysis, and also for an ability to concentrate on subjects for a long time without sleep.
> Napoleon (1769–1821), *Correspondance de Napoléon Ier, publié par ordre de l'Empereur Napoléon III*, No. 337, Vol. I, 1858–1870.

You must not needlessly fatigue the troops.
> Napoleon, 29 July 1806, *Correspondance de Napoléon Ier, publié par ordre de l'Empereur Napoléon III*, No. 10563, Vol. XII, 1858–1870.

It is at night when a commander must work: if he tires himself to no purpose during the day, fatigue overcomes him at night... A commander is not expected to sleep.
> Napoleon, conversation with Gaspard Gourgaud, *Journal inédit de 1815 à 1818*, Vol. II, n.d.

I was obliged to do so [sleep] when I fought battles that lasted three days; Nature was also to have her due: I took advantage of the smallest intervals, and slept where and when I could.
> Napoleon (1769–1821) quoted in Emmanuel de Las Casas, *Memorial de St. Hélène*, II, 1816.

Fatigue, exertion, and privation constitute a separate destructive factor in war – a factor not essentially belonging to combat, but more or less intricately involved in it, and pertaining

FATIGUE AND REST

especially to the realm of strategy. This factor is also present in tactical situations, and possibly in its most intense form; but since tactical actions are of shorter duration, the effects... will be limited. On the strategic plane, however, where the dimensions of time and space are enlarged, the effects are always perceptible, and often decisive.
>Major-General Carl von Clausewitz, On War, 3.12, 1832, tr. Michael Howard and Peter Paret, 1976.

I have just read your despatch about sore tongued and fatigued horses. Will you pardon me for asking what the horses of your army have done since the battle of Antietam that fatigues anything?
>President Abraham Lincoln, 24 October 1862, letter to Major-General George B. McClellan, about his slow pursuit of General Lee after the Battle of Antietam, *The Portable Lincoln*, 1992.

I often sent small squads at night to attack and run in the pickets along a line of several miles. Of course, these alarms were very annoying, for no human being knows how sweet sleep is but a soldier. I wanted to use and consume the Northern cavalry in hard work. I have often thought that their fierce hostility to me was more on account of the sleep I made them lose than the number we killed or captured.
>Colonel John S. Mosby, *Mosby's War Reminiscences*, 1887.

No human being knows how sweet sleep is but a soldier.
>Colonel John S. Mosby, *Mosby's War Reminiscences*, 1887.

Before eleven o'clock the general-in-chief remarked to the staff: 'We shall have a busy day to-morrow, and I think we had better get all the sleep we can to-night. I am a confirmed believer in the restorative qualities of sleep, and always like to get at least seven hours of it, though I have often been compelled to put up with much less.' 'It is said,' remarked Washburne, 'that Napoleon often indulged in only four hours of sleep, and still preserved all the vigor of his mental faculties.' 'Well, I, for one, never believed those stories,' the general replied. 'If the truth were known, I have no doubt it would be found that he made up for his short sleep at night by taking naps during the day.'
>General of the Army Ulysses S. Grant, as recounted by Brigadier-General Horace Porter, *Campaigning With Grant*, 1897.

That men should be prepared to fight to the last is neither surprising nor unusual; it is counted upon, in fact in establishing any plan of operations; but that men who have been retreating continuously for ten days and are lying about half-dead with exhaustion can pick up their arms and go into the attack with bugles blowing is something that had never entered into our calculations, never been taught as even a remote possibility in our military schools.
>General Heinrich von Kluck, 1920, on the Anglo-French counter-attack of the First Battle of the Marne, 1914, quoted in Georges Blond, *The Marne*, 1967.

I saw about twenty-seven of twenty-nine divisions in battle. There were more failures, more crushed careers of officers of considerable rank, which grew out of physical exhaustion more than any other one cause. One acquired judgement with the years, but lost the resiliency of tendons and muscles. Leadership in the field depends to an important extent on one's legs, stomach, nervous system, and on one's ability to withstand hardships and lack of sleep and still be disposed energetically and aggressively to command men and dominate men on the battlefield. In World War I, many men had to be relieved because their spirit –

FEAR

their tenacity of purpose, their power of leadership over tired men – was broke through physical fatigue. They became pessimistic.
> General of the army George C. Marshall (1880–1959), quoted in Bernard Loeffke, *From Warrior to Healer*, 1998.

The tanks traveled, in a straight line, perhaps 570 kilometers... But, Andreya Lavrentevich, their speedometers show more than 2,000. A man has no speedometer and nobody knows what wear and tear has taken place.
> General Mikhail Ye. Katukov, 1945, quoted in Cornelius Ryan, *The Last Battle*, 1966.

Men of goodwill saddled with the fate of others need great courage to be idle when only rest can clear their fuddled wits.
> Lord Moran, *The Anatomy of Courage*, 1945.

Fatigue makes cowards of us all.
> General George S. Patton, Jr., *War As I Knew It*, 1947.

In battle, whatever wears out the muscles reacts on the mind and whatever impairs the mind drains physical strength.
Tired men take fright more easily. Frightened men swiftly tire.
The arrest of fear is as essential to the recovery of physical vigor by men as is rest to the body which has been spent by hard marching or hard work.
We are therefore dealing with a chain reaction. Half of control during battle comes of the commander's avoiding useless expenditure of the physical resources of his men while taking action to break the hold of fear. The other half comes from sensible preparation beforehand.
> Brigadier-General S.L.A. Marshall, *The Soldier's Load*, 1980.

FEAR

Fear makes men forget, and skill which cannot fight is useless.
> Brasidas of Sparta, 429 BC, to the Peloponnesian Fleet before action in the Crisaean Gulf, quoted in Thucydides, *The Peloponnesian War*, c. 404 BC, tr. Jowett, 1881.

Fear is apparently a formidable ally for a guard.
> Xenophon, *How To Be A Good Cavalry Commander*, c. 360 BC (*Hiero the Tyrant and Other Treatises*, tr. Robin Waterfield, 1997).

Darius and his army remained during the night marshalled in their original order; for they had no proper camp surrounding them, and they also feared that the enemy would make a night attack. It did more harm than anything else to the Persian cause at this crisis, that they stood so long under arms and that the fear, which usually proceeds great dangers, was not produced suddenly by a crisis, but cultivated for a long period, til it mastered their minds.
> Arrian, *Anabasis of Alexander*, 3.11, c. AD 125, tr. P.A. Brunt, 1976; the night before the Battle of Gaugamela (Arbela), 331 BC.

He who is afraid is half beaten.
> Field Marshal Prince Aleksandr V. Suvorov, *The Science of Victory*, 1799.

On the battlefield the real enemy is fear and not the bayonet or bullet.
> Robert Jackson, *A Systematic View of the Formation, Discipline, and Economy of Armies*, 1804.

War is not at all such a difficult art as people think... In reality it would seem that he is vanquished who is afraid of his adversary and that the whole secret of war is this.
> Napoleon, 1807, comment to Alexander I at Tilsit.

All soldiers run away, madam.
> The Duke of Wellington (1769–1852), when asked whether British soldiers ever ran away.

FEAR

Fear is the beginning of wisdom.
  General of the Army William T. Sherman (1820–1891).

The exceptional man may not feel fear, but the great mass of men do – their nervous control alone stands between them and a complete yielding to fear.

This nervous control may be upset in two principal ways. It may be worn thin by a long-continued strain – it may be shattered in a single instant by sudden shock. Usually it gives under a combination of these influences. The control is worn away imperceptibly by the anxiety and suspense of waiting for the enemy's blow, by the noise and concussive effect of shellfire, and by loss of the sleep that renovates the tired will. Then without warning the shock of a sudden surprise danger snaps the fine drawn thread of the will to resist. Stubborn resistance changes in a moment to panic-stricken flight. Fear becomes uncontrollable terror. (July 1921.)
  Captain Sir Basil Liddell Hart, *Thoughts on War*, 1944.

Fear unhinges the will, and by unhinging the will it paralyses the reason; thoughts are dispersed in all direction in place of being concentrated on one definite aim. Fear, again, protects the body; it is the barometer of danger; is danger falling or rising, is it potent or weak? Fear should answer these questions, especially physical fear, and, thus knowing that danger confronts us, we can secure ourselves against it. Whilst moral fear is largely overcome by courage based on reason, physical fear is overcome by courage based on physical means.
  Major-General J.F.C. Fuller, *The Foundations of the Science of War*, 1926.

You are not all going to die. Only two percent of you here, in a major battle would die. Death must not be feared. Every man is frightened at first in battle. If he says he isn't he is a God damn liar. Some men are cowards, yes, but they fight just the same or get the hell scared out of them watching men who do fight and who are just as scared as they. The real hero is the man who fights even though he is scared. Some get over their fright in a few minutes under fire, some take hours, some take days. The real man never lets the fear of death over-power his honor, his duty to his country and his innate manhood.
  General George S. Patton, Jr., June 1944, speech to the Third Army.

By cowardice I do not mean fear. Fear is the response of the instinct of self-preservation to danger. It is only morbid, as Aristotle taught, when it is out of proportion to the degree of danger. In invincible fear – 'fear stronger than I am' – the soldier has to struggle with a flood of emotions; he is made that way. But fear even when morbid is not cowardice. That is a label we reserve for something that a man does. What passes through his mind is his own affair.
  Lord Moran, *The Anatomy of Courage*, 1945.

When the infantryman's mind is gripped by fear, his body is captured by inertia, which is fear's Siamese twin. 'In an attack half of the men on a firing line are in terror and the other half are unnerved.' So wrote Major General J.F.C. Fuller when a young captain. The failure of the average soldier to fire is not in the main due to conscious recognition of the fact that the act of firing may entail increased exposure. It is a result of a paralysis which comes of varying fears. The man afraid wants to do nothing; indeed, he does not care even to think of taking action.

Getting him on his way to the doing of one positive act – the digging of a foxhole or the administering of first-aid to a comrade – persuading him to make any constructive use of his muscle

FIGHTING

power, and especially putting him at a job which he can share with other men, may become the first step toward getting him to make appropriate use of his weapons under combat conditions. Action is the great steadying force. It helps clear the brain. The man who finds that he can still control his muscles will shortly begin to use them. But if he is to make a rapid and complete recovery, he requires help from others.
  Brigadier S.L.A. Marshall, Men Against Fire, 1947.

Everybody gets frightened. This is basic. I do not believe that many soldiers are frightened of death. Most people are frightened of dying and everybody is frightened of being hurt. The pressures of noise, of weariness, of insecurity lower the threshold of a man's resistance to fear. All these sources of stress can be found in battle, and others too – hunger, thirst, pain, excess of heat or cold and so on. Fear in war finds victims fattened for the sacrifice.
  General Sir John Hackett, The Profession of Arms, 1983.

Never look a frightened man in the eye if you want to keep him going. Of course, it occurs to me now that this can operate the other way round too. One can also be found out oneself in this way.
  General Sir John Hackett, The Profession of Arms, 1983.

I've been scared in every war I've ever been in.
  General H. Norman Schwarzkopf, 1991, quoted in Peter David, Triumph in the Desert, 1991.

FIGHTING
War – I know it well, and the butchery of men.
Well I know, shift to the left, shift to the right
my tough tanned shield. That's what the real drill,
defensive fighting means to me. I know it all,
how to charge in the rush of plunging horses –
I know how to stand and fight to the finish,
twist and lunge in the War-god's deadly dance.
On guard!
  Homer, The Iliad, 7.271–281, 800 BC, tr. Robert Fagles, 1990; Hector to Telemonian Ajax as they began their duel.

Italian commander: 'If you are a great general, come down and fight me.'
Marius: 'If you are a great general, come and make me fight you.'
  Gaius Marius (c. 157–86 BC), in F.E. Adcock, The Roman Art of War Under the Republic, 1940.

You see my sword dripping with blood and my horse with sweat. It is thus that the Moors are beaten in the field of battle.
  El Cid (Rodrigo Diaz de Vivar), 1094, after routing the Almoravides before Valencia, quoted in Cronica del Cid Ruy Diaz, 1498.

You must know, then, that there are two methods of fighting, the one by law, the other by force: the first method is that of men, the second of beasts; but as the first method is often insufficient, one must have recourse to the second. It is therefore necessary for a prince to know how to use both the beast and the man.
  Niccolò Machiavelli, The Prince, 1513.

The object of a good general is not to fight, but to win. He has fought enough if he gains a victory.
  The Duke of Alva, c. 1560.

To imagine that it is possible to perform great military deeds without fighting is just empty dreams.
  Napoleon (1769–1821).

Hard pounding, this, gentlemen; try

FIGHTING

who can pound the longest.
> The Duke of Wellington, 18 June 1815, to his staff at the Battle of Waterloo, quoted in Elizabeth Longford, *Years of the Sword*, 1970.

Essentially war is fighting, for fighting is the only effective principle in the manifold activities generally designated as war. Fighting, in turn, is a trial of moral and physical forces through the medium of the latter. Naturally moral strength must not be excluded, for psychological forces exert a decisive influence on the elements involved in war.
> Major-General Carl von Clausewitz, *On War*, 2.1, 1832, tr. Michael Howard and Peter Paret, 1976.

Jackson: The business of a soldier is to fight. Armies are not called out to dig trenches, or throw up breastworks, and live in camps, but to find the enemy, and strike him; to invade his country, and do him all possible damage in the shortest possible time.
Graham: This would involve great destruction of both life and property.
Jackson: Yes, while it lasted; but such a war would of necessity be of brief continuance, and so would be an economy of prosperity and life in the end. To move swiftly, strike vigorously, and secure all the fruits of the victory is the secret of successful war.
> Lieutenant-General Thomas J. 'Stonewall' Jackson, 1861, conversation with the Rev. James Graham at Lexington, quoted in Graham, 'Some Reminiscences of Stonewall Jackson', *Things and Thoughts*, I, 1901.

I have not word of encouragement to give!... The fact is that people have not yet made up their minds that we are at war with the South. They have not buckled down to the determination to fight this war through; for they have got the idea into their heads that we are going to get out of this fix somehow by strategy! That's the word – *strategy*! General McClellan thinks he is going to whip the Rebels by strategy; and the army has got the same notion... The people *have not* made up their minds that we are at war, I tell you! They think there is a royal road to peace, and that General McClellan is to find it. The army has not settled down into the conviction that we are in a terrible war that has got to be fought out – no; and the officers have not either...
> Abraham Lincoln, shortly after the Battle of Antietam, September 1862, quoted in Sandburg, *Abraham Lincoln*, II, pp. 194–195.

War means fighting, and fighting means killing.
> Lieutenant-General Nathan Bedford Forrest (1821–1877).

I do not advise rashness, but I do desire resolute and actual fighting, with necessary casualties.
> Major-General Philip H. Sheridan, 23 September 1864, message to Brigadier-General William W. Averall.

This is a very suggestive age. Some people seem to think that an army can be whipped by waiting for rivers to freeze over, exploding powder at a distance, drowning out troops, or setting them to sneezing; but it will always be found in the end that the only way to whip an army is to go out and fight it.
> General of the Army Ulysses S. Grant, to Brigadier-General Horace Porter, *Campaigning With Grant*, 1897.

I have seen more than most boys my age – probably more than any. I am not squeamish, but I have seen acts of great barbarity perpetrated at Omdurman and have thoroughly sickened of human blood. I shall always be glad that I was one of those who took these brave men on with weapons little better than theirs and with only our discipline to back against their numbers. All the

FIREPOWER

rest of the army merely fed out death by machinery.
> Sir Winston S. Churchill, letter to his cousin, the Ninth Duke of Marlborough, 29 September 1898, quoted in *The Gazette*, Library of Congress, Vol. 15. No 8, 27 February 2004, pp. 6–7.

The object of fighting is to kill without getting killed. Don't disperse your force; you can't punch with an open hand; clench your fist; keep your command together.

Fight when holding, advancing or retiring: always fight or be ready to fight.

Aim at surprise; see without being seen. If you meet a man in a dark room, you jump; you should always try to make your enemy jump, either by day or night. A jumping man can't hit.

Never remain halted without a lookout. Sentries must be posted, no matter what troops are supposed to be in front of you.

Guard your flanks and keep touch with neighboring units. Try to get at the enemy's flanks.

Send information back to your immediate Commander. Negative information is as important as positive. State time and place of your message. You cannot expect assistance from your superiors unless you tell them where you are and how you are situated.

Hold what you have got and what you gain. Never withdraw from a position until ordered to do so.
> WHEN IN DOUBT FIGHT IT OUT. (1918)
> Major-General J.F.C. Fuller, *Memoirs of an Unconventional Soldier*, 1936.

You can tell the general and the admiral we're going to stick and fight it out.
> General David A. Shoup, 20 November 1943, as commander (colonel) of the Second Marine Regiment during the desperate first hours of the landing on Tarawa, message to the commander of the Second Marine Division, Major-General Julian Smith, and commander of the invasion task force, Admiral Harry Hill; quoted in *Shoup Journal*, Marine Corps Historical Center.

You've never lived until you've almost died. For those who fight for it, life has a meaning the protected will never know.
> Popular saying in SOG (Studies and Operations Group – clandestine special operations) units in Vietnam, quoted in Leigh Wade, *The Protected Will Never Know*, 1998.

If you have to close with the enemy, particularly on foot at close quarters, there is no more emotive piece of kit than the rifle.
> General Sir Michael Jackson, Chief of the Army General Staff, quoted in *Preview – The Journal of the Defence Procurement Agency*, Page 20, November 2002

Actually it's a lot of fun to fight. You know, it's a hell of a hoot... It's fun to shoot some people. I'll be right up front with you. I like brawling. You go into Afghanistan, you got guys who slap women around for five years because they don't got no manhood left anyway, so it's a hell of a lot of fun to shoot them.
> Lieutenant-General James Mattis, 1 February 2005, 'Marine General Counseled Over Comments,' *Associated Press*, 4 February 2005.

FIREPOWER

Battles are won by superiority of fire.
> Frederick the Great, *Military Testament*, 1768.

By concentrating our howitzers and cannon at a single point we will gain the local superiority, and perhaps be able to beat them. The real difficulty is to make the hole in the enemy line, but once we have done that we will overcome the remaining obstacles soon enough.
> Frederick the Great, 11 June 1778, letter to

FIRST BATTLE

Prince Henry, *Politische Correspondenz Friedrichs des Grossen*, 1879–1939.

Fire opens the gates of victory.
Field Marshal Prince Aleksandr V. Suvorov, *The Science of Victory*, 1796, quoted in Bragin, *Field Marshal Kutuzov*, 1944.

Maxim 92. In battle like in siege, skill consists in converging a mass of fire on a single point: once the combat is opened, the commander who is adroit will suddenly and unexpectedly open fire with a surprising mass of artillery on one of these points, and is sure to seize it.
Napoleon, *The Military Maxims of Napoleon*, tr. Burnod, 1827.

The best protection against the enemy's fire is a well directed fire from our own guns.
Admiral David G. Farragut, 14 March 1863, General Order for the attack on Port Hudson.

Fire kills.
Marshal of France Henri-Phillipe Pétain (1856–1951).

There is but one means to extenuate the effects of enemy fire: it is to develop a more violent fire oneself.
Marshal of France Ferdinand Foch, *Precepts and Judgments*, 1919.

It is fire-power, and fire-power that arrives at the right time and place, that counts in modern war – not man-power. (March 1924.)
Captain Sir Basil Liddell Hart, *Thoughts on War*, 1944.

The tendency towards under-rating fire-power... has marked every peace interval in modern military history.
Captain Sir Basil Liddell Hart, *Thoughts on War*, 1944.

Battles are won by frightening the enemy. Fear is induced by inflicting death and wounds on him. Death and wounds are produced by fire. Fire from the rear is more deadly and three times more effective than fire from the front, but to get fire behind the enemy, you must hold him by frontal fire and move rapidly around his flank.
General George S. Patton, Jr., *War As I Knew It*, 1947.

FIRST BATTLE

Then fell on Polydore his vengeful rage,
The youngest hope of Priam's stooping age:
(Whose feet for swiftness in the race surpast)
Of all his sons, the dearest, and the last.
To the forbidden field he takes his flight
In the first folly of a youthful knight,
To vaunt his swiftness, wheels around the plain,
Vaunts not long, with all his swiftness slain.
Struck where the crossing belts unite behind,
And golden rings the double back-plate join'd:
Forth thro' the navel burst the thrilling steel;
And on his knees with piercing shrieks he fell;
The rushing entrails pour'd upon the ground
His hands collect; and darkness wraps him round.
Homer, *The Iliad*, 20.471–484, c. 800 BC, tr. Alexander Pope, 1743; the death of Prince Polydorus, the youngest of the Trojan princes, who had disobeyed his father and gone into his first battle to die at the hands of Achilles.

Impatient to flesh his virgin sword.
Homer, *The Odyssey*, 20, c. 800 BC, tr. Alexander Pope, 1725–1726.

Let the boy win his spurs.
King Edward III, 26 August 1346, of his

FIRST BATTLE

son, the Black Prince, at the Battle of Crécy, when others urged the king to send help to his sorely pressed flank.

That shall be my music in the future!
> Charles XII, King of Sweden, 1700, at Copenhagen, the first time under fire as he listened to the bullets' whistle.

Only a fool will claim that he is as calm in his first battle as in his tenth... I know that my heart was pounding when reveille sounded on the morning of that memorable day.
> Frederick the Great, referring to the Battle of Mollwitz, 10 April 1741, quoted in Duffy, *The Military Life of Frederick the Great*, 1985.

I heard the bullets whistle, and believe me, there is something charming in the sound.
> General George Washington, 3 May 1754, letter to his mother about his first action.

As to the way in which some of our ensigns and lieutenants braved danger – the boys just come out from school – it exceeds all belief. They ran as at cricket.
> The Duke of Wellington (1769–1852), of the first action of new officers during the Peninsular War (1809–1813).

War loses a great deal of its romance after a soldier has seen his first battle. I have a more vivid recollection of the first than the last one I was in. It is a classical maxim that it is sweet and becoming to die for one's country; but whoever has seen the horrors of a battlefield feels that is far sweeter to live for it.
> Colonel John S. Mosby, *Mosby's War Reminiscences*, 1887.

When first under fire and you're wishful to duck,
Don't look or take heed at the man that is struck,
Be thankful you're living and trust to your luck,

And march to your front like a soldier.
> Rudyard Kipling (1865–1936), 'The Young British Soldier'.

Nothing is more exhilarating than to be shot at without result.
> Sir Winston S. Churchill, *The Malakand Field Force*, 1898.

I must have been unconscious for sometime. When I came to, Sergeant Bentele was working over me. French shell and shrapnel were striking intermittently in the vicinity... A quarter of an hour later, buglers sounded 'regimental call' and 'assembly.' From all sides parts of the regiment worked their way toward the area west of Bleid. One after the other the different companies came in. There were many gaps in their ranks. In its first fight the regiment had lost twenty-five per cent of its officers and fifteen per cent of its men in dead, wounded, and missing. I was deeply grieved to learn that two of my best friends had been killed...
> Field Marshal Erwin Rommel, of the first battle of the 124th Infantry Regiment, August 1914, *Infantry Attacks*, 1944.

The first fight, the first contact with war! I do not know what this held for others, but for me it had as much moving poetry as my first youthful love affair, my first kisses. (1918)
> Marshal Jospeh Pilsudski, *The Memoirs of a Polish Revolutionary and Soldier*, 1931.

Many of you have not been fortunate enough to have engaged in combat, and owing to the foolish writings of sob-sisters and tear-jerkers, you may have erroneous ideas of what battle is like. You will read of men – imaginary men – who on the eve of battle sit around the camp fire and discuss their mothers, and their sisters, and their sweet-hearts, and talk regretfully of their past life and fear foolishly for their future. No one has a higher or more respectful devotion to

FLEXIBILITY

women than I have; but the night before battle you do not sit around a fire... You go to sleep and have to be kicked in the butt in the morning so as to start the war. You have not dreamed of dying or worried about your boyhood. You have slept the sleep of fighting males eager for the kill.

General George S. Patton, December 1941, *The Patton Papers*, II, 1974.

Far as training went, our pilots were coming directly from basic trainers right into B-17's. They knew nothing about formation flying. We merely prayed to get 'em off the ground and get 'em down again.

You'd go to bed at night and think: *How could anybody every have the gall to bring a rabble like this into battle?*

And then you'd say to yourself: *You, too. How will you stand up? You've never been shot at, you don't know how you'll feel. Maybe you know more about the business of being a pilot; maybe you know more about navigation and bombardment, and even gunnery, than your men. But what do you know about how it feels to be in combat? Will you stand up? Will you have the nerve to ask them to stand up to it?*

Things like that you'd keep thinking. Then you'd be too exhausted to go to sleep.

General Curtis LeMay, *Mission With LeMay*, 1965.

FLEXIBILITY

One must not always use the same modes of operation against the enemy, even though they seem to be working out successfully. Often enough the enemy will become used to them, adapt to them, and inflict disaster on us.

The Emperor Maurice, *The Strategikon*, c. AD 600 (*Maurice's Strategikon*, tr. George Dennis, 1984).

Strategy is an art, and one who practices it must be supple and cunning and know how to make a timely alteration at every turn of it. For there is a time when it is not shameful to flee, if the occasion allows, and again to pursue relentlessly, each according to one's advantage; where success would seem more by cunning than by force, risking everything is to be deprecated. Since many and various matters lead toward one end, victory, it is a matter of indifference which one one uses to reach it.

General John Kinnamos, *Deeds of John and Manuel Comnenus*, 1976.

The Duke being asked how it was that he had succeeded in beating Napoleon's Marshals, one after another, said 'I will tell you. They planned their campaigns just as you might make a splendid set of harness. It looks very well; and answers very well; until it gets broken; and then you are done for. Now I made my campaigns of ropes. If anything went wrong, I tied a knot; and went on.'

The Duke of Wellington (1769–1852), quoted in William Fraser, *Words on Wellington*, 1889.

Strategic deployment can, and must be, planned far ahead. Battles in a war of positions demand similar treatment, but in the war of movement and the actions incidental to it the situations which the commander has to visualize follow one another in motley succession. He has to decide in accordance with his instinct. Thus soldiering becomes an art, and the soldier a strategist.

General Erich Ludendorff, *My War Memories, 1914–1918*, 1919.

The success of a commander does not arise from following rules or models. It consists in absolutely new comprehension of the dominant facts of the situation at the time, and all the forces at work. Every great operation of war is unique. What is wanted is a profound appreciation of the actual

event. There is no surer road to disaster than to imitate the plans of bygone heroes and fit them to novel situations.
>Sir Winston S. Churchill, *Marlborough*, 1933–1938.

Let us now discuss flexibility. What is flexibility? It is the concrete realization of the initiative in military operations; it is the flexible employment of armed force... the central task in directing a war, a task most difficult to perform well. In addition to organizing and educating the army and the people, the business of war consists in the employment of troops in combat, and all these things are done to win the fight. Of course it is difficult to organize an army, etc., but it is even more difficult to employ it, particularly when the weak are fighting the strong. To do so requires subjective ability of a very high order and requires the overcoming of the confusion, obscurity and uncertainty peculiar to war and the discovery of order, clarity and certainty in it; only thus can flexibility in command be realized.
>Mao Tse-tung, *On Protracted War*, May 1938.

Ensure that both plan and dispositions are flexible – adaptable to circumstances. Your plan should foresee and provide for a next step in case of success or failure, or partial success – which is the most common case in war. Your dispositions (or formation) should be such as to allow this exploitation or adaptation in the shortest possible time.
>Captain Sir Basil Liddell Hart, *Strategy*, 1954.

In battle, the art of command lies in understanding that no two situations are ever the same; each must be tackled as a wholly new problem to which there will be a wholly new answer.
>Field Marshal Viscount Montgomery of Alamein, *The Memoirs of Field Marshal Montgomery*, 1958.

I have used the word 'flexibility' for the last time. It seems that it is a large general purpose tent under which chaos, confusion, and incompetency are kept well hidden from the public.
>General Creighton Abrams (1914–1974), quoted in *Military Review*, 4/1985.

THE FOG OF WAR

Frederick: That was a diabolical day. Did you understand what was going on?
Catt: Your Majesty, I had a good grasp of the preliminary march, and the first arrangements for the battle. But all the rest escaped me. I could make no sense of the various movements.
Frederick: You were not the only one, my dear friend. Console yourself, you weren't the only one!
>Frederick the Great, speaking of the Battle of Zorndorff, 1758, quoted in Catt, *Unterhaltungen mit Friedrich dem Grossen*, 1884.

The general unreliability of all information presents a special problem in war: all action takes place, so to speak, in a kind of twilight, which like fog or moonlight, often tends to make things seem grotesque and larger than they really are.

Whatever is hidden from full view in this feeble light has to be guessed at by talent, or simply left to chance. So once again for lack of objective knowledge one has to trust to luck.
>Major-General Carl von Clausewitz, *On War*, 2.2, 1832, tr. Michael Howard and Peter Paret, 1976.

In war obscurity and confusion are normal. Late, exaggerated or misleading information, surprise situations and counterorders are to be expected.
>General of the Army George C. Marshall, *Infantry in Battle*, 1939.

Errors of judgment there must be in war, and few would cavil at them, especially those due to the fog of war. But it is different when the fog is self-

FOLLOW-UP

created by confused thought and limited study.
  Captain Sir Basil Liddell Hart, *Thoughts on War*, 1944.

The Vietnam War is the foggiest in my own personal experience. Moreover, it is the first war I know of wherein the fog of war is thicker away from the scene of conflict than on the battlefield.
  General Earle G. Wheeler, 18 December 1967, address to the Economic Club of Detroit, quoted in *Joint Force Quarterly*, Autumn–Winter 1997–1998.

FOLLOW-UP

Many generals believe that they have done everything as soon as they have issued orders, and they order a great deal because they find many abuses. This is a false principle; proceeding in this fashion, they will never reestablish discipline in an army in which it has been lost or weakened. Few orders are best, but they should be followed up with care; negligence should be punished without partiality and without distinction of rank or birth; otherwise you will make yourself hated. One can be exact and just, and be loved at the same time as feared. Severity must be accompanied with kindness, but this should not have the appearance of pretense, but of goodness.
  Field Marshal Maurice Comte de Saxe, *My Reveries*, 1732, tr. Phillips, 1940.

The principal task of the general is mental, large projects and major arrangements. But since the best dispositions become useless if they are not executed, it is essential that the general should be industrious to see whether his orders are executed or not.
  Frederick the Great, *Instructions to His Generals*, 1747, tr. Phillips, 1940.

It is not enough to give orders, they must be obeyed.
  Napoleon, 11 June 1806, advice to his stepson, Prince Eugene, *Correspondance de Napoléon Ier, publié par ordre de l'Empereur Napoléon III*, No. 10350, XII, 1858–1870.

If you want anything done well, do it yourself.
  The Duke of Wellington, quoted in William Fraser, *Words on Wellington*, 1889.

There is but one way; – to do as I did – to have A HAND OF IRON. The moment there was the slightest neglect in any department I was down on them.
  The Duke of Wellington, cited in Elizabeth Longford, *Wellington: Years of the Sword*, 1970.

No man of action, no commander, has finished when he has taken his decision and embodied it in an order. He remains to the last moment responsible for its execution in the way he intended and for the manifestation of his will in every stage of its accomplishment... One evening before a battle I was taking steps to discover whether our order had reached all the quarters concerned, and I received the brief answer in an honest Berlin accent, '*Ich greife an*,'* He had understood, and that was the essential thing.
Ich greife an, I attack.
  Colonel-General Hans von Seekt, *Thoughts of a Soldier*, 1930.

Commanders must remember that... the issuance of an order, or the devising of a plan, is only about 5 percent of responsibility of command. The other 95 percent is to insure, by personal observation, or through interposing of staff officers that the order is carried out.
  General George S. Patton, Jr, quoted in *Military Review*, 9/1980.

It is a mistake to think that once an order is given there is nothing more to be done; you have got to see that it is carried out in the spirit which you intended. Once he has decided on his outline plan and how he will carry it out, the commander should himself draft the initial operational order or directive, and not allow his staff to do so. His staff and subordinates then begin their more detailed work, and this is based on the written word of the commander himself. Mistakes are thus reduced to a minimum. This was my method, beginning from the days when I commanded a battalion.

> Field Marshal Viscount Montgomery of Alamein, *The Memoirs of Field Marshal Montgomery*, 1958.

You may have heard this story about General Pershing in World War I. While inspecting a certain area, he found a project that was not going too well, even though the second lieutenant in charge seemed to have a pretty good plan. General Pershing asked the lieutenant how much pay he received, and when the lieutenant replied: '$141.67 per month, Sir,' General Pershing said, 'Just remember that you get $1.67 for making your plan and issuing the order, and $140.00 for seeing that it is carried out.'

> General of the Army Omar N. Bradley, 'Leadership', *Parameters*, Winter 1972.

FOLLY

I was in a furious rage. The Turks should never, by the rules of sane generalship, have ventured back to Tafileh at all. It was simple greed, a dog-in-the-manger attitude unworthy of a serious enemy, just the sort of hopeless thing a Turk would do. How could they expect a proper war when they gave us no chance to honour them? Our morale was continually being ruined by their follies, for neither could our men respect their courage, nor our officers respect their brains. Also, it was an icy morning, and I had been up all night and was Teutonic enough to decide that they should pay for my changed mind and plan.

> Colonel T.E. Lawrence, *The Seven Pillars of Wisdom*, 1935.

Another lesson of history is that the most dangerous folly of all is man's failure to recognize his own aptitude for folly. That failure is a common affliction of authority.

> Captain Sir Basil Liddell Hart, *Through the Fog of War*, 1938.

One never attributes folly to his enemy – but then, of such stuff are surprises made.

> Lieutenant-General Phillip B. Davidson, writing of the rashness of the 1968 Tet Offensive by the North Vietnamese, *Vietnam at War*, 1988.

FORCE

Force has no place, there is need of skill.

> Herodotus, *The History*, c. 450 BC.

Nothing is so weak and unstable as a reputation for power not based on force.

> Cornelius Tacitus, *Annals*, AD 116.

In every country, force yields to civic qualities. Bayonets are lowered before the priest who speaks in the name of Heaven and before the scholar who inspires respect for his science. I have predicted to certain military men, who had scruples, that military rule will never take root in France, unless the nation were stultified by fifty years of ignorance. All attempts will fail, and their authors will be the victims. I do not govern in the capacity of a general but by virtue of the civil qualities which in the eyes of the nation qualify me for the government. If the nation did not hold this opinion, the government would not last. I knew what I was doing when, as army commander, I styled myself 'Member of the Institute'; I was sure of being understood by the last drummer boy.

FORCE

Napoleon, 1802, meeting of the Council of State, quoted in Christopher Herold, ed., *The Mind of Napoleon*, 1955.

Do you know what I admire most in this world? It's the total inability of force to organize anything. There are only two powers in the world – the sword and the spirit. By spirit I understand the civil and religious institutions... In the long run, the sword is always beaten by the spirit.
Napoleon, 1808, conversation with Fontanes, quoted in Christopher Herold, ed., *The Mind of Napoleon*, 1955.

People are very apt to believe that enthusiasm carried the French through their revolution, and was the parent of those exertions which nearly conquered the world; but if the subject is nicely examined, it will be found that enthusiasm was the name only, but that force was the instrument which brought forward those great resources... and that a perseverance in the same system of applying every individual and every description of property to the service of the army, by force, has since conquered Europe.
The Duke of Wellington, 22 August 1809 at Miajadas, letter to Lord Castlereagh, commenting on how the Spanish leadership attempted to govern with enthusiasm, but without regularity and order, quoted in Julian Rathbone, *Wellington's War: His Peninsular Dispatches*, 1984.

Force, to counter opposing force, equips itself with the inventions of art and science. Attached to force are certain self-imposed, imperceptible limitations hardly worth mentioning, known as international law and custom, but they scarcely weaken it. Force – that is, physical force, for moral force has no existence save as expressed in the state and the law – is thus the *means* of war; to impose our will on the enemy is its object. To secure that object we must render the enemy powerless; and that, in theory, is the true aim of warfare.
Major-General Carl von Clausewitz, *On War*, 1.1, 1832, tr. Michael Howard and Peter Paret, 1976.

Not by speeches and decisions of majorities will the greatest problems of the time be decided – that was the mistake of 1848–49 – but by iron and blood.
Prince Otto von Bismarck, 29 September 1862, while ambassador to France, to a group of Prussian ministers.

An efficient military body depends for its effect in war – and in peace – less upon its position than upon its concentrated force.
Rear Admiral Alfred Thayer Mahan, *Naval Administration and Warfare*, 1908.

Not believing in force is the same as not believing in gravitation.
Leon Trotsky (1879–1940).

The one means that wins the easiest victory over reason: terror and force.
Adolf Hitler, *Mein Kampf*, 1.2, 1924.

Against naked force the only possible defense is naked force. The aggressor makes the rules for such a war; the defenders have no alternative but matching destruction with more destruction, slaughter with greater slaughter.
President Franklin Roosevelt, 21 August 1941, message to young Democrats Convention, Louisville, Kentucky.

The more I reflect on the experience of history the more I come to see the instability of solutions achieved by force, and to suspect even those instances where force has had the appearance of resolving difficulties.
Captain Sir Basil Liddell Hart, *Thoughts on War*, 1944.

FORESIGHT/ANTICIPATION

Once you decide to use force, you had better make sure you have plenty of it. If you need a battalion to do a job, it's much better to have the strength of a division. You probably won't suffer any casualties at all in that way.
General of the Army Dwight D. Eisenhower, *The New York Times*, 10 May 1965.

Bush and Blair don't understand anything but the power of force. Every time they kill us, we kill them, so the balance of terror can be achieved. It is the duty of every Muslim to fight. Killing Jews is top priority.
Osama bin Ladin, undated video, reported on 11 November 2001 by the *Daily Telegraph*

FORESIGHT/ANTICIPATION

After agreeing upon a treaty or a truce with the enemy, the commander should make sure that his camp is guarded more strongly and more closely. If the enemy chooses to break the agreement, they will only gain a reputation for faithlessness and the disfavor of God, while we shall remain in safety and be true to our word. A general should not have to say: 'I did not expect it.'
The Emperor Maurice, *The Stratigikon*, c. AD 600 (*Maurice's Strategikon*, tr. George Dennis, 1984).

As it turned out, they wished to do it all in safety and to fall upon us by night, when we were not watching. And as I was so much on my guard, they found me standing in front of their thoughts.
Hernan Cortes, 1519, on his entry into the Valley of Mexico, in Madariaga, *Hernan Cortes*, 1942.

Think to a finish!
Field Marshal Viscount Allenby of Meggido, 1902, upon taking command of the Fifth Lancers.

The vital point in actual warfare is to apply to the enemy what we do not wish to be applied to ourselves and at the same time not to let the enemy apply it to us. Therefore, it is most important that what we consider would embarrass the enemy we should apply to them before they can do the same to us; we must always forestall them.
Admiral Marquis Heihachiro Togo, February 1905, to the officers of the Japanese Fleet, quoted in Ogasawara, *Life of Admiral Togo*, 1934.

Dangers which are warded off by effective precaution and foresight are never even remembered.
Sir Winston S. Churchill (1874–1965).

Want of foresight, unwillingness to act when action would be simple and effective, lack of clear thinking, confusion of counsel until emergency comes, until self-preservation strikes its jarring gong – these are the features which constitute the endless repetition of history.
Sir Winston S. Churchill, speech in the House of Commons, 2 May 1935, *Winston S. Churchill: His Complete Speeches 1897–1963*, VI, 1974, p. 6499.

A commander must think two stages ahead.
Field Marshal Viscount Montgomery of Alamein, *The Memoirs of Field Marshal Montgomery*, 1958.

While Rommel was leading his troops in person against our strongly-held defensive positions on the Alam Halfa ridge, Montgomery was planning the battle of Alamein. That was the difference between the two.
Lieutenant-General Sir Brian Horrocks, *Escape to Action*, 1961.

It is a cardinal responsibility of a commander to foresee insofar as possible where and when crises affecting his command are likely to occur. It

FORTUNE

starts with his initial estimate of the situation – a continuing mental process from the moment of entering the combat zone until his unit is pulled out of the line. Ask yourself these questions. What are the enemy's capabilities? What shall I do, or what could I do, if he should exercise that one of his capabilities which would be most dangerous to me, or most likely to interfere with the accomplishment of my mission?
General Matthew B. Ridgway, 'Leadership', Military Review, 10/1966.

The job of a commander is to think ahead. A GHQ must be dealing with events foreseen two and preferably three or more days away. The imagination and foresight with which he manages that is one of the inescapable tests of a commander: because if there is one rule about a battle it is, 'What is possible today may not be possible tomorrow.'
General Saad El Shazly, The Crossing of the Suez, 1980.

It is easier to get into something than to get out of it.
Secretary of Defense Donald Rumsfeld, Rumsfeld's Rules, Revised edition 2000.

FORTIFICATIONS

When somebody... asked why Sparta lacked fortification walls, he pointed to the citizens under arms and said: 'These are the Spartans' walls.'
Plutarch, quoting Agesilaus, Eurypontid King of Sparta (400–360 BC), The Lives, c. AD 100 (Plutarch on Sparta, tr. Richard Talbert, 1988).

By God's throat, even if that castle were all built of butter and not of iron and stone, I have no doubt it would defend me against him [King of France] and all his forces.
Richard I, 'The Lion Heart', King of England, 1198–1199, of his new castle, Gaillard.

The art of defending fortified places consists in putting off the moment of their reduction.
Frederick the Great, Instructions to His Generals, 1747, tr. Phillips, 1940.

The only advantage a fortified line may afford is to render the situation of the enemy so difficult as to induce him to operate incorrectly, which in turn may cause him to be defeated with numerically inferior forces. Or if one is confronted by a capable general, the fortifications may have the effect of compelling him methodically to negotiate the obstacles that one has created with due deliberation. In this manner time is gained.
Napoleon, 1807, 'Notes sur la defence de l'Italie', Correspondance de Napoléon Ier, publié par ordre de l'Empereur Napoléon III, No. 14704, Vol. XVIII, 1858–1870.

FORTUNE

I who never yet feared anything that was human, have amongst such as were divine, always had a dread of Fortune as faithless and inconstant...
Lucius Aemilius Paullus, 216 BC, quoted in Plutarch, Plutarch's Lives, c. AD 100, tr. John Dryden.

It is difficult for a man to whom fortune has never proved false to reflect upon its uncertainties.
Hannibal, 202 BC, to Scipio Africanus on the day before the Battle of Zama, which ended the Second Punic War in the defeat of Carthage, quoted in Livy, Annals of the Roman People, c. 25 BC–AD 17 (Livy: A History of Rome, 30, tr. Moses Hadas and Joe Poe, 1962).

In war we must always leave room for strokes of fortune and accidents that cannot be foreseen.
Polybius, Universal History, c. 125 BC (The Rise of the Roman Empire, tr. Ian Scott-Kilvert, 1987).

FREEDOM

Go ahead, my friend. Be bold and fear nothing. You have got Caesar and Caesar's fortune with you in your boat.
> Julius Caesar, 28 November 49 BC, during the Civil War, when the master of the boat in which he was travelling from Brindisium to Dyrrhaccium was so terrified of the rough sea that he was about to turn about, quoted in Plutarch, *The Lives*, c. AD 100 (*The Fall of the Roman Republic*, tr. Rex Warner, 1972).

For I know well the terror and misfortune which have laid hold upon the enemy and compel them to become less brave, for the one fills them with fear because of what has already happened, and the other brushes aside their hope of success. For Fortune, once seen to be bad, straightway enslaves the spirit of those who have fallen in her way.
> Count Belisarius, December AD 533, address to his army before the Battle of Tricamarum in North Africa, during which he destroyed the last hope of the Vandals, quoted in Procopius, *History of the Wars*, 4.1, c. AD 560, tr. H.B. Dewing, 1929.

It is better to subdue an enemy by famine than by sword, for in battle, *fortuna* has often a much greater share than *virtu*.
> Niccolò Machiavelli, *The Art of War*, 1521.

Indeed the fate of states and the reputation of generals rest on the most trifling incidents. A few seconds are enough to determine their fortune.
> Frederick the Great, *Oeuvres de Frederic le Grand*, (1846–1857).

Rule fortune.
> Field Marshal Prince Aleksandr V. Suvorov (1729–1800), quoted in Blease, *Suvorof*, 1920.

In warfare every opportunity must be seized; for fortune is a woman; if you miss her to-day, you need not expect to find her tomorrow.
> Napoleon, 30 July 1800, quoted in R.M. Johnston, ed., *The Corsican*, 1910.

I propose to get into fortune's way…
> The Duke of Wellington, 10 December 1812, to Lord Beresford, *The Dispatches of Field Marshal the Duke of Wellington, During His Various Campaigns in India, Denmark, Portugal, Spain, the Low Countries, and France*, IX, 1834–1838, p. 609.

…The contingent element inseparable from the waging of war [which] gives to that activity both its difficulty and its grandeur.
> General Charles de Gaulle, *The Edge of the Sword*, 1932.

I am more than ever impressed with the part that good or bad fortune sometimes plays in tactical engagements. The authors give us credit, where no credit is due, for being able to choose the exact time for our attack on the Japanese carriers when they were at the great disadvantage – flight decks full of aircraft fueled, armed and ready to go. All I can claim credit for, myself, is a very keen sense of the urgent need for surprise and a strong desire to hit the enemy carriers with our full strength as early as we could reach them.
> Admiral Raymond A. Spruance, foreword to *Midway: The Battle That Doomed Japan*, 1955; surely a description of how 'fortune favours the bold'.

FREEDOM

What stands if Freedom fall?
> Rudyard Kipling, 'For all we have and are,' verse, 1940.

The winning of freedom is not to be compared to the winning of a game – with the victory recorded forever in history. Freedom has its life in the heart, the actions, the spirit of men and so it must be daily earned and refreshed – else like a flower cut from its life-giving roots, it will wither and die.

FRIENDLY FIRE/FRATRICIDE

General of the Army Dwight D. Eisenhower, 12 June 1945, The Guildhall Speech, London, in John S.D. Eisenhower, *General Ike: A Personal Reminiscence*, p. 233.

Those we lost were last seen on active duty. Their final act on this earth was to fight a great evil, and bring liberty to others. All of you – all in this generation of our military – have taken up the highest calling of history… And wherever you go, you carry a message of hope – a message that is ancient, and ever new. In the words of the prophet Isaiah: 'To the captives, 'Come out,' – and to those in darkness, 'Be free.'

President George W. Bush, 1 May 2003, speech from the deck of the USS *Abraham Lincoln* announcing the end of military operations in the liberation of Iraq.

With those attacks, the terrorists and their supporters declared war on the United States. And war is what they got… Our commitment to liberty is in America's tradition, declared at our founding, affirmed in Franklin Roosevelt's Four Freedoms, asserted in the Truman Doctrine and in Ronald Reagan's challenge to an evil empire. We are committed to freedom in Afghanistan, in Iraq and in a peaceful Palestine. The advance of freedom is the surest strategy to undermine the appeal of terror in the world… You all can know, friend and foe alike, that our nation has a mission. We will answer threats to our security and we well defend the peace…The war on terror is not over, yet it is not endless. We do not know the day of final victory, but we have seen the turning of the tide.

President George W. Bush, 1 May 2003, speaking from the deck of the aircraft carrier, USS *Abraham Lincoln*, announcing the end of military operations in the liberation of Iraq.

FRICTION

…The effect of reality on ideas and intentions in war.

Major-General Carl von Clausewitz, to Marie von Brühl, 29 September 1806.

Everything in war is very simple, but the simplest thing is difficult. The difficulties accumulate and end by producing a kind of friction that is inconceivable unless one has experienced war… Countless minor incidents – the kind you can never really foresee – combine to lower the general level of performance, so that one always falls far short of the intended goal. Iron willpower can overcome friction; it pulverizes every obstacle, but of course it wears down the machine as well.

Friction is the only concept that more or less corresponds to the factors that distinguish real war from war on paper. The military machine… is composed of individuals, every one of whom retains his potential for friction… the least important of whom may chance to delay things or somehow make them go wrong.

Major-General Carl von Clausewitz, *On War*, 1.7, 1832, tr. Michael Howard and Peter Paret, 1976.

Friendly relations contribute a lot to diminish the friction of war. Comradeship is the oil in the machine. If 'Bill' appeals to his friend 'Dick' he is more sure of good and prompt support than if an overtired O.C. 9th Battalion sends a formal message to an overburdened O.C. 999th Battery. It is thus important to lose no opportunity of creating a cooperative atmosphere by making touch. The military saying that 'time spent in reconnaissance is rarely wasted' applies just as clearly to the more personal kind of reconnaissance within your own front. (July 1933.)

Captain Sir Basil Liddell Hart, *Thoughts on War*, 1944.

FRIENDLY FIRE/FRATRICIDE

True, but this is a necessary evil which

FUNERALS

we must endure, to arrest impending destruction.

> General Marquis Cornwallis, 15 March 1781, to Brigadier-General O'Hara, of the necessity of firing on his own troops mixed with the advancing Americans at the Battle of Guilford Courthouse, quoted in Henry Lee, *Memoir of the War in the Southern Department*, 1869.

Friendly fire, isn't.

> US Navy SEAL list of quips published just before Operation Just Cause in Panama, December 1989.

I detest the term 'friendly fire.' Once a bullet leaves a muzzle or a rocket leaves an airplane, it is not friendly to anyone. Unfortunately, fratricide has been around since the beginning of war. The very chaotic nature of the battlefield, where quick decisions make the difference between life and death, has resulted in numerous incidents of troops being killed by their own fire.

> General H. Norman Schwarzkopf, *It Doesn't Take a Hero*, 1992.

FUNERALS

As he was dying, he said, looking towards his companions, 'I see that my funeral rites will be imposing.'

> Alexander the Great, anticipating the wars among his successors, quoted in Plutarch, *Moralia*, Vol. III, 181, 2nd century AD, p.67

The funeral pyre was raised and gold was brought up from the hoard in payment of wergild for Hneaef. Upon the pyre were placed the dead warriors in bloodied corslets, gilded boar glittering above the helmets. Then Hildeburh in double grief ordered the body of her slain son to be placed by the side of her brother Hneaef, nephew at uncle's shoulder. She keened her sorrow as the bodies mounted to the pyre.

The fire swirled up to the heavens and roared above the mound. Heads melted and the wounded bodies burst with spring blood. Then the flames, greediest of spirits, swallowed up those taken away in battle, their glory gone with the smoke.

> *Beowulf*, c. AD 1000 (*My Name is Beowulf*, tr. Frederick Rebsamen, 1971).

There should be no real neglect of the dead, because it has a bad effect on the living; for each soldier values himself as highly as though he were living in a good house at home.

> General of the Army William T. Sherman, *Memoirs of W.T. Sherman*, 1875.

There was a tenderness under the apple trees as powder-grimed officers and men brought in the dead; a tenderness for lost comrades, who had fought together so often and so well, that went beyond reverence and compassion... Funerals in the field, rough and ready though they be, seem less bleak than those performed with funeral rites, as though the soldier whose calling deals with sudden death can find a way to stand easy in its shadow.

> Brigadier Lord Lovat, *March Past*, 1978.

G

GENERALS/GENERALSHIP

There are five dangerous faults which may affect a general: (1) Recklessness, which leads to destruction; (2) cowardice, which leads to capture; (3) a hasty temper that can be provoked by insults; (4) delicacy of honor that is sensitive to shame; (5) over-solicitude for his men, which exposes him to worry and trouble. These are the five besetting sins of a general, ruinous to the conduct of war.

When an army is overthrown and its leader slain, the cause will surely be found among the five dangerous faults. Let them be a subject of meditation.
  Sun Tzu, *The Art of War*, 8, c. 500 BC, tr. Giles, 1910.

For a general must also be capable of furnishing military equipment and providing supplies for the men; he must be resourceful, active, careful, hardy, and quick-witted; he must be both gentle and brutal, at once straightforward and designing, capable of both caution and surprise, lavish and rapacious, generous and mean, skilful in defence and attack; and there are many other qualifications, some natural, some acquired, that are necessary to one who would succeed as a general. It is well to understand tactics too; for there is a wide difference between right and wrong disposition of the troops, just as stones, bricks, timber and tiles flung together anyhow are useless, whereas when the materials that neither rot nor decay, that is the stones and tiles, are placed at the bottom and the top, and the bricks and timber are put together in the middle, as in building, the result is something of great value, a house, in fact.
  Socrates (470–399 BC), quoted by Xenophon, *Memorabilia*, 3.6–7, c. 360 BC (tr. E.C. Marchant, 1923).

He used to say that a general needs to show daring towards his opponents, goodwill towards his subordinates and a cool head in crisis.
  Plutarch, paraphrasing Agesilaus, Eurypontid King of Sparta (400–360 BC), *The Lives*, c. AD 100, *(Plutarch on Sparta*, tr. Richard Talbert, 1988).

When people said of Scipio Africanus that he lacked aggressiveness, he is reported to have answered: 'My mother bore me a general, not a warrior.'
  Scipio Africanus (236–184 BC), quoted in Frontinus, *Stratagems*, 4.7, c. AD 84–96, tr. Charles E. Bennett, 1925.

A good general not only sees the way to victory; he also knows when victory is not possible.
  Polybius, *Histories*, 1, c. 125 BC.

It was Caesar's hope that he could finish the business without fighting or casualties because he had cut his adversaries off from food supply. Why lose men, even for victory? Why expose soldiers who deserved so well of him to wounds? Why even tempt fortune. Victory through policy is as much a mark of the good general as victory by the sword.
  Julius Caesar, *The Civil War*, c. 45 BC, (*The Gallic Wars and other Writings by Julius Caesar*, tr. Moses Hadas, 1957).

GENERALS/GENERALSHIP

There are five skills and four desires involved in generalship.

The five skills are: skill in knowing the disposition and power of enemies, skill in knowing the ways to advance and withdraw, skill in knowing how empty or how full countries are, skill in knowing nature's timing and human affairs, and skill in knowing the features of terrain.

The four desires are: desire for the extraordinary and unexpected in strategy, desire for thoroughness in security, desire for calm among the masses, and desire for unity of hearts and minds.

> Zhuge Liang (AD 180–234), *The Way of the General* (*Mastering the Art of War*, tr. Thomas Cleary, 1989).

A ship cannot cross the sea without a helmsman, nor can one defeat an enemy without tactics and strategy. With these and the aid of God it is possible to overcome not only an enemy force of equal strength but even one greatly superior in numbers. For it is not true, as some inexperienced people believe, that wars are decided by courage and numbers of troops, but, along with God's favor, by tactics and generalship, and our concern should be with these rather than wasting our time in mobilizing large numbers of men. The former provide security and advantage to men who know how to use them well, whereas the other brings trouble and financial ruin.

> The Emperor Maurice, *The Strategikon*, c. AD 600 (*Maurice's Strategikon*, tr. George Dennis, 1984).

It is safer and more advantageous to overcome the enemy by planning and generalship than by sheer force; in the one case the results are achieved without loss to oneself, while in the other some price has to be paid.

> The Emperor Maurice, *The Strategikon*, c. AD 600 (*Maurice's Strategikon*, tr. George Dennis, 1984).

The commanding general, as well as the general, the military governor, the ruling general: his office [is] warfare. [He is] the maneuverer of troops – a courageous warrior, one whose mission is to go to his death.

The good commanding general, [or] general, [is] able, prudent, a holder of vigil, a maneuverer of forces. He devises strategy; he declares, he assumes the responsibility of war. He distributes, he supervises the arms; he distributes, commands, supervises the provisioning. He lays out, he searches out the roads [to the foe]; he tracks [them]. He establishes the war huts, the prisons, the market places in enemy lands. He places the sentries, posts the chosen ones, stations the spies, the hidden ones, the concentrated ones. He interrogates them; he discovers the places where the enemy will approach.

The stupid commanding general, [or] general, causes trouble, causes death, leads one into danger.

> Aztec account of generalship, c. 1500, quoted in Fray Bernardino de Sahagún, *General History of the Things of New Spain, Book 10 – The People*, 1961.

Pay well, command well, hang well.

> General Sir Ralph Hopton (1596–1652), *Maxims for the Management of an Army*, 1643.

The knowledge of a real general must be varied. He must have an accurate idea of politics in order to be informed of the intention of princes and the forces of states and of their communications; to know the number of troops and the princes and their allies can put in the field; and to judge the condition of their finances. Knowledge of the country where he must wage war serves as the base for all strategy. He must be able to imagine himself in the enemy's shoes in order to anticipate all the obstacles that are likely to be placed in his way. Above all, he must train his mind to furnish him with a multitude of expedients, ways and means in case of need. All this requires study and exercise. For those who are destined for the military profession, peace must be a time of meditation, and war the period where one puts his ideas into practice.

> Frederick the Great, 'Reflexions sur la projets de campagne', Oeuvres de Frederic le Grand, 1846–1856, in Jay Luvaas, ed., *Frederick the Great on the Art of War*, 1966.

Maxim 81. It is exceptional and difficult to find all the qualities of a great general combined in one man. What is most desirable and distinguishes the exceptional man, is the balance of intelligence and ability with character or courage. If courage is predominant, the general will hazard far beyond his conceptions; and on the contrary, he will not dare to accomplish his conceptions if his character or his courage are below his intelligence.

> Napoleon, *The Military Maxims of Napoleon*, tr. Burnod, 1827.

I have seen the order of the day signed by you that makes you the laughing stock of Germany, Austria, and France. Have you no friend who will tell you the truth? You are King and brother of the Emperor, – ridiculous title in warfare! You must be a soldier, and again a soldier, and always a soldier! You must bivouac with your outposts, spend night and day in the saddle, march with your advance guard so as to get information, or else remain in your seraglio. You wage war like a satrap. By Heaven! Is it from me you have learned that! – from me, who with an army of 200,000 men live with my skirmishers?

> Napoleon, 17 May 1809, to his brother Jérôme, quoted in R.M. Johnston, ed., *The Corsican*, 1910.

To be a good general a man must know mathematics; it is of daily help in straightening one's ideas. Perhaps I owe my success to my mathematical conceptions; a general must never imagine things, that is the most fatal of all. My great talent, the thing that marks me most, is that I see clearly; it is the same with my eloquence, for I can distinguish what is essential in a question from every angle. The great art in battle is to change the line of operations during the course of the engagement; that is an idea of my own, and quite new.

The art of war does not require complicated maneuvers; the simplest are the best, and common sense is fundamental. From which one might wonder how it is generals make blunders; it is because they try to be clever. The most difficult thing is to guess the enemy's plan, to sift the truth from all the reports that come in. The rest merely requires common sense; it is like a boxing match, the more you punch the better it is. It is also necessary to read the map well.

> Napoleon, 29 January 1818, St Helena, quoted in R.M. Johnston, ed., *The Corsican*, 1910.

The most essential qualities for a general will always be: *first*, a high moral courage, capable of great resolution; *second*, a physical courage which takes no account of danger. His scientific or military acquirements are secondary to these. It is not necessary that he should be a man of vast erudition; his knowledge may be limited but it should be thorough, and he should be perfectly grounded in the principles at the base of the art of war.

Next in importance come the qualities of his personal character. A man who is gallant, just, firm, upright, capable of esteeming merit in others instead of being jealous of it, and skillful in making his merit add to his own glory, will always be a good general and may even pass for a great man... *finally*, the union of wise theory with great character will constitute the great general.

GENERALS/GENERALSHIP

Lieutenant-General Antoine-Henri Baron de Jomini, *Summary of the Art of War*, 1838, tr. Mendell and Craighill, 1862.

A general has much to bear and needs strong nerves. The civilian is too inclined to think that war is only like the working out of an arithmetical problem with given numbers. It is anything but that. On both sides it is a case of wrestling with powerful, unknown physical and psychological forces, a struggle which inferiority in numbers makes all the more difficult. It means working with men of varying force of character and with their own views. The only quantity that is known and constant is the will of the leader.
General Erich Ludendorff, *My War Memories, 1914–1918*, 1919.

These, then, are the three pillars of generalship – courage, creative intelligence, and physical fitness; the attributes of youth rather than of middle age.
Major-General J.F.C. Fuller, *Generalship: Its Diseases and Their Cure*, 1933.

What I hold to be the first essential of a general [is] the quality of robustness, the ability to stand the shocks of war.
Field Marshal Viscount Wavell of Cyrenaica, 1939, 'Lees Knowles Lectures', Trinity College, Cambridge.

I don't mind being called tough, since I find in this racket it's the tough guys who lead the survivors.
General Curtis LeMay, 1943.

Here are six points which are important for successful generalship. A general must:
(a) Know his 'stuff' thoroughly.
(b) Be known and recognized by the troops.
(c) Ensure the troops are given tasks that are within their powers. Success means mutual confidence – failure the reverse.
(d) See that subordinate commanders are disturbed as little as possible in carrying out their tasks.
(e) Command by personal contact.
(f) Be human and study the human factor.
Major-General Sir Francis de Guingand, *Operation Victory*, 1947.

Generals are often prone, if they have the chance, to choose a set-piece battle, when all is ready, at their own selected moment, rather than to wear down the enemy by continued unspectacular fighting. They naturally prefer certainty to hazard. They forget that war never stops, but burns on from day to day with ever-changing results not only in one theatre but in all.
Sir Winston S. Churchill (1874–1965).

I have always believed that a motto for generals must be 'No Regrets,' no crying over spilt milk.
Field Marshal Viscount Slim of Burma, *Defeat Into Victory*, 1956.

To exercise high command successfully one has to have an infinite capacity for taking pains and for careful preparation; and one has also to have an inner conviction which at times will transcend reason. Having fought, possibly over a prolonged period, for the advantage and gained it, there then comes the moment for boldness. When that moment comes, will you throw your bonnet over the mill and soar from the known to seize the unknown? In the answer to this question lies the supreme test of generalship in high command.
Field Marshal Viscount Montgomery of Alamein, *The Memoirs of Field Marshal Montgomery*, 1958.

Let me give you some advice, Lieutenant. Don't become a general. Don't ever become a general. If you become a general you just plain have too much to worry about.

GENERAL STAFF

General of the Army Dwight D. Eisenhower, 9 May 1965.

GENERAL STAFF

Normally it is not possible for an army simply to dismiss incompetent generals. The very authority which their office bestows upon generals is the first reason for this. Moreover, the generals form a clique, tenaciously supporting each other, all convinced that they are the best possible representatives of the army. But we can at least give them capable assistants. Thus the General Staff officers are those who support incompetent generals, providing the talents that might otherwise be wanting among leaders and commanders.

General Gerhard von Scharnhorst (1755–1813), quoted in T.N. Dupuy, *A Genius for War*, 1977.

The most highly trained general staff, with the best of ideas and principles cannot guarantee good leadership in an army by itself. It must be backed by a great general who serves as its leader and counterbalance, who from time to time prevents the staff from entangling itself in its own red tape. A good staff on the other hand is an indispensable aid for a general.

Major-General Carl von Clausewitz (1780–1831), *The Campaign of 1799 in Italy and Switzerland*, 1906.

The difficulty of always selecting a good general has led to the formation of a good general staff, which being near the general may advise him and thus exercise a beneficial influence over the operations. A well-instructed general staff is one of the most useful organizations, but care must be observed to prevent the introduction into it of false principles.

Lieutenant-General Antoine-Henri Baron de Jomini, *Summary of the Art of War*, 1838, tr. Mendell and Craighill, 1862.

Great achievements, small display: more reality than appearance.

Field Marshal Helmuth Graf von Moltke (1800–1891), motto of the German General Staff, quoted in Heinz Guderian, *Panzer Leader*, 1953.

The General Staff officer was, so to speak, a man apart. As the war became more technical, his duties became more arduous. It was no longer sufficient for him to have a general knowledge of all arms and their employment. He had to be a good artilleryman and, in addition, to possess a sound knowledge of the use of aircraft, signalling, supply questions and a thousand other things, while he had to master many details which the divisional commander had no time to settle. In spite of every effort to keep them brief, the orders which had to draft grew ever longer and more complicated. The more technical the war became, the more did these orders grow into veritable works of art, involving infinite skill and knowledge. There was no other way, if things were to go smoothly. The variety of his functions often compelled the General Staff officer to keep many things in his own hands. Care had to be taken that the independence of other services did not suffer on this account, and that the commander, too, was not 'shelved.' I could never have allowed either of these developments.

General Erich Ludendorff, *My War Memories, 1914–1918*, 1919.

The General Staff was certainly one of the most remarkable structures within the framework of our German Army. Side by side with the distinctly hierarchical form of the commands it constituted a special element which had its foundation in the great intellectual prestige of the Chief of Staff of the Army, Field Marshal Count von Moltke. The peace training of the General Staff officer offered a guarantee that in case of

GENIUS

war all the commanders in the field should be controlled from a single source, and all their plans governed by a common aim. The influence of the General Staff on those commanders was not regulated by any binding order. It depended far more on the military and personal qualities of the individual officer. The first requirement of the General Staff officer was that he should keep his own personality and actions entirely in the background. He had to work out of sight, and therefore be more than he seemed.
> Field Marshal Paul von Hindenburg, *Out of My Life*, 1920.

General Staff officers have no names.
> Colonel-General Hans von Seekt, quoted in Hermann Foertsch, *The Art of Modern Warfare*, 1940.

No soldier can doubt the immense value of a general staff if it is the general's servant, and not the general's gaoler. I have said that the staff has no responsibilities; it has none, though it has duties; because it has no powers of decision or command. It can suggest, but it has no responsibility for actions resulting; the general alone is responsible, therefore the general alone should and must decide, and, more than this, he must elaborate his own decisions and not merely have them thrust upon him by his staff like a disc upon a gramophone.
> Major-General J.F.C. Fuller, *Generalship: Its Diseases and Their Cure*, 1933.

My general staff training revolted against an operation of this sort. If old von Moltke thought I had planned this offensive he would turn in his grave.
> Field Marshal Gerd von Rundstedt, November 1944, referring to the plan for the Ardennes Offensive of December 1944.

Prerequisites for appointment to the General Staff were integrity of character and unimpeachable behavior and way of life both on and off duty. Next came military competence; a man had to have proved himself at the front, had to have an understanding for tactical and technical matters, a talent for organization and powers of endurance both physical and mental; he had also to be industrious, of sober temperament, and determined.

In selecting officers from this point of view it is possible that intellectual ability was sometimes overvalued in comparison to strength of character and particularly to warmth of personality; but these last two qualities are much less easily estimated, particularly since they do not by their very nature tend to be spectacular.
> Colonel-General Heinz Guderian, *Panzer Leader*, 1953.

The German General Staff was a human institution, and it, too, had human frailties and weaknesses, which contributed, at least in part, to its two major defeats in this century. Yet the performance of that General Staff, and of the Army it designed and built in both of those disastrous wars, was comparable in terms of military excellence to Napoleon and Hannibal at their best. Perhaps, in this sense, it is not too much to say that in striving to institutionalize excellence in military affairs, the German General Staff can be said to have institutionalized military genius itself.
> Colonel T.N. Dupuy, *A Genius for War: The German General Staff and the German Army*, 1977.

GENIUS

Great geniuses have a sort of intuitive knowledge; they see at once the cause, and its effect, with the different combinations, which unite them: they do not proceed by common rules, successively from one idea to another, by slow languid steps, no: the *Whole*, with all its

GENIUS

circumstances and various combinations, is like a picture, all together present to their mind; they want no geometry, but an age produces few of this kind of men: and in the common run of generals, geometry, and experience, will help them to avoid gross errors.

Major-General Henry Lloyd, *History of the Late War in Germany*, 1766.

Of the conquerors and eminent military characters who have at different times astonished the world, Alexander the Great and Charles the Twelfth of Sweden are two of the most singular; the latter of whom was the most heroic and the most extraordinary man of whom history has left any record. An army which had Alexander or Charles in its eye was different from itself in its simple nature, it imbibed a share of their spirit, became insensible of danger, and heroic in the extreme.

Robert Jackson, *A Systematic View on the Formation, Discipline, and Economy of Armies*, 1804.

Achilles was the son of a goddess and a mortal: he symbolizes the genius of war. Its divine half is composed of all those elements which depend on moral factors – on character, on talent, on the interests of your adversary, on public opinion, and on the morale of your troops, who are either strong and victorious or weak and beaten, depending on which they think they will be. Armaments, entrenchments, positions, order of battle, everything pertaining to the combinations of material factors – these make up the earthly half.

Napoleon, dictation at St Helena (1815–1821), quoted in Christopher Herold, ed., *The Mind of Napoleon*, 1955.

Genius is not extravagant; it is ardent, and it conceives great projects; but it knows beforehand how to attain the result, and it uses the simplest means, because its faculties are essentially calculating, industrious, and patient. It is creative, because its knowledge is vast; it is quick and preemptory, not because it is presumptuous, but because it is well-prepared.

General Sir William Napier (1782–1853), quoted in George F.R. Henderson, *Stonewall Jackson*, 1898.

...The realm of genius, *which rises above all rules*.

Pity the soldier who is supposed to crawl among these scraps of rules, and not good enough for genius, which genius can ignore, or laugh at. No; what genius does is the best rule, and theory can do no better than show how and why this should be the case.

Major-General Carl von Clausewitz, *On War*, 2.2, 1832, tr. Michael Howard and Peter Paret, 1976.

Towering genius disdains a beaten path. It seeks regions hitherto unexplored.

Abraham Lincoln, 27 January 1838, speech in Springfield, Illinois.

The stroke of genius that turns the fate of a battle? I don't believe it. A battle is a complicated operation, that you prepare laboriously. If the enemy does this, you say to yourself I shall do that. If such and such happens, these are the steps I shall take to meet it. You think out every possible development and decide on the way to deal with the situation created. One of these developments occurs; you put in operation your pre-arranged plan, and everyone says, 'What genius to have thought of that at the critical moment!' Whereas the credit is really due to the labour of preparation done beforehand.

Marshal of France Ferdinand Foch, *Daily Mail* (London), 19 April 1919.

Despite every wish to give full weight to the genius of the Commander, we must

GENIUS

fain recognize that there are immutable bounds set to what any individual can achieve. We know what a Lee can do against the good, ordinary General; yet – there are the appointed limits – as Gettysburg showed. A thrust is well-timed and deadly; the over-tasked blade breaks in the hand of the master. The stroke remains a master stroke; *it did not penetrate*, that was all!

General Sir Ian Hamilton, *The Soul and Body of an Army*, 1921.

Opposed to genius stand the men of logarithms, who, by their 'if-he-does-this-I-shall-do-that,' attempt to exhaust the cache of chances which God keeps stocked far away in starland. Genius is God's secret, that is all. Foch thinks he won the war because he 'calculated,' as the Yankees say. He forgets that the Germans equally pride themselves on calculations. He either forgets, or he is too modest to tell us, that he has in him like a burning fire a passion for sheer fighting, a fire of passion which burnt up all his sums and figures when the moment came. But if the great Marshal had only told us this, we should have known what weight to attach to his 'if-he-does-this-I-shall-do-that.'

General Sir Ian Hamilton, *Soul and Body of an Army*, 1921.

The military genius is he who can produce original combinations out of the forces of war; he is the man who can take all these forces and so attune them to the conditions which confront him that he can produce startling and, frequently, incomprehensible results. As an animal cannot explain the instincts which control it, neither can a man of genius explain the powers which control him. He acts on the spur of the moment, and he acts rightly, because this power is in his control. That some explanation exists cannot be doubted, but so far science has not revealed it, though the psychologist is working towards its fringe.

Major-General J.F.C. Fuller, *The Foundations of the Science of War*, 1926.

As genius is a personal gift, so is imitation a collective instinct. One man possessed of genius may alter the course of history, in fact, such a man has always altered the course of history, when alteration has been rapid. Three men of genius, working as a committee, could not do this, and still less so a crowd of normal men.

Major-General J.F.C. Fuller, *The Foundations of the Science of War*, 1926.

Generals have been legion; artists of war few. Many more evidences of military genius can be traced in the scantier records of irregular and guerrilla forces. The natural explanation is that the natural gifts of these leaders who have emerged straight from the womb of conflict, instead of a professional incubation chamber, have not been cramped or warped by convention. . . . it does cast a reflection upon the customary method of training leaders in organized armies. It is significant that in the American Civil War, which was the most fruitful field of generalship in modern times, most of the outstanding leaders were men who after an apprenticeship in the Regular Army had gone into some field of civil activity, which in varying degree freed their minds from the fetters of military convention and routine, while they retained a useful experience of the functioning of the military machine. (March 1937.)

Captain Sir Basil Liddell Hart, *Thoughts on War*, 1944.

'Genius' is a tiresome and misleading word to apply to the military art, if it suggests, as it does to many, one so gifted by nature as to obtain his successes by inspiration rather than through study. Nor does the definition

of genius attributed to Carlyle as 'an infinite capacity for taking pains' suit the great commander, as it suggests the pedant or martinet. Good generals, unlike poets, are made rather than born, and will never reach the first rank without much study of their profession; but they must have certain natural gifts – the power of quick decision, judgment, boldness and, I am afraid, a considerable degree of toughness, almost callousness, which is harder to find as civilization progresses.
> Field Marshal Viscount Wavell of Cyrenaica, *Soldiers and Soldiering*, 1953.

THE GENTLEMAN

I feel every thing that hurts the Sensibility of a Gentleman.
> General George Washington, to the Marquis de Lafayette, 1 September 1778, *The Writings of George Washington*, XII, p. 382, John C. Fitzgerald, ed., 1934.

Recollect that you must be a seaman to be an officer; and also, that you cannot be a good officer without being a gentleman.
> Admiral Viscount Nelson (1758–1805), to his midshipmen, quoted in Robert Southey, *Life of Nelson*, 1813.

The forbearing use of power does not only form a touchstone, but the manner in which an individual enjoys certain advantages over others is a test of a true *gentleman*.

The power which the strong have over the weak, the magistrate over the citizen, the employer over the employed, the educated over the unlettered, the experienced over the confiding, even the clever over the silly – the forbearing or inoffensive use of all this power or authority, or a total abstinence from it when the case admits it, will show the gentleman in a plain light. The gentleman does not needlessly and unnecessarily remind an offender of a wrong he may have committed against him. He cannot only forgive, he can forget; and he strives for the nobleness of self and mildness of character which impart sufficient strength to let the past be but the past. *A true gentleman of honor feels humbled himself when he cannot help humbling others.*
> General Robert E. Lee, a note found in his satchel after his death, quoted in William Jones, *Personal Reminiscences, Anecdotes, and Letters of Gen. Robert E. Lee*, 1874.

We want to make our nation entirely one of gentlemen, men who have a strong sense of honour, of chivalry towards others, of playing the game bravely and unselfishly for their side, and to play it with a sense of fair play and happiness in all.
> Lieutenant-General Baden-Powell, *Headquarters Gazette*, 1914.

The identity of the officer with the gentleman should persist in his relations with men of all degree. In the routine of daily direction and disposition, and even in moments of exhortation, he had best bring courtesy to firmness. The finest officers that one has known are not occasional gentlemen, but in every circumstance: in commissioned company and, more importantly in contact with those who have no recourse against arrogance.
> Brigadier-General S.L.A. Marshall, *The Armed Forces Officer*, 1950.

GLORY

Who is the King of glory? The Lord of hosts, he is the King of glory.
> Psalms 24:10.

Mother tells me,
the immortal goddess Thetis with her
    glistening feet,
that two fates bear me on to the day of
    death.
If I hold out here and I lay siege to Troy,
my journey home is gone, but my glory

GLORY

never dies.
If I voyage back to the fatherland I love,
my pride, my glory dies...
true, but the life that's left me will be long,
the stroke of death will not come on me quickly.
> Homer, *The Iliad*, 9.497–505, *c.* 800 BC, tr. Robert Fagles, 1990; the 'choice of Achilles' between a long, dull life or a short one sweet with everlasting glory.

So now I meet my doom. Well let me die –
but not without struggle, not without glory, no,
in some great clash of arms that even men to come
will hear of down the years!
> Homer, *The Iliad*, 22.359–362, *c.* 800 BC, tr. Robert Fagles, 1990; Hector aware that doom was rushing upon him in the shape of brilliant Achilles.

The nearest way to glory – a short cut, as it were – is to strive to be what you wish to be thought to be.
> Socrates (469–399 BC), quoted in Cicero, *De Officiis*, 2.12.43, 44 BC, tr. Walter Miller.

My own assessment of myself is based on the extent not of my life but of my glory. I could have been content with my father's inheritance, and within Macedonia's bounds have enjoyed a life of ease as I awaited an old age without renown or distinction (though even inactive men cannot control their destiny, and those who believe a long life is the only good are often overtaken by a premature death). But no – I count my victories, not my years and, if I accurately compute fortune's favours to me, my life has been long.
> Alexander the Great, quoted in Curtius, *The History of Alexander*, *c.* AD 41, tr. John Yardley, 1984.

Stand firm; for well you know that hardship and danger are the price of glory, and that sweet is the savour of a life of courage and of deathless renown beyond the grave.
> Alexander the Great, July 326 BC, in India when the Macedonians refused to go on, quoted in Arrian, *The Campaigns of Alexander*, 5.26, *c.* AD 150, tr. Aubrey de Sélincourt, 1971.

From where the eagles are resting,
from where the tigers are exhalted,
the sun is invoked.
Like a shield that descends,
so does the Sun set,
In Mexico night is falling,
war rages on all sides,
Oh Giver of Life!
war comes near...
Proud of itself
is the city of Mexico-Tenochtitlan.
Here no one fears to die in war.
This is our glory.
This is Your Command,
oh Giver of Life!
Have this in mind, oh princes,
do not forget it.
Who could conquer Tenochtitlan?
Who could shake the foundation of heaven?
With our arrows,
with our shields,
the city exists.
Mexico-Tenochtitlan remains.
> Aztec poem, *c.* 1500, exhalting the glory of their great capital and its martial people, quoted in Miguel Leon-Portilla, *Pre-Columbian Literatures of Mexico*, 1969.

There must be a beginning of any great matter, but the continuing unto the end until it be thoroughly finished yields the true glory.
> Sir Francis Drake, 17 May 1587, to Lord Walsingham.

My greatest happiness is to serve my gracious King and Country, and I am envious only of glory; for if it be a sin to covet glory, I am the most offending soul alive.

GLORY

Admiral Viscount Nelson, 18 February 1800, letter to Lady Hamilton.

Where is the republic, ancient or modern, that has not granted honours? Call them trifles if you like, but it is by trifles that men are influenced. I would not utter such a sentiment as this in public, but here, among statesmen and thinkers, things should be spoken of as they are. In my opinion the French do not care for liberty and equality; they must be given distinctions. Do you suppose you can persuade men to fight by a process of analysis? Never; the process is valid only for the man of science in his study. The soldier demands glory, distinction, rewards.
Napoleon, 14 May 1802, in Paris, quoted in R.M. Johnston, ed., *The Corsican*, 1910.

Die young, and I shall accept your death – but not if you have lived without glory, without being useful to your country, without leaving a trace of your existence: for that is not to have lived at all.
Napoleon, 1802, letter to Jérôme, quoted in Christopher Herold, ed., *The Mind of Napoleon*, 1955.

Glory can only be won where there is danger.
Napoleon, 14 November 1806.

How glorious it is to be a Frenchman!
Marshal Ney, June 1807, after the Battle of Friedland, cited in Raymond Horricks, *Marshal Ney*, 1982.

A new Prometheus, I am nailed to a rock to be gnawed by a vulture. Yes, I have stolen the fire of Heaven and made a gift of it to France. The fire has returned to its source, and I am here!
The love of glory is like the bridge that Satan built across Chaos to pass from Hell to Paradise: glory links the past with the future across a bottomless abyss. Nothing to my son, except my name!
Napoleon, a note found written in his hand after his death on 5 May 1821, quoted in Christopher Herold, ed., *The Mind of Napoleon*, 1955.

The exceeding brightness of military glory – that attractive rainbow, that rises in showers of blood – that serpent's eye, that charms to destroy.
President Abraham Lincoln, 'Speech in the United States House of Representatives on the War with Mexico', 12 January 1848, *Collected Works of Abraham Lincoln*, I, 1953.

There is a true glory and a true honor, the glory of duty done, the honor of the integrity of principle.
General Robert E. Lee, in Early, *Southern Historical Society Papers*, XI.

A nation without glory is like a man without courage, a woman without virtue. It takes the first place in our human estimate of national fame. All States long for it, and certainly it is a big factor in the consciousness of national strength which commands the respect of both friends and enemies. It is a national heirloom of priceless value to the people to whom the world accord it and who are ready to fight rather than risk its loss. When the nation to whom it was once universally conceded begins to sneer at it as unimportant, and to ridicule its worth, the tide of that nation's greatness has surely turned...
  Glory to the nation is what sunlight is to all human beings. Without it the State dwindles in size and grows weak in strength, as the man in a dark dungeon becomes daily whiter, until at last his whiteness passes into the colourlessness of death.
Field Marshal Viscount Wolseley, *The Story of a Soldier's Life*, II, 1903.

What a glory shines on the brave and true.
Sir Winston S. Churchill (1874–1965).

THE GOLDEN BRIDGE

THE GOLDEN BRIDGE

Do not thwart an enemy returning home.
To surround an enemy you must leave a way of escape
Do not press an enemy at bay.
> Sun Tzu, *The Art of War*, c. 500 BC, tr. Samuel Griffith, 1969.

Scipio Africanus used to say that a road not only ought to be afforded the enemy for flight, but that it ought even to be paved.
> Frontinus, *Stratagems*, 4.7, c. AD 84–96, tr. Charles E. Bennett, 1925.

When the enemy is surrounded, it is well to leave a gap in our lines to give them the opportunity to flee, in case they judge that flight is better than remaining and taking their chances.
> The Emperor Maurice, *The Strategikon*, c. AD 600 (*Maurice's Strategikon*, tr. George Dennis, 1984).

In God's name! they go. Let them depart. And go we to give thanks to God. We shall not follow them farther, for it is Sunday. Seek not to harm them. It suffices me that they go.
> Joan of Arc, 8 May 1429, to her soldiers as the English raised the siege of Orléans, *Joan of Arc in Her Own Words*, 1996.

When an enemy army is in flight, you must either build a golden bridge for it or stop it with a wall of steel.
> Napoleon (1769–1821), quoted in Christopher Herold, ed., *The Mind of Napoleon*, 1955.

When the two sides are too evenly matched to offer a reasonable chance of early success to either, the statesman is wise who can learn something from the psychology of strategy. It is an elementary principle of strategy that, if you find your opponent in a strong position too costly to force, you should leave him a line of retreat – as the quickest way of loosening his resistance. It should, equally, be a principle of policy, especially in war, to provide your opponent with a ladder by which he can climb down.
> Captain Sir Basil Liddell Hart, *Strategy*, 1954.

GREAT CAPTAINS /GREATNESS

A consecutive series of great actions never is the result of chance and luck; it always is the product of planning and genius. Great men are rarely known to fail in their most perilous enterprises... Is it because they are lucky that they became great? No, but being great, they have been able to master luck.
> Napoleon, 14 November 1816, conversation on St Helena. Emmanuel Las Casas, *Mémorial de St. Hélène*, quoted in Christopher Herold, ed., *The Mind of Napoleon*, 1955.

Many a time has the soldier's calling exhausted strong characters, and that surprisingly quickly. The fine intellect and resolute will of one year give place to the sterile imaginings and faint heart of the next. That is perhaps the tragedy of military greatness.
> Field Marshal Paul von Hindenburg, *Out of My Life*, 1920.

If I may hazard to set down the qualifications of the great captain, then I should say that they are:
(i) Imagination operating through reason.
(ii) Reason operating through audacity.
(iii) And audacity operating through rapidity of movement.

The first creates unsuspected forms of thought; the second establishes original forms of action; and the third impels the human means at the disposal of the commander to accomplish his purpose with the force and rapidity of a thunderbolt. From the mind, through the soul, we thus gain our ends by

GREAT CAPTAINS/GREATNESS

means of the body.
> Major-General J.F.C. Fuller, *The Foundations of the Science of War*, 1926.

It may well be that the greatest soldiers have possessed superior intellects, may have been thinkers; but this was not their dominant characteristic... [they] owed their successes to indomitable wills and tremendous energy in execution and achieved their initial hold upon the hearts of their troops by acts of demonstrated valor... the great leaders are not our responsibility, but God's.
> General George S. Patton, Jr., 27 October 1927, lecture, 'Why Men Fight', *The Patton Papers*, I, 1972.

History shows that rather than resign himself to a direct approach, a Great Captain will take even the most hazardous indirect approach – if necessary, over mountains, deserts, or swamps, with only a fraction of his force, even cutting himself from his communications. Facing, in fact, every unfavourable condition rather than the risk of stalemate. (Oct. 1928.)
> Captain Sir Basil Liddell Hart, *Thoughts on War*, 1944.

He must accept the loneliness which according to Faguet, is the 'wretchedness of superior beings.' Contentment and tranquility and the simple joys which go by the name of happiness are denied to those who fill positions of great power. The choice must be made, and it is a hard one: whence that vague sense of melancholy which hangs about the skirts of majesty, in things no less than in people. One day somebody said to Napoleon, as they were looking at an old and noble monument: 'How sad it is!' 'Yes,' came the reply, 'as sad as greatness.'
> General Charles de Gaulle, *The Edge of the Sword*, 1932.

The commander with the imagination – the genius, in fact – to use the new forces may have his name written among the 'great captains.' But he will not win that title lightly or easily; consider for a moment the qualifications he will require. On the ground he will have to handle forces moving at a speed and ranging at a distance far exceeding that of the most mobile cavalry of the past; a study of naval strategy and tactics as well as those of cavalry (tanks) will be essential to him. Some ideas on his position in battle and the speed at which he must make his decisions may be derived from the battle of Jutland; not much from Salisbury Plain or the Long Valley. Needless to say, he must be able to handle Air Forces with the same knowledge as forces on land.
> Field Marshal Viscount Wavell of Cyrenaica, *Generals and Generalship*, 1941.

One mark of a great man is the power of making lasting impressions upon the people he meets.
> Sir Winston S. Churchill (1874–1965).

It was because he really understood war that he became so good at securing peace. He was the least militaristic of soldiers, and free from the lust of glory. It was because he saw the value of peace that he became so unbeatable in war. For he kept the end in view, instead of falling in love with the means. Unlike Napoleon, he was not infected by the romance of war, which germinates illusions and self-deceptions. That was how Napoleon failed, and Wellington prevailed.
> Captain Sir Basil Liddell Hart, 'Wellington', quoted in Barrett Parker, *Famous British Generals*, 1951.

H

HEALTH

The soldier's health must come before economy or any other consideration.
> Napoleon, 17 June 1813, letter to Daru, *Correspondance de Napoléon Ier, publié par ordre de l'Empereur Napoléon III*, XXV, p. 396, 1858–1870.

If the facilities for washing were as great as those for drink, our Indian army would be the cleanest body of men in the world.
> Florence Nightingale, 1863, commenting as a member of the Royal Commission on the sanitary state of the Indian army.

Good health and a robust constitution are invaluable in a general... In a sick body, the mind cannot possibly remain permanently fresh and clear. It is stunted by the selfish body from the great things to which it should be entirely devoted.
> Field Marshal Colmar Baron von der Goltz, *The Nation in Arms*, 1887.

Health in a general is, of course, most important, but it is a relative quality only. We would all of us, I imagine, sooner have Napoleon sick on our side than many of his opponents whole. a great spirit can rule in a frail body, as Wolfe and others have showed us. Marlborough during his great campaigns would have been ploughed by most modern medical boards.
> Field Marshal Viscount Wavell of Cyrenaica, *Soldiers and Soldiering*, 1953.

HERO/HEROISM

They say: 'Exhalted is the Sun of our land! We were lost in the land daily in the darkness, which King Rameses III has expelled. The lands and countries are stripped, and brought to Egypt as slaves; gifts gathered together for her gods' satiety, provisions, supplies, are a flood in the Two Lands. The multitude rejoices in this land, none is sad, for Amon has established his son upon his throne, all the circuit of the sun is united in his grasp; the vanquished Asiatics and the Tehenu. Taken are those who were spoiling the condition of Egypt. The land had been exposed in continual extremity, since the former kings. They were desolated, the gods as well as all people. There was no hero to seize them when they retreated. Lo, there was a youth like a gryphon, like a bull ready for battle upon the field. His horses were like hawks... roaring like a lion terrible in rage... His name is a flame, the terror of him is in the countries.'
> Rameses III, Egyptian Pharaoh, c. 1193 BC, description of his victory in the First Libyan War, quoted in James Breasted, *Ancient Records of Egypt*, 1906.

The action was one that rewarded itself.
> Scipio Africanus (236–184 BC), quoted in Polybius, *The Histories*.

Though the actions of our hero shine with great brilliancy, they must not be imitated, except with peculiar caution. The more resplendent they are, the more easily they seduce the youthful, headlong, and angry warrior, to whom we cannot often enough repeat that valor without wisdom is insufficient,

HISTORY

and that the adversary with a cool head, who can combine and calculate, will finally be victorious over the rash individual.
>   Frederick the Great, 'Reflections on Charles XII', *Posthumus Works of Frederick II, King of Prussia*, tr. Holcroft, 1789; in Jay Luvaas, ed., *Frederick the Great on the Art of War*, 1966.

Permit me, Your Highness, to report that heroes are to be found also in the lower ranks.
>   Field Marshal Prince Aleksandr V. Suvorov, report to Prince Potemkin of the Battle of Kinburn during the 1787 Kherson campaign against the Turks, quoted in K. Ossipov, *Suvorov*, 1945.

War is, or anyhow should be, an heroic undertaking; for without heroism it can be no more than an animal conflict, which in place of raising man through an ideal, debases him through brutality.
>   Major-General J.F.C. Fuller, *Generalship: Its Diseases and Their Cure*, 1933.

Soldiers, all men in fact, are natural hero worshippers. Officers with a flare for command realize this and emphasize in their conduct, dress and deportment the qualities they seek to produce in their men... the influence one man can have on thousands is a never ending source of wonder to me.
>   General George S. Patton, Jr., 6 June 1944, *The Patton Papers*, II, 1974.

On a trip to the West Coast, President Kennedy was asked by a little boy, 'Mr. President, how did you become a war hero?'
'It was involuntary. They sank my boat.'
>   President John F. Kennedy, quoted in Adler, *The Kennedy Wit*, 1964.

It is in disaster, not success, that the heroes and the bums really get sorted out.
>   Admiral James B. Stockdale, 'In War, In Prison, In Antiquity', *Parameters*, 12/1987.

It doesn't take a hero to order men into battle. It takes a hero to be one of those men who goes into battle.
>   General H. Norman Schwarzkopf, *It Doesn't Take a Hero*, 1992.

HISTORY
General

My purpose is not to relate at length every motion, but only such as were conspicuous for excellence or notorious for infamy. This I regard as history's highest function, to let no worthy action be uncommemorated, and to hold out the reprobation of posterity as a terror to evil words and deeds.
>   Cornelius Tacitus, *Annals*, 4.65, AD 116, (*The Complete Works of Tacitus*, tr. Alfred Church and William Brodribb, 1942).

What will history say – what will posterity think?
>   Napoleon, his favourite phrase according to de Bourriene, his confidant and secretary, 1797–1802, quoted in Louis de Bourriene, *Memoirs of Napoleon Bonaparte*, I, 1855, p. xxiv.

Everything on earth is soon forgotten, except the opinion we leave imprinted upon history.
>   Napoleon, 27 November 1802, to General Leclerc, *Correspondance de Napoléon Ier, publié par ordre de l'Empereur Napoléon III*, VIII, p. 113, 1858–1870, quoted in Christopher Herold, ed., *The Mind of Napoleon*, 1955.

My usual reading fare, when I go to bed, is the old chronicles of the third, fourth, and fifth centuries. I read them or have them translated for me. Nothing is more curious, or less known, than the change from the old way of life to the new, than the transition from the ancient political bodies to the new ones that were founded on their ruins.

HISTORY

Napoleon, 1806, at a meeting of the Council of State, quoted in Christopher Herold, ed., *The Mind of Napoleon*, 1955.

Fellow citizens, we cannot escape history... We will be remembered in spite of ourselves.
Abraham Lincoln, 1 December 1862, second annual message to Congress.

It is history that teaches us to hope.
General Robert E. Lee, c. 1866, letter to Colonel Charles Marshall.

People get the history they deserve.
General Charles de Gaulle, 1920, lecture to cadets at St Cyr.

For the first time in my life I have seen 'History' at close quarters, and I know that its actual process is very different from what is presented to posterity.
General Max Hoffman, *War Diaries and Other Papers*, 1929.

There is a modern, and too common, tendency to regard history as a specialist subject. On the contrary, it is the corrective to specialization. Viewed right, it is the broadest of studies, embracing every aspect of life. It lays the foundation of education by showing how mankind repeats its follies, and what those follies are... It is universal experience – infinitely longer, wider, and more varied than any individual's experience. How often do we hear people claim a knowledge of the world and of life because they are sixty or seventy years old. Most of them might be described as a 'young sixty, or seventy'. There is no excuse for any literate person if he is less than three thousand years old in mind.
Captain Sir Basil Liddell Hart, *Through the Fog of War*, 1938.

The history of free men is never really written by chance but by choice – their choice.
President Dwight D. Eisenhower, 9 October 1956, address in Pittsburgh, Pennsylvania.

Literature is history with the truth left in. I believe we can profit from the study of the classical texts as never before. The veneer of civilization, so recent and fragile, is being stripped from much of the world. The old problems are today's problems – and tomorrow's. If we want to know 'Who is our enemy?' we must look within.
Lieutenant-Colonel Ralph Peters, 'Our New Old Enemies', *Parameters*, Summer 1999.

Let us say one thing. If we are wrong we will have destroyed a threat that, at its least, is responsible for inhuman carnage and suffering. That is something I am confident history will forgive.
Prime Minister Tony Blair, Address to a joint session of the US Congress, 17 July 2003.

Studying Military History

[History] hath triumphed over time, which besides it, nothing but Eternity hath triumphed over...
  In a word, we may gather out of history a policy no less wise than eternal; by the comparison and application of other men's forepassed miseries with our own like errors and ill deservings.
Sir Walter Raleigh, *History of the World*, 1614.

It is not only Experience and Practice which maketh a Soldier worthy of his Name: but the knowledge of the manifold Accidents which arise from the variety of humane actions, which is best, and most speedily learned by reading History: for upon the variety of Chances that are set forth, he may meditate on the Effects of other men's adventures, that their harms may be his

HISTORY

warnings, and their happy proceedings his fortunate Directions in the Art Military.
>   Anonymous Captain J.S. (probably John Smythe), *Military Discipline – Or the Art of War*, 1689.

It is not probable that any similar chain of causes should, in a short time, produce the same circumstances as those under which we were... generals are never placed in exactly similar situations... past facts are good to store in the imagination and the memory; they furnish a repository of ideas whence a supply of materials may be obtained, but which ought to be purified by passing through the strainer of judgment.
>   Frederick the Great, *The History of the Seven Years War, Posthumus Works*, II, pp. ix, xi, xii, quoted in Azer Gat, *The Origins of Military Thought: from the Enlightenment to Clausewitz*, 1989.

I had it from Xenophon, but our friends here are astonished at what I have done because they have read nothing.
>   Major-General Sir James Wolfe, of his use of light infantry at the siege of Louisburg, 1758, quoted in Liddell Hart, *Great Captains Unveiled*, 1927.

History I conquered rather than studied: that is to say, I wanted from it and retained of it only what could add to my ideas, I spurned what was of no use, and I seized upon certain conclusions that pleased me.
>   Napoleon, conversation with Mme de Rémusat, quoted in Christopher Herold, ed., *The Mind of Napoleon*, 1955.

Maxim 78. Peruse again and again the campaigns of Alexander, Hannibal, Caesar, Gustavus Adolphus, Turenne, Eugene and Frederick. Model yourself upon them. This is the only means of becoming a great captain, and of acquiring the secret of the art of war.

Your own genius will be enlightened and improved by this study, and you will learn to reject all maxims foreign to the principles of the great commanders.
>   Napoleon, *The Military Maxims of Napoleon*, tr. George D'Aguilar, 1831 (David Chandler, ed., 1987).

If we read history with an open mind, we cannot fail to conclude that, among all the military virtues, the energetic conduct of war has always contributed most to glory and success.
>   Major-General Carl von Clausewitz, *On War*, 4.3, 1832, tr. Michael Howard and Peter Paret, 1976.

Military history, accompanied by sound criticism, is indeed the true school of war.
>   Lieutenant-General Antoine-Henri Baron de Jomini, *Summary of the Art of War*, 1838, tr. Mendell and Craighill, 1862.

...The most effective means of teaching war during peace.
>   Field Marshal Helmuth Graf von Moltke (1800–1891).

The smallest detail, taken from the actual incident in war, is more instructive to me, a soldier, than all the Thiers and Jominis in the world. They speak, no doubt, for the heads of states and armies but they never show me what I wish to know – a battalion, a company, a squad in action.
>   Colonel Charles Ardant du Picq, *Battle Studies*, 1880, tr. Greely, 1957.

It is now accepted with naval and military men who study their profession, that history supplies the raw material from which they are to draw their lessons, and reach their working conclusions. Its teachings are not, indeed, pedantic precedents; but they are the illustrations of living principles.
>   Rear Admiral Alfred Thayer Mahan, *From Sail to Steam*, 1907.

HISTORY

To every one who desires to become a commander, there is available a book entitled 'military history'... This reading matter, I must admit, is not always highly flavored. One will have to work through a mass of ingredients anything but palatable. But behind it all one arrives just the same at the facts, often most gratifying facts; at the bottom lies the knowledge as to how everything happened, how it had to happen, and how it will happen again.
> Field Marshal Alfred Graf von Schlieffen, 15 October 1910, speech at the centenary celebration of the *Kriegsakademie*, quoted in von Freytag-Loringhoven, *Generalship in the Great War*, 1920.

History. – To keep the brain of an army going in time of peace, to direct it continually towards its task of war, there is no book more fruitful to the student than that of history. If war, in its just aspect, is but a struggle between two wills more or less powerful and more or less informed, then the accuracy of decisions arrived at in war will always depend upon the same considerations as those of the past. The same errors reappear, leading to the same checks. The art of war is always to be drawn from the same sources.
> Marshal of France Ferdinand Foch, *Precepts and Judgments*, 1919.

The road to high command leads through a long path called the history of war. Like all long roads the scenery is not always interesting; there are desert stretches of prosaic facts. But now and again the traveler reaches eminences where he sees the most sublime panoramas ever vouchsafed to mortals – the deathless deeds of the great who have passed to Valhalla, which is death but not oblivion.
> General George S. Patton, Jr, note card, c. 1920, entitled 'Training', from the Patton Collection, Library of Congress.

Unless history can teach us how to look at the future, the history of war is but a bloody romance.
> Major-General J.F.C. Fuller, *British Light Infantry in the Eighteenth Century*, 1925.

Do make it clear that generalship, at least in my case, came of understanding, of hard study and brain-work and concentration. Had it come easy to me I should not have done it so well.
 So please, if you see me that way and agree with me, do use me as a text to preach for more study of books and history, a greater seriousness in military art. With 2,000 years of examples behind us we have no excuse, when fighting, for not fighting well.
> Colonel T.E. Lawrence, 26 April 1933, letter to Captain Sir Basil Liddell Hart, quoted in *The Liddell Hart Memoirs*, I, 1965.

The value of the knowledge acquired by study must not be overestimated. The soldier faced with the necessity for independent decision must not mentally search the pages of his professional encyclopaedia nor seek to remember how the great generals of history, from alexander to Zieten, would have acted in a similar case. Such knowledge as that derived from the study of the history of war is only of living practical value when it has been digested, when the permanent and the important has been extracted from the wealth of detail and has been incorporated with a man's own mental resources – and it is not every man who has the gift for this.
> Colonel-General Hans von Seekt, *Thoughts of a Soldier*, 1930.

The real way to get value out of the study of military history is to take particular situations, and as far as possible get inside the skin of the man who made a decision, realize the conditions in which the decision was made, and then see in what way you

HISTORY

could have improved upon it.
> Field Marshal Viscount Wavell of Cyrenaica, 1930, lecture to officers at Aldershot.

More than most professions the military is forced to depend upon intelligent interpretation of the past for signposts charting the future. Devoid of opportunity, in peace, for self-instruction through actual practice of his profession, the soldier makes maximum use of historical record in assuring the readiness of himself and his command to function efficiently in emergency. The facts derived from historical analysis he applies to conditions of the present and the proximate future, thus developing a synthesis of appropriate method, organization, and doctrine.
> General of the Army Douglas MacArthur, *Annual Report of the Chief of Staff*, June 30, 1935.

The discovery of uncomfortable facts had never been encouraged in armies, who treated their history as a sentimental treasure rather than a field of scientific research. (June 1936.)
> Captain Sir Basil Liddell Hart, *Thoughts on War*, 1944.

In my view, the single most important foundation for any leader is a solid academic background in history. That discipline gives perspective to the problems of the present and drives home the point that there is really very little new under the sun. Whenever a policymaker starts his explanation of how he intends to handle a problem with such phrases as 'We are at the takeoff point of a new era...,' you know you are heading for trouble. Starting by ignoring the natural yardstick of 4,000 years of recorded history, busy people, particularly busy opportunists, have a tendency to see their dilemmas as so unique and unprecedented that they deserve to make exceptions to law, custom, or morality in their own favor to get around them...
> Admiral James B. Stockdale, 'Educating Leaders', *The Washington Quarterly*, Winter 1983.

Writing Military History

I shall be content if those shall pronounce my History useful who desire to have a view of events as they did really happen, and as are very likely, in accordance with human nature, to repeat themselves at some future time – if not exactly the same, yet very similar.
> Thucydides, *The Peloponnesian War*, c. 404 BC, tr. Benjamin Jowett, 1881.

Few persons will understand me, but I write for the connoisseurs, trusting that they will not be offended by the confidence of my opinions.
> Field Marshal Maurice Comte de Saxe, *My Reveries*, 1732, tr. Phillips, 1940.

Write the history of a battle? As well write the history of a ball.
> The Duke of Wellington, quoted in Richard Holmes, *Acts of War: Human Behavior in Battle*, 1985.

My ambition was to write a book that would not be forgotten in two or three years, and which anyone interested in the subject would certainly take up more than once.
> Major-General Carl von Clausewitz, *On War*, 1832, p. 63, tr. Michael Howard and Peter Paret, 1976; notes on the genesis of his early manuscript, 1818.

Ten million pities that G.B.S. [George Bernard Shaw] didn't raise the Arab Revolt and write the history of it. There is a book of books gone to waste, because nobody made him a Brigadier-General! It's easy to be wise afterwards, I suppose. If I'm Prime Minister for the next war, all independent commands

shall be given to the finest writers of the generation. We may lose the war... but think of the glories to be published as the Peace rolls on!
> Colonel T.E. Lawrence, letter to Mrs. Charlotte Shaw, 15 October 1924, in *The Letters of T.E. Lawrence*, 1988.

If truth is many-sided, mendacity is many-tongued.

History cannot proceed by silences. The chronicler of ill-recorded times has none the less to tell his tale. If facts are lacking, rumor must serve. Failing affidavits, he must build with gossip.
> Sir Winston S. Churchill, *Marlborough*, 1933.

A military philosopher may recognize the inevitability of gradual evolution, but a military historian must point out how often it has proved fatal to nations. The money spent on armies that failed to adapt themselves to changing conditions has proved too literally a sinking fund – acting as a millstone around the investor's neck when he was plunged into the deep waters of war. (March 1934.)
> Captain Sir Basil Liddell Hart, *Thoughts on War*, 1944.

Naval officers prefer to make history rather than write it – because of which preference they probably do a better job of the former.
> Admiral Ernest J. King, 1942, quoted in *Naval War College Review*.

History will bear me out, particularly as I shall write that history myself.
> Winston S. Churchill (1874–1965), quoted in Timothy Garton Ash, 'In the Churchill Museum,' *New York Review of Books*, 7 May 1987

Some readers may think I have at times, whether as a subaltern or as a general, treated too lightly, even when in this minor key, the grim and tragic business of war. If they do, I can only plead guilty to a fault – or is it a virtue? – that has for centuries marked the British soldier. There would be fewer battle honours emblazoned on his colours had he lacked it.
> Field Marshal Viscount Slim of Burma, *Unofficial History*, 1959.

Most battles are more like a schoolyard in a rough neighborhood at recess time than a clash between football giants in the Rose Bowl. They are messy, inorganic, and uncoordinated. It is only much later, after the clerks have tidied up their reports and the generals have published their memoirs, that the historian with his orderly mind professes to discern an understandable pattern in what was essentially catch-as-catch-can, if not chaotic, at the time.
> Brigadier-General S.L.A. Marshall, *Ambush*, 1969.

HONOUR

When asked why the best men prefer an honourable death to a life without honour, he said: 'Because they regard the latter as the gift of Nature, and the former as being in their own hands.'
> King Leonidas of Sparta, 480 BC, quoted in Plutarch, *The Lives*, c. AD 100 (*Plutarch on Sparta*, tr. Richard Talbert, 1988).

Thus choosing to die resisting, rather than to live submitting, they fled only from dishonour, but met danger face to face, and after one brief moment, while at the summit of their fortune, left behind them not their fear, but their glory.
> Pericles, 431/430 BC, funeral oration for the Athenian fallen, quoted in Thucydides, *History of the Peloponnesian War*, c. 404 BC, 2.42, tr. William Crawley, 1910, Robert Strassler, ed., *The Landmark Thucydides*, 1996.

Learning of his approach, Alexander rode and met him in advance of the line with a few Companions; halting his

HONOUR

horse, he admired the stature of Porus, who was approximately over five cubits in height, his beauty, and the appearance he gave of spirit not enslaved, but of one man of honour meeting another after a fine struggle against another king for his own kingdom. Then Alexander spoke to him first and urged him to say what he desired to be done with him. Porus is said to have replied: 'Treat me, Alexander, like a king,' and Alexander, pleased with the reply, answered: 'that you shall have, Porus, for my own sake; now demand what you would wish for yours.' He replied that everything was comprised in this one request. Alexander was all the more pleased with this reply, and gave Porus the government of his Indians and added still further territory even greater in extent to his old realm. In this way he himself acted like a king in his treatment of a man of honour, while in Porus he found from this time entire fidelity.
Arrian, *Anabasis of Alexander*, 5.19, c. AD 150, tr. P.A. Brunt; the meeting of Alexander the Great and Porus after the Battle of the Hydaspes, June 326 BC.

The best memorial for a mighty man is to gain honour ere death.
Beowulf, c. AD 1000.

War must be carried on systematically, and to do it you must have men of character activated by principles of honor.
General George Washington (1732–1799).

I would lay down my life for America, but I cannot trifle with my honor.
Admiral John Paul Jones, 4 September 1777, letter to A. Livingston.

The Neapolitan officers did not lose much honour, for, God knows, they had not much to lose; but they lost all they had.
Admiral Viscount Nelson, writing of the rout of the Neapolitan Army by the French, 15 December 1798, quoted in Robert Southey, *Life of Nelson*, 1813.

The bravest man feels an anxiety '*circa praecordia*' as he enters the battle; but he dreads disgrace more.
Admiral Viscount Nelson, quoted in Alfred Thayer Mahan, *The Life of Nelson*, 1897.

Unfit as my ship was, I had nothing left for the honour of our country but to sail, which I did in two hours afterward.
Admiral Viscount Nelson, 16 September 1793, letter from Naples.

What I want you to preserve is honour, and not a few planks of wood.
Napoleon, 15 September 1804, letter to the Minister of Marine, *Correspondance de Napoléon Ier, publié par ordre de l'Empereur Napoléon III*, IX, p. 527, 1858–1870.

One obtains everything from men by appealing to their sense of honour.
Napoleon, 1 March 1805, to Fouché, *Correspondance de Napoléon Ier, publié par ordre de l'Empereur Napoléon III*, VI, p. 178, 1858–1870.

It is no dishonour to be defeated by my army.
Napoleon, 1805, to a captured Russian officer after the Battle of Austerlitz.

Whoever prefers death to ignominity will save his life and live in honour, but he who prefers life will die and cover himself with disgrace.
Napoleon, 1 October 1809, letter to General Clarke, *Correspondance de Napoléon Ier, publié par ordre de l'Empereur Napoléon III*, XIX, p. 542, 1858–1870, quoted in Christopher Herold, ed., *The Mind of Napoleon*, 1955.

The British Army is what it is because it is officered by gentlemen; men who would... scorn to do a dishonourable

HOPE

thing and who have something more at stake than a reputation for military smartness.
> The Duke of Wellington (1769–1852), quoted in *The Autobiography of Sir Harry Smith*, I, 1901.

The soldier's trade, if it is to mean anything at all, has to be anchored to an unshakeable code of honor. Otherwise, those of us who follow the drums become nothing more than a bunch of hired assassins walking around in gaudy clothes... a disgrace to God and mankind.
> Major-General Carl von Clausewitz, *On War*, 1832, tr. Michael Howard and Peter Paret, 1976.

Honour is manly decency. The shame of being found wanting in it means everything to us.
> Alfred Comte de Vigny, *Servitude and Grandeur of Arms*, 1835.

What is life without honor? Degradation is worse than death. We must think of the living and of those who are come after us, and see that with God's blessing we transmit to them the freedom we have ourselves inherited.
> Lieutenant-General Thomas J. 'Stonewall' Jackson, August 1862, to an officer who requested leave to visit a dying member of his family.

The muster rolls on which the name and oath were written were pledges of honor redeemable at the gates of death. And they who went up to them, knowing this, are on the lists of heroes.
> Major-General Joshua L. Chamberlain, *The Passing of the Armies*, 1915.

Never give in, never give in, never, never, never – in nothing great or small, large or petty – never give in except in convictions of honour and good sense.
> Sir Winston S. Churchill, 29 October 1941, address at Harrow School.

No nation can safely trust its martial honor to leaders who do not maintain the universal code which distinguishes between those things that are right and those things that are wrong.
> General of the Army Douglas MacArthur, February 1946, describing General Homma's conduct in The Philippines, 1942, in *Reminiscences*, 1962.

The choice of death or dishonour is one which has always faced the professional fighting man, and there must be no doubt in his mind what his answer must be. He chooses death for himself so that his country [or comrades] may survive, or on a grander scale so that the principles for which he is fighting may survive.
> Lieutenant-General Sir John Crowley, quoted in John D. Luck, *To Fight With Intrepidity*, 1998.

HOPE

When he had shared out or signed away almost all the property of the crown, Perdiccas asked him, 'But your majesty, what are you leaving for yourself?' 'My hopes!' replied Alexander.
> Alexander the Great, 334 BC, as he prepared to cross the Hellespont to conquer the Persian Empire on an empty purse, quoted in Plutarch, *The Lives*, c. AD 100, (*The Age of Alexander*, tr. Ian Scott-Kilvert, 1972).

Hope encourages men to endure and attempt everything; in depriving them of it, or in making it too distant, you deprive them of their very soul.
> Field Marshal Maurice Comte de Saxe, *My Reveries*, 1732, tr. Phillips, 1940.

I have... had my tiny dog hole where I sleep in the line smashed up by a shell wh [sic] had it... detonated perfectly w'd [sic] have been the end of my chequered fortunes. One becomes quite reconciled to the idea of annihilation & death seems to be divested of any element of

HUMANITY

tragedy. The only thing to dread is some really life wrecking wound which left one a cripple, an invalid, or an idiot. But that one must hope is not on the agenda of the Fates.
> Sir Winston S. Churchill, letter to his cousin, the Ninth Duke of Marlborough, 12 January 1916, quoted in *The Gazette*, Library of Congress, Vol 15. No 8, 27 February 2004, p. 7.

Loss of hope, rather than loss of life, is the factor that really decides wars, battles, and even the smallest combats. The all-time experience of warfare shows that when men reach the point where they see, or feel, that further effort and sacrifice can do no more than delay the end they commonly lose the will to spin it out, and bow to the inevitable.
> Captain Sir Basil Liddell Hart, *Defence of the West*, 1950.

HORSES

Hast thou given the horse strength? hast thou clothed his neck with thunder?

Canst thou make him afraid as a grasshopper? the glory of his nostrils is terrible.

He paweth in the valley, and rejoiceth in his strength; he goeth on to meet the armed men.

He mocketh at fear, and is not afrighted; neither turneth he back from the sword.

The quiver rattleth against him, the glittering spear and the shield.

He swalloweth the ground with fierceness and rage: neither believeth he that *it is* the sound of the trumpet.

He saith among the trumpets, Ha, Ha; and he smelleth the battle afar off, the thunder of the captains, and the shouting.
> Job 39:19–25.

But good horses with competent riders will manage to escape even from hopeless situations.
> Xenophon, *On Horsemanship*, c. 360 BC (*Hiero the Tyrant and Other Treatises*, tr. Robin Waterfield, 1997).

The disappearance of the horse from war does not suggest the withdrawal of riding from an officer's education in peace. It suggests exactly the opposite.
> General Sir John Hackett, 1960, lecture to the Royal United Services Institute (RUSI).

It was the horse, of course, that distinguished the cavalryman from the other branches, even the artillery, because of the manner in which they used them. It made the cavalry an elite branch of the service, and it was in horsemanship that most of the legends of the school originated. The horse doubled the work of training cavalrymen in comparison with that of training the dismounted branches, and it was the horse that increased the sense of responsibility of cavalrymen, they could not forget him for even one day.
> General Lucian C. Truscott, Jr., *The Twilight of the Cavalry*, 1989; speaking of the US Army Cavalry School at Fort Riley, Kansas.

HUMANITY

Kind-hearted people might of course think there was some ingenious way to disarm or defeat an enemy without too much bloodshed, and might imagine this is the true goal of the art of war. Pleasant as it sounds, it is a fallacy that must be exposed: war is such a dangerous business that the mistakes which come from kindness are the very worst.
> Major-General Carl von Clausewitz, *On War*, 1.1, 1832, tr. Michael Howard and Peter Paret, 1976.

Men who take up arms against one another in public do not cease on this account to be moral beings, responsible to one another and to God.
> US Army General Order No. 100, 1863.

HUMOUR

The greatest kindness in war is to bring it to a speedy conclusion.
> Field Marshal Helmuth Graf von Moltke, 11 December 1880.

Quite contrary to trivial opinion, all professional military men do not walk blind and brutal. I have known some who demonstrated as much pity as they did courage, and they showed a lot of that.

When you are dealing firsthand with the quivering element of life and death – When you are trying to figure out the best manner in which to save certain lives as well as to take others, and in the same operation – You do not necessarily become calloused. Neither does a surgeon.
> General Curtis LeMay, *Mission With LeMay*, 1965.

HUMOUR

What impressed me most at the time, and survives as an inspiring memory, was the steadiness and fortitude of the British private soldier combined with his queer ironic humour in days of deep privation and adversity.
> Herbert Asquith, *Moments of Memory*, 1937; referring to the crisis of the German *Friedensturm* of 1918 and the 'backs to the wall' resolve of the British Army.

Don't be careless about yourselves – on the other hand not too careful. Live well but do not flaunt it. Laugh a little and teach your men to laugh – get good humour under fire – war is a game that's played with a smile. If you can't smile, grin. If you can't grin, keep out of the way until you can.
> Sir Winston S. Churchill (1874–1965).

To speak of the importance of a sense of humor would be futile, if it were not that what cramps so many men isn't that they are by nature humorless as that they are hesitant to exercise what humor they possess. Within the military profession, this is as unwise as to let the muscles go soft or to spare the mind the strain of original thinking. Great humor has always been in the military tradition.
> Brigadier-General S.L.A. Marshall, *The Armed Forces Officer*, 1950.

HUNGER

Achilles, godlike and brave, send not the Achaeans thus against Ilium to fight the Trojans fasting, for the battle will be no brief one, when it is once begun, and heaven has filled both sides with fury; bid them first take food, both bread and wine, by the ships, for in this there is strength and stay. No man can do battle the livelong day to the going down of the sun if he is without food. However much he may want to fight his strength will fail him before he knows it; hunger and thirst will find him out, and his limbs will grow weary under him. But a man can fight all day if he is full fed with meat and wine; his heart beats high, and his strength will stay till he has routed all his foes.
> Homer, *The Iliad*, 19, c. 800 BC, tr. Samuel Butler; Odysseus urging Achilles, who was eager to avenge the death of Patroclus, to taste his revenge slightly cold.

Gaius Caesar used to say that he followed the same policy towards the enemy as did many doctors when dealing with physical ailments, namely, that of conquering the foe by hunger rather than steel.
> Frontinus, *Stratagems*, 4.7, c. AD 84–96, tr. Charles E. Bennett, 1925.

Famine makes greater havoc in an army than the enemy, and is more terrible than the sword.
> Flavius Vegetius Renatus, *Military Institutions of the Romans*, c. AD 378, tr. Clark, 1776.

The general achieves the most who tries to destroy the enemy's army more by

HUNGER

hunger than by force of arms.
  The Emperor Maurice, *The Strategikon*, c. AD 600 (*Maurice's Strategikon*, tr. George Dennis, 1984).

Few victories are won on an empty belly.
  Sir John Hawkwood (1320–1394).

The greatest secret of war and the masterpiece of a skillful general is to starve his enemy. Hunger exhausts men more surely than courage, and you will succeed with less risk than by fighting.
  Frederick the Great, *Instructions to His Generals*, 1747, tr. Phillips, 1940.

The first art of a military leader is to deprive the enemy of subsistence.
  Field Marshal Prince Aleksandr V. Suvorov (1729–1800).

A starving army is actually worse than none. The soldiers lose their discipline and their spirit. They plunder even in the presence of their officers.
  The Duke of Wellington, August 1809, letter from Spain.

Food, food, food! Without it, there is no limit to the horrors this undisciplined mass of men may bring upon the town.
  Napoleon, 29 November 1812, letter to M. Maert, Minister for Foreign Affairs, of the approach of his starving army to Vilna on its retreat from Moscow, *Correspondance de Napoléon Ier, publié par ordre de l'Empereur Napoléon III*, XXIV, 19362, 1858–1870.

There is an enemy I dread much more than the hostile Creek… that meagre-monster 'Famine'.
  Major-General Andrew Jackson, 23 October 1813, letter to Leroy Pope, *Correspondence*, I, p. 336.

The waning *moral* at home was intimately connected with the food situation. In the daily food the human body did not receive the necessary nourishment, especially albumen and fats, for the maintenance of physical and mental vigour. In wide quarters a certain decay of bodily and mental powers of resistance was noticeable, resulting in an unmanly and hysterical state of mind which under the spell of enemy progaganda encouraged the pacifist leanings of many Germans. In the summer of 1917 my first glimpse of this situation gave me a great shock. This state of mind was a tremendous element of weakness. It was all a question of human nature. It could be eliminated to some extent by strong patriotic feeling, but in the long run could only be finally overcome by better nourishment. More food was needed.
  General Erich Ludendorff, *My War Memories, 1914–1918*, 1919.

Nothing undermines morale more decisively than hunger; quickest of all is the effect of any digestive upset. That was strikingly demonstrated in the late summer of 1918, when the moral slump of the German troops became most marked at a moment when stomach disorders, due to bad food, were rife among them. The old saying that 'an army marches on its stomach' has a wider and deeper application than has yet been given to it. An army fights on its stomach, and falls if its stomach is upset. (May 1939.)
  Captain Sir Basil Liddell Hart, *Thoughts on War*, 1944.

I–J–K

IDEALISM/IDEALS

It is, indeed, an observable fact that all leaders of men, whether as political figures, prophets, or soldiers, all those who can get the best out of others, have always identified themselves with high ideals, and this has given added scope and strength to their influence. Followed in their lifetime because they stand for greatness of mind rather than self-interest, they are later remembered less for the usefulness of what they have achieved than for the sweep of their endeavors. Though sometimes reason may condemn them, feeling clothes them in an aura of glory.
General Charles de Gaulle, *The Edge of the Sword*, 1932.

Men who are infused with a faith, even a false one, will beat men who have no faith; only a good one can withstand the impact. Those who complain of the younger generation's lack of patriotism should, rather, reproach themselves for their failure to define and teach patriotism in higher terms than the mere preservation of a geographical area, its inhabitants, and their material interests. Such material appeal offers no adequate inspiration, nor cause for sacrifice to the young. Those who are concerned with practical questions of defence ought to realize the practical importance of ideals. (August 1936.)
Captain Sir Basil Liddell Hart, *Thoughts on War*, 1944.

IGNORANCE

One must obey these old fellows who, never having studied their profession, obsessed by an antiquated routine which they call experience, and taking advantage of a long existence which they consider a long life, set out to traduce, pull to pieces, and ridicule budding genius which they detest, because they are compelled to value it more than themselves.
Major-General Henry Lloyd, *Introduction of the History of the War in Germany*, 1766.

There is an enemy greater than the hospital: the damned fellow who 'doesn't know'. The hint-dropper, the riddle-poser, the deceiver, the word-spinner, the prayer-skimper, the two-faced, the mannered, the incoherent. The fellow who 'doesn't know' has caused a great deal of harm... One is ashamed to talk about him. Arrest for the officer who 'doesn't know' and house arrest for the field or general officer.
Field Marshal Prince Aleksandr V. Suvorov, *The Science of Victory*, 1796.

The ignorant suspect no difficulties. They want to solve a problem of transcendental mathematics by means of a second-degree formula. All questions of grant tactics are indeterminate physico-mathematical equations that are incapable of solution by formulas of elementary mathematics.
Napoleon, dictation at St Helena, *Correspondance de Napoléon Ier, publiée par ordre de l'Empereur Napoléon III*, XXXI, p. 338, 1858–1870, quoted in Christopher Herold, ed., *The Mind of Napoleon*, 1955.

An ignorant officer is a murderer. All brave men confide in the knowledge he

IMAGINATION

is supposed to possess; and when the death-trial comes their generous blood flows in vain. Merciful God! How can an ignorant man charge himself with so much bloodshed? I have studied war long, earnestly and deeply, but yet I tremble at my own derelictions.
    General Sir William Napier (1782–1853).

The greatest difficulty I find is in causing orders and regulations to be obeyed. This arises not from a spirit of disobedience, but from ignorance.
    General Robert E. Lee, quoted in Montross, *War Through the Ages*, 1960.

I have known him for a long time. I am very fond of him, but he has not yet gone as far as Caesar's Commentaries in studying the history of war since he forgot the history he learned at West Point.
    General of the Armies John J. Pershing, 1918, comments on a deficient officer, quoted in Frank Vandiver, *Black Jack*, 1977.

An ignorant man cannot be a good soldier. He may be brave and audacious, and, in the hand-to-hand struggles of the past, his ignorance may have appeared but a small defect, since he could rapidly clinch with danger. But to-day this defect has grown big; the stout arm of Cannae, of Crécy, or even of Inkerman, demands at least a cunning brain. Fighting intervals and distances have increased, and there is more room for ignorance to display its feathers, and the corridors of fear are long and broad.
    Major-General J.F.C. Fuller, *The Foundations of the Science of War*, 1926.

Wisdom begins at the point of understanding that there is nothing shameful about ignorance; it is shameful only when a man would rather remain in that state than cultivate other men's knowledge.

Brigadier-General S.L.A. Marshall, *The Armed Forces Officer*, 1950.

IMAGINATION

What a thing is imagination! Here are men who don't know me, who have never seen me, but who only knew of me, and they are moved by my presence, they would do anything for me! And this same incident arises in all centuries and in all countries! Such is fanaticism! Yes, imagination rules the world. The defect of our modern institutions is that they do not speak to the imagination. By that alone can a man be governed; without it he is but a brute.
    Napoleon, 17 June 1800, at Milan, after being cheered by German and Hungarian prisoners taken at the Battle of Marengo three days previously, men whom Napoleon had taken prisoner before in the campaigns of 1796 and 1797, quoted in R.M. Johnston, ed., *The Corsican*, 1910.

Without that great gift [imagination] only a very inferior order of ambition in any walk of life can be satisfied, and certainly without it no one can ever become a renowned leader of armies. How largely it was possessed by Moses, Xenophon, Hannibal, Caesar, Turenne, Marlborough, Napoleon, and Wellington! It is said to rule the world… And yet, whilst imagination may convert into a poet the man of poor physique, it would not of itself make an able general of him. He who aspires to lead soldiers in war should be not only a thorough master of the soldier's science, but he must possess a healthy strength of body, an iron nerve, calm determination, and that instinctive electric power which causes men to follow the leader who possesses it, as readily, as surely, as iron filings do the magnet… It is the necessity for this rare, this exceptional combination of mental gifts with untiring physical power and stern resolution that accounts for the fact that the truly great

IMPEDIMENTA

commander is rare indeed among God's creations.
> Field Marshal Viscount Wolseley, *The Story of a Soldier's Life*, 1869.

Imagination forms one of the four parts of genius and has itself two clearly marked attributes: the one, fancy or the power of ornamenting facts as, for instance, making the wolf speak to Little Red Riding Hood or describing the Battle of Le Cateau; the other, inventions; not fairy tales but machines. Inventions do not often make their first bow to armies on the battle-field. They have been in the air for some time; hawked about the ante-chambers of the men of the hour; spat upon by common sense; cold-shouldered by interests vested in what exists; held up by stale functionaries to whom the sin against the Holy Ghost is to 'make a precedent' – until, one day, arrives a genius who by his imagination sees; by his enthusiasm moves; by his energy keeps moving; by his courage cuts the painter of tradition.
> General Sir Ian Hamilton, *The Body and Soul of an Army*, 1921.

The most indispensable attribute of the great captain is imagination.
> General of the Army Douglas MacArthur, 1959, letter to B.H. Liddell Hart.

IMPEDIMENTA

First of all, then, I propose that we burn the wagons we have, that the baggage-train may not be our captain, and we may go where is best for the army. Next, burn the tents too: for these are a nuisance to carry, they don't help us to fight or to find provisions. Further, let us get rid of superfluous baggage, keeping only what we need for battle or eating or drinking; so the greatest number of us will be under arms, and as few as possible carrying baggage. You know, of course, that if we are beaten, someone else owns all our goods; if we win, we may call the enemy our porters.
> Xenophon, *Anabasis*, 3.2, c. 360 BC (*The March Up Country*, tr. W.H.D. Rouse, 1959); 401 BC, encouraging the stranded Greek Army to abandon its impedimenta to allow escape from the Persians.

For the purpose of limiting the number of pack animals, by which the march of the army was especially hampered, Gaius Marius had his soldiers fasten their utensils and food up in bundles and hang these on forked poles, to make the burden easy and to facilitate rest; whence the expression 'Marius's Mules.'
> Frontinus, *Stratagems*, 4.6, c. AD 84–96, tr. Charles E. Bennett, 1925.

We can get along without anything but food and ammunition. The road to glory cannot be followed with much baggage.
> Lieutenant-General Richard S. Ewell, 1862, orders during the Valley campaign.

An army is efficient for action and motion exactly in the inverse ratio of its *impedimenta*.
> General of the Army William T. Sherman, *Memoirs of General W.T. Sherman*, II, 1875.

A serious handicap on mobility lies in the tendency of armies to increase their requirement. The desire for stronger armament causes a demand for more transport; the desire for better communications leads to more complex communications; the desire for superior organization creates an additional organization to deal with the development. In building up its powers to overcome the enemy's resistance an army is apt to set up an internal resistance to its advance. Thus what has been gained by the advent of mechanized means of movement is largely offset by the growth of impedimenta. It is not a necessary growth, but military evolution has always tended to be a process of accretion, tacking new means on to the old body. The idea of

INCOMPETENCE

remodelling design is foreign to tradition. And the result to-day is the paradoxical one that armies have become less mobile as their limbs have become more mobile. The hare dons the shell of the tortoise.
    Captain Sir Basil Liddell Hart, *Thoughts on War*, 1944.

INCOMPETENCE

Many badly conceived enterprises have the luck to be successful because the enemy has shown an even smaller degree of intelligence.
    Thucydides, *The History of the Peloponnesian War*, c. 404 BC.

The General has no plan, or even an idea of a plan nor do I believe he knows the meaning of the word Plan. I entered fully into a discussion upon the situation of the army the day before yesterday, in which I pointed out the inutility of augmenting his army here and the danger of the measure if he should not – and he could not – augment his means of supply, and I gave him papers to read upon the subject to his future operations. He had not uttered one word to me upon that subject from that time to this. And Moore told me last night that he had said to him, 'You may either land your corps or not as you think proper,' as if it was a matter of perfect indifference whether this army should have 10,000 men in addition or whether 15,000 additional mouths should be fed by means calculated for half the number. The people are really more stupid and incapable than any I have yet met with, and, if things go on in this disgraceful manner, I must quit them.
    The Duke of Wellington, 26 August 1808, after being superseded in command of the British forces in Portugal by incompetent general officers due to the iron rule of seniority, quoted in Arthur Bryant, *The Great Duke*, 1972.

I don't know what effect these men will have on the enemy, but, by God, they frighten me.
    The Duke of Wellington, 1809, impression of senior British officers sent to the Army in Portugal.

When I reflect on the characters and attainments of some of the general officers of this army... I tremble; and, as Lord Chesterfield said of the generals of his day, 'I only hope that when the enemy reads the list of their names he trembles as I do.'
    The Duke of Wellington, 20 August 1810, letter to Lt Col. Henry Torrens.

When I was at Tilsit with the Emperor Alexander and the King of Prussia, I was the most ignorant of the three in military affairs! These two sovereigns, especially the King of Prussia, were completely *au fait* as to the number of buttons there ought to be in front of a jacket, how many behind, and the manner in which the skirts ought to be cut. Not a tailor in the army knew better than Frederick how many measures of cloth it took to make a jacket. In fact, I was nobody in comparison with them. They continually tormented me with questions about matters belonging to tailors, of which I was entirely ignorant, though, in order not to affront them, I answered just as gravely as if the fate of an army depended upon the cut of a jacket. The King of Prussia changed his fashion every day. He was a tall, dry looking fellow, and would give a good idea of Don Quixote. At Jena, his army performed the finest and most showy manoeuvres possible, but I soon put a stop to their *coglionerie*, and taught them that to fight and to execute dazzling manoeuvres and wear splendid uniforms were very different affairs. If the French army had been commanded by a tailor, the King of Prussia would certainly have gained the day, from his superior knowledge and art!

INDEPENDENCE OF COMMAND

Napoleon, 16 May 1817, on St Helena, cited in R.M. Johnston, ed., *The Corsican*, 1910.

There are field-marshals who would not have shone at the head of cavalry regiments and vice versa.
    Major-General Carl von Clausewitz, *On War*, 1832, tr. Michael Howard and Peter Paret, 1976.

The incompetence of the authorities and of G.H.Q. is greater on the other side than with us; and that is saying a good deal.
    General Max Hoffmann, November 1915, *War Diaries and Other Papers*, I, 1929.

Presumably victory is our object. This war is a business proposition; it therefore requires ability. Nevertheless, our Army is crawling with 'duds'; though habitual offenders, they are tolerated because of the camaraderie of the old Regular Army: an Army so small as to permit of all its higher members being personal friends. Good-fellowship ranks with us above efficiency; the result is a military trade union which does not declare a dividend.
    Major-General J.F.C. Fuller, 1918, *Memoirs of an Unconventional Soldier*, 1936.

More and more does the 'system' tend to promote to *control*, men who have shown themselves efficient *cogs* in the machine. It is especially so in the Army to-day. There are few commanders in our higher commands. And even these, since their chins usually outweigh their foreheads, are themselves outweighted by the majority of commanders who are essentially staff officers. These tend to be *desperately* conventional in organizing the Army and moulding its doctrines for war. And in war they would almost certainly prove *recklessly* cautious. (May 1936.)
    Captain Sir Basil Liddell Hart, *Thoughts on War*, 1944.

INDEPENDENCE OF COMMAND

There are... commands of the sovereign which must not be obeyed.
    Sun Tzu, *The Art of War*, 8, c. 500 BC, tr. Giles, 1910.

Maxim 72. A general-in-chief has no right to shelter his mistakes in war under cover of his sovereign, or of a minister, when they are both distant from the scene of operation, and must consequently be either ill informed or wholly ignorant of the actual state of things.
    Hence it follows that every general is culpable who undertakes the execution of a plan which he considers faulty. It is his duty to represent his reasons, to insist upon a change of plan; in short, to give in his resignation rather than allow himself to become the instrument of his army's ruin. Every general-in-chief who fights a battle in consequence of superior orders, with the uncertainly of losing, is equally blamable.
    Napoleon, *The Military Maxims of Napoleon*, tr. George D'Aguilar, 1831 (David Chandler, ed., 1987).

To Field Marshal Rommel: In the situation in which you find yourself there can be no other thought but to stand fast and throw every gun and every man into the battle. The utmost efforts are being made to help you. Your enemy, despite his superiority, must also be at the end of his strength. It would not be the first time in history that a strong will has triumphed over the bigger battalions. As to your troops, you can show them no other road than that to victory or death.
                                ADOLF HITLER

This order demanded the impossible. Even the most devoted soldier can be killed by a bomb.
    Erwin Rommel, Hitler's order from East Prussia to stand fast at El Alamein in North

THE INDIRECT APPROACH

Africa and Rommel's comment, *The Rommel Papers*, 1953.

THE INDIRECT APPROACH

In all fighting, the direct method may be used for joining battle, but indirect methods will be needed in order to secure victory.

Indirect tactics, efficiently applied, are inexhaustible as Heaven and Earth, unending as the flow of rivers and streams; like the sun and moon, they end but to begin anew; like the four seasons, they pass but to return once more.
  Sun Tzu, *The Art of War*, 5, c. 500 BC, tr. Giles, 1910.

If you would shoot a general, first destroy his horse.
  Japanese military proverb, quoted in Inoguchi and Nakajima, *The Divine Wind*, 1958.

Three men behind the enemy are worth more than fifty in front of him.
  Frederick the Great (1712–1786).

Maxim 16. It is an approved maxim of war, never do what the enemy wishes you to do, for this reason alone, that he desires it. A field of battle, therefore, which he has previously studied and reconnoitered, should be avoided, and double care should be taken where he has had time to fortify or entrench. One consequence deducible from this principle is, never to attack a position in front which you can gain by turning.
  Napoleon, *The Military Maxims of Napoleon*, tr. George D'Aguilar, 1831 (David Chandler, ed., 1987).

To see correctly a general must understand the nature of the changes which take place in war. The enemy does not attack him physically, but mentally; for the enemy attacks his ideas, his reason, his plan. The physical pressure directed against his men reacts on him through compelling him to change his plan, and changes in his plan react on his men by creating a mental confusion which weakens their morale. Psychologically, the battle is opened by a physical blow which unbalances the commander's mind, which in its turn throws out of adjustment the morale of his men, and leads to their fears impeding the flow of his will. If the blow is a totally unexpected one, the will of the commander may cease altogether to flow, and, the balance in the moral sphere being utterly upset, self-preservation fusing with self-assertion results in panic.
  Major-General J.F.C. Fuller, *The Foundations of the Science of War*, 1926.

Throughout history the direct approach has been the normal form of strategy, and a purposeful indirect approach the rare exception. It is curious how often generals have adopted the latter, not as their initial strategy, but as a last resource. Yet it has brought them a decision where the direct approach had brought them failure – and thereby left them in a weakened condition to attempt the indirect. A decisive success obtained in such deteriorated conditions acquires all the greater significance. (Oct. 1928.)
  Captain Sir Basil Liddell Hart, *Thoughts on War*, 1944.

The art of the indirect approach can only be mastered, and its full scope appreciated, by study of and reflection upon the whole history of war. But we can at least crystallize the lessons into two simple maxims, one negative, and the other positive. The first is that in face of the overwhelming evidence of history no general is justified in launching his troops to a *direct* attack upon an enemy *firmly* in position. The second, that instead of seeking to upset the enemy's equilibrium by one's attack, it must be upset *before* a real attack is, or can be, successfully launched… (Nov. 1928.)

INFANTRY

Captain Sir Basil Liddell Hart, *Thoughts on War*, 1944.

Battle should no longer resemble a bludgeon fight, but should be a test of skill, a manoeuvre combat, in which is fulfilled the great principle of surprise by striking 'from an unexpected direction against an unguarded spot.'

Captain Sir Basil Liddell Hart, *Thoughts on War*, 1944.

You can tell a man's character by the way he makes advances to a woman. Men like you, for example – when the fleet's in port and you go off to have a good time, you seem to have only two ways of going about things. First, you put it straight to the woman: 'Hey, how about a lay?' Now, any woman, even the lowest whore, is going to put up at least a show of refusing if she's asked like that. So what do you do next? You either act insulted and get rough, or you give up immediately and go off to try the same thing on the next woman. That's all you're capable of. But take a look at Western men – they're quite different. Once they've set their sights on a woman, they invite her out for a drink, or to dinner, or to go dancing. In that way they gradually break down her defenses until, in the end, they get what they want, and in style at that. Where achieving a particular aim is concerned, that's surely a far wiser way of going about things. At any rate, they're the kind of men you'd be dealing with if there were a war, so you'd better give it some thought.

Admiral Isoroku Yamamoto, 1939, to his junior officers, quoted in Agawa, *The Reluctant Admiral*, 1979.

Use steamroller strategy; that is, make up your mind on course and direction of action, and stick to it. But in tactics, do not steamroller. Attack weakness. Hold them by the nose and kick them in the pants.

General George S. Patton, Jr., *War As I Knew It*, 1947.

The enemy wanted to concentrate their forces. We compelled them to disperse. By successfully launching strong offensives on the points they had left relatively unprotected, we obliged them to scatter their troops all over the place in order to ward off our blows, and thus created favourable conditions for the attack at Dien Bien Phu, the most powerful entrenched camp in Indo-China, considered invulnerable by the Franco-American general staff. We decided to take the enemy by the throat at Dien Bien Phu.

General Vo Nguyen Giap, *People's War, People's Army*, 1961.

I was not going to have a Grozny or a Stalingrad. I needed a more cunning way and to build up a web of courage among the ordinary people.

Major-General Robin Brims, Commander, British 1st Armoured Division, as he surrounded Basra and Az Zubay without besieging them and picked up intelligence from people as they came out, quoted in Charles Moore, *Daily Telegraph*, 28 April 2003.

INFANTRY

The infantry must ever be valued as the very foundation and nerve of an army.

Niccolò Machiavelli, *Discourses*, 1517.

There, it all depends upon that article whether we do the business or not. Give me enough of it, and I am sure.

The Duke of Wellington, May 1815, pointing to a British infantryman on the eve of the Waterloo campaign, Creevy, *The Creevy Papers*, 1934.

It was the most desperate business I ever was in. I never took so much trouble about any Battle [Waterloo], & never was so near being beat. Our loss is immense particularly in the best of All

INFANTRY

Instruments, British Infantry. I never saw the Infantry behave so well.
> The Duke of Wellington, 19 June 1815, the day after Waterloo, quoted in Brett-James, *The Hundred Days*, 1964.

My opinion is that the best troops we have, probably the best in the world, are the British infantry, particularly the old infantry that has served in Spain. This is what we ought to keep up; and what I wish above all others to retain.
> The Duke of Wellington, advice to British ministers intent on reducing the army after 1815, *The Dispatches of Field Marshal the Duke of Wellington During His Various Campaigns in India, Denmark, Portugal, Spain, the Low Countries, and France*, XII, 1834–1838, p. 668.

Infantry is the Queen of Battles.
> General Sir William Napier (1785–1860); later adopted as the motto of the US Infantry.

Maxim 93. The better the infantry is, the more it should be used carefully and supported with good batteries. Good infantry is, without doubt, the sinew of an army; but if it is forced to fight for a long time against a very superior artillery, it will become demoralized and will be destroyed.
> Napoleon, *The Military Maxims of Napoleon*, tr. Burnod, 1827.

It is admitted by all military men that infantry is the great lever of war, and that the artillery and cavalry are only indispensable accessories…
  Two essential conditions constitute the strength of infantry: –
  That the men be good walkers and inured to fatigue.
  That the firing be well-executed.
  The physical constitution, and the national composition of the French armies, fulfill the former most advantageously; the vivacity and intelligence of the soldiers ensure the success of the latter.
> Marshal of France Michel Ney, Duc d'Elchingen, Prince de la Moskova, *Memoirs of Marshal Ney*, 1834.

In the end of ends, infantry is the deciding factor in every battle. I was in the infantry myself and was body and soul an infantry man. I told my sons to join the infantry. They did so, but, as happened to so many our our young men, the freedom of the air drew them from the trenches. But the fine saying of the old 'Directions for Infantry Exercise,' remain true in war: 'The infantry bears the heaviest burden of a battle and requires the greatest sacrifice; so also it promises the greatest renown.'
> General Erich Ludendorff, *My War Memories, 1914–1918*, 1919.

The infantry, the infantry with mud behind their ears.
The infantry, the infantry that drinks up all the beers.
The cavalry, the artillery, and the goddamned engineers –
They couldn't beat the infantry in a hundred million years!
> Old US Infantry marching song, quoted in Geoffrey Parret, *Eisenhower*, 1999, p. 53.

When the smoke cleared away, it was the man with the sword, or the crossbow, or the rifle, who settled the final issue on the field.
> General of the Army George C. Marshall, 3 February 1939, speech before the National Rifle Association.

I love the infantryman because they are the underdogs. They are the mud-rain-and-wind boys. They have no comforts, and they learn to live without the necessities. And in the end they are the guys that wars can't be won without.
> Ernie Pyle, *New York World Telegram*, 5 May 1943.

Look into an infantryman's eyes and

INITIATIVE

you can tell how much war he has seen... I don't make the infantryman look noble, because he couldn't look noble even if he tried. Still there is a certain nobility and dignity in combat soldiers and medical aid men with dirt in their ears. They are rough and their language gets coarse because they live a life stripped of convention and niceties. Their nobility and dignity come from the way they live unselfishly and risk their lives to help each other.
   Sergeant Bill Mauldin, *Up Front*, 1945, p. 42, 11–15.

Another element [of leadership] to be considered is the Man to be led, and with whose morale we are concerned. I am constantly reminded of this point by a cartoon which hangs over my desk at home which depicts an infantryman with his rifle across his knees as he sits behind a parapet. Above him is the list of the newest weapons science has devised and the soldier behind the parapet is saying: 'But still they haven't found a substitute for ME.'
   General of the Army Omar N. Bradley, 'Leadership', *Parameters*, Winter 1972, p. 2.

...The least spectacular arm of the Army, yet without them you cannot win a battle. Indeed, without them, you can do nothing. Nothing at all, nothing.
   Field Marshal Viscount Montgomery of Alamein, quoted in *Military Review*, 5/1981.

Quite frankly, the average British infantryman is far better [than American infantrymen]. They're a tribe of feral monsters, but they're highly disciplined monsters. You don't want to get in their way.
   Anonymous British officer, quoted by Paul Martin, *The Washington Times*, 3 April 2003.

INITIATIVE
The Operational Dynamic
Whoever is first in the field and awaits the coming of the enemy, will be fresh for the fight; whoever is second in the field and has to hasten to the battle, will arrive exhausted.
   Therefore the clever combatant imposes his will on the enemy, but does not allow the enemy's will to be imposed on him.
   By holding out advantages to him, he can cause the enemy to approach of his own accord; or by inflicting damage, he can make it impossible for the enemy to draw near.
   Sun Tzu, *The Art of War*, 6, c. 500 BC, tr. Giles, 1910.

The art of war is, in the last result, the art of keeping one's freedom of action.
   Xenophon (431–352 BC).

It is an incontestable truth that it is better to forestall the enemy, than to find yourself anticipated by him.
   Frederick the Great, 'Anti-Machiavelli', *Posthumous works of Frederick II, King of Prussia*, 1789, tr. Holcroft, quoted in Jay Luvaas, *Frederick the Great on the Art of War*, 1966.

We must get the upper hand, and if we once have that, we shall keep it with ease, and shall certainly succeed.
   The Duke of Wellington, 17 August 1803, dispatch in India.

Instead of being on the defensive, I would be on the offensive; instead of guessing at what he [the enemy] means to do, he would have to guess at my plans. The difference in war is 25 per cent.
   General of the Army William T. Sherman, 11 October 1864, letter to Grant urging his attack toward Atlanta and the sea, *The War of the Rebellion: A Compilation of the Official Records of the Union and Confederate Armies*, Series I, Vol. XXXIX, Part 3, p. 202, US Government Printing Office, 1880–1901.

In any fight, it's the first blow that counts; and if you keep it up hot

INITIATIVE

enough, you can whip 'em as fast as they come up.
> Lieutenant-General Nathan Bedford Forrest (1821–1877).

It was to be considered that hostile troops would allow themselves to be held if we attacked, or at least threatened, the enemy at some really sensitive point. It was further to be remembered that, with the means available, protection of the Colony could not be ensured even by purely defensive tactics, since the total length of land frontier and coast-line was about equal to that of Germany. From these considerations it followed that it was necessary, not to split up our small available forces in local defence, but, on the contrary, to keep them together, to grip the enemy by the throat and force him to employ his forces for self-defence…
> Field Marshal Paul von Lettow-Vorbeck, *East African Campaigns*, 1957, describing the defence of German East Africa in WWI, during which German forces of fewer than 15,000 drew into the theatre a third of a million Allied troops.

The initiative… means an army's freedom of action as distinguished from an enforced loss of freedom. Freedom of action is the very life of an army and, once it is lost, the army is close to defeat or destruction.
> Mao Tse-tung, *On Protracted War*, 1938.

I determined to make no-man's land our land. We set out to besiege the besiegers.
> Lieutenant-General Leslie Morshead, his principle of defence as Commander of the Tobruk garrison and the Ninth Australian Division, 1941, quoted in Horner, *The Commanders*, 1984.

When so much was uncertain, the need to recover the initiative glared forth.
> Winston S. Churchill, after Dunkirk, 1940, *Their Finest Hour*, 1949.

I have never given a damn what the enemy was going to do or where he was. What I have known is what I have intended to do and then have done it. By acting in this manner I have always gotten to the place he expected me to come about three days before he got there.
> General George S. Patton, Jr., August 1944, *The Patton Papers*, II, 1974.

During war the ball is always kicking around loose in the middle of the field and any man who has the will may pick it up and run with it.
> Brigadier-General S.L.A. Marshall, *The Armed Forces Officer*, 1950.

He who shoots first laughs last.
> General Aleksandr I. Lebed's favourite Russian military maxim, *Sobesednik* (Moscow), March 1998.

The Personal Quality

True, when the man leaps in the breach
    that way
no one can blame or disobey him, no
    Achaean,
not when he spurs the troops and gives
    commands.
> Homer, *The Iliad*, 10.150–153, c. 800 BC, tr. Robert Fagles, 1990; King Nestor of Pylos speaking of Menalaus' initiative to organise resistance when the Trojans stormed the Greek camp.

Tell the king that after the battle my head is at his disposal, but meantime I hope he will permit me to exercise it in his service!
> General F.W. Seydlitz, to Frederick the Great who was continually ordering an ill-advised manoeuvre from the opposite end of the battlefield at Zorndorf, 25 August 1758, quoted in Christopher Duffy, *The Military Life of Frederick the Great*, 1986.

It certainly was so, and if ever you commit such a breach of your orders I will forgive you also.

INITIATIVE

Admiral Lord St Vincent, 14 February 1797, on the evening of the Battle of Cape St Vincent when an officer complained that Nelson's action was a breach of orders, quoted in David Walder, *Nelson*, 1978.

When the signal, No. 39, was made, the Signal Lieutenant reported to him. He continued his walk, and did not appear to take notice of it. The lieutenant meeting his Lordship at the next turn asked, 'whether he should repeat it?' Lord Nelson answered, 'No, acknowledge it.'... his Lordship called after him, 'Is No. 16 [for close action] still hoisted?' The lieutenant answering in the affirmative, Lord Nelson said, 'Mind you keep it so.' He now walked the deck considerably agitated, which was always known by his moving the stump of his right arm. after a turn or two, he said to me, in a quick manner, 'Do you know what's shown on board the Commander-in-Chief, No. 39?' On asking him what that meant, he answered, 'Why, to leave off action.' 'Leave off action!' he repeated, and then added, with a shrug, 'Now damn me if I do.' He also observed, I believe, to Captain Foley, 'You know, Foley, I have only one eye – I have a right to be blind sometimes:' and with an archness peculiar to his character, putting the glass to his blind eye, he exclaimed, 'I really do not see the signal.'

> Admiral Viscount Nelson's refusal to abandon a successful action despite orders to do so during the Battle of Copenhagen, 2 April 1801, as described by Colonel William Parker, quoted in Alfred Thayer Mahan, *The Life of Nelson*, 1897.

When I am without orders, and unexpected occurrences arise, I shall always act as I think the honour and glory of my King and Country demand.

> Admiral Viscount Nelson, November 1804, letter to Hugh Elliot.

But, in case signals can neither be seen or perfectly understood, no captain can do very wrong if he places his ship alongside that of the enemy.

> Admiral Viscount Nelson, 9 October 1805, memorandum to the fleet before Trafalgar.

All in all it seems... that an unusual spirit, of independence of those above and acceptance of responsibility, has grown up throughout the Prussian officer corps as it has in no other army... Prussian officers will not stand for being hemmed in by rule and stereotypes as happens in Russia, Austria or Britain... We follow the more natural course of giving scope to every individual's talent, of using a looser rein. We back up every success as a matter of course – even when it runs counter to the intentions of the commander-in-chief... the subordinate commander exploits every advantage by taking initiatives off his own bat, without his superior's knowledge or permission.

> Frederick III, 1860, essay, quoted in Richard Simpkin, *Race to the Swift*, 1985.

That individuality of action which so strongly characterizes the conduct of our troops in battle, if unguided or misdirected, can but produce confusion. But let the same idea control the mind of every man, let them apply these general principles to the incidents of battle as they arise, and success is certain.

> Major-General James E.B. Stuart, General Order No. 26, Cavalry Tactics, 30 July 1863, *Letters of Major General James E.B. Stuart*, 1990.

A favourable situation will never be exploited if commanders wait for orders. The highest commander and the youngest soldier must always be conscious of the fact that omission and inactivity are worse than resorting to the wrong expedient.

> Field Marshal Helmuth Graf von Moltke (1800–1891).

INITIATIVE

My aim was to turn highly-disciplined troops into responsible men possessed of initiative. Discipline is not intended to kill character, but to develop it. The purpose of discipline is to bring about the uniformity in co-operating for the attainment of the common goal, and this uniformity can only be obtained when each one sets aside the thought of his own personal interests. This common goal is – Victory.
General Erich Ludendorff, *My War Memories, 1914–1918*, 1919.

The magazine rifle with its low trajectory and smokeless powder, spoke volumes to the captain of 1899–1902. It told him he could still conduct his company into the zone of aimed fire, but that, having got them there, he must:
(1) Keep his direct command at the cost of double losses.
(2) Let each little group understand the common objective. Then leave them to the promptings of their own consciences of what was right rather than to the dread of what was wrong.
General Sir Ian Hamilton, *The Soul and Body of an Army*, 1921.

If, in the opinion of the leader, the plan has, through change in conditions, become inoperative, then he ceases to be a leader and becomes, for the time being, an independent commander and must act as if he were a general-in-chief. That is to say, he must replace the inoperative plan by an operative one – that is, one which will permit of the economical expenditure of force. To carry on a plan which manifestly has failed is the act of a fool, whether he be the general-in-chief or a private soldier. Once again we come back to our starting-point, namely, intelligence.
Major-General J.F.C. Fuller, *The Foundations of the Science of War*, 1926.

The capacity for independent action of low-level infantry sub-units has long been our weakest point. The cry of 'I'm waiting for orders,' which really meant 'I'm doing nothing,' was the real scourge of our activities in the field.
Marshal of the Soviet Union Mikhail N. Tukhachevskiy, 1931–1932, quoted in Richard Simpkin, *Deep Battle*, 1987.

There are cases of... initiative exaggerated to the point of violating discipline and interfering with the concentration of the combined effort... In the last analysis, an exaggerated display of initiative is due in the first place to an absence or weakness of decision higher up. A spirit of enterprise in a commander is never a danger in itself.

...so it comes about that the authorities dread any officer who has the gift of making decisions and cares nothing for routine and soothing words. 'Arrogant and undisciplined' is what the mediocrities say of him, treating the thoroughbred with a tender mouth as they would a donkey which refuses to move, not realizing that asperity is, more often than not, the reverse side of a strong character, that you can only lean on something that offers resistance, and that resolute and inconvenient men are to be preferred to easy-going natures without initiative.
General Charles de Gaulle, *The Edge of the Sword*, 1932.

Eisenhower, the Department is filled with able men who analyze the problems well but feel compelled always to bring them to me for final solution. I must have assistants who will solve their own problems and tell me later what they have done.
General George C. Marshall, quoted in Stephen E. Ambrose, *Eisenhower: Soldier, General of the Army, President-Elect*, 1963.

Never tell people *how* to do things. Tell them *what* to do and they will surprise you with their ingenuity.

INNOVATION

General George S. Patton, Jr., *War As I Knew It*, 1947.

Initiative I gave as the third of the qualities of an officer. Initiative simply means that you do not sit down and wait for something to happen. In war, if you do, it will happen all right, but it will be almighty unpleasant. Initiative for you means that you keep a couple of jumps ahead, not only of the enemy, but of your own men.

Field Marshal Viscount Slim of Burma, *Courage and Other Broadcasts*, 1957.

It is better to struggle with a stallion when the problem is how to hold it back, than to urge on a bull which refuses to budge.

General Moshe Dayan (1915–1981).

Commanders at all levels had to act more on their own; they were given greater latitude to work out their own plans to achieve what they knew was the Army Commander's intention. In time they developed to a marked degree the flexibility of mind and a firmness of decision that enabled them to act swiftly to take advantage of sudden information or changing circumstances without reference to superiors... This acting without orders, in anticipation of orders, or without waiting for approval, yet always within the over-all intention, must become second nature in any form of warfare where formations do not fight closely *en cadre*, and must go down to the smallest units. It requires in the higher command a corresponding flexibility of mind, confidence in its subordinates, and the power to make its intentions clear right through the force.

Field Marshal Viscount Slim of Burma, *Defeat Into Victory*, 1956; of the command relationship within the British 14th Army in the Burma-India-China theatre in WWII.

To make perfectly clear that action contrary to orders was not considered as disobedience or lack of discipline, German commanders began to repeat one of Moltke's favorite stories, of an incident observed while visiting the headquarters of Prince Frederick Charles. A major, receiving a tongue-lashing from the Prince for a tactical blunder, offered the excuse that he had been obeying orders, and reminded the Prince that a Prussian officer was taught that an order from a superior was tantamount to an order from the King. Frederick Charles promptly responded: 'His Majesty made you a major because he believed you would know when not to obey his orders.' This simple story became guidance for all following generations of German officer.

Colonel T.N. Dupuy, *A Genius For War: The Army and the German General Staff, 1807–1945*, 1977.

INNOVATION

In war novelties of an atavistic nature are generally horrible; nevertheless, in the public mind, their novelty is their crime; consequently, when novelties of a progressive character are introduced on the battlefield, the public mind immediately anathematizes them, not necessarily because they are horrible but because they are new. Nothing insults a human being more than an idea his brains are incapable of creating. Such ideas detract from his dignity for they belittle his understanding. In April 1915, a few hundred British and French soldiers were gassed to death; gas being a novelty, Europe was transfixed with horror. In the winter of 1918–1919, the influenza scourge accounted for over 10,000,000 deaths, more than the total casualties killed throughout the Great War; yet the world scarcely twitched an eyelid, though a few people went so far as to sniff eucalyptus.

Major-General J.F.C. Fuller, *The Reformation of War*, 1923.

Accusing as I do without exception all the

INSPIRATION

great Allied offensives of 1914, 1916 and 1917, as needless and wrongly conceived operations of infinite cost, I am bound to reply to the questions – what else could have been done? And I answer it, pointing to the Battle of Cambrai, 'This could have been done'. This in many variants, this in larger and better forms ought to have been done, and would have been done if only the generals had not been content to fight machine-gun bullets with the breasts of gallant men, and think this was waging war.

> Sir Winston S. Churchill, on the first use of tanks on the battlefield, *The World Crisis*, 1923.

Looking back on the stages by which various fresh ideas gained acceptance, it can be seen that the process was eased when they could be presented, not as something radically new, but as the revival in modern terms of a time-honoured principle or practice that had been forgotten. This required not deception, but care to trace the connection – since 'there is nothing new under the sun'. A notable example was the way that the opposition to mechanization was diminished by showing the mobile armoured vehicle – the fast moving tank – was fundamentally the heir of the armoured horseman, and thus the natural means of reviving the decisive role which cavalry had played in past ages.

> Captain Sir Basil Liddell Hart, *Sherman*, 1929.

Prejudice against innovation is a typical characteristic of an Officer Corps which has grown up in a well-tried and proven system.

> Field Marshal Erwin Rommel, *The Rommel Papers*, 1953.

INSPIRATION

The first city we shall see was built by Alexander. Our every step will evoke memories of the past worthy of emulation by Frenchmen.

> Napoleon, 22 June 1798, proclamation to his troops at sea on the expedition to Egypt, quoted in R.M. Johnston, ed., *The Corsican*, 1910.

It is this animation which so largely constitutes the art of war, and of which it is so difficult to write. It is not one soul lighting another – this is mere fanaticism – but rather one mind illuminating many minds, by one heart causing thousands to beat in rhythm which, like a musical instrument, accompanies the mind in control. It is a union between intelligence and heart; between the will of the general and the willingness of his men; that fusion of the mental and moral spheres.

> Major-General J.F.C. Fuller, *The Foundations of the Science of War*, 1926.

While war is terribly destructive, monstrously cruel, and horrible beyond expressions, it nevertheless causes the divine spark in men to glow, to kindle, and to burst into a living flame, and enables them to attain heights of devotion to duty, sheer heroism, and sublime unselfishness that in all probability they would never have reached in the prosecution of peaceful pursuits.

> Major-General John A. Lejeune, *The Reminiscences of a Marine*, 1929.

Study the human side of history... To learn that Napoleon in 1796 with 20,000 beat combined forces of 30,000 by something called economy of force or operating on interior lines is a mere waste of time. If you can understand how a young unknown man inspired a half-starved, ragged, rather Bolshie, crowd; how he filled their bellies; how he outmarched, outwitted, and out-bluffed and defeated men who had studied war all their lives and waged it according to the text-books of their time, you will have learnt something worth knowing.

INSTINCT/INTUITION

I have never accepted what many people have kindly said, namely that I inspired the nation. It was the nation and the race dwelling around the globe that had the lion heart. I had the luck to be called upon to give the roar.
Sir Winston S. Churchill, 30 November 1954, speech on his 80th birthday at Westminster Hall.

When all the units had been briefed by their officers, I would stand out on the porch of the headquarters and watch the preparations. Soldiers would be going and coming, checking their weapons and equipment, loading trucks, talking to each other and their officers. The camp would be a beehive of activity, alive with purpose. Each one knew precisely what his job was, and how he was going to do it. Each had been readied by months of the hardest training. I could see the determination in their eyes, and invariably I would feel a surge of assurance. It was a reciprocal process, a flow of confidence from them to me and from me to them. A commander has to inspire his men, but it was always clear to me that they inspire him as well.
General Ariel Sharon, *Warrior*, 1989.

INSTINCT/INTUITION

As with a man of the world, instinct becomes almost habit so that he always acts, speaks and moves appropriately, so only the experienced officer will make the right decision in major and minor matters – at every pulsebeat of war. Practice and experience dictate the answer. 'This is possible, that is not.'
Major-General Carl von Clausewitz, *On War*, 1, 1832, tr. Michael Howard and Peter Paret, 1976.

The consecutive achievements of a war are not premeditated but spontaneous and guided by military instinct.
Field Marshal Viscount Wavell of Cyrenaica (1883–1950).
Field Marshal Helmuth Graf von Moltke, *Instructions for the Commanders of Large Formations*, 1869.

The greatest commander is he whose intuitions most nearly happen.
Colonel T.E. Lawrence, 'The Science of Guerrilla Warfare', *Encyclopaedia Britannica*, 1929.

Great war leaders have always been aware of the importance of instinct. Was not what Alexander called his 'hope,' Caesar his 'luck' and Napoleon his 'star' simply the fact that they knew they had a particular gift of making contact with realities sufficiently closely to dominate them? For those who are greatly gifted, this facility often shines through their personalities. There may be nothing in itself exceptional about what they say or their way of saying it, but other men in their presence have the impression of a natural force destined to master events.
General Charles de Gaulle, *The Edge of the Sword*, 1932.

The fighting instinct is necessary to success on the battlefield – although even here the combatant who can keep a cool head has an advantage over the man who 'sees red' – but should always be ridden on a tight rein. The statesman who gives that instinct its head loses his own; he is not fit to take charge of the fate of nations.
Captain Sir Basil Liddell Hart, *Strategy*, 1954.

INTEGRITY

Commanders must have integrity; without integrity, they have no power. If they have no power, they cannot bring out the best in their armies. Therefore, integrity is the hand of warriorship.
Sun Bin, *The Lost Art of War*, c. 350 BC, tr. Thomas Cleary.

INTEGRITY

If you choose godly, honest men to be captains of horse, honest men will follow them.
> Oliver Cromwell, September 1643, letter to Sir William Springe.

By God, Mr. Chairman, at this moment I stand astonished at my own moderation!
> Robert Lord Clive, 1773, in reply to a Parliamentary inquiry into charges of corruption during his conquests in India.

I would sacrifice Gwalior, or every frontier of India, ten times over, in order to preserve our credit for scrupulous good faith, and the advantages and honor we gained by the late war and the peace; and we must not fritter them away in arguments, drawn from over-strained principles of the laws of nations, which are not understood in this country. What brought me through many difficulties in the war, and the negotiations for peace? The British good faith, and nothing else.
> The Duke of Wellington, 1804, *The Dispatches of Field Marshal the Duke of Wellington, During His Various Campaigns in India, Denmark, Portugal, Spain, the Low Countries, and France*, III, 1834–1838.

I have just been offered two hundred fifty thousand dollars and the most beautiful woman I have ever seen to betray my trust. I am depositing the money with the Treasury of the United States and request immediate relief from this command. They are getting close to my price.
> Major-General Arthur MacArthur (1845–1912), quoted by his son, Douglas MacArthur, *Reminiscences*, 1964.

In order to be a leader, a man must have followers. And to have followers, a man must have their confidence. Hence the supreme quality for a leader is unquestionably integrity. Without it, no real success is possible, no matter whether it is on a section gang, a football field, in an army, or in an office. The first great need, therefore, is integrity and high purpose.
> General of the Army Dwight D. Eisenhower, quoted in *Airpower*, Summer 1996.

Integrity. And so to the greatest of the virtues on my list, one without which the leader is lost. Integrity, of course, embraces much more than just simple honesty. It means being true to your men, true to your outfit, and above all true to yourself. Integrity of purpose, loyalty upward and loyalty downward, humanity, unselfishness – these are its components. They come more easily to a man of conscience.
> Lieutenant-General Sir James Glover, 'A Soldier and His Conscience', *Parameters*, 9/1983.

Integrity is one of those words that many people keep in that desk drawer labeled 'too hard.' It is not a topic for the dinner table or the cocktail party. When supported with education, one's integrity can give a person something to rely on when perspective seems to blur, when rules and principles seem to waver, and when faced with a hard choice of right and wrong. To urge people to develop it is not a statement of piety but of practical advice. Anyone who has lived in an intense extortion environment [PoW] realizes that the most potent weapon an adversary can bring to bear is manipulation, the manipulation of a prey's shame. A clear conscience is one's only protection.
> Admiral James B. Stockdale, 'Educating Leaders', *Washington Quarterly*, Winter 1983.

The integrity of society is approximately equal to the lowest common denominator of its people. If developing individuals observe that people with known moral defects, or people known to be crooked or liars, are accepted by society without penalty,

INTELLECT/INTELLIGENCE

they may conclude that integrity is not worth their effort.
> Admiral Arleigh A. '31 Knot' Burke, 'Integrity', US Naval Institute, *Proceedings*, 10/1985.

INTELLECT/INTELLIGENCE

To those who congratulated him on his victory in the battle against the Arcadians, he said: 'It would be better if our intelligence were beating them rather than our strength.'
> Archidamus III (d. 338 BC), Eurypontid King of Sparta, quoted in Plutarch, *The Lives*, c. AD 100 (*Plutarch on Sparta*, tr. Richard Talbert, 1988).

Men with sharpness of mind are to be found only among those with a penchant for thought.
> Shiba Yoshimasa, *The Chikubasho*, 1380 (*Ideals of the Samurai*, tr. William Wilson, 1982).

What distinguished a man from a beast of burden is thought, and the faculty of bringing ideas together... a pack mule can go on the campaigns of Prince Eugene of Savoy, and still learn nothing of tactics.
> Frederick the Great (1712–1786).

The discipline of the mind is as requisite as that of the body to make a good soldier.
> General Sir John Moore, 1804, letter to Robert Brownrigg.

We only wish to represent things as they are, and to expose the error of believing that a mere bravo without intellect can make himself distinguished in war.
> Major-General Carl von Clausewitz, *On War*, 1.3, 1832, tr. Michael Howard and Peter Paret, 1976.

Marked intellectual capacity is the chief characteristic of the most famous soldiers. Alexander, Hannibal, Caesar, Marlborough, Washington, Frederick, Napoleon, Wellington, and Nelson were each and all of them something more than fighting men. Few of their age rivalled them in strength of intellect. It was this, combined with the best qualities of Ney and Blücher, that made them masters of strategy, and lifted them high above those who were tacticians and nothing more.
> Colonel George F.R. Henderson, *Stonewall Jackson and the American Civil War*, 1898.

It is true that military men, exaggerating the relative powerlessness of the intelligence, will sometimes neglect to make use of it at all. Here the line of least resistance comes into its own. There have been examples of commanders avoiding all intellectual effort and even despising it on principle. Every great victory is usually followed by this kind of mental decline. The Prussian Army after the death of Frederick the Great, for instance of this. In other cases, the military men note the inadequacy of knowledge and therefore trust to inspiration alone or to the dictates of fate. That was the prevalent state of the French Army at the time of the Second Empire: 'We shall muddle through, somehow.'
> General Charles de Gaulle, *The Edge of the Sword*, 1932.

There is one word in our language which I believe to be the most potent word in the dictionary. So that you do not forget it, I have written it on the blackboard – it is the word WHY. Whatever I say to you, whatever your instructors say to you, whatever you read, whatever you think, ask yourself the reason why. If you do not do so, however much you may strive to learn, you will be mentally standing at ease. Remember this: your brain is not a museum for the past, or a lumber room for the present; it is a laboratory for the future, even if the future is only five minutes ahead of you; a creative centre

… in which new discoveries are made and progress is fashioned.
> Major-General J.F.C. Fuller, lecture to students at the Camberley staff college, January 1933, *Memoirs of an Unconventional Soldier*, 1936.

The influence of thought on thought is the most influential factor in history.
> Captain Sir Basil Liddell Hart, *The Ghost of Napoleon*, 1934.

The Army for all its good points, is a cramping place for a *thinking man*. As I have seen too often, such a man chafes and goes – or else decays. And the root of the trouble is the Army's root fear of the truth. Romantic fiction is the soldierly taste. The heads of the Army talk much of developing character yet deaden it – frowning on younger men who show promise of the personality necessary for command, and the originality necessary for surprise. (October 1935.)
> Captain Sir Basil Liddell Hart, *Thoughts on War*, 1944.

There has been no illustrious captain who did not possess taste and feeling for the heritage of the human mind. At the root of Alexander's victories one will always find Aristotle.
> General Charles de Gaulle, *The Army of the Future*, 1941.

INTREPIDITY

O strange! What will you do with this man, who can bear neither good nor bad fortune? He is the only man who neither suffers us to rest when he is victor, nor rests himself when he is overcome. We shall have, it seems, perpetually to fight with him; as in good success his confidence, and in ill success his shame, still urges him to some further enterprise.
> Hannibal, 209 BC, on the intrepidity of Marcus Claudius Marcellus, Roman proconsul, at the Battle of Asculum,
quoted in Plutarch, *The Lives*, c. AD 100, tr. John Dryden.

Remember, that even intrepidity must be restrained within certain moderate limits, and, when it becomes pernicious, ceases to be honourable.
> Count Belisarius, (c. AD 505–565), urging restraint upon his troops, quoted in Mahon, *Life of Belisarius*, 1829.

To fight the enemy bravely with a prospect of victory is nothing; but to fight with intrepidity under the constant impression of defeat, and inspire irregular troops to do it, is a talent peculiar to yourself.
> Major-General Nathaniel Greene, letter to Colonel Francis 'Swamp Fox' Marion after his capture of Fort Watson, South Carolina, 23 April 1781.

Few men can trust their judgment when in physical danger. On the other hand men's hearts go to the leader who is ready – as Marlborough and Wellington were – to risk his life. The intrepid General has gone more than half way to win the affections of his men.
> Lord Moran, *The Anatomy of Courage*, 1945.

INVENTION

On seeing an arrow shot from a catapult when it was first brought from Sicily, he exclaimed: 'By Heracles, man's valour is done for.'
> Plutarch, quoting Archidamus III (d. 338 BC), Eurypontid King of Sparta, *The Lives*, c. AD 100 (*Plutarch on Sparta*, tr. Richard Talbert, 1988).

I was told by Mr. Charles King, when President of Columbia College, that he had been present in company with [Admiral Stephen] Decatur at one of the early experiments in steam navigation. Crude as the appliances still were, demonstration was conclusive; and Decatur, whatever his prejudices,

JUDGEMENT

was open to conviction. 'Yes,' he said gloomily, to King, 'it is the end of our business; hereafter any man who can boil a tea-kettle will be as good as the best of us.'
Rear Admiral Alfred Thayer Mahan, *From Sail to Steam*, 1907.

The past had its inventions and when they coincided with a man who staked his shirt on them the face of the world changed. Scythes fixed to the axles of war chariots; the moving towers which overthrew Babylon; Greek fire; the short bow, the cross-bow, the Welsh long bow and ballistra; plate armour, the Prussian needle gun; the Merrimac and Ericsson's marvellous coincidental reply. The future is pregnant with invention...
General Sir Ian Hamilton, *The Soul and Body of an Army*, 1921.

JUDGEMENT

There are roads which must not be followed, armies which must not be attacked, towns which must not be besieged, positions which must not be contested, commands of the sovereign which must not be obeyed.
Sun Tzu, *The Art of War*, 8, c. 500 BC, tr. Giles, 1910.

Judgment, and not headlong courage, is the true arbiter of war.
Count Belisarius (c. AD 505–565).

That quality which I wish to see the officers possess, who are at the head of the troops, is a cool, discriminating judgment when in action, which will enable them to decide with promptitude how far they can go and ought to go, with propriety; and to convey their orders, and act with such vigour and decision, that the soldier will look to them with confidence in the moment of action, and obey them with alacrity.
The Duke of Wellington, General Order, 15 May 1811.

In reviewing the whole array of factors a general must weigh before making his decision, we must remember that he can gauge the direction and value of the most important ones only by considering numerous other possibilities – some immediate, some remote. He must *guess*, so to speak: guess whether the first shock of battle will steel the enemy's resolve and stiffen his resistance, or whether, like a Bologna flask, it will shatter as soon as its surface is scratched; guess the extent of debilitation and paralysis that the drying up of particular sources of supply and the severing of certain lines of communication will cause in the enemy; guess whether the burning pain of an injury he has dealt will make the enemy collapse with exhaustion or, like a wounded bull, arouse his rage; guess whether the other powers will be frightened or indignant, and whether and which political alliances will be dissolved or formed. When we realize that he must hit upon all this and much more by means of his discreet judgment, as a marksman hits a target, we must admit that such an accomplishment of the human mind is no small achievement. Thousands of wrong turns running in all directions tempt his perception; and if the range, confusion and complexity of the issues are not enough to overwhelm him, the dangers and responsibilities may.
Major-General Carl von Clausewitz, *On War*, 7.22, 1832, tr. Michael Howard and Peter Paret, 1976.

If you take a chance, it usually succeeds, presupposing good judgment.
Lieutenant-General Sir Giffard Martel (1889–1958).

Judgment comes from experience and experience comes from bad judgment.
General Simon Bolivar Buckner, quoted by General of the Army Omar N. Bradley, 'Leadership', *Parameters*, Winter 1972.

JUNIOR OFFICERS

JUNIOR OFFICERS

A young officer should never think he does too much; they are to attend to the looks of the men, and if any are thinner or paler than usual, the reason of their falling off may be enquired into.

Major-General Sir James Wolfe (1727–1759), *Instructions for Young Officers*.

There are three things, young gentleman, which you are constantly to bear in mind. First, you must always implicitly obey orders, without attempting to form any opinion of your own respecting their propriety. Secondly, you must consider every man your enemy who speaks ill of your king; and thirdly, you must hate a Frenchman as you do the devil.

Admiral Viscount Nelson, 1792, to a midshipman aboard HMS *Agamemnon*, quoted in Robert Southey, *Life of Nelson*, 1813.

As to the way in which some of our ensigns and lieutenants braved danger – the boys just come out from school – it exceeds all belief. They ran as at cricket.

The Duke of Wellington, quoted in Rodgers, *Recollections*, 1859.

As the war dragged on into the autumn of 1916, the German Army, like our own, had need of underpinning. But the lynch-pins were missing. A German soldier writes:

> The tragedy of the Somme battle was that the best soldiers, the stoutest-hearted men were lost; their numbers were replaceable, their spiritual worth never could be.

Our own forces were in no better plight. Most of those who were meant by nature to lead men had been struck down by two years of war.

That is the story of all wars. A national army is held together by young leaders. They are most needed when weariness has dulled the first purpose of such a force in the third or fourth year of war. But as time passes they have become scarce, for the best go first and have gone first since war began.

Dead like the rest, for this is true;
War never chooses an evil man but the good.

It is the theme of the *Iliad*. It runs through the humblest diary of the war.

Lord Moran, *The Anatomy of Courage*, 1945.

Modern battles are fought by platoon leaders. The carefully prepared plans of higher commanders can do no more than project you to the line of departure at the proper time and place, in proper formation, and start you off in the right direction.

General of the Army George C. Marshall, *Selected Speeches and Statements*, 1945.

Never in the history of warfare has individual leadership meant so much. Land fighting has almost inevitably broken down into mobile action by small units, and the ability of junior officers, non-commissioned officers and even privates to assume responsibility and initiative in a pinch has carried us through to victory...

We have paid no small price. By their very leadership, men have multiplied their risks. An unusually heavy percentage of our outstanding young officers were killed in action on Iwo Jima – most of them shot down in front of their men. Such men will be sorely missed.

General Alexander A. Vandegrift, 1945.

Said a newly arrived lieutenant to an old sergeant of the 12th Cavalry: 'You've been here a long time, haven't you?' 'Yes sir,' replied the sergeant. 'The troop commanders, they come and go, but it don't hurt the troop.'

Brigadier-General S.L.A. Marshall, *The Armed Forces Officer*, 1950.

JUSTICE

The pious Greek, when he had set up altars to all the great gods by name, added one more altar, 'To the Unknown God'. So whenever we speak and think of the great captains and set up our military altars to Hannibal and Napoleon and Marlborough and suchlike, let us add one more altar, 'To the Unknown Leader', that is, to the good company, platoon, or section leader who carries forward his men or holds his post, and often falls unknown. It is these who in the end do most to win wars.
> Field Marshal Viscount Wavell of Cyrenaica, *Soldiers and Soldiering*, 1953.

The most terrible warfare is to be a second lieutenant leading a platoon when you are on the battlefield.
> General Dwight D. Eisenhower, 17 March 1954, remark.

JUSTICE

When asked once which of the two virtues was finer, courage or justice, he declared: 'Courage has no value if justice is not in evidence too; but if everyone were to be just, then no one would need courage.'
> Agesilaus, Eurypontid King of Sparta (400–360 BC), quoted in Plutarch, *The Lives*, c. AD 100 (*Plutarch on Sparta*, tr. Richard Talbert, 1988).

Commanders must be just; if they are not just, they will lack dignity. If they lack dignity, they will lack charisma; and if they lack charisma, their soldiers will not face death for them. Therefore justice is the head of warriorship.
> Sun Bin, c. 350 BC, *The Lost Art of War*, tr. Thomas Cleary.

I shall not consider anyone of you a fellow-soldier of mine, no matter how terrible he is reputed to be to the foe, who is not able to use clean hands against the enemy. For bravery cannot be victorious unless it be arrayed along with justice.
> Count Belisarius, AD 533, to his troops, at the beginning of the Vandalic War, who complained when he executed two soldiers who had murdered another, quoted in Procopius, *History of the Wars*, 3.12, c. AD 560, tr. H.B. Dewing.

There is no authority without justice.
> Napoleon, 12 March 1803, remark.

It depends on justice, freedom from corruption, and unswerving truth to one's word and to every obligation one has undertaken.
> The Duke of Wellington, 1798, defining the bases on which British rule in India would be maintained, cited in Arthur Bryant, *The Great Duke*, 1972.

KILLING

At the word of Assur, the great lord, my lord, on flank and front I pressed upon the enemy like the onset of raging storm. With the weapons of Assur, my lord, and the terrible onset of my attack, I stopped their advance, I succeeded in surrounding them, I decimated the enemy host with arrow and spear. All of their bodies I bored through like a sieve... The chariots and their horses, whose riders had been slain at the beginning of the terrible onslaught, and who had been left to themselves, kept running back and forth... I put an end to their riders' fighting... They abandoned their tents and to save their lives they trampled the bodies of their fallen soldiers, they fled like young pigeons that are pursued. They were besides themselves; they held back their urine, but let their dung go into their chariots. In pursuit of them I dispatched my chariots and horses after them. Those among who had escaped, who had fled for their lives, wherever my charioteers met them, they cut them down with the sword.
> Sennacherib (705–681 BC), King of

KILLING

Assyria, quoted in Daniel Luckenbill, *Ancient Records of Assyria and Babylonia*, 1926.

Do not put a premium on killing.
Li Ch'uan, Tang Dynasty commentator on Sun Tzu, *The Art of War*, 3, c. 500 BC, tr. Samuel Griffith, 1963.

The overfaithful sword returns the user His heart's desire at price of his heart's blood.
Rudyard Kipling, 'The Proconsuls', 1905.

Christmas [1944] dawned clear and cold; lovely weather for killing Germans, although the thought seemed somewhat at variance with the spirit of the day.
General George S. Patton, Jr., *War As I Knew It*, 1947.

War is a savage business, a form of legalized murder, a business where killing one's fellow men without mercy is a duty and sometimes a form of sport. Nobody enjoys killing their fellow creatures, but in war one's likes and dislikes must take second place to defeat and survival. In this book there is much killing and I make no excuse for recording cases with satisfaction and often with relish, whilst on reflection one is shocked at depriving others, just as good as oneself, of their lives and for no better reason than that both of us are obeying orders and performing an unpleasant duty.
Colonel Richard Meinertzhagen, *Army Diary 1899–1925*, 1960.

Youths only want one thing, to kill you so they can go to paradise.
Osama bin Laden in *fatwa* issued in 1996.

The purpose of the military is to kill, and if you cannot stomach that, you should not have a military.
Lieutenant-Colonel Ralph Peters, *Fighting for the Future*, 1999.

Of all the notions I have advanced over the years, the only one that has met with consistent rejection is the statement that men like to kill. I do not believe that all men like to kill. At the extreme, there are those saintly beings who would rather sacrifice their own lives before taking the life of another. The average man will kill if compelled to, in uniform in a war or in self-defense, but has no evident taste for it. Men react differently to the experience of killing. Some are traumatized. Others simply move on with their lives. But there is at least a minority of human beings – mostly male – who enjoy killing. That minority may be small, but it does not take many enthusiastic killers to trigger the destruction of a fragile society. Revolutions, pogroms, genocides, and civil wars are made not by majorities. The majority may gloat or loot, but the killing minority drives history.
Lieutenant-Colonel Ralph Peters, *Fighting for the Future*, 1999.

They have no moral inhibition on the slaughter of the innocent. If they could have murdered not 7,000 but 70,000 does anyone doubt they would have done so and rejoiced in it?
Prime Minister Tony Blair, speech to Labour Party Conference 2 Oct 2001.

If avenging the killing of our people is terrorism then history should be a witness that we are terrorists. Yes, we kill their innocents and this is legal religiously and logically.
Osama bin Ladin, undated video, reported on 11 November 2001 by the *Daily Telegraph*.

L

LAST WORDS

Remember my last saying: show kindness to your friends, and then shall you have it in your power to chastise your enemies.
> Cyrus the Great, c. 529 BC, quoted in Xenophon, *The Education of Cyrus*, tr. H.G. Dakyns, 1992.

I have lived long enough for I die unconquered.
> Epaminondas, who waited until he had heard the Thebans he commanded had defeated the Spartans on the battlefield of Mantinea, to order the javelin withdrawn from his body and then died, quoted in Cornelius Nepos, *Life of Epaminondas*, c. 30 BC.

To the best (or strongest).
> Alexander the Great, 323 BC, on his deathbed in Babylon when asked to whom he bequeathed his empire, quoted in Curtius, *The history of Alexander*, c. AD 41, tr. John Yardley, 1984.

Let me free the Roman people from their long anxiety, since they think it tedious to wait for an old man's death... they have sent an ambassador to suggest to Prusias (King of Bithynia) the crime of murdering a guest.
> Hannibal, 183 BC, upon poisoning himself as the Romans pursued him even into exile.

Ungrateful country, you will not possess even my bones.
> Scipio Africanus, 183 BC, inscription to be engraved on his tomb in Campania, after he had been tried for bribery in Rome, quoted in Valerius Maximus, *De Dictis Factisque*, III.

Thou too, my son?
> Julius Caesar, 15 March 44 BC, to Marcus Junius Brutus, thought to be his son, as he stabbed Caesar in the Senate House in Rome, quoted in Suetonius, *The Lives of the Twelve Caesars*, AD 121, tr. Philemon Holland, 1606.

Oh wretched valour thou wert but a name, and yet I worshipped thee as real indeed. But now it seems thou wert but fortune's slave.
> Marcus Junius Brutus, 42 BC, as he fell upon his sword after the defeat of the assassins of Julius Caesar at the Battle of Philippi.

Come, give it to me, if we have anything to do.
> Emperor Septimius Severus, AD 211, still trying to accomplish imperial business as he expired, quoted in Dio Cassius, *Romaika*, 76.15.2, c. AD 230.

I desire to leave to the men that come after me a remembrance of me in good works.
> King Alfred the Great, AD 901.

Forth we go in our lines
Without our armour, against the blue
  blades.

The helmets glitter: I have no armour.
Our shrouds are down in those ships.
> Harald Hardraade, 25 September 1066, poem composed as the Battle of Stamford Bridge began, quoted in David Howarth, *The Year of the Conquest*, 1978; the shrouds referred to are the byrnies his men left in their ships.

LAST WORDS

God has promised me a great victory after my death.
>El Cid, 1094, after the rout of the Almoravides before Valencia, quoted in *Cronica Particular del Cid*, 1512.

Now let the rest go as it will, I care no more.
>Henry II, 6 July 1189, as he learned of the betrayal of his son, quoted in Marion Meade, *Eleanor of Aquitaine*, 1977.

Youth, I forgive thee. Take off his chains, give him 100 shillings, and let him go.
>Richard I, 1199, to the man who had caused his mortal wound with a crossbow bolt. Richard's wishes were not honoured: the archer was flayed alive after the King's death.

Carry my bones before you on your march. For the rebels will not be able to endure the sight of me, dead or alive.
>Edward I, 1307.

Rejoice my departed soul with a beautiful series of victories.
>Osman I, 1326, quoted in Lord Kinross, *The Ottoman Centuries*, 1979.

Make my skin into drumheads for the Bohemian cause.
>Jan Ziska, 1424.

I pray you, go to the nearest church, and bring me the cross, and hold it up level with my eyes until I am dead. I would have the cross on which God hung be ever before my eyes while life lasts in me.
  Jesu, Jesus!
>Joan of Arc, 30 May 1431, at the stake, *Joan of Arc in Her Own Words*, 1996.

The city is taken, and I am still alive!
>Emperor Constantine XI, 29 May 1453, before plunging to his death in the midst of Turkish troops who were pouring through a breach in the walls of Constantinople, quoted in Lord Kinross, *The Ottoman Centuries*, 1979.

The fleeting pomp of the world is like the green willow
…but at the end a sharp axe destroys it, a north wind fells it…
>King Nezahualcoyotl of Texcoco, 1472, quoted in Frances Gilmore, *Flute of the Smoking Mirror*, 1949.

Pity me not: I die as a man of honour ought, in the discharge of my duty: they indeed are objects of pity who fight against their king, their country and their oath.
  God and my country!
>Pierre du Terrail, Chevalier de Bayard, 1524, to the Genoese rebel who offered him commiseration for his mortal wound; also 'Let me die facing the enemy'.

I knew what it was to trust to your false promises, Malinche; I knew that you had destined me to this fate, since I did not fall by my own hand when you entered my city of Tenochtitlan. Why do you slay me so unjustly? God will demand it of you!
>Emperor Cuahtémoc of the Aztecs, Lent 1525, just before Cortes had him hanged on trumped-up charges, quoted in William Prescott, *The History of the Conquest of Mexico*, 1843.

What have I done, oh my children, that I should meet such a fate? And from your hands, too, you, who have met with friendship and kindness from my people, with whom I have shared my treasures, who have received nothing but benefits from my hands!
>Emperor Atahualpa of the Incas, 26 July 1533, just before Pizarro had him garrotted, quoted in William Prescott, *History of The Conquest of Peru*, 1847.

What ho! traitors! have you come to kill me in my own house? …Jesu!

LAST WORDS

Francisco Pizarro, 26 June 1541, when the assassins broke into his house to murder him; he killed three before being overwhelmed, quoted in William Prescott, *History of the Conquest of Peru*, 1847.

It has long been a question whether one can conscientiously hold property of Indian slaves. Since this point has not yet been determined, I enjoin it on my son Martin and his heirs that they spare no pains to come to an exact knowledge of the truth; as a matter which deeply concerns the conscience of each of them, no less than mine.
>Hernan Cortes, the final element of his will, which he executed shortly before his death on 2 December 1547 in Seville, quoted in William Prescott, *The History of the Conquest of Mexico*, 1843.

I have enough, save thyself brother!
>Gustavus II Adolphus, 1632, to the Duke of Lauenberg as they were overtaken by Imperialist cavalry at the Battle of Lützen.

They don't want to listen to me when I'm alive, so why should they obey me when I'm dead.
>King John III of Poland (Jan Sobieski), 1696.

I have loved war too well.
>King Louis XIV of France, 1 September 1715.

That was once a man.
>The Duke of Marlborough, shortly before his death on 16 June 1722, 'One day he paced with failing steps the state rooms of his palace, and stood long and intently contemplating his portrait by Kneller. Then he turned away with the words above', quoted in Churchill, *Marlborough*, 1933.

No, not quite naked; I shall have my uniform on. Herr Jesu, to Thee I live; Herr Jesu, to Thee I die: in life and death, Thou art my gain.
>King Frederick William I of Prussia, 1740, upon hearing the passage from the *Book of Job*, 'Naked I came into the world and naked shall I go.'

Now God be praised, I will die in peace.
>Major-General James Wolfe, 13 September 1759, upon hearing that his forces had triumphed over the French at Quebec, quoted in Churchill, *A History of the English Speaking Peoples: The Age of Revolution*, 1957.

Oh that I had served my Lord as faithfully as I have served my King.
>Major-General Louis-Joseph Marquis de Montcalm, 14 September 1759, the day after his defeat at the hands of Wolfe on the Plains of Abraham outside Quebec, quoted in Parkman, *Montcalm and Wolfe*, 1884.

We have crossed the mountain; things will go better now.
>King Frederick the Great of Prussia, 17 August 1786, to the sun at his palace of Sans Souchi.

Waiting are they, waiting are they? Well, let 'em wait!
>Ethan Allen, 1789, upon being told that the angels were waiting for him.

It is well. I die hard, but am not afraid to go.
>General George Washington, 14 December 1799.

Genoa... Battle... Forward!
>Field Marshal Aleksandr Suvorov, May 1800, quoted in Phillip Longworth, *The Art of Victory*, 1965.

I am satisfied. Thank God, I have done my duty.
>Admiral Viscount Nelson, 21 October 1805, in the cockpit of HMS *Victory* as the Battle of Trafalgar waned and the British victory became apparent, quoted in Rober Southey, *The Life of Nelson*, 1813.

LAST WORDS

Are the French beaten?
[When told yes] I hope the people of England will be satisfied. I hope that my country will do me justice.
    Lieutenant-General Sir John Moore, 16 January 1809, to his surgeon after losing an arm to a cannon-ball at the Battle of Corunna in Portugal, quoted in Carola Oman, *Sir John Moore*, 1953.

Napoleon: 'You will live, my friend:'
Lannes: 'I trust I may, if I can still be of use to France and your Majesty.'
    Marshal of France Jean Lannes, 21 May 1809, mortally wounded at Aspern-Essling.

Good God! To survive a hundred fields and die like this...
    Marshal of France Guillaume Marie Brunne, 1 August 1815, when about to be torn to pieces by a Royalist mob after Waterloo, quoted in R.F. Delderfield, *Napoleon's Marshals*, 1966.

Spare the face, straight to the heart!
    Marshal of France Murat to his firing squad, 13 October 1815, quoted in R.F. Delderfield, *Napoleon's Marshals*, 1966.

Soldiers, when I give the command to fire, fire straight at my heart. Wait for the order. It will be my last to you. I protest against my condemnation. I have fought a hundred battles for France, and not one against her... Soldiers, Fire!
    Marshal of France Michel Ney, after he had insisted on commanding his own firing squad, 7 December 1815, quoted in Raymond Horricks, *Marshal Ney: The Romance and the Real*, 1982.

Nostitz, you have learned many things from me. Now you are to learn how peacefully a man can die.
    Field Marshal Prince Gebhard von Blücher, 1819.

Tête d'Armée! (Head of the Army!)
    Napoleon, 5.30 pm, 5 May 1821, St Helena.

Hau! You, too, Mbopa, son of Sitayi, you too, are killing me.
    Shaka Zulu, 1828, to one of his assassins, quoted in E.A. Ritter, *Shaka Zulu*, 1955.

Let us go – let us go – this people don't want us in this land! Let's go, boys. Take my luggage on board the frigate.
    Simon Bolivar, 17 December 1830, after resigning in disillusionment as supreme chief of Columbia.

Order A.P. Hill to prepare for action! Pass the infantry to the front rapidly! Tell Major Hawks... (leaving the sentence unfinished he stopped, then a smile came over his face) Let us now cross over the river and rest under the shade of the trees.
    Lieutenant-General Thomas J. 'Stonewall' Jackson, 10 May 1863, quoted in John Selby, *Jackson As Military Commander*, 1968.

Forward men, forward for God's sake and drive those fellows out of those woods...
    Major-General John Fulton Reynolds, 1 July 1863, while leading the Iron Brigade of his First Corps on the first day of the Battle of Gettysburg, quoted in Edward Coddington, *Gettysburg: A Study in Command*, 1968.

Give them the cold steel, men!
    Brigadier-General Lewis Armistead, 3 July 1863, as he climbed over the stone wall at Gettysburg, the 'high tide of the Confederacy'.

Doctor, I suppose I am going fast now. It will soon be over. But God's will be done. I hope I have fulfilled my destiny to my country and my duty to God... I am resigned; God's will be done.
    Major-General James E.B. Stuart, 12 May 1864, as he lay dying from wounds suffered at Yellow Tavern, quoted in Douglas Southall Freeman, *Lee's Lieutenants*, 1944.

THE LEADER

Tell A.P. Hill he must come up...
  General Robert E. Lee, 12 October 1870, quoted in Brigadier-General Fitzhugh Lee, *General Lee*, 1894.

Now mark this, if the Expeditionary Force, and I ask for no more than two hundred men, does not come in ten days, the town may fall; and I have done my best for the honour of our country. Goodbye.
  General Charles George Gordon, 14 December 1884, last diary entry while waiting for the relief of Khartoum.

I am a great sufferer all the time... all that I can do is pray that the prayers of all these good people may be answered so far as to have us all meet in another and better world. I cannot speak even in a whisper.
  General Ulysses S. Grant, 23 July 1885, written as he lay dying of throat cancer.

Let me go! Let me go!
  Clara Barton, 12 April 1912, on her deathbed, quoted in Percy Epler, *The Life of Clara Barton*, 1915.

It must come to a fight. Only keep the right wing strong.
  Field Marshal Alfred Count von Schlieffen, 1913.

This I would say, standing as I do in view of God and Eternity: I realize that patriotism is not enough; I must have no hatred and bitterness towards anyone.
  Edith Cavell, 12 October 1915, to the chaplain attending her before her execution by German firing squad.

Gentlemen, the battle is done. The victory is ours!
  Admiral George Dewey, 1917.

Don't you think I'll be back?
  Rittmeister Manfred Freiherr von Richthofen, the 'Red Baron', 21 April 1918, to his ground crew as he took off on his last mission.

Have you been dreaming about all this?
  Marshal of the Soviet Union N.T. Tukhachevskiy, early June 1937, to a fellow prisoner confessing to fantastic, trumped-up charges during Stalin's Great Purges, quoted in Robert Conquest, *The Great Terror*, 1990.

Long Live the Party! Long Live Stalin!
  General I.E. Yakir, as he was being led to the firing squad during Stalin's Great Purges, 11 June 1937, quoted in Robert Conquest, *The Great Terror*, 1990.

To die by the hand of one's own people is hard.
  Field Marshal Erwin Rommel, 14 October 1944, to his son before he took the poison brought by Hitler's emissaries.

It's too dark... I mean too late.
  General George S. Patton, Jr, 21 December 1945, quoted in Carlo d'Este, *Patton: A Genius For War*, 1995.

I want to go. God take me.
  General of the Army Dwight D. Eisenhower, to his son, John Eisenhower, *Strictly Personal*, 12, 1974

I couldn't sleep last night – I had great difficulty. I can't have very long to go now. I've got to meet God – and explain all those men I killed at Alamein...
  Field Marshal Montgomery, among his last words, late February 1975, quoted in Nigel Hamilton, *Monty – the Last Years of the Field Marshal*, 1983.

THE LEADER

The leader must himself believe that willing obedience always beats forced obedience, and that he can get this only by really knowing what should be done. Thus he can secure obedience from his men because he can convince them that he knows best, precisely as a good

THE LEADER

doctor makes his patients obey him. Also he must be ready to suffer more hardships than he asks of his soldiers, more fatigue, greater extremes of heat and cold.
   Xenophon, *Cyropaedia*, c. 360 BC.

When he was asked how someone might most surely earn people's esteem, he replied: 'By the best words and the finest actions.'
   Plutarch, quoting Agesilaus, Eurypontid King of Sparta (400–360 BC), *The Lives*, c. AD 100, (*Plutarch on Sparta*, tr. Richard Talbert, 1988).

An army of deer led by a lion is more to be feared than an army of lions led by a deer.
   Attributed to Chabrias (c. 420–357/6 BC) and to Philip of Macedon (382–336 BC).

A leader is a dealer in hope.
   Napoleon (1769–1821).

The more the leader is in the habit of demanding from his men, the surer he will be that his demands will be answered.
   Major-General Carl von Clausewitz, *On War*, 1832, tr. Michael Howard and Peter Paret, 1976.

The moral equilibrium of the man is tremendously affected by an outward calmness on the part of the leader. The soldier's nerves, taut from anxiety of what lies ahead, will be soothed and healed if the leader sets an example of coolness. Bewildered by the noise and confusion of battle, the man feels instinctively that the situation cannot be so dangerous as it appears if he sees that his leader remains unaffected, that his orders are given clearly and deliberately, and that his tactics show decision and judgment. However 'jumpy' the man feels, the inspiration of his leader's example shames him into swallowing his own fear. But if the leader reveals himself irresolute and confused, the more even than if he shows personal fear, the infection spreads instantly to his men. (July 1921.)
   Captain Sir Basil Liddell Hart, *Thoughts on War*, 1944.

To be a successful leader, a commander must not only gain contact with his men, but he must give them the impression of fearlessness.
   Major-General John A. Lejeune, *The Reminiscences of a Marine*, 1929.

What, above all else, we look for in a leader is the power to dominate events, to leave his mark on them, and to assume responsibility for the consequences of his actions. The setting up of one man over his fellows can be justified only if he can bring to the common task the drive and certainty which comes of character. But why, for that matter, should a man be granted, free gratis and for nothing, the privilege of domination, the right to issue orders, the pride of seeing them obeyed, the thousand and one tokens of respect, unquestioning obedience, and loyalty which surround the seat of power? To him goes the greater part of the honor and glory. But that is fair enough, for he makes the best repayment that he can by shouldering risks. Obedience would be intolerable if he who demands it did not use it to produce effective results, and how can he do so if he does not possess the qualities of daring, decision, and initiative?
   General Charles de Gaulle, *The Edge of the Sword*, 1932.

The truly great leader overcomes all difficulties, and campaigns and battles are nothing but a long series of difficulties to be overcome.
   General of the Army George C. Marshall, 18 September 1941, Fort Benning, Georgia.

No man is a leader until his

appointment is ratified in the minds and hearts of his men.
Anonymous, *The Infantry Journal*, 1948.

The good military leader will dominate the events which surround him; once he lets events get the better of him he will lose the confidence of his men, and when that happens he ceases to be of value as a leader.
Field Marshal Viscount Montgomery of Alamein, *The Memoirs of Field Marshal Montgomery*, 1958.

When all is said and done the leader must exercise an effective influence, and the degree to which he can do this will depend on the personality of the man – the 'incandescence' of which he is capable, the flame which burns within him, the magnetism which will draw the hearts of other men toward him. What I personally would want to know about a leader is:

Where is he going?

Will he go all out?

Has he the talents and equipment, including knowledge, experience and courage? Will he take decisions, accepting full responsibility for them, and take risks where necessary?

Will he then delegate and decentralize, having first created an organization in which there are definite focal points of decision so that the master plan can be implemented smoothly and quickly?
Field Marshal Viscount Montgomery of Alamein, *The Memoirs of Field Marshal Montgomery*, 1958.

Every great leader I have known has been a great teacher, able to give those around him a sense of perspective and to set the moral, social, and motivational climate among his followers.
Admiral James B. Stockdale, 'Educating Leaders', *The Washington Quarterly*, 30 March 1983.

I've made the points that leaders under pressure must keep themselves absolutely clean morally (the relativism of the social sciences will never do). They must lead by example, must be able to implant high-mindedness in their followers, must have competence beyond status, and must have earned their followers' respect by demonstrating integrity.
Admiral James B. Stockdale, 'Machiavelli, Management, and Moral Leadership', *Military Ethics*, 1987.

A leader establishes the ethical... environment in which the entire operation is going to be accomplished.
General H. Norman Schwarzkopf, 1991, quoted in Peter David, *Triumph in the Desert*, 1991.

There are three types of leaders: Those who make things happen; those that watch things happen; and those who wonder what happened!
American military saying.

LEADERSHIP

There is no command without leadership.
Sun Bin, *The Lost Art of War*, c. 350 BC, tr. Thomas Cleary.

The quality of leadership needs, above all, spirit, intelligence, and sympathy. Spirit is needed to fire men to self-sacrificing achievements; intelligence, because men will only respect and follow a leader whom they feel knows his profession thoroughly; sympathy, to understand the mentality of each individual in order to draw out the best that is in him. Given these qualities, men will conquer fear to follow a leader. (July 1921.)
Captain Sir Basil Liddell Hart, *Thoughts on War*, 1944.

At first push one would scarcely expect to find the behavior of a piece of cooked spaghetti an illustration of

LEADERSHIP

successful leadership in combat... it scarcely takes demonstration to prove how vastly more easy it is to PULL a piece of cooked spaghetti in a given direction along a major axis than it is to PUSH it in the same direction. Further, the difficulty increases with the size of either the spaghetti or the command....
   General George S. Patton, Jr., January 1928, *The Patton Papers*, I, 1972.

In the World War, nothing was more dreadful to witness than a chain of men starting with a battalion commander and ending with an army commander sitting in telephone boxes... talking, talking, talking in place of leading, leading, leading.
   Major-General J.F.C. Fuller in *Generalship: Its Diseases and Their Cure*, 1936

For ten years I had burrowed into the essence of the work of leadership, the element – as Clausewitz says – of danger, the element of uncertainty, finally, as I put it, the element of eternal contradictions, which are not to be solved, but are cut like the Gordian knot by the sword of decision, the sword of command.
   Marshal Joseph Pilsudski, *The Memoirs of a Polish Revolutionary and Soldier*, 1931.

Leadership in war is an art, a free creative activity based on a foundation of knowledge. The greatest demands are made on the personality.
   Generals Werner von Fritsch and Ludwig Beck, *Troop Leadership*, 1933.

Let us set up a standard around which the brave and loyal can rally.
   Sir Winston S. Churchill (1874–1965).

A few men had the stuff of leadership in them, they were like rafts to which all the rest of humanity clung for support and hope.
   Lord Moran, 12 May 1943, preface to *The Anatomy of Courage*, 1945.

War is much too brutal a business to have room for brutal leading; in the end, its only effect can be to corrode the character of men, and when character is lost, all is lost. The bully and the sadist serve only to further encumber an army; their subordinates must waste precious time clearing away the wreckage that they make. The good company has no place for an officer who would rather be right than loved, for the time will quickly come when he walks alone, and in battle no man may succeed in solitude.
   Brigadier-General S.L.A. Marshall, *Men Against Fire*, 1947.

In the British Army there are no good units and no bad units – only good and bad officers and NCOs. They make or break the unit. Today we cannot afford anything but the good ones. No man can be given a more honourable task than to lead his fellow countrymen in war. We, the officers and NCOs, owe it to the men we command and to our country that we make ourselves fit to lead the best soldiers in the world, that in peace the training we give them is practical, alive and purposeful, and that in war our leadership is wise, resolute and unselfish.
   Leaders are made more often than they are born. You all have leadership in you. Develop it by thought, by training and by practice...
   Field Marshal Viscount Slim of Burma, 1949, Foreword to the first edition of the *British Army Journal*, later renamed *British Army Review*.

I have been very lucky in my service. In getting on for forty years I have commanded everything from a squad of six men to an Army Group of a million and a quarter, and, believe me, whether you command ten men or ten million, the essentials of leadership are the same. Leadership is that mixture of example, persuasion and compulsion which makes men do what you want

them to do. If I were asked to define leadership, I should say it is the Projection of Personality. It is the most intensely personal thing in the world, because it is just plain you.
Field Marshal Viscount Slim of Burma, *Courage and Other Broadcasts*, 1957.

No leader, however great, can long continue unless he wins victories.
Field Marshal Viscount Montgomery of Alamein, *The Memoirs of Field-Marshal Montgomery*, 6, 1958.

Leadership can be taught, but not character. The one arises from the other. A successful leader of men must have character, ability and be prepared to take unlimited responsibility. Responsibility can only be learned by taking responsibility; you cannot learn the piano without playing on one. Leadership is the practical application of character. It implies the ability to command and to make obedience proud and free. One of the questions demanded by the War Office before 1914 of officers recommended for the Staff College was: Does this officer give commands in a manner likely to secure cheerful obedience? My Commanding Officer – Du Maurier, whose habits were slightly unorthodox – replied: 'I do not know if this officer's orders would be cheerfully obeyed, but I can guarantee that they would be carried out.'
The War Office were satisfied.
Colonel Richard Meinertzhagen, *Army Diary 1899–1925*, 1960.

I hold that leadership is not a science, but an art. It conceives an ideal, states it as an objective, and then seeks actively and earnestly to attain it, everlastingly persevering, because the records of war are full of successes coming to those leaders who stuck it out just a little longer than their opponents.
General Matthew B. Ridgway, 'Leadership', *Military Review*, 9/1966.

The final test of a leader is the feeling you have when you leave his presence after a conference or interview. Have you a feeling of uplift and confidence? Are you clear s to what is to be done, and what is your part of the task? Are you determined to pull your weight in achieving the object? Or is your feeling the reverse?
Field Marshal Viscount Montgomery of Alamein, *The Path to Leadership*, 1, 1961.

A certain ruthlessness is essential, particularly with inefficiency and also with those who would waste his time. People will accept this provided the leader is ruthless with himself.
Field Marshal Viscount Montgomery of Alamein, *The Path to Leadership*, 15, 1961.

Having decided on his policy, his objective, he must not be led off his target by the faint-hearted; he will do well to discard the faint-hearted once they are discovered.
Field Marshal Viscount Montgomery of Alamein, *The Path to Leadership*, 15, 1961.

One final aspect of leadership is the frequent necessity to be a philosopher, able to understand and to explain the lack of a moral economy in this universe, for many people have a great deal of difficulty with the fact that virtue is not always rewarded nor is evil always punished. To handle tragedy may indeed be the mark of an educated man, for one of the principal goals of education is to prepare us for failure. To say that is not to encourage resignation to the whims of fate, but to acknowledge the need for forethought about how to cope with undeserved reverses. It's important that our leadership steel themselves against the natural reaction of lashing out or withdrawing when it happens. The test of character is not 'hanging in there' when the light at the end of the tunnel is expected but performance of duty

and persistence of example when the situation rules out the possibility of the light ever coming.
>Admiral James B. Stockdale, *Naval War College Review*, July–August 1979.

We never lost sight of the reality that people, particularly gifted commanders, are what make units succeed. The way I like to put it, leadership is the art of accomplishing more than the science of management says is possible.
>General Colin Powell, *My American Journey*, 1995.

LINES OF COMMUNICATION
The secret of war lies in the secret of the lines of communication... Strategy does not consist of making half-hearted dashes at the enemy's rear areas; it consists in really mastering his communications, and then proceeding to give battle.
>Napoleon, 1806, comment to Jomini, quoted in David Chandler, *The Campaigns of Napoleon*, 1966.

The line that connects an army with its base of supplies is the heel of Achilles – its most vital and vulnerable point.
>Colonel John S. Mosby, *Mosby's War Reminiscences*, 1887.

Free supplies and open retreat are two essentials to the safety of an army or a fleet.
>Rear Admiral Alfred Thayer Mahan, *Naval Strategy*, 1911.

Nine times out of ten an army has been destroyed because its supply lines have been severed.
>General of the Army Douglas MacArthur, 23 August 1950, to the Joint Chiefs of Staff.

In planning any stroke against the enemy's communications, either by manoeuvre round his flank or by rapid penetration of a breach in his front, the question will arise as to the most effective point of aim – whether it should be directed against an immediate rear of the opposing force, or further back.

In general, the nearer to the force that the cut is made, the more immediate the effect; the nearer to the base, the greater the effect...

A further consideration is that while a stroke close in rear of the enemy force may have more effect on the minds of the enemy troops, a stroke far back tends to have more effect on the mind of the enemy commander.
>Captain Sir Basil Liddell Hart, *Strategy*, 1954.

LOGISTICIANS
Nobody ever heard of a quartermaster in history.
>Major-General Nathaniel Greene, 1778, letter to General George Washington upon receiving his appointment as Quartermaster General of the Continental Army.

I have prevailed upon Lord Clive to appoint a Commissary of Stores... Matters can now be brought into some shape and we shall know what we are about, instead of trusting to the vague calculations of a parcel of blockheads who know nothing and have no data.
>The Duke of Wellington, 28 and 31 October 1798, in India to his brother, Sir Henry Wellesley, *Supplementary Despatches, Correspondence, and Memoranda of Field Marshal Arthur Duke of Wellington, K.G., 1794–1818*, I, 1858, pp. 124, 84, quoted in Elizabeth Longford, *Wellington: The Years of the Sword*, 1969.

Every man of you should be mentioned by name, soldiers of Verdun, soldiers in the line and soldiers in the rear. For if I give the place of honor, as is meet, to those who fell in the front of the battle, still I know that their courage would have availed nothing without the patient toil, continued day and night, to the last limit of their strength, on the part of the men to whose efforts were

LOGISTICS

due the regular arrival of the reinforcements, of munitions, and of food, and the evacuation of wounded: the truck drivers along the Sacred Way, the railroad engineers, the ambulance force.
Marshal of France Henri-Phillipe Pétain, 1927, at the dedication of the Ossuaire at Verdun, *Verdun*, 1930.

An adequate supply system and stocks of weapons, petrol and ammunition are essential conditions for any army to be able to stand successfully the strain of battle. Before the fighting proper, the battle is fought and decided by the Quartermasters.
Field Marshal Erwin Rommel, quoted in Archibald Wavell, *Soldiers and Soldiering*, 1953.

LOGISTICS

Master Sun said: 'A city might have walls of iron and be surrounded by moats of boiling water, but if it is inadequately provisioned, even a Chiang T'ai-kung or a Mo Ti would be unable to defend it.'
Sun Tzu, c. 500 BC, a passage recovered from later works not found in *The Art of War*, quoted in *Sun Tzu: The Art of Warfare*, tr. Roger Ames, 1993.

Without supplies neither a general nor a soldier is good for anything.
Clearchus, 401 BC, speech to the Ten Thousand, quoted in Xenophon, *Anabasis*, 1.3, c. 360 BC (*The Persian Expedition*, tr. Rex Warner, 1949).

On the third day of his campaign Philip arrived at Onocarsis, a place in Thrace with a beautifully prepared grove; it was in every regard a pleasant place to spend time, especially in summer. He was going to camp in this beautiful area, but he learned that there was no feed for the animals.

'What kind of life are we leading,' he said, 'if we have to live at the pleasure of an ass?'
King Philip II of Macedon, quoted in Alfred S. Bradford, *Philip II of Macedon: A Life from the Ancient Sources*, 1992.

An army cannot preserve good order unless its soldiers have meat in their bellies, coats on their backs and shoes on their feet.
The Duke of Marlborough, 1703, letter to Colonel Cadogan.

Our greatest difficulty is that of making our bread follow us, for the troops I have the honour to command cannot subsist without it, and the Germans that are used to starve cannot advance without us.
The Duke of Marlborough, July 1704, during the manoeuvering before the battle of Blenheim, quoted in Winston Churchill, *Marlborough*, II, 1967, p. 325.

Without supplies no army is brave.
Frederick the Great, *Instructions to his Generals*, 1747, tr. Phillips, 1940.

The troops will have observed the extreme difficulty of supplying them with bread in this part of the country, and the necessity that exists, that they should take care of that which is issued to them, and make it last for the time specified in General Orders; for want of attention to this object, and care of their bread, the best operations are necessarily relinquished.
The Duke of Wellington, 17 May 1809, at Ruivaes in Portugal, *Selections from the General Orders of Field Marshal the Duke of Wellington During His Various Commands*, 1847.

The mere readiness of the troops is nothing in comparison with the preparations required for the departments of the Service [logistics].
The Duke of Wellington, quoted in Arthur Bryant, *The Great Duke*, 1972.

An army marches on its stomach.

LOGISTICS AND GENERALSHIP

Napoleon; source cited in *Oxford Dictionary of Quotations* as 'anonymous' and also attributed 'to Napoleon in *Windsor Magazine*, 1904, p. 268; probably from a long passage in Las Casas…'.

The whole of military activity must relate directly or indirectly to the engagement. The end for which a soldier is recruited, clothed, armed, and trained, the whole object of his sleeping, eating, drinking, and marching is simply that he should fight at the right place and the right time.
  Major-General Carl von Clausewitz, *On War*, 1.2, 1832, tr. Michael Howard and Peter Paret, 1976.

Victory is the beautiful, bright-colored flower. Transport is the stem without which it never could have blossomed.
  Sir Winston S. Churchill, *The River War*, 1899.

Mobility is the true test of a supply system.
  Captain Sir Basil Liddell Hart, *Thoughts on War*, 1944.

Good logistics is combat power.
  Lieutenant-General William G. Pagonis, *Moving Mountains*, 1992.

LOGISTICS AND GENERALSHIP

The commander who fails to provide his army with necessary food and other supplies is making arrangements for his own defeat, even with no enemy present.
  The Emperor Maurice, *The Strategikon*, c. AD 600 (*Maurice's Strategikon*, tr. George Dennis, 1984).

When the Duke of Cumberland has weakened his army sufficiently, I shall teach him that a general's first duty is to provide for its welfare.
  Field Marshal Maurice Comte de Saxe, 1747, before Laufeld.

Understand that the foundation of an army is the belly. It is necessary to procure nourishment for the soldier wherever you assemble him and wherever you wish to conduct him. This is the primary duty of a general.
  Frederick the Great, *Instructions to His Generals*, 1747, tr. Phillips, 1940.

What I want to avoid is that my supplies should command me.
  Field Marshal Francois Comte de Guibert, *Essai Général de la Tactique*, 1770.

It is very necessary to attend to all this detail and to trace a biscuit from Lisbon into a man's mouth on the frontier and to provide for its removal from place to place by land or by water, or no military operations can be carried out.
  The Duke of Wellington, 1811, *The Dispatches of Field Marshal the Duke of Wellington During His Various Campaigns in India, Denmark, Portugal, Spain, the Low Countries, and France*, VII, 1834–1838, p. 406.

What makes the general's task so difficult is the necessity of feeding so many men and animals. If he allows himself to be guided by the supply officers he will never move and his expedition will fail.
  Napoleon, *Maxims of War*, 1831.

The 'feeding' of an army is a matter of the most vital importance, and demands the earliest attention of the general intrusted with a campaign. To be strong, healthy, and capable of the largest measure of physical effort, the soldier needs about three pounds gross of food per day, and the horse or mule about twenty pounds. When a general first estimates the quantity of food and forage needed for any army of fifty to one hundred thousand men, he is apt to be dismayed, and here a good staff is indispensable, though the general cannot throw off on them the responsibility. He must give the subject his

LOOTING/PILLAGE/PLUNDER/RAPINE

personal attention, for the army reposes in him alone, and should never doubt the fact that their existence overrides in importance all other considerations. Once satisfied of this, and that all has been done that can be, the soldiers are always willing to bear the largest measure of privation.

General of the Army William T. Sherman, *Memoirs of General W.T. Sherman*, 1875,

I believe that the task of bringing the force to the fighting point, properly equipped and well-formed in all that it needs is at least as important as the capable leading of the force in the fight itself... In fact it is indispensable and the combat between hostile forces is more in the preparation than the fight.

General Sir John Monash (1865–1931), quoted in Horner, *The Commanders*, 1984.

The more I see of war, the less I think that general principles of strategy count as compared with administrative problems and the gaining of intelligence. The main principles of strategy, e.g., to attack the other fellow in the flank or rear in preference to the front, to surprise him by any means in one's power and to attack his morale before you attack him physically are really things that every savage schoolboy knows. But it is often outside the power of the general to act as he would have liked owing to lack of adequate resources and I think that military history very seldom brings this out, in fact, it is almost impossible that it should do so without a detailed study which is often unavailable. For instance, if Hannibal had another twenty elephants, it might have altered his whole strategy against Italy.

Field Marshal Viscount Wavell of Cyrenaica, 1942, letter to Liddell Hart.

[Logistics]... the crux of generalship – superior even to tactical skill.

Field Marshal Viscount Wavell of Cyrenaica (1883–1950).

Many generals have failed in war because they neglected to ensure that what they wanted to achieve operationally was commensurate with their administrative resources; and some have failed because they over-insured in this respect. The lesson is, there must always be a nice balance between the two requirements.

Field Marshal Viscount Montgomery of Alamein, *The Memoirs of Field Marshal Montgomery*, 1958.

LOOTING/PILLAGE/PLUNDER/RAPINE

Then his majesty prevailed against them at the head of his army, and when they saw his majesty prevailing against them they fled headlong to Megiddo in fear, abandoning their horses and their chariots of gold and silver... Now, if only the army of his majesty had not given their heart to plundering the things of the enemy, they would have captured Megiddo at this moment when the wretched foes of Kadesh and the wretched foe of this city were hauled up in haste to bring them into this city.

Pharaoh Thutmose III, 1439 BC, at the Battle of Megiddo, quoted in James Breasted, *Ancient Records of Egypt*, 1906.

War without fire is like sausages without mustard.

Henry V (attributed), quoted in John Gillingham, *The Wars of the Roses*, 1981, p. 24.

The city and the buildings are mine; but I resign to your valour the captives and the spoil, the treasures of gold and beauty; be rich and be happy. Many are the provinces of my empire: the intrepid soldier who first ascends the walls of Constantinople shall be rewarded with the government of the fairest and most wealthy; and my gratitude shall accumulate his honours and fortunes above all measure of his

LOOTING/PILLAGE/PLUNDER/RAPINE

own hopes.
>Mehmed II, exhortation to his army at the beginning of the siege of Constantinople (1453), quoted in Gibbon, *The Decline and Fall of the Roman Empire*, III, 1776–1788.

Do you think my men are nuns?
>General Count Johann Tilly, 1621, to the complaints of the rapine committed by his Imperialist troops, quoted in Theodore Ayrault Dodge, *Gustavus Adolphus*, 1895.

God be my witness, you are yourselves the destroyers, wasters and spoilers of your Fatherland. My heart is sickened when I look at you.
>Gustavus II Adolphus, King of Sweden, September 1632, to the officers of his German Protestant allies who had stolen cattle during his invasion of Bavaria.

I fear the Turkish Army less than their camp.
>Field Marshal Prince Eugene, quoted in Chandler, *Atlas of Military Strategy*, 1980; referring to the breakdown in discipline that would occur in the looting of the Turkish camp.

It is our business to give protection, and support, to the poor, distressed inhabitants; not to multiply and increase their calamities.
>General George Washington, 21 January 1777, General Orders, *The Writings of George Washington*, VII, p. 47, John C. Fitzgerald, ed., 1932.

Fire authorizes looting, which the soldier permits himself in order to save the rest from the remnants of the fire.
>Napoleon, 20 September 1812, letter to Tsar Aleksandr I complaining of the burning of Moscow by incendiaries, *Correspondance de Napoléon Ier, publiée par ordre de l'Empereur Napoléon III*, XXIV, 19213, 1858–1870.

We started with the army in the highest order, and up to the day of battle nothing could get on better; but that event has, as usual, totally annihilated all order and discipline. The soldiers of the army have got among them about a million sterling in money... the night of the battle, instead of being passed in getting rest and food to prepare them for the pursuit of the following day, was passed by the soldiers in looking for plunder. The consequence was, that they were totally knocked up... This is the consequence of the state of discipline of the British army. We may gain the greatest victories; but we shall do no good until we shall so far alter our system, as to force all ranks to perform their duty. The new regiments are, as usual, the worst of all.
>The Duke of Wellington, 29 June 1813, to Bathurst after the Battle of Vittoria, *The Dispatches of Field Marshal the Duke of Wellington During His Various Campaigns in India, Denmark, Portugal, Spain, the Low Countries, and France*, X, 1834–1838, p. 473.

I do not care whether I command a large or a small army; but large or small, I will be obeyed and I will not suffer pillage.
>The Duke of Wellington, early 1814, during the invasion of southern France when he suppressed looting by his army by frequent hangings in an effort to win the good will of the population, quoted in Arthur Bryant, *The Great Duke*, 1972.

Happily, however, policy and morality are equally opposed to the system of pillage. I have meditated much on this subject: and have often been urged to gratify my soldiers in this manner. But nothing is so certain to disorganize and completely ruin an army. A soldier loses all discipline as soon as he gets an opportunity to pillage; and, if by pillage he enriches himself, he immediately becomes a bad soldier, and will not fight.
>Napoleon, 1 July 1816, conversation with Count Emmanuel Las Casas on St Helena,

LOOTING/PILLAGE/PLUNDER/RAPINE

The Life, Exile, and Conversations of the Emperor Napoleon, II, 1890.

Disrepute having been brought upon our brave soldiers by the bad conduct of some of their numbers, showing on all occasions, when marching through territory occupied by sympathizers of the enemy, a total disregard of rights of citizens, and being guilty of wanton destruction of private property the Genl. commanding, desires and intends to enforce a change in this respect.

Interpreting Confiscation Acts by troops themselves, has a demoralizing effect, weakens them in exact proportion to the demoralization and makes open and armed enemies of many who, from opposite treatment would become friends or at worst non-combatants.

It is ordered, therefore, that the severest punishment, be inflicted upon every soldier, who is guilty of taking or destroying private property, and any commissioned officer guilty of like conduct, or of countenancing it shall be deprived of his sword and expelled from the camp, not to be permitted to return.

General Ulysses S. Grant, General Order No. 3, 13 January 1862, Head Quarters District of Cairo.

Ah, General Hood, when you Texans come about the chickens have to roost mighty high.

General Robert E. Lee, to General John B. Hood when he commanded the Texas Brigade, *Advance and Retreat*, 1880.

The commanding general considers that no greater disgrace could befall the army, and through it our whole people, than the perpetration of the barbarous outrages upon the unarmed, and defenceless and the wanton destruction of private property that have marked the course of the enemy in our own country.

Such proceedings not only degrade the perpetrators and all connected with them, but are subversive of the discipline and efficiency of the army, and destructive of the ends of our present movement.

It must be remembered that we make war only upon armed men, and that we cannot take vengeance for the wrongs our people have suffered without lowering ourselves in the eyes of all whose abhorrence has been excited by the atrocities of our enemies, and offending against Him to whom vengeance belongeth, without whose favor and support our efforts must all prove in vain.

The commanding general therefore earnestly exhorts the troops to abstain with most scrupulous care from unnecessary or wanton injury to private property, and he enjoins upon all officers to arrest and bring to summary punishment all who shall in any way offend against the orders on this subject.

General Robert E. Lee, General Order No. 73, issued at Chambersburg, Pennsylvania, 27 June 1863, during the Gettysburg campaign, *The Wartime Papers of Robert E. Lee*, 1961.

Insult to a soldier does not justify pillage, but it takes from the officer the disposition he would otherwise feel to follow up the inquiry and punish the wrong-doers.

General of the Army William T. Sherman, letter to the editor of the *Bulletin*, 21 September 1862, *Memoirs*, I, 1875.

Sherman was right behind us with an army, an army that was no respecter of ducks, chickens, pigs, or turkeys, for they used to say of one particular regiment in Sherman's corps that it could catch, scrape, and skin a hog without a soldier leaving the ranks.

Admiral David D. Porter, *Incidents and Anecdotes of the Civil War*, 1885.

War looses violence and disorder; it

inflames passions and makes it relatively easy for the individual to get away with unlawful actions. But it does not lessen the gravity of the offense or make it less necessary that constituted authority put him down. The main safeguard against lawlessness and hooliganism in any armed body is the integrity of the officers. When men know that their commander is absolutely opposed to such excesses, and will take forceful action to repress any breach of discipline, they will conform. But when an officer winks at any depradation by his men, it is no different than if he had committed the act.

Brigadier-General S.L.A. Marshall, *The Armed Forces Officer*, 1950.

Theft is a crime no matter where or from whom. It is often from small beginnings that the rot sets in: a soldier 'lifts' a few oranges from an orchard and gets away with it; an officer, held in the highest esteem, takes a revolver as booty from the enemy, carries off a few 'finds' from an historic site, or in supposedly innocent mischief sends his wife a souvenir 'from a house in occupied territory' or 'found on an infiltrator'. In this way the dam of army morals is burst asunder and the officer is powerless to prevent the spread of corruption to which he has himself succumbed. No distinction can be drawn here between the penny and the pound; petty or large-scale, crime can be called by no other name. No commander can demand discipline and honesty from his men if he himself does not serve as an example. And plainly the higher the rank the graver the crime – and the more rigorous punishment there must be.

General Yigal Allon, *The Making of Israel's Army*, 1970.

LOYALTY

Those who would be military leaders must have loyal hearts, eyes and ears, claws and fangs. Without people loyal to them, they are like someone walking at night, not knowing where to step. Without eyes and ears, they are as though in the dark, not knowing where to proceed. Without claws and fangs, they are like hungry men eating poisoned food, inevitably they die... Therefore good generals always have intelligent and learned associates for their advisors, thoughtful and careful associates for their eyes and ears, brave and formidable associates for their claws and fangs.

Zhuge Liang (AD 180–234), *The Way of the General*, (*Mastering the Art of War*, tr. Thomas Cleary, 1989).

In all the military works it is written: To train samurai to be loyal separate them when young, or treat them according to their character. But it is no use to train them according to any fixed plan, they must be educated by benevolence. If the superior loves benevolence then the inferior will love his duty.

Tokugawa Ieyasu (1543–1616), quoted in A.L. Sadler, *The Maker of Modern Japan: The Life of Tokugawa Ieyasu*, 1937.

This faith gave you victory at Shiloh and Vicksburg. Also, when you have completed your best preparations, you go into battle without hesitation as at Chattanooga – no doubts, no reserve; and, I tell you that it was this that made us act with confidence. I knew wherever I was that you thought of me and if I got into a tight place you would come, if alive.

General of the Army William T. Sherman, 10 March 1864, near Memphis, letter to General Grant.

...A monstrous theory, viz., that a regimental officer who is the only officer present with a party of soldiers actually and seriously engaged with the enemy can, under any pretext whatever, be justified in deserting them... The

LOYALTY

more helpless the position in which an officer finds his men, the more it is his bounden duty to stay and share their fortune, whether for good or ill.
>Field Marshal Sir Garnet Wolseley (1833–1913), quoted in Donald Morris, *The Washing of the Spears*, 1968.

Loyalty is the marrow of honor.
>Field Marshal Paul von Hindenburg, *Out of My Life*, 1920.

Those who are naturally loyal say little about it, and are ready to assume it in others. In contrast, the type of soldier who is always dwelling on the importance of 'loyalty' usually means loyalty to his own interests. (Sept. 1932.)
>Captain Sir Basil Liddell Hart, *Thoughts on War*, 1944.

An officer... should make it a cardinal principle of life that by no act of commission or omission on his part will he permit his immediate superior to make a mistake.
>General Malin Craig, 12 June 1937, address to the graduating class of West Point.

We learn from history that those who are disloyal to their own superiors are most prone to preach loyalty to their subordinates. Loyalty is a noble quality, so long as it is not blind and does not exclude the higher loyalty to truth and decency. But the word is much abused. For 'loyalty,' analyzed, is too often a polite word for what would be more accurately described as 'a conspiracy for mutual inefficiency.' In this sense it is essentially selfish... 'loyalty' is not a quality we can isolate – so far as it is real, and of intrinsic value, it is implicit in the possession of other virtues.
>Captain Sir Basil Liddell Hart, *Through the Fog of War*, 1938.

There is a great deal of talk about loyalty from the bottom to the top. Loyalty from the top down is even more necessary and much less prevalent.
>General George S. Patton, Jr., *War As I Knew It*, 1947.

Loyalty is the big thing, the greatest battle asset of all. But no man ever wins the loyalty of troops by preaching loyalty. It is given him by them as he proves his possession of the other virtues. The doctrine of a blind loyalty to leadership is a selfish and futile military dogma except in so far as it is ennobled by a higher loyalty in all ranks to truth and decency.
>Brigadier-General S.L.A. Marshall, *Men Against Fire*, 1947.

The very touchstone of loyalty is that just demands will be put upon it.
>Brigadier-General S.L.A. Marshall, *The Armed Forces Officer*, 1950.

I also learnt that the discipline demanded from the soldier must become loyalty in the officer.
>Field Marshal Viscount Montgomery of Alamein, *The Memoirs of Field Marshal Montgomery*, 1958.

The essence of Loyalty is the courage to propose the unpopular, coupled with a determination to obey, no matter how distasteful the ultimate decision. And the essence of leadership is the ability to inspire such behavior.
>Lieutenant-General Victor H. 'The Brute' Krulak, 'A Soldier's Dilemma', *Marine Corps Gazette*, 11/1986.

I explained my idea of loyalty. 'When we are debating an issue, loyalty means giving me your honest opinion, whether you think I'll like it or not. Disagreement, at this stage, stimulates me. But once a decision has been made, the debate ends. From that point on, loyalty means executing the decision as if it were your own.'
>General Colin Powell, *My American Journey*, 1995.

LUCK

LUCK

The state benefits more from a lucky general than from a brave one. The first achieves his results with little effort, whereas the other does so at some risk.
  The Emperor Maurice, *The Strategikon*, c. AD 600 (*Maurice's Strategikon*, tr. George Dennis, 1984).

As I rely very much upon Providence, so I shall be ready of approving all occasions that may offer.
  The Duke of Marlborough, 4 May 1710, letter to the Earl of Godolphin.

One can see that a general should be skillful and lucky and that no one should believe so fully in his star that he abandons himself to it blindly. If you are lucky and trust in luck alone, even your success reduces you to the defensive; if you are unlucky you are already there.
  Frederick the Great, *Instructions to His Generals*, 1747, tr. Phillips, 1940.

The affairs of war, like the destiny of battles, as well of empires, hang upon a spider's thread.
  Napoleon, *Political Aphorisms*, 1848.

Luck in the long run is given only to the efficient.
  Field Marshal Helmuth Graf von Moltke (1800–1891).

Luck is like a sum of gold, to be spent.
  Field Marshal Viscount Allenby of Meggido (1861–1936).

The harder I work, the luckier I get.
  Lieutenant-General Victor H. 'The Brute' Krulak, *Life Magazine*, 11/1965, a favourite motto.

That's the way it is in war. You win or lose, live or die – and the difference is just an eyelash.
  General of the Army Douglas MacArthur, *Reminiscences*, 1964.

There is still another ingredient in this formula for a great leader… and that is LUCK. He must have opportunity. Then, of course, when opportunity knocks, he must be able to rise and open the door.
  General of the Army Omar N. Bradley, 'Leadership', *Parameters*, Winter 1972.

M

MAN

The man whose profession is arms should calm his mind and look into the depths of others. Doing so is likely the best of the martial arts.
  Shiba Yoshimasa, *Chikubasho*, 1380 (*Ideals of the Samurai*, tr. William Wilson, 1982).

The same troops, who if attacking would have been victorious, may be invariably defeated in entrenchments. Few men have accounted for it in a reasonable manner, for it lies in human hearts and one should search for it there. No one has written of this matter which is the most important, the most learned, and the most profound of the profession of war. And without a knowledge of the human heart, one is dependent upon the favor of fortune, which sometimes is very inconstant.
  Field Marshal Maurice Comte de Saxe, *My Reveries*, 1732, tr. Phillips, 1940.

...The human heart – the groundwork of genius in the art of war.
  Major-General Henry 'Light Horse Harry' Lee, *Memoir of the War in the Southern Department*, 1869.

The human character is the subject of the military officer's study; for it is upon man that his trials are made. He must, therefore, know, in the most precise manner, what man can do, and what he cannot do; he must also know the means by which his exertions are to be animated to the utmost extent of exertion. The general's duty is consequently an arduous duty; the capacity of learning it is the gift of Nature; the school is in the camp and the cottage rather than the city and the palace; for a man cannot know things in their foundations till he sees them without ; as he cannot judge of the hardships of service till he has felt them in experience...
  Robert Jackson, *A Systematic View of the Formation, Discipline, and Economy of armies*, 1804.

It needs a long experience of war to perceive its principles; one must have undertaken many offensive operations to realize how the slightest incident means encouragement or discouragement, brings about one result or another. In warfare men are nothing, a man is everything.
  Napoleon, 30 August 1808, a note on Spanish affairs, quoted in R.M. Johnston, ed., *The Corsican*, 1910.

Sentiment rules the world, and he who fails to take that into account can never hope to lead.
  Napoleon (1769–1821).

If you wish to study men, learn how far their patience can stretch.
  Napoleon, 1816, conversation on St Helena, quoted in Christopher Herold, ed., *The Mind of Napoleon*, 1955.

Remember also that one of the requisite studies for an officer is *man*. Where your analytical geometry will serve you once, a knowledge of men will serve you daily. As a commander, to get the right man in the right place is one of the questions of success or defeat.

MAN

There is a soul to an army as well as to the individual man, and no general can accomplish the full work of his army unless he commands the souls of his men, as well as their bodies and legs.
General of the Army William T. Sherman, *Memoirs of General W.T. Sherman*, 1875.

It is the leader who reckons with the human nature of his troops, and of the enemy, rather than with their mere physical attributes, numbers, armament and the like, who can hope to follow in Napoleon's footsteps.
Colonel G.F.R. Henderson, quoted in Marshall, *The Armed Forces Officer*, 1950.

I will again sound the note of warning against that plausible cry of the day which finds all progress in material advance, disregarding that noblest sphere in which the mind and heart of man, in which all that is god-like in man, reign supreme; and against the temper which looks not to man, but to his armor.
Rear Admiral Alfred Thayer Mahan, *Naval Administration and Warfare*, 1908.

We will have to say that in any case the decisive role does not belong to technology – behind technology there is always a living person without whom the technology is dead.
Mikhail Frunze (1885–1925), quoted in M. Gareyev, *Frunze, Military Theorist*, 1985.

Wars may be fought with weapons, but they are won by men. It is the spirit of the men who follow and the man who leads that gains the victory.
General George S. Patton, Jr., in *The Cavalry Journal*, 9/1933.

If I had time and anything like your ability to study war, I think I should concentrate entirely on the 'actualities of war' – the effects of tiredness, hunger, fear, lack of sleep, weather... The principles of strategy and tactics, and the logistics of war are really absurdly simple: it is the actualities that make war so complicated and so difficult, and are usually so neglected by historians.
Field Marshal Viscount Wavell of Cyrenaica to Liddell Hart, quoted in John Connell, 'Writing About Soldiers', *Journal of the Royal United Services Institute*, 8/1965.

A commander should have a profound understanding of human nature, the knack of smoothing out troubles, the power of winning affection while communicating energy, and the capacity for ruthless determination where required by circumstances. He needs to generate an electrifying current and to keep a cool head in applying it. (Oct. 1933.)
Captain Sir Basil Liddell Hart, *Thoughts on War*, 1944.

This is the so-called theory that 'weapons decide everything', which constitutes a mechanical approach to the question of war and a subjective and one-sided view. Our view is opposed to this; we see not only weapons but also people. Weapons are an important factor in war, but not the decisive factor; it is people, not things, that are decisive. The contest of strength is not only a contest of military and economic power, but also a contest of human power and morale.
Mao Tse-tung, *On Protracted War*, May 1938.

The art of leading, in operations large or small, is the art of dealing with humanity, of working diligently on behalf of men, of being sympathetic with them, but equally, of insisting they make a square facing toward their own problems.
Brigadier-General S.L.A. Marshall, *Men Against Fire*, 1947.

Admiral David G. Farragut, 13 October 1864, letter to his son.

MANOEUVRE

An army must be as hard as steel in battle and can be made so; but, like steel, it reaches its finest quality only after much preparation and only provided the ingredients are properly constituted and handled. Unlike steel, an army is a most sensitive instrument and can easily become damaged; its basic ingredient is men and, to handle an army well, it is essential to understand human nature. Bottled up in men are great emotional forces which have got to be given an outlet in a way which is positive and constructive, and which warms the heart and excites the imagination. If the approach to the human factor is cold and impersonal, then you achieve nothing. But if you gain the confidence and trust of your men, and they feel their best interests are safe in your hands, then you have in your possession a priceless asset and the greatest achievements become possible.
> Field Marshal Viscount Montgomery of Alamein, *The Memoirs of Field Marshal Montgomery*, 1958.

It is essential to understand that battles are won primarily in the hearts of men.
> Field Marshal Viscount Montgomery of Alamein, *The Memoirs of Field Marshal Montgomery*, 1958.

It is people who win wars, not technology. If servicemen are not properly trained and motivated to use the equipment they have, they will achieve very little with it.
> General Sir Peter de la Billière, *Storm Command*, 1992.

MANOEUVRE
Principle of War

Maneuvering with an army is advantageous; with an undisciplined multitude, most dangerous.
> Sun Tzu, *The Art of War*, 7, c. 500 BC, tr. Giles, 1910.

...For as a ship, if you deprive it of its steersman, falls with all its crew into the hands of the enemy; so, with an army in war, if you outwit or out-maneuver its general, the whole will often fall into your hands.
> Polybius, *Histories*, c. 125 BC.

The enemy should see me marching when he believes me to be fettered by calculations of subsistence; this new kind of war must astonish him, must give him no chance to breathe anywhere, and make plain at his cost this constant truth – that there is scarcely any position tenable in face of a well-constituted, temperate, patient and manoeuvring army.
> Field Marshal François Duc de Guibert, *Essai Général de Tactique*, 1770.

Aptitude for maneuver is the supreme skill in a general; it is the most useful and rarest of gifts by which genius is estimated.
> Napoleon (1769–1821), quoted in Commandant J. Colin, *The Transformation of War*, 1895.

[Victory/winning battles] ...more by the movement of troops than by fighting.
> General of the Army William T. Sherman (1820–1891).

Maneuvers are threats; he who appears most threatening wins.
> Colonel Charles Ardant du Picq, *Battle Studies*, 1880, tr. Greely, 1957.

The fundamental characterization of a strategy of maneuver is not the formal offensive but initiative and action.
> Leon Trotsky, 1921.

Nearly all the battles which are regarded as masterpieces of the military art, from which have been derived the foundation of states and the fame of commanders, have been battles of manoeuvre.
> Sir Winston S. Churchill, *The World Crisis*, II, 1923.

MARCH TO THE SOUND OF THE GUNS

Battles are won by slaughter and manoeuvre. The greater the general, the more he contributes in manoeuvre, the less he demands in slaughter.
 Sir Winston S. Churchill, *The World Crisis*, II, 1923.

He ...saw that there was no point in fighting for a piece of sand. As he put it, it would be 'rather like a sailor fighting for a wave or an airman for a cloud.' A patch of sand, he decided, was an irrelevance. If some foolish Iraqi wished to dig a hole and sit in it, the best way of defeating him would be to go round behind him and deprive him of his water. As long as the Iraqis chose to defend sand, he reckoned 'they had neutered themselves', and he set about training to fight a fast-moving battle in depth, as he put it, 'fundamentally to alter the way we did business, to accommodate the reality of the circumstances.'
 General Sir Peter de la Billière, quoting Major-General Rupert Smith commanding the British First Armoured Division in its preparation for Desert Storm, December 1990, *Storm Command*, 1992.

MANHOOD/MANLINESS

Be men, my friends! Discipline fill your hearts,
maintain your pride in the eyes of other men!
Remember, each of you, sons, wives, wealth, parents –
are mother and father dead or alive? No matter,
I beg you for their sakes, loved ones far away –
now stand and fight, no turning back, no panic.
 Homer, *The Iliad*, 15.769–774, c. 800 BC, tr. Robert Fagles, 1990; Nestor, King of Pylos, appealing to the manhood of the panicking Greeks as the Trojans pressed them back upon their ships.

When someone inquired of him what children should learn, he said: 'What they will also use when they become men.'
 Plutarch, quoting Agesilaus, Eurypontid King of Sparta (400–360 BC), *The Lives*, c. AD 100 (*Plutarch on Sparta*, tr. Richard Talbert, 1988).

Before a man can be manly the gift which makes him so must be there, collected by him slowly, unconsciously as his bones, his flesh, his blood.
 Field Marshal Earl Roberts, *Diary*, 1876.

Arm and prepare to acquit yourselves like men, for the time of your ordeal is at hand.
 Field Marshal Earl Roberts, *Message to the Nation*, 1912.

Our children don't need to be coddled, and they shouldn't be condemned. Above all, for heaven's sake, let your sons alone and let them grow up to be men.
 Lieutenant-General Lewis B. 'Chesty' Puller (1898–1971), comment at a meeting of the Parent-Teacher Association, Saluda, Virginia, after his retirement in 1955.

War makes everyone indecently naked. The essence of everyone is spread out on the palm of fate in three days for all to see: either you are a man and soldier or you have gotten male characteristics through a misunderstanding.
 Lieutenant-General Aleksandr I. Lebed, *I Hurt For This Great Country*, 1995.

MARCH TO THE SOUND OF THE GUNS

The safest way of achieving victory is to seek it among the enemy's battalions.
 Field Marshal Prince Aleksandr V. Suvorov (1729–1800), quoted in K. Ossipov, *Suvorov*, 1945.

March to the sound of the guns.
 The Duke of York, 1793, to his commanders during the Dunkirk campaign.

MARCHING

Marshal Grouchy had only left his camp at Gemlouz at ten in the morning, and was halfway to Wavres between noon and one o'clock. He heard the dreadful cannonade of Waterloo. No experienced man could have mistaken it: it was the sound of several hundred guns, and from that moment two armies were hurling death at each other. General Excelmans, commanding the cavalry, was profoundly moved by it. He went up to the Marshal and said to him: 'The Emperor is at grips with the English army; there can be no doubt about it, such a furious fire can be no skirmish. Monsieur le Marechal, we must march towards the sound of the guns. I am an old soldier of the Army of Italy; I have heard General Bonaparte preach this principle a hundred times. If we turn to the left we shall be on the battlefield in two hours'....

Napoleon (1769–1821), quoted in Somerset de Chair, ed., *The Waterloo Campaign*, 1957.

MARCHING

But on the armies came
as if the whole earth were devoured by wildfire, yes,
and the ground thundered under them, deep as it does
for Zeus who loves the lightning, Zeus in all his rage
when he lashed the ground around Typhoeus in Arima,
there where they say the monster makes his bed of pain –
so the earth thundered under their feet, armies trampling,
sweeping through the plain at blazing speed.

Homer, *The Iliad*, 2.887–895, c. 800 BC, tr. Robert Fagles, 1990.

An army is exposed to more danger on marches than in battles. In an engagement the men are properly armed, they see their enemies before them and are prepared to fight. But on a march the soldier is less on his guard, has not his arms always ready and is thrown into disorder by a sudden attack or ambuscade. A general, therefore, cannot be too careful and diligent in taking necessary precautions to prevent a surprise on the march.

Flavius Vegetius Renatus, *Military Institutions of the Romans*, c. AD 378, tr Clark, 1776.

We march straight on; we march to victory.

Harold Godwinson, King of England, 1066, refusing William of Normandy's offer of parley before the Battle of Hastings.

Marches are war.

Napoleon, quoted in David Chandler, *The Campaigns of Napoleon*, 1966.

I have accomplished my object; I have destroyed the Austrian army by simple marching. I have made 60,000 prisoners, taken 120 guns, more than 90 flags, and more than 30 generals.

Napoleon, 19 October 1805, to Josephine, of the Ulm campaign, quoted in R.M. Johnston, ed., *The Corsican*, 1910.

Maxim 9. The strength of an army, like power in mechanics, is reckoned by multiplying the mass by the rapidity. A rapid march increases the morale of an army, and increases its means of victory. Press on!

Napoleon, *The Military Maxims of Napoleon*, tr. George D'Aguilar, 1831 (David Chandler, ed., 1987).

The hardships of forced marches are more often painful than the dangers of battle... I would rather lose one man marching than five in fighting...

Lieutenant-General Thomas J. 'Stonewall' Jackson, 1862, quoted in Henderson, *Stonewall Jackson*, 1898.

MARINES

That twelve hundred land Souldjers be

MARKSMANSHIP

forthwith raysed, to be in readiness, to be distrubuted into his Mats Fleets prepared for sea Service...
  Charles II, 28 October 1664, Order in Council establishing the Royal Marines, the first permanent body of marines.

I never knew an appeal made to them for honor, courage, or loyalty that they did not more than realize my highest expectations. If ever the hour of real danger should come to England they will be found the Country's Sheet Anchor.
  Admiral Lord St Vincent, 1802, of the Royal Marines.

How much might be done with a hundred thousand soldiers such as these.
  Napoleon, 15 July 1815, while inspecting the Royal Marine guard aboard HMS *Bellerophon*, following his surrender after Waterloo.

The Marines are properly the garrisons of His Majesty's ships, and upon no pretence ought they to be moved from a fair and safe communication with the ships to which they belong.
  The Duke of Wellington, 21 April 1837, to the House of Lords.

A ship without Marines is like a garment without buttons.
  Rear Admiral David Dixon Porter, 1863, letter to Colonel John Harris, USMC.

The Marines have landed and the situation is well in hand.
  Richard Harding Davis (1864–1916) – attributed.

'SOLDIER AN' SAILOR TOO'
(The Royal Regiment of Marines)
As I was spittin' into Ditch aboard o' the *Crocodile*,
I seed a man on a man-o'-war got up in the Reg'lars style
'E was scrapin' the paint from off of 'er plates, an' I sez to 'im, "Oo are you?"
Sez 'e 'I'm a Jolly – 'Er Majesty's Jolly – soldier an' sailor too!'
Now 'is work begins by Gawd knows when, and 'is work is never through;
'E isn't one o' the reg'lar Line, nor 'e isn't one of the crew.
'E's a kind of giddy harumfrodite – soldier an' sailor too!

An', after, I met 'im all over the world, a-doin' all kinds of things,
Like landin' 'isself with a Gatlin' gun to talk to them 'eathen kings;
'E sleeps in an 'ammick instead of a cot, an' 'e drills with the deck on a slew;
An' 'e sweats like a Jolly – 'Er Majesty's Jolly – soldier an' sailor too!
For there isn't a job on the top o' the earth the beggar don't know, nor do –
  Rudyard Kipling (1865–1936).

Just rejoice at the news and congratulate our forces and the marines...
Rejoice, rejoice.
  Prime Minister Margaret Thatcher, 26 April 1982, on the capture of the island of South Georgia during the Falklands War.

MARKSMANSHIP

In archery we have three goals: to shoot accurately, to shoot powerfully, to shoot rapidly.
  Anonymous Byzantine general, *On Strategy (Peri Strategias)* c. AD 527–565 (*Three Byzantine Military Treatises*, tr. George Dennis, 1985.

Every English archer beareth under his girdle twenty-four Scots.
  English proverb, alluding to the 24 arrows each longbowman carried in his belt and the deadly accuracy of his marksmanship.

Aim for their shoelaces!
  Oliver Cromwell (1599–1658), during the English Civil War; an early recognition of the fact that most men shoot high.

Great Care should be observed in

MASS

choosing active Marsksmen; the manifest Inferiority of inactive Persons, unused to Arms, in this kind of Service, although equal in Numbers, to lively Persons who have practised Hunting is inconceivable. The Chance against them is more than two to one.
> General George Washington, 16 April 1756, to Robert Dinwiddie, *The Writings of George Washington*, I, p. 313, John C. Fitzgerald, ed., 1931.

It is not sufficient that the soldier must shoot, he must shoot well.
> Napoleon (1769–1821).

Good marksmanship is always the most important thing for the infantry.
> General Gerhard von Scharnhorst, 1812, memorandum.

You don't hurt 'em if you don't hit 'em.
> Lieutenant-General Lewis 'Chesty' Puller, quoted in Burke Davis, *Marine*, 1962.

MASS
Principle of War

That the impact of your army may be like a grindstone dashed against an egg – that is effected by the science of weak points and strong.
> Sun Tzu, *The Art of War*, 5, c. 500 BC, tr. Giles, 1910.

Use the most solid to attack the most empty.
> The 'Martial' Emperor Ts'ao Ts'ao, AD 155–220, commentary to Sun Tzu, *The Art of War*, 5, c. 500 BC, tr. Samuel Griffith, 1963.

One should not be overly fond of famous swords and daggers. For even if one has a sword valued at 10,000 cash, he will not overcome 100 men carrying spears valued at 100 cash.
> Asakura Toshikage (1428–1481), *The 17 Articles of Asakura Toshikage* (*Ideals of the Samurai*, tr. William Wilson, 1982).

Move upon the enemy in one mass on one line so that when brought to battle you shall outnumber him, and from such a direction that you compromise him.
> Napoleon (1769–1821).

The art of making the weight of all one's forces successively bear on the resistances one may meet.
> Marshal of France Ferdinand Foch, *Principles of War*, 1913.

Do not let my opponents castigate me with the blather that Waterloo was won on the playfields of Eton, for the fact remains geographically, historically and tactically, whether the great Duke uttered such nonsense or not, that it was won on fields in Belgium by carrying out a fundamental principle of war, the principle of mass; in other words by marching on to those fields three Englishmen, Germans or Belgiums to every two Frenchmen.
> Major-General J.F.C. Fuller, 'Principles of War with Reference to the Campaigns of 1914–1915', *Royal United Services Institution Journal*, LXI: 1, 1916.

We must not replace crushing strikes against the enemy with pinpricks... It is necessary to prepare an operation which will be like an earthquake.
> Marshal of the Soviet Union Georgi K. Zhukov, 12 May 1944.

The theory of human mass dominated the military mind from Waterloo to the World War. This monster was the child of the French Revolution by Napoleon. The midwife who brought it into the military world was the Prussian philosopher of war, Clausewitz, cloudily profound. He unfortunately died while his own thought was still fermenting – leaving his papers in sealed packets, with the significant note: 'Should the work be interrupted by my death, then what is found can only be called a mass of formless conceptions ...open to endless

MEDALS AND DECORATIONS

misconceptions.' So they proved.
Captain Sir Basil Liddell Hart, *Thoughts on War*, 1944.

MEDALS AND DECORATIONS

At the storming of a city the first man to scale the wall is awarded a crown of gold. In the same way those who have shielded and saved one of their fellow-citizens or their allies is honoured with gifts from the consul, and the men whose lives they have saved present them of their own free will with a crown; if not, they are compelled to do so by the tribunes who judge the case. Moreover, a man who has been saved in this way reveres his rescuer as a father for the rest of his life and must treat him as if he were a parent. And so by means of such incentives even those who stay at home feel the impulse to emulate such achievements in the field no less than those who are present and see and hear what takes place. For the men who receive these trophies not only enjoy great prestige in the army and soon afterwards in their homes, but they are also singled out for precedence in religious processions when they return. On these occasions nobody is allowed to wear decorations save those who have been honoured for their bravery by the consuls, and it is the custom to hang up the trophies they have won in the most conspicuous places in their houses, and to regard them as proofs and visible symbols of their valour. So when we consider this people's almost obsessive concern with military rewards and punishments, and the immense importance which they attach to both, it is not surprising that they emerge with brilliant success from every war in which they engage.
Polybius, *Histories*, 6.39, c. 125 BC (*The Rise of the Roman Empire*, tr. Ian Scott-Kilvert, 1987).

Honor them with titles, present them with goods, and soldiers willingly come join you. Treat them courteously, inspire them with speeches, and soldiers willingly die. Give them nourishment and rest so that they do not become weary, make the code of rules uniform, and soldiers willingly obey. Lead them into battle personally, and soldiers will be brave. Record even a little good, reward even a little merit, and the soldiers will be encouraged.
Zhuge Liang (AD 180–234) commentary to Sun Tzu, *The Art of War* (*Mastering the Art of War*, tr. Thomas Cleary, 1989).

It is not titles that honor men, but men that honor titles.
Niccolò Machiavelli, *Discourses*, 1517.

Drinkwater: As for you, Commodore, they will make you a baronet.
Nelson: No, no. if they want to mark my services, it must not be in that manner.
Drinkwater: Oh, you wish to be made a Knight of the Bath?
Nelson: Yes, if my services have been of any value, let them be noticed in a way that the public may know me – or them.
Admiral Viscount Nelson, 14 February 1797, after the victory at Cape St Vincent, preferring the ribbon and star of a Knight of the Bath to display his valour, rather than the higher honour of being made a baronet.

Show me a republic, ancient or modern, in which there have been no decorations. Some people call them baubles. Well, it is by such baubles that one leads men.
Napoleon, 19 May 1802, on establishing the *Légion d'honneur* (Legion of Honour).

In honour I gained them, and in honour I will die with them.
Admiral Viscount Nelson, 21 October 1805, at the Battle of Trafalgar, when urged not to wear his brilliant decorations, which

ultimately attracted the aim of a French marksman who wounded him mortally, quoted in Robert Southey, *Life of Nelson*, 1813.

A soldier will fight long and hard for a bit of colored ribbon.
  Napoleon, 15 July 1815, to the captain of HMS *Bellerophon*, upon entering exile.

I'd rather be a second lieutenant with a DSC (Distinguished Service Cross) than a general without it.
  General George S. Patton, Jr., 1918, quoted in *The Washington Post*, 18 May 1996, p. C1.

A medal glitters, but it also casts a shadow.
  Sir Winston S. Churchill, 22 March 1944, speech in the House of Commons.

They recommended me for a Silver Star for that action, and back in Corps Headquarters at Noumea some jerk reduced it to a Bronze Star. What right have those people got to put their cotton-picking hands into things like that? They didn't see the action, and have no way on earth to judge. Wouldn't you think they could see what it does to morale? I can stand it. I've got enough damned medals. But what it does to the young kids is inexcusable.
  Lieutenant-General Lewis 'Chesty' Puller, quoted in Burke Davis, *Marine*, 1962.

MEDICAL CORPS

War is the only proper school of the surgeon.
  Hippocrates, *Wounds of the Head*, c. 415 BC.

Eight or ten of the less-skilled soldiers in each tagma should be assigned as medical corpsmen to each bandon, especially in the first battle line. They should be alert, quick, lightly clothed, and without weapons. Their duty is to follow about a hundred feet to the rear of their own *tagma*, to pick up and give aid to anyone seriously wounded in the battle, or who has fallen off his horse, or is otherwise out of action, so they may not be trampled by the second line. For each person so rescued the corpsman should receive from the treasury one *nomisma* over above his pay...
  The Emperor Maurice, *The Strategikon*, c. AD 600 (*Maurice's Strategikon*, tr. George Dennis, 1984).

You medical people will have more lives to answer for in the other world than even we do.
  Napoleon, quoted in Barry O'Meara, *Napoleon in Exile*, 1822.

You Gentlemen of England who sit at home in all the well-earned satisfaction of your successful cases, can have little idea from reading the newspapers of the Horror and Misery of operating upon these dying, exhausted men.
  Florence Nightingale, 1854, letter from the British hospital at Scutari during the Crimean War to Dr Brown.

A corps of Medical officers was not established soley for the purpose of attending the wounded and sick ...the labors of Medical officers cover a more extended field. The leading idea, which should be constantly kept in view, is to strengthen the hands of the Commanding General by keeping his army in the most vigorous health, thus rendering it, in the highest degree, efficient for enduring fatigue and privation, and for fighting. In this view, the duties of such a corps are of vital importance to the success of an army, and commanders seldom appreciate the full effect of their proper fulfilment [sic].
  Major Jonathan Letterman (1824–1872), Medical Director of the Army of the Potomac.

If I were to speak of war, it would not be to show you the glories of conquering armies but the mischief and misery they strew in their track; and how, while

they march on with tread of iron and plumes proudly tossing in the breeze, some one must follow closely in their steps, crouching to the earth, toiling in the rain and darkness, shelterless like themselves, with no thought of pride or glory, fame or praise, or reward; hearts breaking with pity, faces bathed in tears and hands in blood. This is the side which history never shows.
> Clara Barton, quoted in Percy Epler, *The Life of Clara Barton*, 1915.

Fanned by the fierce winds of war, medical science and surgical art have advanced unceasingly, hand in hand. There has certainly been no lack of subjects for the treatment. The Medical profession at least cannot complain of unemployment through lack of raw material.
> Sir Winston S. Churchill, 10 September 1947, speech to the International Congress of Physicians at the Guildhall, London.

MEETING ENGAGEMENT

As the Thebans were retreating from Orchomenus towards Tegyrae, the Spartans, at the same time marching from Locris, met them. As soon as they came in view, advancing through the straits, one told Pelopidas, 'We are fallen into our enemy's hands;' he replied, 'And why not they into ours?'
> Pelopidas, at the Battle of Tegyrae, 375 BC, as told by Plutarch, *Plutarch's Lives*, c. AD 100, tr. John Dryden.

MEMOIRS/DIARIES

What the layman gets to know of the course of military events is usually nondescript. One action resembles another, and from a mere recital of events it would be impossible to guess what obstacles were faced and overcome. Only now and then, in the memoirs of generals or of their confidents, or as the result of close historical study, are some of the countless threads of the tapestry revealed. Most of the arguments and clashes of opinion that precede a major operation are deliberately concealed because they touch political interests, or they are simply forgotten, being considered as scaffolding to be demolished when the building is complete.
> Major-General Carl von Clausewitz, *On War*, 1832, 1.3, tr. Michael Howard and Peter Paret, 1976.

This book should make a contribution towards perpetuating those experiences of the bitter war years; experiences often gained at the cost of great deprivations and bitter sacrifices.
> Field Marshal Erwin Rommel, *Infantry Attacks*, 1937.

The last word in the Foreword to this book is 'truth.' I have tried to write the truth. I suppose everyone claims that about his memoirs! Most official accounts of past wars are deceptively well written, and seem to omit many important matters – in particular, anything which might indicate that any of our commanders ever made the slightest mistake. They are therefore useless as a source of instruction. They remind me of the French general's reply to a British protest in 1918, when the former directed the British to take over a sector from the French which had already been overrun by the Germans forty-eight hours previously. The French general said: '*Mais, mon ami, ca c'est pour l'histoire.*'
> Field Marshal Viscount Montgomery of Alamein, *The Memoirs of Field Marshal Montgomery*, 1958.

I hear my generals are selling themselves dearly.
> Sir Winston S. Churchill, of the flood of memoirs after WWII, quoted in Liddell Hart, *The Sword and the Pen*, 1976.

A general who has taken part in a campaign is by no means best fitted to write its history. That, if it is to be

complete and unbiased, should be the work of someone less personally involved. Yet such a general might write something of value. He might, as honestly as he could, tell of the problems he faced, why he took the decisions he did, what helped, what hindered, the luck he had, and the mistakes he made. He might, by showing how one man attempted the art of command, be of use to those who later may themselves have to exercise it. He might even give, to those who have not experienced it, some impression of what it feels like to shoulder a commander's responsibilities in war. These things I have tried to do in this book.
Field Marshal Viscount Slim of Burma, *Defeat Into Victory*, 1956.

I mean to record the course of events as I saw them. I shall be as objective as I feel possible to be, but I have no intention whatever of departing, for any reason, from my own honest opinion as to events and personalities. So often, people make great play about being completely unprejudiced. Frankly, I am completely prejudiced, and I accept as a guide and as warning, Goethe's saying:
'I can promise to be upright but not to be unprejudiced.'
Air Marshal Lord Tedder, *With Prejudice: The War Memoirs of Marshal of the Royal Air Force Lord Tedder*, G.C.B., 1966.

The worst thing that anybody can do ...is to keep an exact diary, because then he will put down every little resentment he had against Bill Smith and Joe Doakes. He will note everything that annoys him. Finally, his editors get hold of this and they say, 'Oh, this is a gold mine.' They take the things that he said and even emphasize them – the frictional type of entry – instead of those things which show that he thought things were going pretty well. I despise daily biographies as showing real history. I don't believe they do.
General of the Army Dwight D. Eisenhower, *General Eisenhower on the Military Churchill*, 1970.

MENTOR/PREFERMENT

To act in concert with a great man is the first of blessings.
Marquis de Lafayette, 1778, letter to George Washington.

It would finish me, could I have taken a sprig of these brave men's laurels. They are, and I glory in them, my darling children; served in my school; and all of us caught our professional zeal and fire from the great and good Earl of St. Vincent.
Admiral Viscount Nelson, March 1800, letter to Lord Keith on the successful sea engagements fought by men he had trained, quoted in David Howarth, *Nelson: The Immortal Memory*, 1986.

I would never have believed this preference could have happened to me, that I could come out before the others in intellectual qualities, but he [Gerhard von Scharnhorst] convinced me he had thought me the most able in the school, and when I believed myself capable of this, I went ahead with confidence.
Major-General Carl von Clausewitz, quoted in Roger Parkinson, *Clausewitz*, 1971.

The notice of others has been the start of many successful men.
General George S. Patton, Jr., 1915, quoted in Robert H. Patton, *The Pattons*, 1994.

Originality never yet led to preferment.
Admiral Sir John Fisher, *Memories*, 1919.

By the way, I have a favor to ask. When you are handing out diplomas at West Point you will see a cadet named Clarence Gooding, about 120 in standing. I would greatly appreciate your making some personal comment to this young fellow when you hand him his diploma. He was my office orderly at

MERCENARIES

Benning – for only a few months. Having just enlisted, I was greatly impressed with his efficiency, bearing and general intelligence. I found out that he had been trying for West Point and had had no success. Finally he enlisted for this purpose and was refused permission to join the candidates class of men being coached. Without his knowledge, I wrote to his congressman in Texas and got the promise of an appointment if a vacancy occurred. The first alternate dropped out at the eleventh hour and Gooding received his appointment. The principal failed and Gooding passed. He is a fine boy and would be electrified by a personal word from you.

> General of the Army George C. Marshall, 2 June 1936, letter to General of the Armies John J. Pershing, *Papers of George C. Marshall: The Soldierly Spirit*, 1981.

The battalion returned to England in 1913 and an officer of our 2nd Battalion was posted to it who had just completed the two-year course at the Staff College at Camberley. His name was Captain Lefroy. He was a bachelor and I used to have long talks with him about the Army and what was wrong with it, and especially how one could get to real grips with the military art. He was interested at once, and helped me tremendously with advice about what books to read and how to study. I think it was Lefroy who first showed me the path to tread and encouraged my youthful ambition. He was killed later in the 1914–18 war and was a great loss to me and to the Army... All this goes to show how important it is for a young officer to come in contact with the best type of officer and the right influences early in his military career.... In my case, the ambition was there, and the urge to master my profession. But is required advice and encouragement from the right people to set me on the road, and once that was forthcoming it was plainer sailing.

> Field Marshal Viscount Montgomery, *The Memoirs of Field Marshal Montgomery*, 1958.

MERCENARIES

The Carthaginians were in the habit of forming their armies of mercenaries drawn together from the different countries; if they did so for the purpose of preventing conspiracies, and of making the soldiers more completely under the control of their generals, they may seem perhaps, in this respect, not to have acted foolishly, for troops of this sort cannot easily unite together in factious counsels. But when we take another view of the question, the wisdom of the proceeding may be doubted, if we consider the difficulty there is to instruct, soften, and subdue the minds of an army so brought together when rage has seized them, and when hatred and resentment have taken root among them, and sedition is actually begun. In such circumstances, they are no longer men, but beasts of prey. Their fury cannot be restricted within the ordinary bounds of human wickedness or violence, but breaks out into deeds the most terrible and monstrous that are to be found in nature.

> Polybius, *Histories*, 1.67, c. 125 BC (*Familiar Quotations from Greek Authors*, tr. Crauford Taut Ramage, 1895).

If any one supports his state by the arms of mercenaries, he will never stand firm or sure, as they are disunited, ambitious, without discipline, faithless, bold amongst friends, cowardly amongst enemies, they have no fear of God, and keep no faith with men. Ruin is only deferred as long as the assault is postponed; in peace you are despoiled by them, and in war by the enemy. The cause of this is that they have no love or other motive to keep them in the field beyond a trifling wage, which is not enough to make them ready to die for you. They are quite willing to be your

MILITARISM

soldiers so long as you do not make war.
> Niccolò Machiavelli, *The Prince*, 1513.

I roared, loved and romped,
And what I lauded most was sin...
Seeking common whores,
Vagabonding, picking quarrels, cursing,
Drinking away money and blood,
Everything was splendidly good.
> Georg Greflinger, a German mercenary and later scholar, quoted in Langer, *The Thirty Years War*, 1980.

Let us therefore animate and encourage each other, and show the whole world that a Freeman contending for *Liberty* on his own ground is superior to any slavish mercenary on earth.
> General George Washington, 2 July 1776, General Order to the Continental Army.

The Greeks in the service of the Great King were not enthusiastic in his cause. The Swiss in French, Spanish, and Italian service were not enthusiastic in their causes. The troops of Frederick the great, mostly foreigners, were not enthusiastic in his cause. A good general, good training, and good discipline make good troops independently of the cause in which they fight. It is true, however, that fanaticism, love of fatherland, and national glory can inspire fresh troops to good advantage.
> Napoleon (1769–1821), quoted in Christopher Herold, ed., *The Mind of Napoleon*, 1955.

Better to lose one's life for the Fatherland, than hang a foreign tassel to one's sword.
> Konstantinos Rhigas' immortal war song, quoted in Kolokotrones, *Theodoros Kolokotrones*, 1892; referring to the fact that many Greeks had joined foreign armies while Greece suffered under the Turkocratia.

MILITARISM

Those who enjoy militarism, however, will perish; and those who are ambitious for victory will be disgraced. War is not something to enjoy, victory is not an object of ambition.
> Sun Bin, *The Lost Art of War*, c. 350 BC, tr. Thomas Cleary.

Prussia was hatched from a cannon ball.
> Napoleon (1769–1821).

When I come in contact with a militarist, his stupidity depresses me and makes me realize the amount of human obtuseness that has to be overcome before we can make much progress towards peace.
> Captain Sir Basil Liddell Hart, *Thoughts on War*, 1944.

MILITARY EDUCATION

The Lacadaemonians made war their chief study. They are affirmed to be the first who reasoned on the events of battles and committed their observations thereon to writing with such success as to reduce the military art, before considered totally dependent on courage or fortune, to certain rules and fixed principles.
> Flavius Vegetius Renatus, *Military Institutions of the Romans*, c. AD 378, tr Clark, 1776.

War is not an affair of chance. A great deal of knowledge, study, and meditation is necessary to conduct it well.
> Frederick the Great, *Instructions to His Generals*, 1747, tr. Phillips, 1940.

Every art has its rules and maxims. One must study them: theory facilitates practice. The lifetime of one man is not long enough to enable him to acquire perfect knowledge and experience. Theory helps to supplement it, it provides a youth with premature experience and makes him skillful also through the mistakes of others. In the profession of war the rules of the art never are violated without drawing

MILITARY EDUCATION

punishment from the enemy, who is delighted to find us at fault. An officer can spare himself many mistakes by improving himself. We even venture to say that he must do it, because the mistakes that he commits through ignorance cover him with shame, and even in praising his courage one cannot refrain from blaming his stupidity.

What an incentive to work hard! What reasons to travel the thorny road to glory! And what loftier and nobler compensation is there than to have one's name immortalized for his pains, and by his work.

Frederick the Great, 'Avant-propos', 1771, *Oeuvres*, Vol. XXIX, 1846–1856, in Jay Luvaas, ed., *Frederick the Great on the Art of War*, 1966.

None other than a Gentleman, as well as seaman, both in Theory and practice is qualified to support the Character of a Commissioned Officer in the Navy, nor is any Man fit to Command a Ship of War, who is not also capable of communicating his Ideas on Paper in Language that becomes his Rank.

Admiral John Paul Jones (1747–1792).

Gneisenau, if I had only learned something, what might not have been made of me! …But I put off everything I should have learned. Instead of studying, I have given myself to gambling, drink, and women; I have hunted, and perpetrated all sorts of foolish pranks. That's why I know nothing now. Yes, the other way I would have become a different kind of fellow, believe me.

Field Marshal Prince Gebhard von Blücher, to Gneisenau, his chief of staff, during the 1813 campaign.

Correct theories, founded upon right principles, sustained by actual events of wars, and added to accurate military history, will form a true school of instruction for generals. If these means do not produce great men, they will at least produce generals of sufficient skill to take rank next after the natural masters of the art of war.

Lieutenant-General Antoine-Henri Baron de Jomini, *Summary of the Art of War*, 1838, tr. Mendell and Craighill, 1862.

It is necessary that the study of the military sciences should be encouraged and rewarded as well as courage and zeal. The military corps should be esteemed and honored; this is the only way of securing for the army men of merit and genius.

Lieutenant General Antoine-Henri Baron de Jomini, *Summary of the Art of War*, 1838, tr. Mendell and Craighill, 1862.

The close of the siege of Vicksburg found us with an army unsurpassed, in proportion to its numbers, taken as a whole of officers and men. A military education was acquired which no other school could have given. Men who thought a company was quite enough for them to command properly at the beginning, would have made good regimental or brigade commanders; most of brigade commanders were equal to the command of a division, and one, Ransom, would have been equal to the command of a corps at least. Logan and Crocker ended the campaign fitted to command independent armies.

General of the Army Ulysses S. Grant, *Personal Memoirs of U.S. Grant*, 1885.

It is evident that a theoretical knowledge is not sufficient to that end; there must be a free, practical, artistic development of the qualities of mind and character, resting of course on a previous military education and guided by experience, whether it be the experience of military history or actual experience in war.

Field Marshal Helmuth Graf von Moltke, quoted in Ferdinand Foch, *Principles of War*, 1913.

MILITARY EDUCATION

I hope the officers of her Majesty's army may never degenerate into bookworms. There is happily at present no tendency in that direction, for I am glad to say that this generation is as fond of danger, adventure, and all manly out-of-door sports as its forefathers were. At the same time, all now recognize that the officer who has not studied war as an applied science, and who is ignorant of modern military history, is of little use beyond the rank of captain.
> Field Marshal Viscount Wolseley, Preface to the English edition of Colmar von der Goltz, *The Conduct of War*, 1897.

Not to promote war but to preserve peace.
> Elihu Root, 1902, motto of the US Army War College, Carlisle Barracks, Pennsylvania.

There can be no better school for a commander of a regiment, or a brigade, or a division than the work of educating the platoon commander... the best training of all is received by a teacher in teaching his pupils.
> Leon Trotksy, 1922, *Military Writings of Leon Trotsky*, 1969.

To-day we meet together as students, and it is only through mutual loyalty to each other that we shall profit by the work which lies before us... As the director of your studies, the ideal I intend to aim at is that we shall teach each other; first, because we all have a vast amount of war experience behind us, and secondly, because, in my opinion, it is only through free criticism of each other's ideas that truth can be thrashed out. Mere swallowing of either food or opinions does not of necessity carry with it digestion, and without digestion swallowing is but labour lost and food wasted... During your course here no one is going to compel you to work, for the simple reason that a man who requires to be driven is not worth the driving... thus you will become your own masters, and until you learn how to teach yourselves, you will never be taught by others.
> Major-General J.F.C. Fuller, from his first lecture at Camberley staff college, 1923, quoted in *Memoirs of an Unconventional Soldier*, 1936.

Though the military art is essentially a practical one, the opportunities of practicing it are rare. Even the largest-scale peace maneuvers are only a feeble shadow of the real thing. So that a soldier desirous of acquiring skill in handling troops is forced to theoretical study of Great Captains.
> Field Marshal Viscount Wavell of Cyrenaica, c. 1930, lecture to officers at Aldershot.

For my strategy, I could find no teachers in the field: but behind me there were some years of military reading, and even in the little that I have written about it, you may be able to trace the allusions and quotations, the conscious analogies.
> Colonel T.E. Lawrence (1888–1935).

This is just not another training pamphlet; it is a magazine and like all good magazines it will be interesting, stimulating and, I hope, at times amusing. In it you will find current military thought, tips on training, and the lessons of war illustrated by experience in battle.

You will be the authors of the articles; you will contribute the ideas and suggestions that will make alive your training and your leadership. We have all got a lot to learn and we have all got something which, out of our own experience and study, we can teach. This magazine is to enable us to share the results of that experience and that study. I want every officer and NCO to read the British Army Journal and I want a lot of you to contribute to it.

MILITARY INTELLIGENCE

Field Marshal Viscount Slim of Burma, 1949, Foreword to the first edition of the *British Army Journal*, later renamed *British Army Review*.

Fundamental to the training of senior officers will be the most comprehensive instruction possible in technical and organizational matters. The object of this instruction will be to induce a certain independence of mind, so that particular value will need to be laid on teaching officers to think critically on questions of basic principle. Respect for the opinion of this or that great soldier must never be allowed to go so far that nobody dares to discuss it. A sure sense of reality must be aroused. Given a well-founded knowledge of basic principles, any man of reasonably cool and logical mind can work out most of the principles for himself, provided he is not inhibited in his thinking.

Field Marshal Erwin Rommel, *The Rommel Papers*, 1953.

It is sometimes thought that when an officer is promoted to the next higher command, he needs no teaching in how to handle it. This is a great mistake. There is a tremendous difference between a brigade and a division, between a division and a corps; when an officer got promotion, he needed help and advice in his new job and it was up to me to see that he got it.

Field Marshal Viscount Montgomery of Alamein, *The Memoirs of Field Marshal Montgomery*, 1958.

Officers, particularly those in positions of command, must at all times be urged to expand the scope of their knowledge; nothing has a more damaging effect on the quality of the army than a hard core of commanders whose minds are narrow and inflexible. Educated, enlightened commanders will produce soldiers equally enlightened and thirsty for knowledge, soldiers who will understand why they were recruited and for what they must fight.

General Yigal Allon, *The Making of Israel's Army*, 1970.

I was always being asked by the Navy brass what a destroyer skipper needs to know about Immanuel Kant; a liberally educated person meets new ideas with curiosity and fascination. An illiberally educated person meets new ideas with fear.

Admiral James B. Stockdale, *Newsweek*, 1 September 1980.

MILITARY INTELLIGENCE

It is essential to know the character of the enemy and of their principal officers – whether they be rash or cautious, enterprising or timid, whether they fight on principle or from chance.

Flavius Vegetius Renatus, *The Military Institutions of the Romans*, c. AD 378, tr. Clark, 1776.

Nothing is more worthy of the attention of a good general than the endeavour to penetrate the designs of the enemy.

Niccolò Machiavelli, *Discourses*, xviii, 1531.

…That no war can be conducted successfully without early and good intelligence, and that such advices cannot be had but at very great expense.

The Duke of Marlborough (1650–1722), quoted in S.J. Watson, *By Command of the Emperor*, 1957.

Great advantage is drawn from knowledge of your adversary, and when you know the measure of his intelligence and character you can use it to play on his weaknesses.

Frederick the Great, *Instructions to His Generals*, 1747, tr. Phillips, 1940.

Have good spies; get to know everything that happens among your

enemies; sow dissension in their midst. To crush tyranny every method is fair.
  Lazare Carnot, 1793, message to General Pichegru.

The most difficult thing is to discern the enemy's plans, and to detect the truth in all reports one receives: the remainder requires only common sense.
  Napoleon (1769–1821), in S.J. Watson, *By Command of the Emperor*, 1957.

To get information, it is necessary to seize the letters in the postal system, to question travelers. In one word, you have to look for it. Intelligence never comes by itself.
  Napoleon (1769–1821), quoted in Chuqet, *Inédits Napoléoniens*, 1913.

To guess at the intention of the enemy; to divine his opinion of yourself; to hide from him both your intentions and opinion; to mislead him by feigned manoeuvres; to invoke ruses, as well as digested schemes, so as to fight under the best conditions – this is and always was the art of war.
  Napoleon, quoted in J.F.C. Fuller, *Memoirs of an Unconventional Soldier*, 1936.

By 'intelligence' we mean every sort of information about the enemy and his country – the basis, in short, of our own plans and operations. If we consider the actual basis of this information, how unreliable and transient it is, we soon realize that war is a flimsy structure that can easily collapse and bury us in its ruins. The textbooks agree, of course, that we should only believe reliable intelligence, and should never cease to be suspicious, but what is the use of such feeble maxims? They belong to that wisdom which for want of anything better scribblers of systems and compendia resort to when they run out of ideas.
  Major-General Carl von Clausewitz, *On War*, 1.6, 1832, tr. Michael Howard and Peter Paret, 1976.

How can any man say what he should do himself if he is ignorant what his adversary is about?
  Lieutenant-General antoine-Henri Baron de Jomini, *Summary of the Art of War*, 1838, tr. Mendell and Craighill, 1862.

Hide what you have and reveal what you haven't.
  J.F.C. Fuller, *A Military History of the Western World*, 3.13, 1956.

Without Combat Intelligence a commander has no right to issue a combat order.
  Anonymous US Army G2 (intelligence officer at division or higher echelon) in WWII, quoted in Stedman Chandler and Robert Robb, ed, *Front Line Intelligence*, 1946.

Expect only five per cent of an intelligence report to be accurate. The trick of a good commander is to isolate the five per cent.
  General of the Army Douglas MacArthur, quoted in Courtney Whitney, *MacArthur*, 1956.

Without an efficient Intelligence organization a commander is largely blind and deaf.
  Field Marshal Earl Alexander of Tunis, *Memoirs*, 1962.

Americans have always had an ambivalent attitude toward intelligence. When they feel threatened, they want a lot of it, and when they don't, they regard the whole thing as somewhat immoral.
  General Vernon A. Walters, *Silent Missions*, 1978.

MILITARY JUSTICE

The commander should be severe and thorough in investigating charges against his men, but merciful in punishing them. This will gain him their good will.
  The Emperor Maurice, *The Strategikon*, c.

MILITARY LIFE

AD 600 (*Maurice's Strategikon*, tr. George Dennis, 1984).

Come here, my son. Better that I punish thee, than that God, for thy sin, visit vengeance on me and the whole army.
    Gustavus II Adolphus (1594–1632), to a petty officer whom he had grasped by the hair and was handing over to the executioners, after discovering that the man had stolen a cow, quoted in Theodore Ayrault Dodge, *Gustavus Adolphus*, 1895.

First, it is very well known to all men of experience and judgment in matters of arms that all such great captains as have been lieutenants general to emperors, kings, or formed commonwealths, or that with regiments of their own nation have served foreign princes as mercenaries, knowing that justice is the prince of all order and government both in war and peace, by which God is honored and served and magistrates and officers obeyed, have at the first forming of their armies or such regiments, by great advice of counsel, established sundry laws both politic and martial, with officers for the superintending and due execution of the same. These laws have been notified to all their men of war, as also, at every encamping or lodging, they have been set, written, or printed in certain tables in convenient places for all soldiers and men of war to behold, to the intent that none might transgress the same through ignorance.
    John Smythe (c. 1580–1631), *Certain Discourses Military*, 1966.

He [the general] should have a good disposition free from caprice and be a stranger to hatred. He should punish without mercy, especially those who are dearest to him, but never from anger. He should always be grieved when he is forced to execute the military rules and should have the example of Manlius constantly before his eyes. He should discard the idea that it is he who punishes and should persuade himself and others that he only administers the military laws. With these qualities, he will be loved, he will be feared, and, without doubt, obeyed.
    Field Marshal Maurice Comte de Saxe, *My Reveries*, 1732, tr. Phillips, 1940.

Measured military punishment, together with a short and clear explanation of the offense, touches the ambitious soldier more than brutality which drives him to despair.
    Field Marshal Prince Aleksandr V. Suvorov (1729–1800), quoted in Philip Longworth, *The Art of Victory*, 1965.

One of the great defects in our military establishment is the giving of weak sentences for military offenses. The purpose of military law is administrative rather than legal. As the French say, sentences are for the purpose of encouraging the others. I am convinced that, in justice to other men, soldiers who go to sleep on post, who go absent for an unreasonable time during combat, who shirk in battle, should be executed; and that Army Commanders or Corps Commanders should have the authority to approve the death sentence. It is utterly stupid to say that General officers, as a result of whose orders thousands of gallant and brave men have been killed, are not capable of knowing how to remove the life of one miserable poltroon.
    General George S. Patton, Jr., *War As I Knew It*, 1947.

MILITARY LIFE

That variety incident to a military life gives our profession some advantage over those of a more even and consistent nature. We have all our passions and affections roused and exercised, many of which must have wanted their proper employment, had not suitable

occasions obliged us to exert them. Few men are acquainted with the degree of their own courage till danger prove them, and are seldom justly informed how far the love of honour and the dread of shame are superior to the love of life... Constancy of temper, patience, and all the virtues necessary to make us suffer with a good grace are likewise parts of our character, and... frequently called in to carry us through unusual difficulties...
> Major-General Sir James Wolfe (1727–1759), quoted in Liddell Hart, *Great Captains Unveiled*, 1927.

I have met many officers of the garrison who were with me in Mexico. You have often heard me say the cordiality and friendship in the army was the great attraction of the service. It is that, I believe, that has has kept me so long, and it is that which now makes me fear to leave it. I do not know where I should meet with so much friendship out of it.
> General Robert E. Lee, letter from Mobile, Alabama, 1849, quoted in Fitzhugh Lee, *General Lee*, 1894, pp. 43–85.

Well, never leave the Corps until you are absolutely unable to continue. It's the greatest life there is. I miss it terribly.
> Lieutenant-General Lewis B. 'Chesty' Puller, 7 February 1962, comment after retirement to visiting Marine NCOs, Marine Corps press release.

Men have joined armed forces at different times for different reasons. I do not see many young men joining for philosophical reasons. Almost always the desire for an active life has been prominent among reasons for taking up the profession of arms, but there have usually been contributory motives. These have often been ephemeral in value, and in kind accidental rather than essential.
> General Sir John Hackett, *The Profession of Arms*, 1963.

MILITARY MUSIC

Noble and manly music invigorates the spirit, strengthens the wavering man, and incites him to great and worthy deeds.
> Homer, *The Iliad*, c. 800 BC, tr. Fagles, 1990.

After this they joined battle, the Argives and their allies advancing with haste and fury, the Spartans slowly and to the music of many flute players – a standing institution in their army, that has nothing to do with religion, but is meant to make them advance evenly, stepping in time, without breaking their order, as large armies are apt to do in the moment of engagement.
> Thucydides, *The History of the Peloponnesian War*, 5.70, c. 404 BC, tr. Richard Crowley, 1910; of the Spartans at the Battle of Mantinea, summer 418 BC.

I don't believe we can have an army without music.
> General Robert E. Lee, quoted in Burke Davis, *Gray Fox*, 1956.

They would march away in the dark, singing to the beat of drums.
> Siegfried Sassoon, *Memoirs of a Fox-Hunting Man*, 1928.

A few days later Colonel Sims came to see me at the command post. Accompanying him was the 7th Marines band leader, who reported his casualties in the following manner: 'I lost my first trumpet, second trombone, one drum, and second chair in the clarinet,' or words to that effect. Like all bandsmen he was a musician at all times and at all places.
> General Merrill B. Twining, *No Bended Knee*, 1994, referring to the losses suffered by the band of the Seventh Marines as it was pressed into battle on Guadalcanal in 1943.

MILITARY SERVICE

MILITARY SCIENCE

War is a science replete with shadows in whose obscurity one cannot move with assured step. Routine and prejudice, the natural result of ignorance, are its foundation and support. All sciences have principles and rules. War has none. The great captains who have written of it give us none. Extreme cleverness is required merely to understand them.
>Field Marshal Maurice Comte de Saxe, *My Reveries*, 1732, tr. Phillips, 1940.

War is a science for those who are outstanding; an art for mediocrities; a trade for ignoramuses.
>Frederick the Great (1712–1786).

...The science of winning...
>Field Marshal Prince Aleksandr V. Suvorov (1729–1800).

Military science consists in first calculating all the possibilities accurately and then in making an almost mathematically exact allowance for accident. It is on this point that one must make no mistake; a decimal more or less may alter everything. Now, this apportioning of knowledge and accident can take place only in the head of a genius, for without it there can be no creation – and surely the greatest improvisation of the human mind is that which gives existence to the nonexistent. Accident thus always remains a mystery to mediocre minds and becomes reality for superior men.
>Napoleon, conversation in the early 1800s, quoted in Christopher Herold, ed., *The Mind of Napoleon*, 1955.

Nothing is better calculated to kill natural genius and to cause error to triumph, than those pedantic theories, based upon the false idea that war is a positive science, all the operations of which can be reduced to infallible calculations.
>Lieutenant-General antoine-Henri Baron de Jomini, *Summary of the Art of War*, 1838, tr. Mendell and Craighill, 1862.

It [strategy] is more than science, it is the translation of science into practical life, the further development of the original governing idea, in accordance with the ever-changing circumstances; it is the art of taking action under the pressure of the most trying conditions.
>Field Marshal Helmuth Graf von Moltke, *On Strategy*, 1871.

The Marxian method is a method of historical and social science. There is no 'science' of war, and there never will be any. There are many sciences war is concerned with. But war itself is not a science; war is practical art and skill. How could it be possible to shape principles of military art with the help of Marxian method? It is as impossible as it is impossible to create a theory of architecture or to write a veterinary textbook with the help of Marxism.
>Leon Trotsky, *How the Revolution Developed its Military Power* (Moscow, 1924).

A study of military history brings ample confirmation of Rebecca West's *mot*: 'Before a war military science seems a real science, like astronomy, but after a war it seems more like astrology.'
>Captain Sir Basil Liddell Hart, *Thoughts on War*, 1944.

MILITARY SERVICE

For there is justice in the claim that steadfastness in his country's battles should be as a cloak to cover a man's other imperfections; since the good action has blotted out the bad, and his merit as a citizen more than outweighed his demerits as an individual.
>Pericles, 431/430 BC, funeral speech for the Athenian dead, quoted in Thucydides, *The Peloponnesian War*, 2.42, c. 404 BC, tr. Richard Crawley, 1910.

MISTAKES

However long you live and whatever you accomplish, you will find that the time you spent in the Confederate army was the most profitably spent portion of your life. Never again speak of having lost time in the army!
  General Robert E. Lee, quoted in Douglas Southall Freeman, *R.E. Lee*, IV, 1934.

It is well, that not only the nation pay this great tribute of respect and gratitude once every year to those who fell in defence of its liberties, – but that those who struggled in the same noble cause, and survived, should meet, and in some manner, live over again the scenes which make, and forever must make up, to them the most important era of their lives.

For there is no true loyal soldier to-day, who served his term of enlistment in the war of the Rebellion, who, if asked for some portion of his past life to be taken out of his record and remembrance, but would say – 'Take whatever three or four years of my existence you will, but leave the old army life untouched, – I did in those days what you never did, and I can never do again – leave that to me.'
  Clara Barton, address on Memorial Day, 30 May 1869, quoted in Percy Epler, *The Life of Clara Barton*, 1915.

If we persuade intelligent youth to hold aloof from the Army in peace, we ought not to complain if we are not properly led in war.
  Lord Moran, *The Anatomy of Courage*, 1945.

The majority of men, so long as they are treated fairly and feel that good use is being made of their powers, will rejoice in a new sense of unity with new companions even more than they will mind the increased separation from their old associations.
  Brigadier-General S.L.A. Marshall, *The Armed Forces Officer*, 1950.

I'm not sure, but I rather think that the military service is a profession more personal and more emotionally charged than most of the others: a professional soldier, sailor, or airman, though he may on some occasions be disgusted with his current job – and his job changes constantly – is more inclined to be sentimentally inclined in respect to his career than would say, a lawyer or a shopkeeper. I don't think lawyers, for instance, have songs about regimental exploits – good, bad, and ridiculous – or about the virtues of dying on the field of honor, or, when they die, going to some special place 'halfway down to hell' where there are a lot of good horses and whiskey. At least I hope they don't.
  General Hamilton H. Howze, *A Cavalryman's Story*, 1996.

MISTAKES

In war there is never any chance for a second mistake.
  Lamachus (465–414 BC), quoted in Plutarch, *Apothegms*.

I am more afraid of our own blunders than of the enemy's devices.
  Pericles, 432 BC, speech to the Athenians encouraging a firm policy against the Spartans, quoted in Thucydides, *History of the Peloponnesian War*, 1.144, c. 404 BC, tr. Richard Crawley, 1910.

Mistakes made in ordinary affairs can generally be remedied in a short while, but errors made in war cause lasting harm.
  The Emperor Maurice, *The Strategikon*, c. AD 600 (*Maurice's Strategikon*, tr. George Dennis, 1984).

When a man has committed no faults in war, he can only have been engaged in it but a short time.
  Marshal of France Vicomte de Turenne, 1645, after the Battle of Marienthal.

Gentlemen, when the enemy is com-

MISTAKES

mitted to a mistake, we must not interrupt him too soon.
> Admiral Viscount Nelson, quoted in Alfred Thayer Mahan, *The Life of Nelson*, 1897.

The greatest general is he who makes the fewest mistakes.
> Napoleon (1769–1821).

If I am to be hanged for it, I cannot accuse a man who I believe has meant well and whose error was one of judgement and not intention. Although my errors, and those of others also, are visited heavily upon me, that is not the way in which any, much less a British army, can be commanded.
> The Duke of Wellington, 1810, of General Craufurd, *Supplementary Despatches, Correspondence and Memoranda of Field Marshal Arthur Duke of Wellington, K.G., 1794–1818*, VI, 1860, p. 564.

After it is all over, as stupid a fellow as I am can see the mistakes that were made. I notice, however, that my mistakes are never told me until it is too late, and you, and all my officers, know that I am always ready and anxious to have their suggestions.
> General Robert E. Lee, in remarks to Major-General Henry Heth, quoted in Early, ed., *Southern Historical Society Papers*, 153, 159–160, cited in Douglas Southall Freeman, *R.E. Lee, A Biography*, 1932.

I draw a distinction between mistakes. There is the mistake which comes through daring – what I call a mistake towards the enemy – in which you must sustain *your* commanders... There are mistakes from the safety-first principle – mistakes of turning away from the enemy; and they require a far more acid consideration.
> Sir Winston S. Churchill (1874–1965).

My God, Senator, that's the reason we do it. I want the mistake down in Louisiana, not over in Europe, and the only way to do this thing is try it out, and if it doesn't work, find out what we need to make it work.
> General of the Army George C. Marshall, to a senator who complained of the number of mistakes made in the Louisiana manoeuvres of 1940, quoted in Forrest Pogue, *George Marshall: Ordeal and Hope: 1939–1942*, 1965.

Nothing is easy in war. Mistakes are always paid for in casualties and troops are quick to sense any blunder made by their commanders.
> General of the Army Dwight D. Eisenhower, *Crusade in Europe*, 1948.

Victory in battle – save where it is brought about by sheer weight of numbers, and omitting all question of the courage of the troops engaged – never comes solely as the result of the victor's planning. It is not only the merits of the victor that decide the issue, but also mistakes on the part of the vanquished.
> Field Marshal Erwin Rommel, *The Rommel Papers*, 1953.

The training of armies is primarily devoted to developing efficiency in the detailed execution of the attack. This concentration on tactical technique tends to obscure the psychological element. It fosters a cult of soundness, rather than surprise. It breeds commanders who are so intent not to do anything wrong, according to 'the book', that they forget the necessity of making the enemy do something wrong. The result is that their plans have no result. For, in war, it is by compelling mistakes that the scales are most often turned.
> Captain Sir Basil Liddell Hart, *Strategy*, 1954.

Happily for the result of the battle – and for me – I was, like other generals before me, to be saved from the

MOBILISATION

consequences of my mistakes by the resourcefulness of my subordinate commanders and the stubborn valour of my troops.
Field Marshal Viscount Slim of Burma, Defeat Into Victory, 1956.

To inquire if and where we made mistakes is not to apologize. War is replete with mistakes because it is full of improvisations. In war we are always doing something for the first time. It would be a miracle if what we improvized under the stress of war should be perfect.
Vice Admiral Hyman G. Rickover, April 1964, testimony before the House Military Appropriations Subcommittee.

You must be able to underwrite the honest mistakes of your subordinates if you wish to develop their initiative and experience.
Lieutenant-General Bruce C. Clarke (1901–1988).

MOBILISATION

The young men shall fight; the married men shall forge weapons and transport supplies; the women will make tents and serve in the hospitals; the children will make up old linen into lint; the old men will have themselves carried into the public squares to rouse the courage of the fighting men, and to preach hatred of kings and the unity of the Republic. The public buildings shall be turned into barracks, the public squares into munitions factories; the earthen floors of cellars shall be treated with lye to extract saltpetre. All suitable firearms shall be turned over to the troops; the interior shall be policed with fowling pieces and with cold steel. All saddle horses shall be seized for the cavalry; all draft horses not employed in cultivation will draw the artillery and supply wagons.
Decree of the Committee of Public Safety, France, 23 August 1793.

When a nation is without establishments and a military system, it is very difficult to organize an army.
Napoleon, Maxims of War, 1831.

In making an army, three elements are necessary – men, weapons and money. There must also be time.
Sir Winston S. Churchill (1874–1965).

Mobilization will inevitably mean war.
German ultimatum to France, 1 August 1914, quoted in Captain Sir Basil Liddell Hart, The Real War, 1930.

The advance of armies formed of millions of men …was the result of years of painstaking work. Once planned, it could not possibly be changed.
General Count Helmuth von Moltke 'The Younger', 2 August 1914, to Kaiser Wilhelm's suggestion that the mobilisation to attack France be abandoned to concentrate instead on Russia, quoted in Captain Sir Basil Liddell Hart, The Real War, 1930.

Everyone will now be mobilized and all boys old enough to carry a spear will be sent to Addis Ababa. Married men will take their wives to carry food and cook. Those without wives will take any woman without a husband. Women with small babies need not go. The blind, those who cannot walk or for any reason cannot carry a spear are exempted. Anyone found at home after receipt of this order will be hanged.
Emperor Haile Selassie's mobilisation order issued after the Italian invasion of Ethiopia in 1935.

I want to make it clear that it is the purpose of the nation to build now with all possible speed every machine, every arsenal, every factory that we need to manufacture our defense material. We have the men – the skill – the wealth – and above all the will.

MOBILTY/MOVEMENT

President Franklin D. Roosevelt, 29 December 1940, 'Fireside Chat' on national security.

The Army used to have all the time in the world and no money; now we've got all the money and no time.
General of the Army George C. Marshall, January 1942, remark on the mobilisation for WWII.

Looking back on those days of mounting tension, I can now discern more clearly the subtleties of our gradual transition from peacetime group of ships brought together for an exercise, to a battle group which was actually going to have to fight, to wage war, and thus accept damage, loss of ships, and loss of lives. Commands became more terse at all levels. They say the first casualty of war is always truth. In our case, I believe it was politeness: 'Get that done right now!' 'Don't just stand there, do it!' Men were more on edge; tasks that once had seemed insignificant now appeared critical. The full range of Navy reasoning and habit which so often in peacetime had appeared petty, pedantic and even churlish, no longer seemed so. Reasons which had once seemed obscure came into sharp focus. It quite remarkable how the prospect of a possible early demise can bring out the best in everyone.
Admiral Sir J.F. 'Sandy' Woodward, *One Hundred Days: The Memoirs of the Falklands Battle Group Commander*, 1992, p.116.

MOBILITY/MOVEMENT

Aptitude for war is aptitude for movement.
Napoleon, *Maxims of War*, 1831.

Force does not exist for mobility, but mobility for force.
Rear Admiral Alfred Thayer Mahan, *Lessons of the War with Spain*, 1899.

Success in war depends upon mobility and mobility upon time. Mobility leads to mass, to surprise and to security. Other things being equal, the most mobile side must win: this is a truism in war as in horse-racing. The tank first of all is a time-saving machine, secondly a shield – it is, in fact, a mechanical horse. If in a given time we can do three times as much as the enemy and lose a third less than he does, our possibilities of success are multiplied by nine.
Major-General J.F.C. Fuller, memorandum of 1917, *Memoirs of an Unconventional Soldier*, 1936.

Movement is the safety-valve of fear; unless we wish to risk a sudden collapse of the man's nerve control, we must open the valve. (July 1921.)
Captain Sir Basil Liddell Hart, *Thoughts on War*, 1944.

We need bold and free flight, we need mobility.
Mikhail Frunze, 1921.

Two things stop the movement of armies: (a) bullets and fragments of shell which destroy the motive power of men, and (b) the confusion of the conflict.
Sir Winston S. Churchill, *The World Crisis*, 1923.

The Principle of Movement. If concentration of weapon-power be compared to a projectile and economy of force to its line of fire, then movement may be looked upon as the propellant and as a propellant is not always in a state of explosive energy, so neither is movement. Movement is the power of endowing mass with momentum; it depends, therefore, largely on security, which, when coupled with offensive power, results in liberty of action. Movement consequently, may be potential as well as dynamic, and, if an army be compared to a machine the power of which is supplied to it by a

MOMENTUM

series of accumulators, should the object of its commander be to maintain movement, he can only accomplish this by refilling one set of accumulators while the other is in the process of being exhausted. The shorter the time available to do this the more difficult will the commander's task be; consequently, one of his most important duties, throughout war, is to increase the motive power of his troops, which depends on two main factors – moral and physical endurance.
Major-General J.F.C. Fuller, *The Reformation of War*, 1923.

It is worth recalling that the Mongols, although their army was entirely composed of mobile troops, found neither the Himalayas nor the far-stretching Carpathians a barrier to progress. For mobile troops there is usually a way around. (Sept. 1927.)
Captain Sir Basil Liddell Hart, *Thoughts on War*, 1944.

MOMENTUM

In battle, momentum means riding on the force of the tide of events. If enemies are on the way to destruction, then you follow up and press them; their armies will surely collapse.
  The rule is 'Use the force of momentum to defeat them...'
Liu Ji (1310–1375), *Lessons of War* (*Mastering the Art of War*, tr. Thomas Cleary, 1989).

Never pull up during an attack.
Field Marshal Prince Aleksandr V. Suvorov, *The Science of Victory*, 1796.

Maxim 9. The strength of an army, like the power in mechanics, is estimated by multiplying the mass by the rapidity; a rapid march augments the morale of an army, and increases all the chances of victory.
Napoleon, *The Military Maxims of Napoleon*, tr. George D'Aguilar, 1831 (ed. David Chandler, 1987).

Two fundamental lessons of war experience are – never to check momentum; never to resume mere pushing.
Captain Sir Basil Liddell Hart, *Thoughts on War*, 1944.

MORAL ASCENDANCY/ MORAL FORCE

Move forward yourself with the cavalry and one battalion, and dash at the first enemy that comes into your neighborhood. You will either cut them up or drive them off... A long defensive war will ruin us... Dash at the first fellows that make their appearance and the campaign will be your own.
The Duke of Wellington, 1800, in India, *The Dispatches of Field Marshal the Duke of Welllington, During His Various Campaigns in India, Denmark, Portugal, Spain, the Low Countries, and France*, II, 1834–1838, pp. 208, 210, 219.

We must get the upper hand, and, if once we have that, we shall keep it with ease and shall certainly succeed.
The Duke of Wellington, 17 August 1803, to Lieutenant-Colonel Close, quoted in Arthur Bryant, *The Great Duke*, 1972.

...They [the British soldiers] have obtained by their spirit and enterprise an ascendency over the French which nothing but great superiority of numbers on their part can get the better of.
General Sir John Moore, 28 December 1808, to Castlereagh, from Benavente, quoted in Charles Oman, *History of the Peninsular War*, I, 1902.

Marching on Landshut I met Bessières retreating. I ordered him to march forward. He objected that the enemy were in force. – Go ahead, – said I, and he advanced. The enemy seeing him take the offensive thought he was stronger

MORAL ASCENDENCE/MORAL FORCE

than they and retreated. In war that is the way everything goes. It is moral force more than numbers that wins victory.
  Napoleon, 14 June 1817, on St Helena, quoted in R.M. Johnston, ed., *The Corsican*, 1910.

The moral elements are among the most important in war. They constitute the spirit that permeates war as a whole, and at an early stage they establish a close affinity with the will that moves and leads the whole mass of force, practically merging with it, since the will is itself a moral quantity. Unfortunately they will not yield to academic wisdom. They cannot be classified or counted. They have to be seen or felt.
  Major General Carl von Clausewitz, *On War*, 3.3, 1832, tr. Michael Howard and Peter Paret, 1976.

In battle, two moral forces, even more than two material forces, are in conflict. The stronger conquers.
  Colonel Charles Ardant du Picq, *Battle Studies*, 1880, tr. Greely, 1957.

The Army of the Tennessee had won five successive victories over the garrison of Vicksburg in the three preceding weeks. They had driven a portion of the army from Port Gibson with considerable loss, after having flanked them out of their stronghold at Grand Gulf. They had attacked another portion of the same army at Raymond, more than fifty miles farther in the interior of the State, and driven them back into Jackson with great loss in killed, wounded, captured and missing, besides loss of large and small arms; they had captured the capital of the State of Mississippi, with a large amount of materials of war and manufactures. Only a few days before, they had beaten the enemy then penned up in the town first at Champion's Hill, next at Big Black River Bridge, inflicting upon him a loss of fifteen thousand or more men …besides large losses in arms and ammunition. The Army of the Tennessee had come to believe that they could beat their antagonist under any circumstances.
  General of the Army Ulysses S. Grant, *Personal Memoirs*, 1885.

They claim that to fire human grapeshot at the enemy without preparation, gives us moral ascendancy. But the thousands of dead Frenchmen, lying in front of the German trenches, are instead those who are giving moral ascendancy to the enemy. If this waste of human material keeps on, the day is not far off when the offensive capacity of our army, already seriously weakened, will be entirely destroyed.
  Lieutenant Abel Ferry, 1916, memorandum to the French Cabinet; Ferry was a Deputy in the French Assembly and a reserve officer in the field.

The most solid moral qualities melt away under the effect of modern arms.
  Marshal of France Ferdinand Foch, *Precepts and Judgments*, 1919.

Moral forces may take a back seat at Committees of Imperial Defense or in War Offices; at the front they are where Joab put Uriah.
  General Sir Ian Hamilton, *The Soul and Body of an Army*, 1921.

Moral force is, unhappily, no substitute for armed force, but it is a very great reinforcement.
  Sir Winston S. Churchill, 21 December 1937, speech in the House of Commons.

When men become fearful in combat, the moral incentive can restore them and stimulate them to action. But when they become hopeless, it is because all moral incentive is gone. Soldiers who have ceased to hope are no longer

MORALE

receptive beings. They have become oblivious to all things, large and small.
Brigadier-General S.L.A. Marshall, *The Armed Forces Officer*, 1950.

Soldiers universally concede the general truth of Napoleon's much-quoted dictum that in war 'the moral is to the physical as three to one'. The actual arithmetical proportion may be worthless, for morale is apt to decline if the weapons are inadequate, and the strongest will is of little use if it is inside a dead body. But although the moral and physical factors are inseparable and indivisible, the saying gains its enduring value because it expresses the idea of the predominance of moral factors in all military decisions. On them constantly turns the issue of war and battle. In the history of war, they form the more constant factors, changing only in degree, whereas the physical factors are different in almost every war and every military situation.
Captain Sir Basil Liddell Hart, *Strategy*, 1954.

MORALE

Note: often spelled 'moral' in the 19th and early 20th centuries

You are all aware that it is not numbers or strength that bring the victories in war. No, it is when one side goes against the enemy with the gods' gift of a stronger morale that their adversaries, as a rule, cannot withstand them.
Xenophon, *The Persian Expedition*, 3.1, c. 360 BC, tr. Rex Warner.

He [Chevalier Folard] supposes all men to be brave at all times and does not realize that the courage of the troops must be reborn daily, that nothing is so variable, and that the true skill of the general consists of knowing how to guarantee it.
Field Marshal Maurice Comte de Saxe, *My Reveries*, 1732, tr. Phillips, 1940.

My lord, I never saw better horses, better clothes, finer belts and accoutrements; but money, which you don't want in England, will buy fine clothes and horses, but it can't buy the lively air, I see in everyone of these troopers.
Field Marshal Prince Eugene, June 1740, quoted in David Chandler, *The Art of War in the Age of Marlborough*, 1976.

A battle is lost less through the loss of men than by discouragement.
Frederick the Great, *Instructions to His Generals*, 1747, tr. Phillips, 1940.

In war, everything depends on morale; and morale and public opinion comprise the better part of reality.
Napoleon (1769–1821).

Morale makes up three quarters of the game; the relative balance of man-power accounts only for the remaining quarter.
Napoleon, 27 August 1808, *Correspondance de Napoléon Ier, publiée par ordre de l'Empereur Napoléon III*, XVII, No. 14276, 1858–1870.

One fights well when the heart is light.
Napoleon, 17 February 1816, on St Helena, to General Gaspard Gourgaud, *Sainte Hélène: Journal inédit de 1815 à 1818*, 1899.

Physical casualties are not the only losses incurred by both sides in the course of the engagement: their moral strength is also shaken, broken and ruined. In deciding whether or not to continue the engagement it is not enough to consider the loss of men, horses and guns; one also has to weigh the loss of order, courage, confidence, cohesion, and plan. The decision rests chiefly on the state of morale, which, in cases where the victor has lost as much as the vanquished, has been the single decisive factor.
Major-General Carl von Clausewitz, *On War*, 4.4, 1832, tr. Michael Howard and Peter Paret, 1976.

MORALE

Note the army organizations and tactical formations on paper are always determined from the mechanical point of view, neglecting the essential coefficient, that of morale. They are almost always wrong.
  Colonel Charles Ardant du Picq, *Battle Studies*, 1880, tr. Greely, 1957.

To General Swinton, too, is due the implanting into all ranks of the fundamental idea of the Tank as a weapon for saving the lives of infantry. This idea was indeed the foundation of the moral of the Tank Corps, for it spread from the fighting personnel to the depots and workshops, and even to the factories.

More than anything else, it was this sentiment which kept men ploughing through the mud of 1917, in the dark days when often the chance of reaching an objective had fallen to ten per cent; which kept workshops in full swing all around the clock on ten and eleven hour shifts for weeks and, once, for months on end; which, finally, secured from the factories an intensive and remarkable output.
  Major-General Hugh Elles, Preface to Clough William-Ellis, *The Tank Corps*, 1919.

First of all you must build the morale of your own troops. Then you must look to the morale of your civilian population. Then, and only then, when these are in good repair should you concern yourself with the enemy's morale. And the best way to destroy the enemy's morale is to kill him in large numbers. There's nothing more demoralizing than that.
  Leon Trotsky (1879–1940), quoted in *RUSI*, 9/1984.

The unfailing formula for production of morale is patriotism, self-respect, discipline, and self-confidence within a military unit, joined with fair treatment and merited appreciation from without. It cannot be produced by pampering or coddling an army, and is not necessarily destroyed by hardship, danger, or even calamity. Though it can survive and develop in adversity that comes as an inescapable indicant of service, it will quickly wither of indifference or injustice on the part of their government or of ignorance, personal ambition, or ineptitude on the part of military leaders.
  General of the Army Douglas MacArthur, *Annual Report of the Chief of Staff, US Army for the Fiscal Year Ending 30 June 1933*.

Morale, only morale, individual morale as a foundation under training and discipline, will bring victory.
  Field Marshal Viscount Slim of Burma, June 1941, to the officers of the Tenth Indian Infantry Division.

Morale is a state of mind. It is steadfastness and courage and hope. It is confidence and zeal and loyalty. It is elan, esprit de corps and determination. It is staying power, the spirit which endures to the end – the will to win. With it all things are possible, without it everything else, planning, preparation, production, count for naught.
  General of the Army George C. Marshall, 15 June 1941, address at Trinity College, Hartford, Connecticut.

Machines are nothing without men. Men are nothing without morale.
  Admiral of the Fleet Ernest J. King, 19 June 1942, graduation address to the US Naval Academy.

Morale is the big thing in war. We must raise the morale of our soldiery to the highest pitch; they must be made enthusiastic, and must enter this battle with their tails high in the air and with the will to win.
  Field Marshal Viscount Montgomery of Alamein, 14 September 1942, instruction

MOTHERS

for Operation Lightfoot, the Battle of El Alamein, quoted in Francis de Guingand, *Operation Victory*, 1947.

To win battles you require good Commanders in the senior ranks, and good senior staff officers; all of these must know their stuff.

You also require an Army in which the morale of the troops is right on the top line. The troops must have confidence in their Commanders and must have the light of battle in their eyes; if this is not so you can achieve nothing.
Field Marshal Viscount Montgomery of Alamein, 10 November 1942, quoted in Nigel Hamilton, *Master of the Battlefield*, 1983.

Do not place military cemeteries where they can be seen by replacements marching to the front. This has a very bad effect on morale, even if it adds to the pride of the Graves Registration Service.
General George S. Patton, *War As I Knew It*, 1947.

Very many factors go into the building-up of sound morale in an army, but one of the greatest is that men be fully employed at useful and interesting work.
Sir Winston S. Churchill, *The Gathering Storm*, 1948.

The morale of the soldier is the greatest single factor in war and the best way to achieve a high morale in war-time is by success in battle. The good general is the one who wins his battles with the fewest possible casualties; but morale will remain high even after considerable casualties, provided the battle has been won and the men know it was not wastefully conducted, and that every care has been taken of the wounded, and the killed have been collected and reverently buried.
Field Marshal Viscount Montgomery of Alamein, *The Memoirs of Field Marshal Montgomery*, 1958.

You must watch your own morale carefully. A battle is, in effect, a contest between two wills – your own and that of the enemy general. If your heart begins to fail you when the issue hangs in the balance, your opponent will probably win.
Field Marshal Viscount Montgomery of Alamein, *The Memoirs of Field Marshal Montgomery*, 1958.

Morale is a state of mind. It is that intangible force which will move a whole group of men to give their last ounce to achieve something, without counting the cost to themselves; that makes them feel they are part of something greater than themselves.
Field Marshal Viscount Slim of Burma, *Serve to Lead*, 1959.

The Pentagon Whiz Kids are, I think, conscientious, patriotic people who are experts at calculating odds, figuring cost-effectiveness and squeezing the last cent out of contract negotiations. But they are heavy-handed butchers in dealing with that delicate, vital thing called 'morale.' This is the stuff that makes ships like the *Enterprise*, puts flags on top of Iwo Jima and wins wars. But I doubt if Mr. McNamara and his crew have any morale settings on their computers.
Admiral Daniel V. Gallery, *Eight Bells and All's Well*, 1965.

The Twin Towers were legitimate targets, they were supporting US economic power. These events were great by all measurement. What was destroyed were not only the towers, but the towers of morale in that country.
Osama bin Ladin, undated video, reported on 11 November 2001 by the *Daily Telegraph*.

MOTHERS

So the voice of the king rang out in tears, the citizens wailed in answer, and noble

MOTHERS

Hecuba
led the wives of Troy in a throbbing
   chant of sorrow:
'O my child – my desolation! How can I
   go on living?
What agonies must I suffer now, now
   you are dead and gone?
You were my pride throughout the city
   night and day –
a blessing to us all, the men and women
   of Troy:
Throughout the city they saluted you
   like a god.
You, you were their greatest glory while
   you lived –
now death and fate have seized you,
   dragged you down!'
   Homer, *The Iliad*, 22.504–512, c. 800 BC, tr.
   Robert Fagles, 1990; Hecuba, Queen of
   Troy, mourning her son, Hector.

When asked by a woman from Attica:
'Why are you Spartan women the only
ones who can rule men?', she said:
'Because we are also the only ones who
give birth to men.'
   Gorgo (born about 506 BC), wife of
   Leonidas, King of Sparta, quoted in
   Plutarch, *The Lives*, c. AD 100 (*Plutarch on
   Sparta*, tr. Richard Talbert, 1988).

Another woman, as she handing her
son his shield and giving him some
encouragement, said: 'Son, [return]
either with this or on this.'
   Anonymous Spartan mother, quoted in
   Plutarch, *The Lives*, c. AD 100 (*Plutarch on
   Sparta*, tr. Richard Talbert, 1988).

Dear Madam, I have been shown in the
files of the War Department a statement of the Adjutant-General of
Massachusetts that you are the mother
of five sons who have died gloriously
on the field of battle. I feel how weak
and fruitless must be any words of
mine which should attempt to beguile
you from the grief of a loss so overwhelming. But I cannot refrain from
tendering to you the consolation that
may be found in the thanks of the
Republic they died to save. I pray that
our heavenly Father may assuage the
anguish of your bereavement, and leave
you only the cherished memory of the
loved and lost, and the solemn pride
that must be yours to have laid so costly a sacrifice on the altar of freedom.
   Abraham Lincoln, letter to Mrs Bixby, 21
   November 1864 (Note: the records were in
   error; Mrs. Bixby had lost two sons for the
   Union despite being a Southern
   sympathiser).

Mothers – wives – and maidens, would
there were some testimonials grand
enough for you, – some tablet that
could show to the world the sacrifice of
American womanhood and American
motherhood in that war! Sacrifices so
nobly and so firmly – but so gently and
beautifully made.
   If like the Spartan mother she did
not send her son defiantly to the field, –
bidding him return only with his shield,
or on it, – if like the Roman matron she
did not take him by the hand and lead
him proudly to the standard of the
Republic; like the True Anglo-Saxon, –
loyal and loving, tender and brave, – she
hid her tears with one hand while with
the other wrung her fond farewell, and
passed him to the state.
   Clara Barton, quoted in Percy H. Epler,
   The Life of Clara Barton, 1905.

GLORY OF WOMEN

You love us when we're heroes, home
   on leave,
Or wounded in a mentionable place.
You worship decorations; you believe
That chivalry redeems the war's
   disgrace.
You make us shells. You listen with
   delight,
By tales of dirt and danger fondly
   thrilled.
You crown our distant ardours while
   we fight,
And mourn our laurelled memories

MUTINY

when we're killed.
You can't believe that British troops 'retire'
When hell's last horror breaks them, and they run,
Trampling the terrible corpses – blind with blood.
O German mother dreaming by the fire,
While you are knitting socks to send your son
His face is trodden deeper in the mud.

   Siegfried Sassoon (1886–1967).

Shame on you. Do not forget that you are the son of a Samurai, that tears are not for you.

   The mother of Saburo Sakai when, as a small boy, he would come home in tears after having been beaten up at school by the bigger boys, quoted in Saburo Sakai, *Samurai!*, 1957.

The motto Papa and I chose when he got his arms as a peer, will guide me as I know it will guide you; 'In Honour Bound'... We who come from an old stock of a privileged family, that has not had to worry over material existence, have inherited that sense of duty towards our fellow men, those especially whose nation we belong to, and who look to us instinctively for example and guidance.
   I know that you feel this too, more than ever in times like these. Let us live or die honourably. I am proud, with the old feelings of our ancestors, that you my son once more are called to such high service. My love for you and my pride in you are too deep for selfish worries or repining.

   Princess Victoria of Battenberg, 1 September 1939, in a letter to her son, later Earl Mountbatten of Burma, quoted in Philip Ziegler, *Mountbatten*, 1985.

Among women, too, it is the young of whom most is demanded. The lot of mothers who lose their sons is hard, but doubly hard is that of the young women who lose the chance of ever becoming wives and mothers. (Feb. 1939.)

   Captain Sir Basil Liddell Hart, *Thoughts on War*, 1944.

MUTINY

How should base mutineers bound together by no rightful obligation, and united only by thirst for pillage and community of crime, display the intrepidity or obtain the success of honourable soldiers? Courage can never exist with the consciousness of guilt and speedily forsakes the man who has forsaken duty.

   Count Belisarius, April AD 536, quoted in Mahon, *The Life of Belisarius*, 1829.

That long official neglect of intolerable grievance, and inexcusable supineness towards measures of progressive improvement had, as they sooner or later infallibly do, at last aroused illegal and exasperated enforcements of redress.

   Jedediah Tucker, of the British naval mutinies of 1797, *Memoirs of Earl St. Vincent*, 1830.

I know some generals are mutinous and preach revolt ...let them take care. I am as high above a general as above a drummer boy and, if necessary, I will have the one shot as soon as the other.

   Napoleon, October 1798, in Egypt, when generals began to complain about his leadership, quoted in S.J. Watson, *By Command of the Emperor*, 1957.

Why does Colonel Grigsby refer to me to learn how to deal with mutineers? He should shoot them where they stand.

   Lieutenant-General Thomas J. 'Stonewall' Jackson, May 1862, upon receiving a report of 12-month volunteers refusing duty.

Where soldiers get into trouble of this nature, it is nearly always the fault of some officer who has failed in his duty.

   Field Marshal Viscount Montgomery of Alamein, quoted in Ahrenfeldt, *Psychiatry in the British Army in the Second World War*, 1958.

N

NAVY

In the [Navy] there was a technology that blocked all my ideas... No man capable of breaking away from its routine and doing creative work ever appeared... I never could find an intermediary between me [and my sailors] who could get them moving.
> Napoleon, quoted in Charles de la Roncière and G. Clerc-Rampal, *Histoire de la Marine Française*, 1934.

Those far distant, storm-beaten ships, upon which [Napoleon's] Grand Army never looked, stood between it and the dominion of the world.
> Rear Admiral Alfred Thayer Mahan, referring to the Royal Navy in *The Influence of Sea Power Upon History*, II, 1892.

The Navy is the 1st, 2nd, 3rd, 4th... *ad infinitum* Line of Defence! If the Navy is not supreme, no Army however large is of the slightest use.
> Admiral Sir John Fisher, speaking before WWI, quoted in Arthur Marder, *The Anatomy of British Sea Power*, 1940.

For the Royal Navy, the watchword should be, 'Carry on, and dread nought.'
> Sir Winston S. Churchill, 6 December 1939, speech in the House of Commons.

Don't talk to me about naval tradition. It's nothing but rum, sodomy, and the lash.
> Sir Winston S. Churchill, during WWII when an admiral stated that a combined operation would be against naval tradition, quoted in Peter Gretton, *Former Naval Person*, 1968.

You seldom hear of the fleets except when there's trouble, and then you hear a lot.
> Admiral John S. McCain, 4 August 1964, *Norfolk Star Ledger*.

NECESSITY

It must be thoroughly understood that war is a necessity, and that the more readily we accept it, the less will be the ardor of our opponents, and that out of the greatest dangers communities and individuals acquire the greatest glory.
> Pericles, 432–431 BC, arguing for war with Sparta, Thucydides, *The Peloponnesian War*, 1.145, c. 404 BC, tr Richard Crawley, 1910.

Necessity hath no law.
> Oliver Cromwell, 1654, address to Parliament.

Necessity, dire necessity, will, nay must, justify my attack.
> General George Washington, 25 December 1776, attributed, before the daring attack on Trenton.

Necessity is the plea for every infringement of human freedom. It is the argument of tyrants; it is the creed of slaves.
> William Pitt the Younger (1759–1806), 18 November 1783, speech in the House of Commons, one month before becoming Prime Minister.

Great men are never cruel without necessity.
> Napoleon (1769–1821), quoted in Christopher Herold, ed., *The Mind of Napoleon*, 1955.

NEUTRALITY

Every officer of the old army remembers how, in 1861, we were hampered with the old blue army-regulations, which tied our hands, and that to do any thing positive and necessary we had to tear it all to pieces – cut red-tape, as it was called – a dangerous thing for an army to do, for it was calculated to bring law and authority into contempt; but war was upon us, and overwhelming necessity overrides all law.
  General of the Army William T. Sherman, *The Memoirs of General W.T. Sherman*, 1875.

NEUTRALITY

A prince is further esteemed when he is a true friend or a true enemy, when, that is, he declares himself without reserve in favor of some one or against another. This policy is always more useful than remaining neutral. For if two neighboring powers come to blows, they are either such that if one wins, you will have to fear the victor, or else not. In either of these two cases it will be better for you to declare yourself openly and make war, because in the first case if you do not declare yourself, you will fall prey to the victor, to the pleasure and satisfaction of the one who has been defeated, and you will have no reason nor anything to defend you and nobody to receive you.
  Niccolò Machiavelli, *The Prince*, 1513.

A neutral is bound to be hated by those who lose and despised by those who win.
  Niccolò Machiavelli, 22 December 1514, letter to Francesco Vettor.

I shall treat neutrality as equivalent to a declaration of war against me.
  Gustavus II Adolphus, King of Sweden, proclamation during the Thirty Years War, 1616–1648.

Neutrality! – what short of talk is that? For or against – there is no middle term!
  Gustavus II Adolphus, to the representatives of the German Protestant party upon his landing in Germany, June 1630, quoted in Nils Ahnlund, *Gustavus Adolphus the Great*, 1940.

The most sincere neutrality is not a sufficient guard against the depradations of nations at war.
  General George Washington, 7 December 1796, Farewell Address.

Never break the neutrality of any port or place, but never consider as neutral any place from whence an attack is allowed to be made.
  Admiral Viscount Nelson, 1804, letter of instruction.

NON-COMBATANTS

Don't hurt civilians, they give us food and drink; a soldier is not a footpad.
  Field Marshal Prince Aleksandr V. Suvorov, *The Science of Victory*, 1796.

It will require the full exercise of the full powers of the Federal Government to restrain the fury of the noncombatants.
  General Winfield Scott, April 1861, after the Confederate attack on Fort Sumter.

Those who live at home in peace and plenty want the duello part of this war to go on: but when they have to bear the burden by loss of property or comforts, they will cry for peace.
  General of the Army Phillip H. Sheridan, 26 November 1864, telegram to Major-General Halleck from Kernstown, Virginia.

When you've shouted 'Rule Britannia',
  when you've sung 'God Save the Queen',
When you've finished killing Kruger with your mouth.
  Rudyard Kipling (1865–1936), 'The absent-Minded Beggar'.

Arras itself was fairly clear, as it was under shell fire; nevertheless, the Hotel

NON-COMMISSIONED OFFICERS

de Commerce, where I pulled up for a drink, was doing good business. There was a dead soldier lying outside with a waterproof sheet over him; but inside, the place was crowded with soldiers drinking, chatting, and joking. The women serving them were now practically shell-proof, anyhow as regards nerves. You cannot terrorize a people for long, especially if there is money to be made.

    Major-General J.F.C. Fuller, note in 1917, *Memoirs of an Unconventional Soldier*, 1936.

NON-COMMISSIONED OFFICERS

As a rule it is easy to find officers, but it is sometimes very hard to find noncommissioned officers.

    Napoleon, 28 January 1809, letter to General Clarke, quoted in Christopher Herold, ed., *The Mind of Napoleon*, 1955.

His Excellency also observes that the non-commissioned officers can do their duty, and can maintain the authority of their situation, only by having the support of the officers belonging to their company given to them upon all occasions, by constantly visiting the soldiers' quarters and by invariable attendance upon the parade, from the moment the soldiers are under arms.

    The Duke of Wellington, 4 April 1810, at Vizeu, *Selections from the General Orders of Field Marshal the Duke of Wellington During His Various Commands 1799–1818*, 1847.

We have in the service the scum of the earth as common soldier; and of late years we have been doing every thing in our power, both by law and by publications, to relax the discipline by which alone such men can be kept in order... As to the non-commissioned officers, as I have repeatedly stated, they are as bad as the men...

    The Duke of Wellington, 2 July 1813, to Lord Bathurst, *The Dispatches of Field Marshal the Duke of Wellington, During His Various Campaigns in India, Denmark, Portugal, Spain, the Low Countries, and France*, X, 1834–1838. p. 496.

All the work is done by the non-commissioned officers of the Guards. It is true that they regularly get drunk once a day – by eight in the evening – and go to bed soon after, but then they always take care to do first what ever they were bid... I am convinced that there would be nothing so intelligent, so valuable, as English soldiers of that rank, if you could get them sober, which is impossible.

    The Duke of Wellington, quoted in Maxwell, *The Life of Wellington*, II, 1899.

We have good corporals and sergeants and some good lieutenants and captains, and those are far more important than good generals.

    General of the Army William T. Sherman (1820–1891).

The backbone of the Army is the non-commissioned man!

    Rudyard Kipling, 'The 'Eathen', 1896.

Even to those who best understand the reasons for the regimenting of military forces, a discipline wrongfully applied is seen only as indiscipline. Invariably it will be countered in its own terms. No average rank-and-file will become insubordinate as quickly, or react as violently, as a group of senior non-commissioned officers, brought together in a body, and then mishandled by officers who are ignorant of the customs of the service and the limits of their own authority. Not only are they conscious of their rights, but they have greater respect for the state of decency and order which is the mark of a proper military establishment than for the insignia of rank. It is this firm feeling of the fitness of things, and his unbounded allegiance to an authority when it is based on character which makes the NCO and

the petty officer the backbone of discipline within the United States fighting establishment.
  Brigadier-General S.L.A. Marshall, *The Armed Forces Officer*, 1950.

My many years in the army have demonstrated that wherever confidence in NCOs is lacking and wherever they are continuously bossed by the officers, you have no real NCOs and no really combatworthy units.
  Marshal of the Soviet Union Georgi K. Zhukov, *Reminiscences and Reflections*, 1974.

NUMBERS

In war it is not numbers that give the advantage. If you do not advance recklessly, and are able to consolidate your own strength, get a clear picture of the enemy's situation, and secure the full support of your men, it is enough.
  Sun Tzu, *The Art of War*, 9, c. 500 BC, tr. Roger Ames, 1993.

When someone said: 'Leonidas, are you here like this, to run such a risk with a few men against many?', he replied: 'If you think that I should rely on numbers, then not even the whole of Greece is enough, since it is a small fraction of their horde; but if I am to rely on courage, then even this number is quite adequate.'
  Leonidas, Agiad King of Sparta, 480 BC, at the Pass of Thermopylae, quoted in Plutarch, *The Lives*, c. AD 100 (*Plutarch on Sparta*, tr. Richard Talbert, 1988).

Once when he heard that the constant campaigning had made the allies discontented – there were large numbers of them following just a few Spartans – [Agesilaus] he was eager to demonstrate the unimportance of their mere numbers. He instructed the allies to sit down together at random, and the Spartans separately on their own. Next he announced that the potters should stand up first, and when they had done so, then the smiths second, then in turn carpenters, builders, and workers in every other craft. Thus virtually all the allies stood up, but not one of the Spartans, since there was a ban on their practising or learning a manual craft. So Agesilaus laughed and said: 'Do you see, gentlemen, how many more soldiers we send out than you?'
  Plutarch, quoting Agesilaus, Eurypontid King of Sparta (400–360 BC), *The Lives*, c. AD 100 (*Plutarch on Sparta*, tr. Richard Talbert, 1988).

Victory in war does not depend entirely upon numbers or mere courage; only skill and discipline will insure it. We find that the Romans owed the conquest of the world to no other cause than continual military training, exact observance of discipline in their camps and unwearied cultivation of the arts of war... A handful of men, inured to war, proceeds to certain victory, while on the contrary numerous armies of raw and undisciplined troops are but masses of men dragged to slaughter.
  Flavius Vegetius Renatus, *Military Institutions of the Romans*, c. Goudy Old Style MT 378, tr. Clark, 1776.

That's all right... The greater the enemy the more they will fall over one another, and the easier it will be for us to cut through. In any case, they're not numerous enough to darken the sun for us.
  Field Marshal Prince Aleksandr V. Suvorov (1729–1800), 1789, before the Battle of Rymnik.

The disparity of generalship often outweighs a difference of numbers, and ...psychological incalculables are apt to make statistical calculations valueless – to the sore perplexity of military pedants, whose minds are fed on 'balance sheet' history.
  Captain Sir Basil Liddell Hart, *Sherman*, 1929.

O

OBEDIENCE

A man submits though his heart breaks with fury.
Better for him by far.
> Homer, *The Iliad*, 1.255–256, c. 800 BC, tr. Robert Fagles, 1990.

The leader must himself believe that willing obedience always beats forced obedience, and that he can get this only by really knowing what should be done. Thus he can secure obedience from his men because he can convince them that he knows best, precisely as a good doctor makes his patients obey him. also he must be ready to suffer more hardships than he asks of his soldiers, more fatigue, greater extremities of cold and heat.
> Xenophon, *Cyropaedia*, c. 360 BC.

The laws of honour cannot necessarily be satisfied by throwing away one's life when it seems convenient. a soldier's duty is to obey.
> Jean de la Valette, Grand Master of the Order of the Knights of St John, June 1565, forbidding his young knights to throw away their lives in a desperate sally against the Turks, who had invested the fort of St Elmo during the siege of Malta.

I speak harshly to no one, but I will have your head off the instant you refuse to obey me.
> Marshal General Vicomte de Turenne (1611–1675), quoted in Maxime Weygand, *Turenne: Marshal of France*, 1930.

Officers must recollect, that to perform their duty well in the field is but a small part of what is required of them; and that obedience to order, regularity, and accuracy in the performance of duties and discipline, are necessary to keep any military body together, and to enable them to perform any military operation with advantage to their country, or service to themselves.
> The Duke of Wellington, 17 August 1812, general order published in Madrid, quoted in Philip Haythornthwaite, *The Armies of Napoleon*, 1994.

I do not believe in the proverb that in order to be able to command one must know how to obey... Insubordination may only be the evidence of a strong mind.
> Napoleon, 1817, letter from St Helena.

Soldiers must obey in all things. They may and do laugh at foolish orders, but they nevertheless obey, not because they are blindly obedient, but because they know that to disobey is to break the backbone of their profession.
> General Sir William Napier (1782–1853).

Men must be habituated to obey or they cannot be controlled in battle and the neglect of the least important order impairs the proper influence of the officer.
> General Robert E. Lee, 1865, to the Army of Northern Virginia.

It is difficult for the non-military mind to realize how great is the moral effort of disobeying a superior, whose order on the one hand covers all responsibility and on the other entails

THE OBJECTIVE/MAINTENANCE OF THE AIM

the most serious personal and professional injury; if violated without due cause, the burden of proving rests with the junior... He has to show, not that he meant to do right, but that he actually did right in disobeying in the particular instance. Under no less vigorous exactions can due military subordination be maintained.
Rear Admiral Alfred Thayer Mahan, *The Life of Nelson*, 1897.

The ugly truth is revealed that fear is the foundation of obedience.
Sir Winston S. Churchill, *The River War*, 1899.

In war, to obey is a difficult thing. For the obedience must be in the presence of the enemy, and in spite of the enemy, in the midst of danger, of varied and unforeseen circumstances, of a menacing unknown, in spite of fatigue of many causes.
Marshal of France Ferdinand Foch, *Principles of War*, 1913.

The mind of the soldier, who commands and obeys without question, is apt to be fixed, drilled, and attached to definite rules.
Field Marshal Viscount Wavell of Cyrenaica, *Generals and Generalship*, 1941.

All through your career of Army life, you men have bitched about what you call this chicken shit drilling. That is all for one reason, instant obedience to orders and it creates alertness. I don't give a damn for a man who is not always on his toes. You men are veterans or you wouldn't be here. You are ready. A man to continue breathing, must be alert at all times. If not, sometimes, some German son of a bitch will sneak up behind him and beat him to death with a sack full of shit.
General George S. Patton, Jr, 6 June 1944, speech to the troops of the Third Army.

All I want is compliance with my wishes, after reasonable discussion.
Sir Winston S. Churchill, to the British Chiefs of Staff, quoted in Dwight D. Eisenhower, *General Eisenhower on the Military Churchill*, 1970.

To know how to command obedience is a very different thing from making men obey. Obedience is not the product of fear, but of understanding, and understanding is based on knowledge.
Brigadier-General S.L.A. Marshall, *The Armed Forces Officer*, 1950.

There are admittedly cases where a senior commander cannot reconcile it with his responsibilities to carry out an order he has been given. Then, like Seydlitz at the Battle of Zorndorf, he has to say: 'After the battle the King may dispose of my head as he will, but during the battle he will kindly allow me to make use of it.' No general can vindicate his loss of a battle by claiming that he was compelled – against his better judgment – to execute an order that led to defeat. In this case the only course open to him is that of disobedience, for which he is answerable with his head. Success usually decides whether he was right or not.
Field Marshal Erich von Manstein, *Lost Victories*, 1957.

THE OBJECTIVE/ MAINTENANCE OF THE AIM
Principle of War

Victory in war is the main object in war. If this is long delayed, weapons are blunted and morale depressed...
Sun Tzu, *The Art of War*, 2, c. 500 BC, tr. Samuel Griffith, 1963.

Whither would you urge me? The most complete and most happy victory is to baffle the force of an enemy without impairing our own and in this favourable situation we are already placed. Is it not wiser to enjoy the

THE OBJECTIVE/MAINTENANCE OF THE AIM

advantages thus easily acquired, than to hazard them in the pursuit of more? Is it not enough to have altogether disappointed the arrogant hopes with which the Persians set out for this campaign and compelled them to a speedy and shameful retreat?
> Count Belisarius, AD 531, when his army demanded an attack of the retreating Persian Army, quoted in Mahon, *The Life of Belisarius*, 1829.

The man who tries to hang onto everything ends up by holding nothing. Your essential objective must be the hostile army...
> Frederick the Great, *General Principles of War*, 1748.

If you start to take Vienna – take Vienna.
> Napoleon (1769–1821).

Many good generals exist in Europe, but they see too many things at once: I see but one thing, and that is the masses; I seek to destroy them, sure that the minor matters will fall of themselves.
> Napoleon (1769–1821).

The loss of Moscow is not the loss of Russia. My first obligation is to preserve the army, to get nearer to those troops approaching as reinforcements, and by the very act of leaving Moscow to prepare inescapable ruin for the enemy... I will play for time, lull Napoleon as much as possible and not disturb him in Moscow. Every device which contributes to this object is preferable to the empty pursuit of glory.
> Field Marshal Prince Mikhail I. Kutuzov, 1812, quoted in Parkinson, *Fox of the North*, 1976.

Pursue one great decisive aim with force and determination – a maxim which should take first place among all causes of victory.
> Major-General Carl von Clausewitz, *Principles of War*, 1812, tr Gatzke, 1942.

The destruction of the enemy forces is always the main object...
> Field Marshal August Wilhelm Graf von Gneisenau (1760–1831).

What do we mean by the defeat of the enemy? Simply the destruction of his forces, whether by death, injury, or any other means – either completely or enough to make him stop fighting. Leaving aside all specific purposes of any particular engagement, the complete or partial destruction of the enemy must be regarded as the sole object of all engagements.
> Field Marshal Carl von Clausewitz, *On War*, 4.3, 1832, tr. Michael Howard and Peter Paret, 1976.

No one starts a war – or rather, no one in his senses ought to do so – without first being clear in his mind what he intends to achieve by that war and how he intends to conduct it. The former is its political purpose; the latter its operational objective. This is the governing principle which will set its course, prescribe the scale of means and effort which is required, and make its influence felt throughout down to the smallest operational detail.
> Field Marshal Carl von Clausewitz, *On War*, 8.2, 1832, tr. Michael Howard and Peter Paret, 1976.

I shall not give my attention so much to Richmond as to Lee's army, and I want all commanders to feel that hostile armies, and not cities, are to be their objectives.
> General of the Army Ulysses S. Grant, 3 May 1864, comments to his staff at the beginning of the 1864 campaign, 3 May 1864, quoted in Porter, *Campaigning With Grant*, 1897.

Why did you come to this war if you can't stand the gaff? War has always been this way... The fat Old Man you

THE OBJECTIVE/MAINTENANCE OF THE AIM

talk about is going to win this campaign. When he does these things will be forgotten. It's the objective that counts, not the incidents.

> General of the Armies John J. Pershing, as a first lieutenant, June 1898, to a fellow officer complaining of the supply problem and leadership of Brigadier-General William R. Shafter, during the campaign in Cuba, quoted in Donald Smyth, 'Pershing in the Spanish-American War', *Military Review*, Spring 1966.

It may be that the enemy's fleet is still at sea, in which case it is the great objective, now as always.

> Rear Admiral Alfred Thayer Mahan, *Naval Strategy*, 1911.

War means fighting for definite results. I am not waging war for the sake of waging war. If I obtain through the armistice the conditions we wish to impose upon Germany, I am satisfied. Once this object has been attained, nobody has the right to shed one drop of blood.

> Marshal of France Ferdinand Foch, *Memoirs of Marshal Foch*, 1931.

The Strategical Objective: Irrespective of the arm employed, the principles of strategy remain immutable, changes in weapons affecting their application only. The first of all strategical principles is 'the principle of the object,' the object being 'the destruction of the enemy's field armies – his fighting personnel.'

Now, the potential fighting strength of a body of men lies in its organization; consequently, if we can destroy this organization, we shall destroy its fighting strength and so have gained our object.

There are two ways of destroying our object.

(i) By wearing it down (dissipating it).

(ii) By rendering it inoperative (unhinging it).

In war the first comprises the killing, wounding, capturing and disarming of the enemy's soldiers – body warfare. The second, the rendering inoperative of his power of command – brain warfare. Taking a single man as an example: the first method may be compared to a succession of slight wounds which will eventually cause him to bleed to death; the second – a shot through the brain.

> Major-General J.F.C. Fuller, memorandum of 24 May 1918 for 'Plan 1919', *Memoirs of an Unconventional Soldier*, 1936.

The true military objective is a mental rather than a physical objective – the paralysis of the opposing command, not the bodies of the actual soldiers. For an army without orders, without co-ordination, without supplies, easily becomes a panic-stricken and famine-stricken mob, incapable of effective action. (March 1925.)

> Captain Sir Basil Liddell Hart, *Thoughts on War*, 1944.

The military object may be expressed in the one word 'conquest,' which presupposes victory in one form or another, and by conquest I understand that condition of success which will admit of a government imposing its will on the enemy's nation, and so attaining the execution of its policy. Conquest may also be considered as the grand strategical military idea, and victory the grand tactical means.

> Major-General J.F.C. Fuller, *The Foundations of the Science of War*, 1926.

The danger of carrying us away from the objective... In an operation, there must be nothing superfluous; it must be the incarnation of aim-directedness. The form of the operation... must recall... not the rococo... but a Greek temple.

> General A.A. Svechin, 1927.

History shows that the unswerving pursuit of any one objective is almost

THE OBJECTIVE/MAINTENANCE OF THE AIM

certain to be barren of result. 'Variability' of objectives, like elasticity of dispositions, is necessary to fulfil an essential principle of war – *flexibility.* (Aug. 1929.)

> Captain Sir Basil Liddell Hart, *Thoughts on War,* 1944.

Here we are dealing with the elementary object of war, war as 'politics with bloodshed', as mutual slaughter by opposing armies. The object of war is specifically 'to preserve oneself and destroy the enemy' (to destroy the enemy means to disarm him or 'deprive him of the power to resist', and does not mean to destroy every member of his forces physically).

> Mao Tse-tung, *On Protracted War,* May 1938.

You ask, What is our policy? I will say: It is to wage war by sea, land, and air, with all our might and with all the strength that God can give us; to wage war against a monstrous tyranny, never surpassed in the dark, lamentable catalogue of human crime. That is our policy. You ask, What is our aim? I can answer in one word: Victory – victory at all costs, victory in spite of all terror; victory, however long and hard the road may be; for without victory there is no survival.

> Sir Winston S. Churchill, 13 May 1940, speech to the House of Commons.

There is no purpose in capturing these manure-filled, water logged villages. The purpose of our operations is to kill or capture the German personnel and vehicles... so that they cannot retreat and repeat their opposition. Straight frontal attacks against villages are prohibited unless after careful study there is no other possible solution.

> General George S. Patton, Jr, 9 December 1944, letter of instruction for the Third Army, *The Patton Papers,* I, 1972.

The alphabet of war, Comrade General, says that victory is achieved not by taking towns but by destroying the enemy. In 1812, Napoleon forgot that. He lost Moscow – and Napoleon was no mean leader of men.

> General Andrei L. Getman, 1945, quoted in Cornelius Ryan, *The Last Battle,* 1966.

In time of war the only value that can be affixed to any unit is the tactical value of that unit in winning the war. Even the lives of those men assigned to it become nothing more than tools to be used in the accomplishment of that mission. War has neither the time nor heart to concern itself with the individual and the dignity of man. Men must be subordinated to the effort that comes with fighting a war, and as a consequence men must die that objectives might be taken. For a commander the agony of war is not in its dangers, deprivations, or the fear of defeat but in the knowledge that with each new day men's lives must be spent to pay the costs of that day's objectives.

> General of the Army Omar N. Bradley, *A Soldier's Story,* 1951.

In discussing the subject of the 'objective' in war it is essential to be clear about, and keep clear in our minds, the distinction between the political and military objective. The two are different but not separate. For nations do not wage war for war's sake, but in pursuance of policy. The military objective is only the means to a political end. Hence the military objective should be governed by the political objective, subject to the basic condition that policy does not demand what is militarily – that, is practically – impossible.

> Captain Sir Basil Liddell Hart, *Strategy,* 1954.

Keep your object always in mind, while adapting your plan to circumstances. Realize that there are more ways than

one of gaining an object, but take heed that every objective should bear on the object. And in considering possible objectives weigh their possibility of attainment with their service to the object attained – to wander down a side-track is bad, but to reach a dead end is worse.
Captain Sir Basil Liddell Hart, *Strategy*, 1954.

OBSTACLES

Country in which there are precipitous cliffs with torrents running between, deep natural hollows, confined places, tangled thickets, quagmires and crevasses, should be left with all possible speed and not approached.

While we keep away from such places, we should get the enemy to approach them; while we face them, we should let the enemy have them on his rear.
Sun Tzu, *The Art of War*, 9, c. 500 BC, tr. Giles, 1910.

Mountains can be crossed wherever goats cross, and winter freezes most rivers.
Frederick the Great, *Instructions to His Generals*, 1747, tr. Phillips, 1940.

Natural hazards, however formidable, are inherently less dangerous and less uncertain than fighting hazards. All conditions are more calculable, all obstacles more surmountable, than those of human resistance. By reasoned calculation and preparation they can be overcome almost to timetable. While Napoleon was able to cross the Alps in 1800 'according to plan', the little fort of Bard could interfere so seriously with the movement of his army as to endanger his whole plan.
Captain Sir Basil Liddell Hart, *Strategy*, 1954.

An obstacle loses 50 per cent of its value if you stand back from it, allowing the enemy to reconnoitre the approaches and subsequently to cross without interference.
Field Marshal Viscount Montgomery of Alamein, *A History of Warfare*, 1968.

OFFENCE AND DEFENCE
The Interrelationship

The general who is skilled in defense hides in the most secret recesses of the earth; he who is skilled in attack flashes forth from the topmost heights of heaven. Thus on the one hand we have ability to protect ourselves; on the other, a victory is complete.
Sun Tzu, *The Art of War*, 4, c. 500 BC, tr. Giles, 1910.

Projects of absolute defense are not practicable because while seeking to place yourself in strong camps the enemy will envelop you, deprive you of your supplies from the rear and oblige you to lose ground, thus disheartening your troops. Hence, I prefer to this conduct the temerity of the offensive with the hazard of losing the battle since this will not be more fatal than retreat and timid defensive. In the one case you lose ground by withdrawing and soldiers by desertion and you have no hope; in the other you do not risk more and, if you are fortunate, you can hope for the most brilliant success.
Frederick the Great, *Instructions to His Generals*, 1747, tr. Phillips, 1940.

Maxim 19. The transition from the defensive to the offensive is one of the most delicate operations in war.
Napoleon, *The Military Maxims of Napoleon*, tr. George D'Aguilar, 1831 (David Chandler, ed., 1987).

A sudden powerful transition to the offensive – the flashing sword of vengence – is the greatest moment for the defense. If it is not in the commander's mind from the start, or rather if it is not an integral part of his idea of defense, he will never be persuaded of the superiority of the defensive from;

OFFENCE AND DEFENCE

all he will see is how much of the enemy's resources he can destroy or capture. But these things do not depend on the way in which the knot is tied, but on the way in which it is untied. Moreover, it is a crude error to equate attack with the idea of assault alone, and therefore, to conceive of defense as merely misery and confusion.
    Major-General Carl von Clausewitz, *On War*, 6.5, 1832, tr. Michael Howard and Peter Paret, 1976.

The best thing for an army on the defensive is to know how to take the offensive at a proper time, and *to take it*.
    Lieutenant-General Antoine-Henri Baron de Jomini, *Summary of the Art of War*, 1838, tr. Mendell and Craighill, 1862.

The offensive knows what it wants… the defensive is in a state of uncertainty.
    Field Marshal Helmuth Graf von Moltke, *Instructions for the Commanders of Large Formations*, 1869.

A clever military leader will succeed in many cases in choosing defensive positions of such an offensive nature from the strategic point of view that the enemy is compelled to attack us in them.
    Field Marshal Helmuth Graf von Moltke (1800–1891).

In war, the defensive exists mainly that the offensive may act more freely.
    Rear Admiral Alfred Thayer Mahan, *Naval Strategy*, 1911.

A purely defensive battle is a duel in which one of the fighters does nothing but *parry*. Nobody would admit that, by so doing, he could succeed in defeating his enemy. On the contrary, he would sooner or later expose himself, in spite of the greatest possible skill, to being touched, to being overcome by one of his enemy's thrusts, even if that enemy were the weaker party.
    Hence the conclusion that the *offensive* form alone, be it resorted to at once or only after the *defensive*, can lead to results, and must therefore *always* be adopted – at least in the end.
    Marshal of France Ferdinand Foch, *Precepts and Judgments*, 1919.

As they were depressed by defence their spirits rose in the offensive. The interests of the Army were best served by the offensive; in defence it was bound gradually to succumb to the ever increasing hostile superiority in men and material. This feeling was shared by everybody. In the West the Army pined for the offensive, and after Russia's collapse expected it with the most intense relief. Such was the feeling of the troops about attack and defence. It amounted to a definite conviction which obsessed them utterly that nothing but an offensive could win the war.
    General Erich Ludendorff, *My War Memories 1914–1918*, 1919.

Where the problem of attack is difficult of solution it is worth considering the possibilities of an alternative form of action which throws the burden of that problem on the enemy, yet keeps the initiative. The alternative is what I would term the 'baited offensive': the combination of offensive strategy with defensive tactics. Throughout history it has proved one of the most effective of moves, and its advantages have increased as modern weapons have handicapped other types of move. By rapidity of advance and mobility of manoeuvre, you may be able to seize points which the enemy, sensitive to the threat, will be constrained to attack. Thus you will invite him to a repulse which in turn may be exploited by a riposte. Such a counterstroke, against an exhausted attacker, is much less difficult than the attack on a defended

THE OFFENSIVE

position. The opportunity for it may also be created by a calculated withdrawal – what one may call the 'luring defensive'. Here is another gambit of future warfare. (Nov. 1935).
    Captain Sir Basil Liddell Hart, *Thoughts on War*, 1944.

The history of wars convincingly testifies... to the constant contradiction between the means of attack and defense. The appearance of new means of attack has always (inevitably) led to the creation of corresponding means of counteraction, and this in the final analysis has led to the development of new methods for conducting engagements, battles, operations (and war in general).
    Marshal of the Soviet Union Nikolai V. Ogarkov, quoted in *Kommunist* (Moscow), 1978.

THE OFFENSIVE
Principle of War

I should say that in general the first of two army commanders who adopts an offensive attitude almost always reduces his rival to the defensive and makes him proceed in consonance with the movement of the former.
    Frederick the Great, *Instructions to his Generals*, 1747, tr. Phillips, 1940.

Pushing off smartly is the road to success.
    Major-General Sir James Wolfe, 1758, letter to a friend.

An offensive, daring kind of war will awe the Indians and ruin the French. Blockhouses and a trembling defensive encourage the meanest scoundrel to attack us.
    Major-General Sir James Wolfe, 1758, letter to General Jeffery Lord Amherst regarding the imminent campaign in North America.

Make war offensively; it is the sole means to become a great captain and to fathom the secrets of the art.
    Napoleon, (1769–1821), *Correspondance de Napoléon Ier, publiée par ordre de l'Empereur Napoléon III*, XXXI, 1858–1870.

Maxim 6. At the start of a campaign, to advance or not to advance is a matter for grave consideration, but when once the offensive has been assumed, it must be sustained to the last extremity.
    Napoleon, *The Military Maxims of Napoleon*, tr. George D'Aguilar, 1831 (David Chandler, ed., 1987).

War, once declared, must be waged offensively, aggressively. The enemy must not be fended off, but smitten down. You may then spare him every exaction, relinquish every gain; but till down he must be struck incessantly and remorselessly.
    Rear Admiral Alfred Thayer Mahan, *The Interest of America in Sea Power*, 1896.

It is this persistent widening and intensifying of the offensive – this pushing vigorously forward on carefully chosen objectives without excessive regard to alignment or close touch – that will give us the best results with the smallest losses...
    Marshal of France Ferdinand Foch, 26 August 1918, letter to Field Marshal Haig, quoted in Liddell Hart, *Foch: Man of Orleans*, 1931.

Hit first! Hit hard! Keep on hitting!
    Admiral Sir John 'Jackie' Fisher, *Memories*, 1919.

In war the only sure defense is offense, and the efficiency of the offense depends on the warlike souls of those conducting it.
    General George S. Patton, Jr, *War As I Knew It*, 1947.

Since I joined the Marines, I have advocated aggressiveness in the field and constant offensive action. Hit quickly, hit hard and keep right on hitting. Give

THE OFFICER

the enemy no rest, no opportunity to consolidate his forces and hit back at you. This is the shortest road to victory.
    General Holland M. 'Howlin' Mad' Smith, *Coral and Brass*, 1949.

THE OFFICER

But it is not enough to cherish them at the moment when you have need of them and when their actions wring cheers from you. It is necessary that in time of peace they enjoy the reputation that they have so justly earned and that they be signally esteemed as men who have shed their blood solely for the honor and the well being of the state.
    Frederick the Great, *Political Testament*, 1752.

War must be carried on systematically, and to do it, you must have good Officers, there are, in my Judgment, no other possible means to obtain them but by establishing your Army upon a permanent basis; and giving your officers good pay; this will induce Gentlemen, and Men of Character to engage; and till the bulk of your Officers are composed of such persons as are actuated by Principles of Honor, and spirit of Enterprize, you have little to expect from them.
    General George Washington, 24 September 1776, letter to the President of Congress.

Be frank with your friends, temperate in your requirements and disinterested in conduct; bear an ardent zeal for the service of your Sovereign; love true fame, distinguish ambition from pride and vainglory; learn early to forgive the faults of others, and never forgive your own...
    Field Marshal Prince Aleksandr V. Suvorov (1729–1800), advice to a young officer cadet, quoted in W. Lyon Blease, *Suvorof*, 1920.

Officers used to serve, not for their starvation pay, but for love of their country and on the off-chance of being able to defend it with their lives... They deliberately, indeed joyously, faced up to a life of adventure, roving action, exile, and poverty because it satisfied and reposed their souls.
    General Sir Ian Hamilton, *The Soul and Body of an Army*, 1921.

The officer is a being apart, a kind of artist breathing the grand air in the brilliant profession of arms, in a uniform that is always seductive... To me the officer is a separate race.
    Mata Hari (1876–1917).

I divide my officers into four classes as follows: the clever, the industrious, the lazy, and the stupid. Each officer possesses at least two of these qualities. Those who are clever and industrious I appoint to the General Staff. Use can under certain circumstances be made of those who are stupid and lazy. The man who is clever and lazy qualifies for the highest leadership posts. He has the requisite nerves and the mental clarity of difficult decisions. But whoever is stupid and industrious must be got rid of, for he is too dangerous.
    Attributed to General Kurt von Hammerstein Equord, c. 1933.

The foundation of our National Defense system is the Regular Army, and the foundation of the Regular Army is the officer. He is the soul of the system. If you have to cut everything out of the National Defense Act the last element should be the officer corps. If you had to discharge every soldier, if you had to do away with everything else, I would still professionally advise you to keep those 12,000 officers. They are the mainspring of the whole mechanism; each one of them would be worth a thousand men at the beginning of a war. They are the

OFFICERS AND MEN

only ones who can take this heterogeneous mass and make it a homogeneous group.
> General of the Army Douglas MacArthur, 1933, statement before Congress defending the officer corps faced with proposed deep cuts, *Reminiscences*, 1964.

We, as officers of the army, are not only members of the oldest of honorable professions, but are also the modern representatives of the demi-gods and heroes of antiquity. Back of us stretches a line of men whose acts of valor, of self-sacrifice and of service have been the theme of song and story since long before recorded history began... In the days of chivalry... knights-officers were noted as well for courtesy and gentleness of behavior as for death-defying courage... From their acts of courtesy and benevolence was derived the word, now pronounced as one, Gentle Man... Let us be gentle. This is, courteous and considerate of the rights of others. Let us be Men. That is, fearless and untiring in doing our duty... Our calling is most ancient and like all other old things it has amassed through the ages certain customs and traditions which decorate and ennoble it, which render beautiful the otherwise prosaic occupation of professional men-at-arms: killers.
> General George S. Patton, Jr (1885–1945), quoted in Fitton, ed., *Leadership*, 1990.

The badge of rank which an officer wears on his coat is really a symbol of servitude – servitude to his men.
> General Maxwell Taylor, quoted in *Army Information Digest*, 1953.

OFFICERS AND MEN

Regard your soldiers as your children, and they will follow you into the deepest valleys; look on them as your own beloved sons, and they will stand by you even unto death.
> Sun Tzu, *The Art of War*, 10, c. 500 BC, tr. Giles, 1910.

Men who think that their officer recognizes them are keener to be seen doing something honorable and more desirous of avoiding disgrace.
> Xenophon (381–c.352 BC).

Tears welled up as they looked at him, and they appeared not as an army visiting its king but one attending a funeral. The grief was especially intense among those at his bedside. Alexander looked at them and said: 'After my death will you find a king who deserves such men?'
> Curtius, of Alexander the Great's farewell to the Macedonians on his deathbed, 323 BC, *The History of Alexander*, 10.5.1, c. AD 45, tr. John Yardley, 1984.

Now as for me, my men, there is not one of you who has not with his own eyes seen me strike a blow in battle; I have watched and witnessed your valour in the field, and your acts of courage I know by heart, with every detail of when and where they took place: and this, surely, is not a thing of small importance. I was your pupil before I was your commander; I shall advance into the line with soldiers I have a thousand times praised and rewarded; and the enemy we shall meet are raw troops with a raw general, neither knowing anything of the other.
> Hannibal, 218 BC, address to his army before meeting the Roman Army under the Consul Cornelius Scipio on the River Ticinus, quoted in Livy, *The History of Rome*, 21.43, c. AD 17 (*The War With Hannibal*, tr. Aubrey de Sélincourt, 1965).

The general's way of life should be plain and simple like that of his soldiers; he should display a fatherly affection toward them; he should give orders in a mild manner; and he should always make sure to give advice and to discuss essential matters with them in person. His concerns ought to be with their safety, their food, and their regular pay. Without these it is impossible to

OFFICERS AND MEN

maintain discipline in an army.
> The Emperor Maurice, *The Strategikon*, c. AD 600 (*Maurice's Strategikon*, tr. George Dennis, 1984).

What makes soldiers in battle prefer to charge ahead rather than retreat even for survival is the benevolence of the military leadership. When soldiers know their leaders care for them as they care for their own children, then the soldiers love their leaders as they do their own fathers. This makes them willing to die in battle, to requite the benevolence of their leaders.
> Liu Ji (1310–1375), *Lessons of War* (*Mastering the Art of War*, tr. Thomas Cleary, 1989).

You must love soldiers in order to understand them, and understand them in order to lead them.
> Marshal of France Vicomte de Turenne (1611–1675).

The kindness of the troops to me has transported me, for I had none in this last action, but such as were with me last year... this gave occasion to the troops... to make me very kind expressions, even in the heat of the action, which I own to you gives me great pleasure, and makes me resolve to endure everything for their sakes...
> The Duke of Marlborough, of the reaction of his troops to his presence at their head at the battle of Elixhem, on 18 July 1705 when he forced the Lines of Brabant, letter to Sarah his wife, quoted in .W. Coxe, *Memoirs of John , Duke of Marlborough*, vol 1, 1736, p.296.

Love the soldier, and he will love you. That is the secret.
> Field Marshal Prince Aleksandr V. Suvorov (1729–1830).

Under such an officer [who understood them], they will conquer or die on the spot, while their action, their hardihood and abstinence enable them to bear up under a severity of fatigue under which larger and apparently stronger men would sink. But it is the principles of integrity and moral correctness that I admire most in Highland soldiers, and this was the trait that first caught my attention. It is this that makes them trustworthy, and makes their courage sure, and not that kind of flash in the pan which would scale a bastion to-day and tomorrow be alarmed at the fire of a picket. You Highland officers may sleep sound at night, and rise in the morning with the assurance that your men, your professional character and honour, are safe.
> Lieutenant-General Sir John Moore, 17 November 1804, letter to Colonel Alexander Napier, quoted in Carola Oman, *Sir John Moore*, 1953.

When asked one day how, after so many years, he could recollect the names and numbers of the units engaged in one of his early combats, Napoleon responded, 'Madam, this is a lover's recollection of his former mistresses.'
> Napoleon, quoted in Count Emmanuel de Las Casas, *Memoirs of the Life, Exile and Conversations of the Emperor Napoleon*, 1836.

The Duke never claimed for one moment credit to himself where he did not feel that it was thoroughly deserved. Someone saying to him 'How do you account, Duke, for your having so persistently beaten the French Marshals?' The Duke replied simply, 'Well, the fact is their soldiers got them into scrapes: mine always got me out.'
> The Duke of Wellington (1769–1852), quoted in William Fraser, *Words of Wellington*, 1889.

For an officer to be overbearing and insulting in the treatment of enlisted men is the act of a coward. He ties the man to a tree with ropes of discipline

and then strikes him in the face, knowing full well that the man cannot strike back.
> Major C.A. Bach, 1917, farewell instructions to graduating student officers at Fort Sheridan, Wyoming.

Good relations between commanders and the rank and file are like all other forms of friendship – if they are to be maintained and bear fruit they must be nourished. Soldiers are human beings – rather more human than other people – and they will never respond whole-heartedly to the commander who treats them as mere automata to be used for his own purpose according to order, and without any thought being given to them as ordinary men. On the other hand, they will always be ready to offer the last ounce of their strength in extricating from any difficulty into which he may have fallen the General in whom they have confidence as a personal friend. Our men are exceedingly accurate judges of an officer's worth and character, and whilst they intensely dislike the officer who does not enter into their feelings and treats them as if they had none, they have unbounded admiration for the one who treats them kindly as well as justly.
> Field Marshal Sir William Robertson, *From Private to Field-Marshal*, 1921.

When the last bugle is sounded, I want to stand up with my soldiers.
> General of the Armies John J. Pershing, quoted in Hale and Turner, *The Yanks are Coming*, 1983.

Sympathy and understanding are vital faculties in a commander – because they evoke a similar response in subordinates, and thus create the basis of co-operation. Arbitrary control commonly breaks down unless it rests on the co-operation of minds attuned to the same key. Without this the first is but voice control. (May 1929.)
> Captain Sir Basil Liddell Hart, *Thoughts on War*, 1944.

Many people think that it is wrong methods that make for strained relations between officers and men and between the army and the people, but I always tell them that it is a question of basic attitude... of having respect for the soldiers and the people.
> Mao Tse-tung, *On Protracted War*, May 1938.

The relationship between a general and his troops is very much like that between the rider and his horse. The horse must be controlled and disciplined, and yet encouraged: he should 'be cared for in the stables as if he was worth 500 and ridden in the field as if he were worth half-a-crown'.
> Field Marshal Viscount Wavell of Cyrenaica, *Generals and Generalship*, 1941.

Officers are responsible, not only for the conduct of their men in battle, but also for their health and contentment when not fighting. An officer must be the last man to take shelter from fire, and the first to move forward. Similarly, he must be the last man to look after his own comfort at the close of a march. He must see that his men are cared for. The officer must constantly interest himself in the rations of the men. He should know his men so well that any sign of sickness or nervous strain will be apparent to him, and he can take such action as may be necessary.
> General George S. Patton, Jr, *War As I Knew It*, 1947.

The example and personal conduct of officers and noncommissioned officers are of decisive influence on the troops. The officer who, in the face of the enemy, is cold-blooded, decisive, and courageous inspires his troops onward. The officer must, likewise, find the way to the hearts of his subordinates and gain their trust through an understanding of their feelings and thoughts, and through never-ceasing care of their needs.

OFFICERS AND MEN

Generals Werner von Fritsch and Ludwig Beck, *Troop Leadership*, 1933, quoted in US War Department, *German Military Training*, 17 September 1942.

Nothing more radical is suggested here than that the leader who would make certain of the fundamental soundness of his operation cannot do better than concentrate his attention on his men. There is no other worthwhile road. They dupe only themselves who believe that there is a brand of military efficiency which consists in moving smartly, expediting papers, and achieving perfection in formations, while at the same time slighting or ignoring the human nature of those whom they command. The art of leading, in operations large or small, is the art of dealing with humanity, of working diligently on behalf of men, of being sympathetic with them, but equally, of insisting that they make a square facing toward their own problems. These are the real bases of a commander's calculations. Yet how often do we hear an executive praised as an 'efficient administrator' simply because he can keep a desk cleared, even though he is despised by everyone in the lower echelons and cannot command a fraction of their loyalty.

Brigadier-General S.L.A. Marshall, *The Armed Forces Officer*, 1950.

The commander must have contact with his men. He must be capable of feeling and thinking with them. The soldier must have confidence in him. There is one cardinal principle which must always be remembered: one must never make a show of false emotions to one's men. The ordinary soldier has a surprisingly good nose for what is true and what is false.

Field Marshal Erwin Rommel, *The Rommel Papers*, 1953.

He [the soldier] will put up with this so long as he knows you are living in relatively much the same way; and he likes to see the C.-in-C. regularly in the forward area, and be spoken to and noticed. He must know that you really care for him and will look after his interests, and that you will give him all the pleasures you can in the midst of his discomforts.

Field Marshal Viscount Montgomery of Alamein, *The Memoirs of Field Marshal Montgomery*, 1958.

The first thing a young officer must do when he joins the Army is to fight a battle, and that battle is for the hearts of his men. If he wins that battle and subsequent similar ones, his men will follow him anywhere; if he loses it, he will never do any real good.

Field Marshal Viscount Montgomery of Alamein, *The Memoirs of Field Marshal Montgomery*, 1958.

The men have a highly developed collective sense. They are almost never mistaken about their commander, and are the people best qualified to judge him. It is possible to mislead them by propaganda and publicity stunts. It is possible to create a cult of admiration for some military figure who rarely comes in contact with them. But it is quite impossible to deceive them as to the character and qualities of their direct commanders with whom they work day by day.

General Yigal Allon, *The Making of Israel's Army*, 1970.

The basis of all good leadership is the relationship built up between a commander and his troops. Napoleon's greatest and, as it turned out, almost his only asset on his return from Elba was the devotion of his army. His physical condition might have deteriorated, his military genius might have waned, but so long as his soldiers were prepared to die for him the French

THE OLD SCHOOL

Army was still a formidable fighting machine.
> Lieutenant-General Sir Brian Horrocks, *Escape to Action*, 1961.

I don't like being separated in comparative luxury from troops who are living hard.
> Field Marshal Sir Claude Auchinleck, 1976, interview with David Dimbleby.

I have a strong belief in what young Marines can achieve if they are properly led. It never ceases to amaze me. In the desert, they built bases and moved vast distances that you wouldn't think anyone would be able to do. And all they ask in return is caring leadership. They are the first to know if you don't really care. And consequently, their performance will reflect that.
> General Walter Boomer, 'Up Front', Al Santoli, ed., *Leading the War: How Vietnam Veterans Rebuilt the US Military*, 1993.

THE OLD SCHOOL

We cannot draw our inspiration indifferently from Turenne, Conde, Prince Eugene, Villars, or Frederick the Great, even less from the tottering theories and degenerate forms of the last century. The best of these doctrines answered a situation and needs which are no longer ours.
> Marshal of France Ferdinand Foch, *The Principles of War*, 1913.

...For if there is anything more characteristic of the Navy than its fighting ability, it is the inertia to change, or conservatism, or the clinging to things that are old because they are old. It must be admitted that this characteristic has been in many things a safeguard; it is also true that in quite as many it has been a drag to progress.
> Admiral of the Fleet Ernest J. King, 'Some Ideas About Organization on Board Ship', prize essay, 1909, *US Naval Institute Proceedings* (written as a lieutenant).

Have we the moral courage to accept the new idea and the energy to carry it out? We doubt it, for before us stands the serried phalanx of the 'old school', who, like the ideal English bishop of Mrs. Browning –
> '...must not love truth too dangerously,
> But prefer the interests of the Church.'

> Major-General J.F.C. Fuller, from his gold-medal-winning essay in *RUSI*, 5/1920.

Cavalry circles loathed the idea of giving up the horse, and thus instinctively decried the tank. They found much support both in the War Office and in Parliament. Wellington's reputed saying that the Battle of Waterloo was won on the playing-fields of Eton is merely a legend, but it is painfully true that the early battles of World War II were lost in the Cavalry Club.
> Captain Sir Basil Liddell Hart, *The Memoirs of Captain Liddell Hart*, I, 1965.

OLD SOLDIERS

But now's the time
for the younger men to lock in rough encounters,
time for me to yield to the pains of old age.
But there was a day I shone among the champions.
> Homer, *The Iliad*, 23.715–719, c. 800 BC, tr. Robert Fagles, 1990; King Nestor of Pylos' reminiscences at the games for Patroclus.

But there are still those moments when I can feel the strength and courage and wisdom move together in my head, and rich evening laughter of good men rises from the benches towards the dozing hawks in the beams above as the poet touches his harp and calls for me. Then I know that Beowulf is my name, and I in my youth with the strength of thirty men in my hand am proud and ready, and will sail again.
> *Beowulf*, c. AD 600 (*Beowulf Is My Name*, tr. Frederick Rebsamen, 1971).

OPINIONS

I dreamed I was back in battle. I waded in blood up to my knees. I saw death as it is on the battle field. The poor boys with arms shot off and legs gone, were lying on the cold ground, with no nurses and no physicians to do anything for them. I saw the surgeons coming, too much needed by all to give special attention to any one. Once again I stood by them and witnessed those soldiers bearing their soldier pains, limbs being sawed off without opiates being taken, or even a bed to lie on. I crept around once more, trying to give them at least a drink of water to cool their parched lips, and I heard them at last speak of mother and wives and sweethearts, but never a murmur or complaint. Then I woke to hear myself groan because I have a stupid pain in my back, that's all. Here on a good bed, with every attention! I am ashamed that I murmur!
>   Clara Barton, 10 April 1912, two days before her death, quoted in Percy Epler, *The Life of Clara Barton*, 1915.

I 'eard the feet on the gravel – the feet o' the men what drill –
An' I sez to my flutterin' 'eart-strings, I sez to 'em, 'Peace, be still!'
>   Rudyard Kipling, 'Back to the Army Again'.

The shadows are lengthening for me. The twilight is here. My days of old have vanished tone and tint; they have gone glimmering through the dreams of things that were. Their memory is one of wonderous beauty, watered by tears, and coaxed and caressed by the smiles of yesterday. I listen vainly, but with thirsty ear, for the witching melody of faint bugles blowing reveille, of far drums beating the long roll. In my dreams I hear again the crash of guns, the rattle of musketry, the strange mournful mutter of the battlefield. But in the evening of my memory, always I come back to West Point. Always there echoes and reechoes in my ears – Duty–Honor–Country.
  Today marks my final call with you. But I want you to know that when I cross the river my last conscious thought will be of the Corps – and the Corps – and the Corps.
  I bid you farewell.
>   General of the Army Douglas MacArthur, 12 May 1964, farewell address to the Corps of Cadets at Westpoint, *Reminiscences*, 1964.

OPERATIONS

No operations plan will ever extend with any sort of certainty beyond the first encounter with the hostile main force. Only the layman believes to perceive in the development of any campaign a consistent execution of a preconceived original plan that has been thought out in all its details and adhered to to the very end.
>   Field Marshal Helmuth Graf von Moltke (1800–1891), quoted in von Freytag-Loringhoven, *Generalship in the World War*, 1920.

An important difference between a military operation and a surgical operation is that the patient is not tied down. But it is a common fault of generalship to assume that he is. (May 1934.)
>   Captain Sir Basil Liddell Hart, *Thoughts on War*, 1944.

OPINIONS

I never suffer my mind to be so wedded to any opinions as to refuse to listen to better ones when they are suggested to me.
>   Henry IV (Henry of Navarre), King of France (1553–1610).

An opinion can be argued with; a conviction is best shot.
>   Colonel T.E. Lawrence (1888–1935).

You have come here to acquire knowledge, to evolve it and to fit it to the

OPPORTUNITY

men you will one day either command or administer. Knowledge is not only acquired in the lecture-room, but also in the mess and in your own private studies. Nothing clarifies knowledge like a free exchange of ideas; consequently, because I happen to be a Major, do not for a moment imagine that rank is a bar to free speech. If you disagree with me or the views of any of your instructors, openly state your disagreement; for we are all students, and the man who cannot change his opinions has mineralized his intellect – he is a walking stone.
Major-General J.F.C. Fuller, *Memoirs of an Unconventional Soldier*, 1936.

I am convinced that the best service a retired general can perform is to turn in his tongue along with his suit and to mothball his opinions.
General of the Army Omar N. Bradley, quoted in James Carlton, *The Military Quotation Book*, 1991.

I can give you my opinion, but remember this is just Al Gray talking: I'm too junior to make policy and too senior to make coffee.
General Alfred M. Gray, *Soldier of Fortune*, 10/1981; while a major-general serving as the USMC's Development Director.

OPPORTUNITY

If the enemy leaves a door open, you must rush in.
Sun Tzu, *The Art of War*, 11, c. 500 BC, tr. Giles, 1910.

...The opportunities of war wait for no man.
Pericles, reply to the Spartan ultimatum, 432/431 BC, quoted in Thucydides, *The History of the Peloponnesian War*, 1.142, c. 404 BC, tr. Richard Crowley, 1910.

The art of certain victory, the mode of harmonizing with changes, is a matter of opportunity. Who but the perspicacious can deal with it? And of all avenues of seeing opportunity, none is greater than the unexpected.
Zhuge Liang (AD 180–234), *The Way of the General* (*Mastering the Art of War*, tr. Thomas Cleary, 1989).

Those designs are best of which the enemy are entirely ignorant till the moment of execution. Opportunity in war is often more to be depended on than courage.
Flavius Vegetius Renatus, *Military Institutions of the Romans*, c. AD 378, tr. Clark, 1776.

In war opportunity is fleeting and cannot be put off at all.
The Emperor Maurice, *The Strategikon*, c. AD 600 (*Maurice's Strategikon*, tr. George Dennis, 1984).

Four things come back not:
The spoken word; the sped arrow;
Time past; the neglected opportunity.
Omar I ibn al-Khattab, Caliph (AD 581–644).

My feeling is that we should confront all the enemy's forces with all the forces of Islam; for events do not turn out according to man's will and we do not know how long a life is left to us, so it is foolish to dissipate this concentration of troops without striking a tremendous blow in the Holy War.
Saladin, July 1187, his decision before the Battle of Hattin to make use of the opportunity presented by concentration, quoted by Ibn al-Athir in Gabrieli, *Arab Historians of the Crusades*, 1957.

I shall last a year, and but little longer: we must think to do good work in that year. Four things are laid upon me: to drive out the English; to bring you to be crowned and anointed at Rheims; to rescue the Duke of Orléans from the hands of the English; and to raise the siege of Orléans.
Joan of Arc, March/April 1429, to Charles,

OPPORTUNITY

Dauphin of France, *Joan of Arc in her Own Words*, 1996.

Nothing is of greater importance in time of war than knowing how to make the best use of a fair opportunity when it is offered.
  Niccolò Machiavelli, *The Art of War*, 1521.

Not only strike while the iron is hot, but make it hot by striking.
  Oliver Cromwell (1599–1658).

It is good to exercise caution, but not to such an extent that all opportunities become lost. Hungry dogs bite best.
  Charles XII, King of Sweden (1682–1718).

Thus, on the day of battle, I should want the general to do nothing. His observations will be better for it, his judgment will be more sane, and he will be in better state to profit from the situations in which the enemy finds himself during the engagement. And when he sees an occasion, he should unleash his energies, hasten to the critical point at top speed, seize the first troops available, advance them rapidly, and lead them in person. These are the strokes that decide battles and gain victories. The important thing is to see the opportunity and to know how to use it.
  Field Marshal Maurice Comte de Saxe, *My Reveries*, 1732, tr. Phillips, 1940.

All great events hang by a single thread. The clever man takes advantage of everything, neglects nothing that may give him some added opportunity; the less clever man, by neglecting one thing, sometimes misses everything.
  Napoleon, 26 September 1797, letter to Tallyrand, *Correspondance de Napoléon Ier, publiée par ordre de l'Empereur Napoléon III*, III, p. 342, 1858–1870, quoted in Christopher Herold, ed., *The Mind of Napoleon*, 1955.

When this movement commences, I shall move out by my left, with all the force I can, holding present intrenched lines. I shall start with no distinct view, further than holding Lee's forces from following Sheridan. But I shall be along myself and will take advantage of anything that turns up.
  General of the Army Ulysses S. Grant, 1865, letter to Major-General Sherman from outside Petersburg, quoted in Liddell Hart, *Sherman*, 1930.

Everything comes to this: to be able to recognize the changed situation and order the foreseeable course and prepare it energetically.
  Field Marshal Helmuth Graf von Moltke, *Ausgewaehlte Werke*, I, 1925.

An advantage is an advantage, however offered or obtained; whether by an enemy's mistake, or by the accidents of the ground that play so large a part in land war; and on either element a skillful defense looks warily for its opportunities to the enemy's mistakes, as well as to other conditions.
  Admiral Alfred Thayer Mahan, *Naval Strategy*, 1911.

The battle that turned the tide of the war in 1914 has given rise to an endless debate among generals and historians as to who was responsible for winning the Battle of the Marne. It is quite as apt to pose the question: 'Was the Marne won by the Allies or lost by the Germans?' The reasonable answer is that most victories in history have been won by seizing opportunities offered by the loser. (June 1931.)
  Captain Sir Basil Liddell Hart, *Thoughts on War*, 1944.

There is no security on this earth; there is only opportunity.
  General of the Army Douglas MacArthur, quoted in Courtney Whitney, *MacArthur: His Rendezvous with History*, 1955.

ORDER

To our men... the jungle was a strange, fearsome place; moving and fighting in it were a nightmare. We were too ready to classify it 'impenetrable'... To us it appeared only as an obstacle to movement; to the Japanese it was a welcome means of concealed manoeuvre and surprise... The Japanese reaped the deserved reward... we paid the penalty.
Field Marshal Viscount Slim of Burma, *Defeat into Victory*, 1956.

ORDER

The Illyrians, to those who have no experience of them, do indeed at first sight present a threatening aspect. The spectacle of their numbers is terrible, their cries are intolerable, and the brandishing of their spears in the air has menacing effect. But in action they are not the men they look, if their opponents will only stand their ground; for they have no regular order, and therefore are not ashamed of leaving any post in which they are hard pressed; to fly and to advance being alike honorable, no imputation can be thrown on their courage. When every man is his own master in battle, he will readily find a decent excuse for saving himself.
Brasidas of Sparta, to the Spartan Army, 423 BC, quoted in Thucydides, *The Peloponnesian War*, 4.126, c. 404 BC, tr. Benjamin Jowett, 1881.

In military operations, order leads to victory. If rewards and penalties are unclear, if rules and regulations are unreliable, and if signals are not followed, even if you have an army of a million strong it is of no practical benefit.

An orderly army is one that is mannerly and dignified, one that cannot be withstood when it advances and cannot be pursued when it withdraws. Its movements are regulated and directed; this gives it security and presents no danger. The troops can be massed but not scattered, can be deployed but not worn out.
Zhuge Liang (AD 180–234), *The Way of the General*, quoted in Thomas Cleary, *Mastering the Art of War*, 1989.

If all is not in order, I will hang you despite my personal high regard.
Field Marshal Prince Aleksandr V. Suvorov, 1799, administrative order to his quartermaster-general during the Italian campaign.

Order marches with weighty and measured strides; disorder is always in a hurry.
Napoleon, *Maxims*, 1804–1815.

It is hard to imagine why outrageous behavior was allowed to persist in the Army during that period [early 1970s]. In the institution, we had kind of lost our bearings in what was right and what was wrong. And what was appropriate discipline and what wasn't. We had misused some of the psychological stuff that management theorists were pushing on us. We started to think that we were running some kind of democratic day camp: Generals sitting around coffee houses talking to troops; troop councils reporting to battalion commanders about what they ought do. It was a breakdown of the command structure. Like inmates had taken over the asylum.

In the 2nd Division, we reinstalled discipline with the attitude: 'Let me explain to you. I'm in charge and you ain't.' That's what the troops needed to hear.
General Colin Powell, 'Fixing the System', Al Santoli, ed., *Leading the Way: How Vietnam Veterans Rebuilt the US Military*, 1993.

ORDERS

When asked another time for what particular reason the Spartans enjoyed notably more success than others, he said: 'Because more than others they train to give orders and to take them.'
Plutarch, quoting Agesilaus, Eurypontid King of Sparta (400–360 BC), *The Lives*, c. AD 100 (*Plutarch on Sparta*, tr. Richard Talbert, 1988).

ORDERS

The orders I have given are strong, and I know not how my admiral will approve of them, for they are, in a great measure, contrary to those he gave me; but the Service requires strong and vigorous measures to bring the war to a conclusion.
> Admiral Viscount Nelson, July 1795, letter to Admiral Collingwood.

To say that an officer is never, for any object, to alter his orders, is what I cannot comprehend. The circumstances of this war so often vary, that an officer has almost every moment to consider, what would my superiors direct did they know what is passing under my nose? But, sir, I find few think as I do. To obey orders is all perfection. To serve my king and destroy the French, I consider as the great order of all, from which little ones spring; and if one of these militate against it (for who can tell exactly at a distance?), I go back, and obey the great order and object, to down – down with the damned French villains! My blood boils at the name of a Frenchman!
> Admiral Viscount Nelson, March 1799, letter from Palermo to the Duke of Clarence, despairing at the army's lack of initiative in pressing the siege of Malta, quoted in Robert Southey, *Life of Nelson*, 1813.

I shall be glad to know who commands this army – I or you? I establish one route, one line of communication – you establish another by ordering up supplies by it. As long as you live, sir, never do that again. Never do anything without my orders.
> The Duke of Wellington, to Sir James McGrigor, who had moved some commissary supplies to Salamanca during the Peninsular campaign, cited in Charles Oman, *Wellington's Army*, 1913.

You must avoid countermanding orders: unless the soldier can see a good reason for benefit, he becomes discouraged and loses confidence.
> Napoleon, 5 August 1806, advice to his stepson, Prince Eugene, *Correspondance de Napoléon Ier, publiée par ordre de l'Empereur Napoléon III*, No. 10699, XIII, 1858–1870.

Give your orders so that they cannot be disobeyed.
> Napoleon, 29 March 1811, to Marshal Berthier, *Correspondance de Napoléon Ier, publiée par ordre de l'Empereur Napoléon III*, No. 17529, XXI, 1858–1870.

I regret to hear from an officer that it is *impossible* to execute an order. If your cavalry will not obey your orders you must *make them* do it, and, if necessary, go out with them yourself. I desire you to go out and post your cavalry where you want them to stay, and arrest any man who leaves his post, and prefer charges and specifications against him, that he may be court-martialed. It will not do to say that men your men cannot be induced to perform their duty. *They must be made to do it…*
> Lieutenant-General Thomas J. 'Stonewall' Jackson, letter to Colonel Sincindiver, 11 February 1862, quoted in William Allan, *History of the Campaign of Gen. T.J. (Stonewall) Jackson in the Shenandoah Valley of Virginia*, 1880.

On *Farragut's quarterdeck*: 'Captain, you begin early in your life to disobey orders. Did you not see the signal flying for near an hour to withdraw from action? …I want none of this Nelson business in my squadron.'

Later in Farragut's cabin: 'I have censured you, sir, on my quarterdeck, for what appeared to be a disregard of my orders. I desire now to commend you and your officers and men for doing what you believe right under the circumstances. Do it again whenever in your judgment it is necessary to carry out your conception of duty.'
> Admiral David Glasgow Farragut, to

ORDERS

Lieutenant Winfield Scott Schley, executive officer of the USS *Richmond*, during the action at Port Hudson, March 1863, quoted in Christopher Martin, *Damn the Torpedos!*, 1970.

I had worded my instructions so strongly, I thought they would wake up a dead man to his true condition.
Major-General Joseph Hooker, April 1863, to his chief of cavalry, Brigadier-General Stoneman, prior to the Battle of Chancellorsville, 1–5 May 1863.

It is the custom of military service to accept instructions of a commander as orders, but when they are coupled with conditions that transfer the responsibility of battle and defeat to the subordinate, they are not orders.
Lieutenant-General James A. Longstreet, comments on Lee's vague orders to Ewell for the second day at Gettysburg, 2 July 1863, *From Manassas to Appomattox*, 1896.

Remember, gentlemen, an order that can be misunderstood will be misunderstood.
Field Marshal Helmuth Graf von Moltke (1800–1891).

Order, counter-order, disorder.
Field Marshal Helmuth Graf von Moltke (1800–1891).

In war the first principle is to disobey orders. *Any fool can obey orders!*
Admiral Sir John Fisher, after the Dogger Bank action, January 1915.

I just gave an order – quite a simple matter unless a man's afraid.
General Sir Ian Hamilton, *Soul and Body of an Army*, 1921.

All orders will have to be as brief as possible. They should be based on a profound appreciation of possibilities and probabilities which... will generally lead to a series of alternatives.

Major-General J.F.C. Fuller (1878–1966).

Petain was once asked what part of an action called for the greatest effort. 'Giving orders,' he replied. There is, indeed, something irrevocable about the intervention of the human will in a sequence of events. Whether useful or not, opportune or misjudged, it is pregnant with unforeseeable consequences. The mere awareness of the audacity of such a proceeding is intimidating. Even in ordinary life there are many who do not find it easy to make decisions, and the number of those prepared to take the initiative is small in comparison with that of the obedient and submissive mass. How much more agonizing must be the call to decisive action in the case of a military commander who knows that so many poor lives depend upon his decision, and that on the highest as well as on the lowest levels it will be judged solely by its results. So heavy is the responsibility in such cases that few men are capable of shouldering the whole unaided. That is why the greatest intellectual qualities are not enough. No doubt they help; so, too, does instinct, but in the last resort, the decision is a moral one.
General Charles de Gaulle, *The Edge of the Sword*, 1932.

In war 'thoughtless' orders often result in the useless sacrifice of men's lives. In peace, they often contribute to the sterilization of men's reason. And they inevitably make the man who transmits the order an accomplice in the crime, however unwilling. Unfortunately, few of the transmitters have the sensitiveness of perception to feel their responsibility. (July 1933.)
Captain Sir Basil Liddell Hart, *Thoughts on War*, 1944.

The issuance of an order is the simplest thing in the world. The important and difficult thing is to see: first, that the

order is transmitted; and, second, that it is obeyed...
> General George S. Patton, Jr, 8 July 1941, speech to the officers of the Second Armored Division, *The Patton Papers*, II, 1974.

I give orders only when they are necessary. I expect them to be executed at once and to the letter and that no unit under my command shall make changes, still less give orders to the contrary or delay execution through unnecessary red tape.
> Field Marshal Erwin Rommel, 22 April 1944, letter of instruction to subordinate commanders.

An officer should not ask a man: 'Would you like to do such-and-such a task?' when he has already made up his mind to assign him to a certain duty. Orders, hesitantly given, are doubtfully received. But the right way to do it is to instill the idea of collaboration. There is something irresistably appealing about such an approach as: 'I need your help. Here's what we have to do.'
> Brigadier-General S.L.A. Marshall, *The Armed Forces Officer*, 1950.

Operations orders do not win battles without the valour and endurance of the soldiers who carry them out.
> Field Marshal Viscount Wavell of Cyrenaica, *Soldiers and Soldiering*, 1953.

Operational command in the battle must be direct and personal, by means of visits to subordinate H.Q. where orders are given verbally. A commander must train his subordinate commanders, and his own staff, to work and act on verbal orders. Those who cannot be trusted to act on clear and concise verbal orders, but want everything in writing are useless.
> Field Marshal Viscount Montgomery of Alamein, *The Memoirs of Field Marshal Montgomery*, 1958.

To grasp the spirit of order is not less important than to accept them cheerfully and keep faith with the contract. But the letter of an instruction does not relieve him who receives it from the obligation to exercise common sense.
> Brigadier-General S.L.A. Marshall, *The Officer as Leader*, 1966.

Order a *naval rating* to 'secure the house' and he'll enter it, close all the doors and windows, and probably throw a line over the roof and lash it down.
Order an *infantryman* to 'secure the house' and he'll enter it, shoot anything that moves, and then probably dig trench about it.
Order an *airman* to 'secure the house' and he'll stroll down to the local estate agent and take out a 7-year lease on it.
> British military adage, quoted in *JFQ Forum*, 95/96 Autumn/Winter

I am not going to start World War III for you.
> Lieutenant-General Michael Jackson, June 1999, refusing the order of NATO commander, General Wesley Clark, to stop by force the Russian *coup de main* aimed at seizing Pristina Airfield in Kosovo, quoted in *Newsweek*, 9 August 1999.

An order is a good basis for discussion.
> Old military saying.

ORGANISATION

Generally, management of the many is the same as management of the few. It is a matter of organization.
> Sun Tzu, *The Art of War*, c. 500 BC, tr. Samuel Griffith, 1969.

The mass needs, and we give it, leaders... We add good arms. We add suitable methods of fighting... We also add a rational decentralization... We animate with passion... An iron discipline... secures the greatest unit... But it depends also on supervision, the mutual

supervision of groups of men who know each other well. A wise organization of comrades in peace who shall be comrades in war... And now confidence appears... Then we have an army.
  Colonel Charles Ardant du Picq, *Battle Studies*, 1880, tr. Greely, 1957.

The primary object of organization is to shield people from unexpected calls upon their powers of adaptability, judgment and decision.
  General Sir Ian Hamilton, *Soul and Body of an Army*, 1921.

Organization is the vehicle of force; and force is threefold in nature; it is mental, moral, and physical. How will the idea affect these spheres of force? This is primarily a question of force and its expenditure. Thus, if the idea is complex, and does not permit of it being readily grasped by others, mistakes are likely to occur; and if its aim is beyond the moral and physical powers of the troops, should it be pushed beyond the limit of their endurance, though organization may for the time being be maintained, ultimately demoralization will set in, and a demoralized organization is one which has become so fragile that a slight blow, especially a surprise blow, will instantaneously shatter it to pieces.
  Major-General J.F.C. Fuller, *The Foundation of the Science of War*, 1926.

ORIGINALITY AND THE CREATIVE MIND

...Qualifications a general ought to possess ...abilities to strike out something new of his own occasionally. For no man excelled in his profession who could not do that, and if a ready and quick invention is necessary and honourable in any profession, it must certainly be so in the art of war above all others.
  Niccolò Machiavelli, *The Art of War*, 1521.

We must hold our minds alert and receptive to the application of unglimpsed methods and weapons. The next war will be won in the future, not in the past. We must go on, or we will go under.
  General of the Army Douglas MacArthur, 1931.

When we study the lives of the great captains, and not merely their victories and defeats, what do we discover? That the mainspring within them was originality, outwardly expressing itself in unexpected actions. It is in the mental past in which most battles are lost, and lost conventionally, and our system teaches us how to lose them, because in the schoolroom it will not transcend the conventional. The soldier who thinks ahead is considered, to put it bluntly, a damned nuisance.
  Major-General J.F.C. Fuller, *Generalship: Its Diseases and Their Cure*, 1933.

In the great battles at Cannae and Lepanto, at Tannenberg and Tsushima, at Cape St. Vincent and Chancellorsville and The Marne, we must acknowledge that a creative mind fashioned the setting for victory in each case. It was a mind that could think beyond the moment of crisis and see beyond the horizon of battle that set the scene for bravery and leadership to work their wonderful magic.
  Lieutenant-General Victor H. 'The Brute' Krulak, 'A Soldier's Dilemma', *Marine Corps Gazette*, 11/1986.

P–Q

PACIFICATION

When entering the territory of an offender do no violence to the shrines of the deities. Do not hunt over the rice fields or damage the earth-works. Do not burn houses or cut down trees. Do not seize domestic animals or grain or agricultural implements. Where you find old people or children allow them to go home unharmed, and do not antagonize even able-bodied men if they do not challenge you. And see that the enemy wounded have medical treatment.
>Ssu Ma Jang Chu, *The Precepts of Ssu Ma Jang Chu*, c. 500 BC, quoted in A.L. Sadler, *Three Military Classics of China*, 1944.

When we conquer our enemies by kind treatment, and by acts of justice, we are more likely to secure their obedience than by a victory in the field of battle. For in the one case they yield to necessity; in the other, it is their own free choice. Besides, how often is the victory dearly bought while the conquest of an enemy by affection may be brought about without expense or loss!
>Polybius, *Histories*, 5.12, c. 125 BC (*Familiar Quotations from Greek Authors*, tr. Crauford Ramage, 1895).

Robbers of the world, having by their universal plunder exhausted the land, they rifle the deep. If the enemy be rich, they are rapacious; if he be poor, they lust for dominion; neither the east nor the west has been able to satisfy them. Alone among men they covet with equal eagerness poverty and riches. To robbery, slaughter, plunder, they give the lying name of empire; they make a solitude and call it peace.
>Cornelius Tacitus, reporting the speech of the British rebel chieftain, Galgacus, during the reconquest of Britain, *Agricola*, 30, AD 98 (*The Complete Works of Tacitus*, tr. Alfred Church and William Brodribb, 1942).

Severity is needed to govern the Turks; I order five or six heads to be sliced off every day in the streets of Cairo. Up till now we have had to behave mildly so as to counteract the reputation of terror that preceded us; at present it is, on the contrary, better to assume the tone that commands obedience with these people, for with them obedience signifies fear.
>Napoleon, 31 July 1798, Cairo, quoted in R.M. Johnston, ed., *The Corsican*, 1910.

We come to give you liberty and equality, but don't lose your heads about it – the first person who stirs without permission will be shot.
>Marshal Pierre-Francois Lefebvre, upon occupying a Franconian town earlier in his career, quoted in David Chandler, *The Campaigns of Napoleon*, 1966.

We can still build a socialist Afghanistan if there are only one million Afghans.
>Soviet comment reported during the war in Afghanistan (c. 1984).

The pacification program in its narrowest sense – the neutralization of the Viet Cong infrastructure in the countryside – was virtually complete by the end of 1970. It would hold firm, with minor shifts due to the combat situation, until the end of the GVN [Government of Vietnam] in 1975. Pacification in its broadest sense – to include reform of the GVN and RVNAF [Republic of Vietnam Armed Forces] – would never be won, but its narrow victory would constitute one of the resounding successes of Indochina War II.

Lieutenant-General Phillip B. Davidson, *Vietnam at War*, 1988.

We're quite keen to get out of Kevlar and into berets and the hearts and minds of the people.

Major-General Robin Brims, Commander, British 1st Armoured Division, signifying the absence of helmets was an indicator of trust in the local Iraqi population, CENTCOM news release, 7 April 2003.

Combat is hard because you are losing soldiers, killing people. But at the end of the day you are destroying things, and we know how to do that. This work (military government) requires inordinate patience. There are incredible frustrations. And you can't just pull a trigger and make it all go away.

Major-General David H. Petraeus, of the military government of the Mosul region of Iraq shortly after the end of military operations in Enduring Freedom, quoted in *The Washington Post*, 16 May 2003.

PACIFISM/ ANTI-WAR ACTIVISM

Once initiated there were but few public men who would have the courage to oppose [war]. Experience proves that the man who obstructs a war in which his nation is engaged, no matter whether right or wrong, occupies no enviable place in life or history. Better for him, individually, to advocate 'war, pestilence, and famine,' than to act as obstructionist to a war already begun. The history of the defeated rebel will be honorable hereafter, compared with that of the Northern man who aided him by conspiring against his government while protected by it. The most favorable posthumous history the stay-at-home traitor can hope for is – oblivion.

General of the Army Ulysses S. Grant, *Personal Memoirs of U.S. Grant*, 1885.

The man who has formed a clear notion of the nature of war, of its necessities, requirements and consequences, to wit, the *soldier*, will take a far more serious view of the potentialities of war than the politician or the business man who coldly weighs its advantages and disadvantages. After all, it is not so difficult to sacrifice one's own life, but the professional duty of risking the lives of others weighs heavily on the conscience. The soldier, having experience of war, fears it far more than the doctrinaire who, being ignorant of war, talks only of peace; for the soldier has gazed into war's bloodshot eyes, he has observed from his point of vantage the battle-fields of a world war, he has had to witness the agonies of nations, his hair has turned grey over the ashes of countless burned homesteads and he has borne the responsibility for the life and death of thousands. The figure of the sabre-rattling, fire-eating general is an invention of poisoned and unscrupulous political strife, a figure welcome to stupid comic papers, a catchword personified. There is no reason why the soldier's attitude towards war should not be called 'pacifism.' It is a pacifism established on knowledge and born of a sense of responsibility, but it is not the pacifism engendered by national abasement or by a hazy internationalism. The soldier will be the first to welcome any effort to diminish the potentialities of war, but he does not march down the street to the slogan of 'No more war!' because he knows that war and peace are decided by higher powers than princes, statesmen, parliaments, treaties and alliances – they are

PANIC

decided by the eternal laws which govern the growth and decay of nations. But the kind of pacifist who would deliberately make his own nation defenceless in such fateful encounters, who prefers to weaken it in alliance with a hostile neighbor rather than support his fellow-countrymen in preparation for legitimate resistance, deserves, as he always did, to be hanged to the nearest lamp-post, were it only a moral one.
> Colonel-General Hans von Seekt, *Thoughts of a Soldier*, 1930.

We are for the abolition of war, we do not want war; but war can only be abolished through war, and to get rid of the gun, we must first grasp it in our hands.
> Mao Tse-tung, *Problems of War and Strategy*, 1938.

The pacifists are at it again. I met a visiting fireman of great eminence who told me this was the 'LAST WAR.' I told him that such statements since 2600 BC had signed the death warrant of millions of young men. He replied with the stock lie – 'Oh yes but things are different now.' My God! Will they never learn?
> General George S. Patton, Jr, 3 March 1944, letter to his wife, *The Patton Papers*, II, 1974.

Rational pacifism must be based on a new maxim – 'If you wish for peace, understand war.'
> Captain Sir Basil Liddell Hart, *Thoughts on War*, 1944.

War is never prevented by running away from it.
> Air Marshal Sir John Slessor, *Strategy for the West*, 1954.

When you guys get home and face an anti-war protester, look him in the eyes and shake his hand. Then, wink at his girlfriend, because she knows she's dating a pussy.
> Lieutenant General James M. Mattis, USMC, commander 1st Marine Division, March 2003 at the beginning of the Iraq campaign.

PANIC

So the Trojans held their watch that
   night but not the Achaeans –
Godsent Panic seized them, comrade of
   bloodcurdling Rout:
All their best were struck by grief too
   much to bear.
As crosswinds chop the sea where the
   fish swarm,
the North Wind and the West Wind
   blasting out of Thrace
in sudden, lightning attack, wave on
   blacker wave, cresting,
heaving a tangled mass of seaweed out
   along the surf –
so the Achaeans' hearts were torn
   inside their chests.
> Homer, *The Iliad*, 9.1–8, c. 800 BC, tr. Robert Fagles, 1990.

Now for a little time it became a hand-to-hand fight, but when the cavalry with Alexander, and Alexander himself, pressed vigorously, shoving the Persians and striking their faces with their spears, and the Macedonian phalanx, solid and bristling with its pikes, had got to close quarters with them, and Darius, who had now long been in a panic, saw nothing but terrors all around, and he was himself the first to turn and flee.
> Arrian, *The Anabasis of Alexander*, 3.14, c. AD 125, tr. P.A. Brunt, 1976; the flight of Darius at the Battle of Gaugamela (or Arbela), 331 BC, which in turn panicked his army.

Every day gives us fresh marks of the great victory; for since my last… we have taken possession of Bruges and Damme, as also Oudenarde… in short, there is so great a panic in the French

army as is not to be expressed.
> The Duke of Marlborough, of the effects of his victory at the battle of Ramillies, on 23 May 1706, letter to his wife Sarah, quoted in W. Coxe, Memoirs of John , Duke of Marlborough, vol 1, 1736, 426.

Both officers and troops must be warned against those sudden panics which often seize the bravest armies when they are not well-controlled by discipline, and when they do not recognize that the surest hope of safety lies in order. An army seized with panic is in a state of demoralization because when disorder is once introduced all concerted action on the part of individuals becomes impossible, the voice of the officers can no longer be heard. No maneuver for resuming the battle can be executed, and there is no course except in ignominious flight.
> Lieutenant-General Antoine-Henri Baron de Jomini, Summary of the Art of War, 1838, tr. Mendell and Craighill, 1862.

The worst enemy a Chief has to face in war is an alarmist.
> General Sir Ian Hamilton, Gallipoli Diary, 1920.

PATIENCE

Victory will be achieved with patience.
> Osama bin Ladin, statement, 12 February 2003, BBC News World Edition.

Our fault has not been impatience.
    The truth is our patience should have been exhausted weeks and months and years ago. Even now, when if the world united and gave him an ultimatum: comply or face forcible disarmament, he might just do it, the world hesitates and in that hesitation he senses the weakness and therefore continues to defy.
    What would any tyrannical regime possessing WMD think viewing the history of the world's diplomatic dance with Saddam? That our capacity to pass firm resolutions is only matched by our feebleness in implementing them.
    That is why this indulgence has to stop. Because it is dangerous. It is dangerous if such regimes disbelieve us.
    Dangerous if they think they can use our weakness, our hesitation, even the natural urges of our democracy towards peace, against us.
    Dangerous because one day they will mistake our innate revulsion against war for permanent incapacity; when in fact, pushed to the limit, we will act. But then when we act, after years of pretence, the action will have to be harder, bigger, more total in its impact. Iraq is not the only regime with WMD. But back away now from this confrontation and future conflicts will be infinitely worse and more devastating.
    Tell our allies that at the very moment of action, at the very moment when they need our determination that Britain faltered. I will not be party to such a course. This is not the time to falter. This is the time for this house, not just this government or indeed this prime minister, but for this house to give a lead, to show that we will stand up for what we know to be right, to show that we will confront the tyrannies and dictatorships and terrorists who put our way of life at risk, to show at the moment of decision that we have the courage to do the right thing.
I beg to move the motion.
> Tony Blair, Speech in the House of Commons 18 Mar 2003, final justification for war.

We have the power to be patient in this, and we're not going to do anything before we are ready... We'll just continue to draw the noose tighter and tighter.
> General Richard B. Myers, Chairman of the Joint Chiefs of Staff, press briefing, 30 March 2003, quoted in The New York Times, 31 March 2003.

PATRIOTISM

This empire has been acquired by men who knew their duty and had the courage to do it, who in the hour of conflict had the fear of dishonor always present to them, and who, if ever they failed in an enterprise, would not allow their virtues to be lost to their country, but freely gave their lives to her as the fairest offering which they could present at her feast.
> Pericles, 431 BC, funeral oration for the Athenian dead, quoted in Thucydides, *The Peloponnesian War, c.* 404 BC, tr. Benjamin Jowett, 1881.

Our country: in her intercourse with foreign nations may she always be in the right; but our country, right or wrong!
> Admiral Stephen Decatur, April 1816, toast at a dinner in Norfolk, Virginia.

Certainly no man has more that should make life dear to him than I have, in the affection of my home; but I do not want to survive the independence of my country.
> Lieutenant-General Thomas J. 'Stonewall' Jackson, quoted in Henderson, *Stonewall Jackson*, II, 1898.

My country, right or wrong; if right, to be kept right; and if wrong, to be set right.
> Major-General Carl Schurz, 29 February 1872, speech.

Patriotism is like a plant whose roots stretch down into race and place subconsciousness; a plant whose best nutrients are blood and tears; a plant which dies down in peace and flowers most brightly in war. Patriotism does not calculate, does not profiteer, does not stop to reason: in an atmosphere of danger the sap begins to stir; it lives; it takes possession of our soul.
> General Sir Ian Hamilton, *The Soul and Body of an Army*, 1921.

The man who is willing to fight for his country is finally the full custodian of its security. If there were no willing man, no power in government could ever rally the masses of the unwilling. But if the spirit and purpose which enable such men to find themselves and to act are to be safeguarded into the future, much more will have to be required of the country than that it point its young people toward the virtues of the production line. There is something almost fatally quixotic about a nation which professes lofty ideas in its international undertakings and yet disdains to talk patriotism to its citizens, as if this were beneath their dignity.
> Brigadier-General S.L.A. Marshall, *Men Against Fire*, 1947.

I admire men who stand up for their country in defeat, even though I am on the other side.
> Sir Winston S. Churchill, *The Gathering Storm*, 1948.

Patriotism is when love for your own people comes first; nationalism, when the hate for people other than your own comes first.
> General Charles de Gaulle, *Life*, 9 May 1969.

PAY

Do not disagree between yourselves. Give the soldiers money and despise everyone else.
> The Emperor Septimius Severus, AD 211, deathbed advice to his sons, quoted in Dio Cassius, *Romaika*, 76.15.2, *c.* AD 230.

To raise any taxes is impracticable, since the provinces are in possession of the enemy, and the long arrear of pay which our soldiers vainly claim loosens every tie of discipline and duty. A debtor is but ill able to command.
> Count Belisarius (*c.* AD 505–565).

Now the chief thing incumbent upon a

PAY

general, in order to maintain his reputation, is to pay well and punish soundly for if he does not pay his men duly, he cannot punish them properly when they deserve it. Suppose, for instance, a soldier should be guilty of a robbery; how can you punish him for that when you give him no pay? And how can he help robbing when he has no other means of subsistence? But if you pay them well and do not punish them severely when they offend, they will soon grow insolent and licentious; then you will become despised and lose your authority; later, tumult and discord will naturally ensue in your army; and will probably end in ruin.

Niccolò Machiavelli, *The Art of War*, 1521.

Without going into the different rates of pay, I shall say only that it should be ample. It is better to have a small number of well-kept and well-disciplined troops than to have a great number who are neglected in these matters. It is not the big armies that win battles; it is the good ones. Economy can be pushed only to a certain point. It has limits beyond which it degenerates into parsimony. If your pay and allowances for officers will not support them decently, then you will only have rich men who serve for debauchery or indigent wretches devoid of spirit.

Field Marshal Maurice Comte de Saxe, *My Reveries*, 1732, tr. Phillips, 1940.

Truly the only good officers are the impoverished gentlemen who have nothing but their sword and their cape; but it is essential that they should be able to live on their pay.

Field Marshal Maurice Comte de Saxe, *My Reveries*, 1732, tr. Phillips, 1940.

There is nothing that gives a man consequence, and renders him fit for command, like a support that renders him independent of everybody but the State he serves.

General George Washington, 24 September 1776, letter to the President of Congress from the Heights of Harlem.

Maxim 40. Every means should be taken to attach the soldier to his colours. This is best accomplished by showing consideration and respect for the old soldier. His pay likewise should increase with his length of service. It is the height of injustice to give a veteran no greater advantages than a recruit.

Napoleon, *The Military Maxims of Napoleon*, tr. George D'Aguilar, 1831 (David Chandler, ed., 1987).

We then related to him a number of bon-mots made by our soldiers during his absence and on his return, with which he was much entertained. But what particularly excited his risability was the answer made by a grenadier at Lyons. A grand review was held there, just after the Emperor had landed on his return from Elba. The Commanding Officer remarked to his soldiers that they were well clothed and well fed, that their pay might be seen upon their persons: – 'Yes, certainly,' replied the grenadier to whom he addressed himself. – 'Well!': continued the officer, with a confident air, 'it was not so under Bonaparte. Your pay was in arrear, he was in your debt?' – 'And what did that signify,' said the grenadier smartly, 'if we chose to give him credit?'

A story told to Napoleon at St Helena, 16 April 1816, quoted in Emmanuel Las Casas, *Memoirs of the Life, Exile, and Conversations of the Emperor Napoleon*, II, 1890.

My present pay is not wholly for present work but is in great part for past services… What money will pay Meade for Gettysburg? What Sheridan for Winchester? What Thomas for Chickamauga?

General of the Army William T. Sherman, 1870, letter on pending legislation to cut Army officers' pay.

PEACE

You cannot pay my Marines enough for what they do for this nation. But you sure can pay them too little.
> General Charles Krulak, testimony before the House Armed Services Committee, quoted in the San Diego Union-Tribune, 23 February 1999.

PEACE

If one has a free choice and can live undisturbed, it is sheer folly to go to war.
> Pericles, at the beginning of the Peloponnesian War, quoted in Thucydides, The Peloponnesian War, c. 404 BC.

Peace is an armistice in a war that is continuously going on.
> Thucydides, The Peloponnesian War, c. 404 BC.

For peace, with justice and honour, is the fairest and most profitable of possessions, but with disgrace and shameful cowardice it is the most infamous and harmful of all.
> Polybius, Histories, 4, c. 125 BC (The Rise of the Roman Empire, tr. Ian Scott-Kilvert, 1987).

The first blessing is peace, as is agreed upon by all men who have even a small share of reason. It follows that if any one should be a destroyer of it, he would be most responsible not only to those near him but also to his whole nation for the troubles which come. The best general, therefore, is that one who is able to bring about peace from war.
> Count Belisarius, AD 530, admonition to the Persian generals not to disrupt peace negotiations, quoted in Procopius, History of the Wars, I, c. AD 560.

I am to salute you, and tell you that the Muslims and Franks are bleeding to death, the country is utterly ruined and goods and lives have been sacrificed on both sides. The time has come to stop this.
> Richard I, 1191, to Saladin, quoted in Baha ad-Din, The Rare and Excellent History of Saladin, 1969 p. 226.

I think it is evident to all men of wisdom and discretion that have read divers notable histories with consideration and judgment, as also that have well considered of this our age, that there are two things of all others that are the greatest enemies to the art and science military and have been the occasion of the great decay, and oftentimes the utter ruin, of many great empires, kingdoms, and commonswealths... the first is long peace, which ensuing after great wars to divers nations that have had notable militias and exercises military in great perfection, they by enjoying long peace have so much given themselves to covetousness, effeminacies, and superfluities that they have either in great part or else utterly forgotten all orders and exercises military.
> John Smythe (c. 1580–1631), Certain Discourses Military, 1590.

Though peace be made, yet it is interest that keeps peace.
> Oliver Cromwell, 4 September 1654, to Parliament.

What my enemies call a general peace is my destruction. What I call peace is merely the disarmament of my enemies. Am I not more moderate than they?
> Napoleon, 1813, quoted in Christopher Herold, The Mind of Napoleon, 1955.

We desire a peace that will be honourable to both parties. And, as I understand this document, we are leaving honour behind us, for we are now not only surrendering our independence, but we are allowing every burgher to be fettered hand and foot. Where is the 'honourable peace' for us? If we conclude peace, we have to do it as men who have to live and die here. We must not agree to a peace which leaves

PEOPLE AND ARMY

behind in the hearts of one party a wound that will never heal.
> General Louis Botha, 19 May 1902, at the peace conference ending the Boer War, quoted in De Wet, *Three Years War*, 1902.

Injustice, arrogance, displayed in the hour of triumph will never be forgotten or forgiven.
> Prime Minister Lloyd George, memorandum written during the Paris peace conference, 1919.

I am not worried about the war; it will be difficult but we shall win it; it is after the war that worries me. Mark you, it will take years and years of patience, courage and faith.
> Field Marshal Jan Christian Smuts, quoted in Tedder, *With Prejudice*, 1948.

If man does find the solution for world peace it will be the most revolutionary reversal of his record we have ever known.
> General of the Army George C. Marshall, 1 September 1945, Biennial Report, Chief of Staff, US Army.

The object of war is a better state of peace – even if only from your own point of view. Hence it is essential to conduct war with constant regard to the peace you desire. That applies both to aggressor nations who seek expansion and to peaceful nations who only fight for self-preservation – although their views of what is meant by a better state of peace are very different.
> Captain Sir Basil Liddell Hart, *Strategy*, 1954.

We seek peace. We strive for peace. And sometimes peace must be defended. A future lived at the mercy of terrible threats is not peace at all.
> President George W. Bush, 28 January 2003, State of the Union Address, *Washington Post*, 29 January 2003, p. A11.

PEOPLE AND ARMY

As commander of the Peloponnese, I led many thousands of men. I undertook expeditions and waged battles, but I confess in all honesty, that during the entire war I did not spend one single penny of my own, for I had none to spend. It was the people who fed me and who kept supplying all that my troops needed, and thanks to my sword I got horses and weapons from the Turks.
> Theodoros Kolokotrones, August 1822, Nemea.

The nation that will insist on drawing a broad line of demarcation between the fighting man and the thinking man is liable to find its fighting done by fools and its thinking done by cowards.
> Lieutenant-General Sir William Butler (1838–1910).

It has always been an article of my creed that Army and people have but one body and one soul, and that the Army cannot remain sound for ever if the nation is diseased.
> General Erich Ludendorff, *My War Memories, 1914–1918*, 1919.

Civilians provide the manpower for our huge armies. Parents provide the sons who fight. They make sacrifices, enormous sacrifices for the cause. Wives lose their husbands, children lose their fathers, families lose their breadwinners. They go short of food, clothing and the comforts they are used to. They pay the heavy taxes required to finance our war effort... [hence] when they know that serious mistakes have been made they want to know why. After all they pay the cost of those mistakes.
> Lieutenant-General H. Gordon Bennett, *Why Singapore Fell*, 1944.

The people are like water and the army is like fish.
> Mao Tse-tung, 'Aspects of China's Anti-Japanese Struggle', 1948.

PERSONAL PRESENCE OF THE COMMANDER

The Vietnamese fighter has always taken care to observe point 9 of its Oath of Honour:

'In contacts with the people, to follow these three recommendations:
 – To respect the people
 – To help the people
 – To defend the people... in order to win their confidence and affection and achieve a perfect understanding between the people and the army.'
General Vo Nguyen Giap, *People's War – People's Army*, 1961.

What a society gets in its armed services is exactly what it asks for, no more and no less. What it asks for tends to be a reflection of what it is. When a country looks at its fighting forces it is looking in a mirror; the mirror is a true one and the face that it sees will be its own.
General Sir John Hackett, *The Profession of Arms*, 1983.

PERSEVERANCE

The merit of the action lies in finishing it to the end.
Genghis Khan (1162–1227).

There must be a beginning of any great matter, but the continuing unto the end until it be thoroughly finished yields the true glory.
Sir Francis Drake, 17 May 1587, letter to Sir Francis Walsingham.

We fight, get beat, rise, and fight again.
Major-General Nathaniel Greene, 22 June 1781, letter to Chevalier de la Luzerne on the campaign in the Carolinas.

I propose to fight it out on this line, if it takes all summer.
General of the Army Ulysses S. Grant, 11 May 1864, dispatch from Spotsylvania Courthouse.

PERSONAL PRESENCE OF THE COMMANDER

It is not fitting for a good king and a brave captain to leave his own encampment.
Ahuítzotl, 'The Lion of Anáhuac', Aztec Emperor, 1487, when a vassal lord offered the comfort of his palace while on campaign.

I see that it is only in my own hand, that my sword is invincible.
Suleyman the Magnificent, late September 1565, upon the return of his defeated army and fleet from the epic siege of Malta, the joint command of which he had delegated to his chief general, Mustapha Pasha, and his chief admiral, Piali, quoted in Ernle Bradford, *The Great Siege: Malta 1565*, 1961.

The presence of a general is necessary: he is the head, he is the all in all of an army. It was not the Roman army conquered Gaul, but Caesar; it was not the Carthaginians made the armies of the Republic tremble at the very gates of Rome, but Hannibal; it was not the Macedonian army marched to the Indus, but Alexander; it was not the French army that carried war to the Weser and to the Inn, but Turenne; it was not the Prussian army that defended Prussia during the seven years against the three strongest Powers of Europe, but Frederick the Great.
Napoleon, 30 July 1800, quoted in R.M. Johnston, ed., *The Corsican*, 1910.

I believe my arrival was most welcome, not only to the Commanders of the Fleet, but to almost every individual in it; and when I came to explain to them the Nelson touch it was like an electric shock. Some shed tears, all approved. 'It was new – it was singular – it was simple!' and from admirals downwards it was repeated, – 'It must succeed if ever they will allow us to get at them. You are, my Lord, surrounded by friends, whom you inspire with confidence.'
Admiral Viscount Nelson, October 1805,

PERSONAL PRESENCE OF THE COMMANDER

upon his return to the fleet before the Battle of Trafalgar when he had described his new tactics, quoted in David Walder, *Nelson*, 1978.

A general who sees with the eyes of others will never be able to command an army as it should be.
   Napoleon, 9 December 1817, to Barry E. O'Meara at St Helena, *Napoleon in Exile*, 1822.

The real reason why I succeeded in my campaign [in India] is because I was always on the spot. I saw everything and did everything myself.
   The Duke of Wellington (1769–1852).

By God! I don't think it would have been done if I had not been there.
   The Duke of Wellington (1769–1852), speaking of the Battle of Waterloo, 1815, quoted in *The Greevey Papers*, I, 1903.

Oh yes – there was nothing like him. He suited a French army so exactly! Depend upon it, at the head of a French army there never was anything like him. In short, I used to say of him that his presence on the field made the difference of forty thousand men.
   The Duke of Wellington (1769–1852), quoted in Earl Stanhope, *Conversations With the Duke of Wellington*, 1888.

The position I am now placed in I feel will prove a trying one, but by having an eye to duty alone I shall hope to succeed. Placed in command of all the Armies as I have been it will be necessary for me to have an office and an A.A. Gen. in Washington, but I will not be there. I shall have Hd.Qrs. in the field and will move from one Army to another so as to be where my presence seems to be most required.
   General of the Army Ulysses S. Grant, letter to T. Lyle Dickey, 15 March 1864, *Papers of USG*, X, 1982, p. 208.

Some men think that modern armies may be so regulated that a general can sit in an office and play on his several columns as on the keys of a piano; this is a fearful mistake. The directing mind must be at the very head of the army – must be seen there, and the effect of his mind and personal energy must be felt by every officer and man present with it, to secure the best results. Every attempt to make war easy and safe will result in humiliation and disaster.
   General of the Army William T. Sherman, *The Memoirs of General W.T. Sherman*, II, 1875.

When the battle becomes hot, they must see their commander, know him to be near. It does not matter even if he is without initiative, incapable of giving an order. His presence creates a belief that direction exists, and that is enough.
   Colonel Charles Ardant du Picq, *Battle Studies*, 1880, tr. Greely, 1957.

On reaching this camp on the 8th of February the Commander-in-Chief immediately proceeded to visit the troops, and by his cheery smile and friendly recognition did much to revive the spirits of those who were feeling disheartened owing to previous failures and disappointments. Lord Roberts possessed an attractive personality, took infinite pains to secure the confidence and esteem of his troops, and to show them that their interests were his – as they undoubtedly were. It is to be regretted that his example is not more frequently followed by other leaders, since the neglect of it greatly reduces the fighting value of the troops and cannot be made good by any other qualities of leadership, with the sole exception, perhaps, of an unbroken string of victories, and this rarely falls to the lot of the commander of whom regimental officers and men know and see little, and for whom they consequently care less.

PERSONAL PRESENCE OF THE COMMANDER

Field Marshal Sir William Robertson, *From Private to Field-Marshal*, 1921.

Perhaps the most damning comment on the plan which plunged the British Army in this bath of mud and blood [Third Battle of Ypres, 1917] is contained in an incidental revelation of the remorse of one who was largely responsible for it. This highly-placed officer from General Headquarters was on his first visit to the battle front – at the end of a four month's battle. Growing increasingly uneasy as the car approached the swamp-like edges of the battle area, he eventually burst into tears, crying 'Good God, did we really send men to fight in that?' To which his companion replied that the ground was far worse ahead. If the exclamation was a credit to his heart it revealed on what a foundation of delusion and inexcusable ignorance his indomitable 'offensiveness' had been based.

Captain Sir Basil Liddell Hart, speaking of Lieutenant-General Sir Lancelot Kiggell, *The Real War, 1914–1918*, 1930.

The most rapid way to shell-shock an army is to shell-proof its generals; for once the heart of an army is severed from its head the result is paralysis. The modern system of command has in fact guillotined generalship, hence modern battles have degenerated into saurian writhings between headless monsters.

Major-General J.F.C. Fuller, *Generalship: Its Diseases and Their Cure*, 1933.

What troops and subordinate commanders appreciate is that a general should be constantly in personal contact with them, and should not see everything simply through the eyes of his staff. The less time a general spends in his office and the more with his troops the better.

Field Marshal Viscount Wavell of Cyrenaica, *Generals and Generalship*, 1941.

Upon going ashore at Sorrento, an officer approached a Ranger in uniform asking, 'Do you know where I can find Colonel Darby?' A slow grin crossed the face of the husky soldier as he answered, 'You'll never find him this far back.'

As told by General William H. Baumer, of Colonel William O. Darby during the Allied landing in Italy on 9 September 1943, quoted in John D. Lock, *To Fight With Intrepidity*, 1998.

The day of a commander in the field is one of constant vigilance. He must continually be getting about to see what is actually happening and be sure that his subordinates are in fact doing what is required of them in the best possible way. This getting about not only keeps the commander informed by first hand knowledge, but the confidence of the troops is raised when they constantly see him amongst themselves. No walk I ever did in New Guinea failed to pay a dividend, either by seeing or improving something or by maintaining the morale of the soldiers by having a chat with those I met.

Major-General George Vasey (1895–1945), quoted in Horner, *The Commanders*, 1984.

A study of the map will indicate where critical situations exist or are apt to develop, and so indicate where the commander should be.

General George S. Patton, Jr., *War As I Knew It*, 1947.

The place of all commanders of armour up to the divisional commander is on the battlefield, and within this wherever they have the best view of the terrain and good communications with the hard core of the tanks. I was always located where I could see and hear what was going on 'in front', that is near the enemy and around myself – namely at the focal point! Nothing and nobody can replace a personal impression.

PERSONAL PRESENCE OF THE COMMANDER

Every experienced commander is familiar with this sort of panic which, in a critical situation, may seize an entire body of troops. Mass hysteria of this type can be overcome only by energetic actions and a display of perfect composure. The example set by a true leader can have miraculous effects. He must stay with his troops, remain cool, issue precise orders, and inspire confidence by his behavior. Good soldiers never desert such a leader. News of the presence of high ranking commanders up front travels like wildfire along the entire front line, bolstering everyone's morale. It means a sudden change from gloom to hope, from imminent defeat to victory.

Major-General Hasso von Manteuffel, c. 1951, unpublished manuscript, quoted in Richard Simpkin, *Tank Warfare*, 1979.

Colonel-General Erhard Raus, *Defense Tactics Against Russian Breakthroughs*, D.A. Pam. No. 20-233, Department of the Army, 1951.

In moments of panic, fatigue, or disorganization, or when something out of the ordinary has to be demanded... the personal example of the commander works wonders, especially if he has the wit to create some sort of legend 'round himself.

Field Marshal Erwin Rommel, *The Rommel Papers*, 1953.

One of the most valuable qualities of a commander is a flair for putting himself in the right place at the vital time.

Field Marshal Viscount Slim of Burma, *Unofficial History*, 1957.

As commander of a division or smaller unit, there will rarely be more than one crisis, one really critical situation facing you at any one time. The commander belongs right at the spot, not at some rear command post. He should be there before the crisis erupts, if possible. If it is not possible, then he should get there as soon as he can after it develops. Once there, then by personal observation of terrain, enemy fires, reactions, and attitudes of his own commanders on the spot – by his eyes, ears, brain, nose, and his sixth sense – he gets the best possible picture of what is happening and can best exercise his troop leadership and the full authority of his command. He can start help of every kind to his hard-pressed subordinates. He can urge commanders to provide additional fire support, artillery, air, and other infantry weapons...

No other means will provide the commander with what his personal perceptions can provide, if he is present at the critical time and place. He can personally intervene, if he thinks that necessary, but only to the extent that such intervention will be helpful and not interfere with his subordinates. He is in a position to make instant decisions, to defend, withdraw, attack, exploit, or pursue.

If, at this time, he is at some rear command post, he will have to rely on reports from others, and time will be lost, perhaps just those precious moments which spell the difference between success and failure. Notwithstanding the console capabilities of future television in combat, I still believe what I have said is true. In any event, keep this time factor in mind. It is the one irretrievable, inextensible, priceless element in war.

General Matthew B. Ridgway, 'Leadership', *Military Review*, 9/1966.

From the practice of the first operations I concluded that those commanders failed most often who did not visit the terrain, where action was to take place, themselves only studied it on the map and issued written orders. The commanders who are to carry out combat missions must by all means know the terrain and enemy battle

PERSONAL RISK OF THE COMMANDER

formations very well in order to take advantage of weak points in his dispositions and direct the main blow there.
  Marshal of the Soviet Union Georgi K. Zhukov, *Reminiscences and Reflections*, 1974.

PERSONAL RISK OF THE COMMANDER

After haranguing the Tenth Legion, Caesar started for the right wing where he saw his men under great pressure. The standards of the Twelfth were huddled in one place and the soldiers so cramped that their fighting was hampered. all the centurions of the fourth cohort were cut down, the standard bearer killed and standard lost, and almost all the centurions of the other cohorts were killed or wounded... the enemy did not remit their pressure... and were pressing in from either flank. Caesar saw that the situation was critical, and there was no reserve to throw in. He snatched a shield from a soldier in the rear... and moved to the front line; he called upon the centurions by name, encouraged the men to advance, and directed them to open their lines out to give freer play with their swords. His coming inspired the men with hope and gave them new heart. Even in a desperate situation each man was anxious to do his utmost when his general was looking on, and the enemy's onset was somewhat slowed down.
  Julius Caesar, 57 BC, when his camp was surprised by the Nervii on the Sambre River, *The Gallic War*, c. 51 BC, tr. Moses Hadas.

I was the first to set a ladder against the fortress on the bridge, and, as I raised it, I was wounded in the throat by a cross-bow bolt. But Saint Catherine comforted me greatly. and I did not cease to ride and do my work.
  Joan of Arc, 7 May 1429, on the capture of the bridge at Orléans, *Joan of Arc in her Own Words*, 1996.

Get to the front with your firing-line. It is no longer a case for acting as in recent years, but you must again put on your boots and your resolution of '93! When the French see your cocked hat with the skirmishers, and see you exposing yourself foremost to the enemy's fire, you can do what you like with them.
  Napoleon, 21 February 1814, letter to Marshal Augereau, quoted in R.M. Johnston, ed., *The Corsican*, 1910.

Jackson is with you... Rally, brave men, and press forward. Your general will lead you. Jackson will lead you. Follow me!
  Lieutenant-General Thomas J. 'Stonewall' Jackson, 9 August 1862, rallying his shattered troops at the battle of Cedar Mountain, quoted in David Martin, *The Second Bull Run Campaign*, 1997.

I long ago made up my mind that it was not a good plan to fight battles with paper orders – that is, for the commander to stand on a hill in the rear and send his aides-de-camp with written orders to the different commanders. My practice has always been to fight in the front rank.
  Major-General Phillip H. Sheridan, 24 October 1864, statement to Charles A. Dana.

In any of these fights, a general officer who does his duty has got to expose himself. Otherwise, he cannot look himself in the face and order men to do things he is afraid to do himself. I am sure that whatever success I have had resulted from my adherence to this belief.
  General George S. Patton, Jr, 9 June 1943, letter to his nephew, Frederick Ayer, quoted in Roger H. Nye, *The Patton Mind*, 1993.

I've always believed no officer's life, regardless of rank, is of such great

PHYSICAL FITNESS

value to his country that he should seek safety in the rear... Officers should be forward with their men at the point of impact. That is where their character stands out and can do the most good... Things work better that way. Men expect you to, and men look to officers and NCOs for example.

Lieutenant-General Lewis 'Chesty' Puller (1898–1971), quoted in *Marine Corps Gazette*, 6/1998.

PHYSICAL FITNESS

No citizen has any right to be an amateur in the matter of physical training. It is part of his profession as a citizen to keep himself in good condition, ready to serve his state at a moment's notice. The instinct of self-preservation demands it likewise: for how helpless is the state of the ill-trained youth in war or in danger! Finally, what a disgrace it is for a man to grow old without ever seeing the beauty and strength of which his body is capable.

Socrates (469–399 BC), according to Xenophon, quoted in Daniel Boorstein, *The Creators*, 1992.

Daily practice of the military exercises is much more efficacious in preserving the health of an army than all the art of medicine.

Flavius Vegetius Renatus, *Military Institutions of the Romans*, c. AD 378, tr. Clark, 1776.

Remember that victory depends on the legs; the hands are only the instruments of victory.

Field Marshal Prince Aleksandr V. Suvorov (1729–1800), quoted in W. Lyon Blease, *Suvorof*, 1920.

In this situation the real leaders of the Army stood forth in bold contrast to the ordinary clay. Men who had sustained a reputation for soldierly qualities, under less trying conditions, proved too weak for the ordeal and became pessimistic calamity howlers. Their organizations were quickly infected with the same spirit and grew ineffective unless a more suitable commander was given charge. It was apparent that the combination of tired muscles, physical discomforts, and heavy casualties weakened the backbone of many. Officers of high rank who were not in perfect physical condition usually lost the will to conquer and took an exceedingly gloomy view of the situation.

General of the Army George C. Marshall, *Memoirs of My Service in the World War*, written 1923–1926, first published 1976.

I want them to be able to march 20 miles, the last five at double time, and then be ready to fight.

Lieutenant-General Lewis B. 'Chesty' Puller, 1951, upon his return from Korea to command a training brigade, quoted by Associated Press, 11 October 1971.

PLANNING/PLANS

For if many ill-conceived plans have succeeded through the still greater fatuity of an opponent, many more, apparently well laid, have on the contrary ended in disgrace. The confidence with which we form our schemes is never completely justified in their execution; speculation is carried on in safety, but, when it comes to action, fear causes failure.

Thucydides, *History of the Peloponnesian War*, 1.121, c. 404 BC, tr. Richard Crawley, 1910.

It is better to see and communicate the difficulties and dangers of an enterprise, and to endeavour to overcome them, than to be blind to everything but success till the moment of difficulty comes, and to despond.

The Duke of Wellington, 1798, *Supplementary Despatches, Correspondence and Memoranda of Field Marshal Arthur Duke of Wellington, K.G.*, 1794–1818, I, 1858, pp. 195–196.

PLANNING/PLANS

There is no greater coward than I when I am drawing up a plan of campaign. I magnify every danger, every disadvantage that can be conceived. My nervousness is painful; but not that I show a cool face to those who are about me. I am like a woman in the throes of childbirth. When once my decision is made, however, I forget all, except what may carry it through to success.
    Napoleon, 27 April 1800, quoted in R.M. Johnston, ed., *The Corsican*, 1910.

Who will attack tomorrow, I or Bonaparte? Bonaparte. Well, Bonaparte has not given me any idea of his projects; and, as my plans will depend upon his, how can you expect me to tell you what mine are? There is one thing certain, Uxbridge, that is, that whatever happens, you and I will do our duty.
    The Duke of Wellington, June 1815, just before the Battle of Waterloo, to General Uxbridge, his nominal second-in-command, quoted in William Fraser, *Words on Wellington*, 1889.

If I always appear prepared, it is because before entering on an undertaking, I have meditated for long and have foreseen what may occur. It is not genius which reveals to me suddenly and secretly what I should do in circumstances unexpected by others; it is thought and meditation.
    Napoleon (1769–1821).

Be audacious and cunning in your plans, firm and persevering in their execution, determined to find a glorious end.
    Major-General Carl von Clausewitz, *Principles of War*, 1812.

Indecision is our bane. A bad plan, in my mind, followed out without wavering, is better than three or four good ones not so dealt with.
    General Charles 'Chinese' Gordon, 1864, quoted in Paul Charrier, *Gordon of Khartoum*, 1965.

I never plan beyond the first battle.
    Field Marshal Helmuth Graf von Moltke, *Ausgewaehlte Werke*, IV, 1925.

Planning is everything – Plans are nothing.
    Field Marshal Helmuth Graf von Moltke (1880–1891), sign above the entrance to the Joint Staff, Department of Defense, the Pentagon.

The main thing is always to have a plan; if it is not the best plan, it is at least better than no plan at all.
    General Sir John Monash, 1918, letter.

Now, as every policy must be plastic enough to admit of fluctuations in national conditions, so must each plan be plastic enough to receive the impressions of war, that is power to change its shape without changing or cracking its substance. This plasticity is determined psychologically by the condition of mentality in the two opposing forces. There is the determination between the commanders-in-chief, and between them and their men, and ultimately, between the two forces themselves.
    Major-General J.F.C. Fuller, *The Reformation of War*, 1923.

When making a plan, try to put yourself in the enemy's mind, and think what course it is least probable he will foresee and forestall. The surest way to success in war is to choose the course of least expectation. (May 1930.)
    Captain Sir Basil Liddell Hart, *Thoughts on War*, 1944.

There is a close analogy between what takes place in the mind of a military commander when planning an action, and what happens to the artist at the moment of conception. The latter does not renounce the use of his intelligence.

PLANNING/PLANS

He draws from it lessons, methods, and knowledge. But his power of creation can operate only if he possesses, in addition, a certain instinctive faculty which we call inspiration, for that alone can give the direct contact with nature from which the vital spark must leap. We can say of the military art what Bacon said of the other arts: 'They are the product of man added to nature.'
General Charles de Gaulle, *The Edge of the Sword*, 1932.

A plan, like a tree, must have branches – if it is to bear fruit. A plan with a single aim is apt to prove a barren pole. (Jan. 1933.)
Captain Sir Basil Liddell Hart, *Thoughts on War*, 1944.

A good plan violently executed Now is better than a perfect plan next week.
General George S. Patton, Jr, *War As I Knew It*, 1947.

On the operational side a C.-in-C. must draw up a master plan for the campaign he envisages and he must always think and plan two battles ahead – the one he is preparing to fight and the next one – so that success gained in one battle can be used as a spring-board for the next. He has got to read the mind of his opponent, to anticipate enemy reactions to his moves, and to take quick steps to prevent enemy interference with his own plans. He has got to be a very clear thinker and able to sort out the essentials from the mass of factors which bear on every problem.
Field Marshal Viscount Montgomery of Alamein, *The Memoirs of Field Marshal Montgomery*, 1958.

But in truth, the larger the command, the more time must go into planning; the longer it will take to move troops into position, to reconnoiter, to accumulate ammunition and other supplies, and to coordinate other participating elements on the ground and in the air. To a conscientious commander, time is the most vital factor in his planning. By proper foresight and correct preliminary action, he knows he can conserve the most precious elements he controls, the lives of his men. So he thinks ahead as far as he can. He keeps his tactical plan simple. He tries to eliminate as many variable factors as he is able. He has a firsthand look at as much of the ground as circumstances render accessible to him. He checks each task in the plan with the man to whom he intends to assign it. Then – having secured in almost every instance his subordinate's wholehearted acceptance of the contemplated mission and agreement on its feasibility – only then does he issue an order.
General Matthew B. Ridgway, *The Korean War*, 1967.

Naturally, in the course of a battle, one would like to fulfill the initial plan… – but what does it mean to plan in war? We plan alone, but we fulfill our plans, if one may do so, together with the enemy, that is, taking account of his counteraction.
Marshal of the Soviet Union Ivan S. Koniev (1897–1973), 1972, quoted in Nathan Leites, *The Soviet Style in War*, 1982.

I grew big with these plans. I had merely to cross a river, capture Brussels and then go on and take the port of Antwerp. And all this in the worst months of the year, December, January, February, through the countryside where snow was waist deep and there wasn't room to deploy four tanks abreast, let alone six armoured divisions; when it didn't get light until eight in the morning and was dark again at four in the afternoon; with divisions that had just been reformed and contained chiefly raw, untried recruits; and at Christmas time.

POLITICS AND THE MILITARY

General Josef 'Sepp' Dietrich (1892–1966), sarcastic comment on the German planning for the Battle of the Bulge, quoted in Charles Messenger, *Hitler's Gladiator*, 1987.

I emphasized meticulous planning not simply because I thought it was the most effective approach, which it is, but because by taking that approach you enforce on your subordinates the same necessity. They have to learn every detail of the topography, every position, every soldier they will be facing. And once they do that, they will be able to decide rationally – not intuitively – on the steps they have to take. They will make their decisions on the basis of knowledge. Experience had also taught me that if you lay your plans in detail before you are under the stress of fighting, the chances are much greater that you will be able to implement at least the outlines of the plans despite the contingencies of battle.
General Ariel Sharon, *Warrior*, 1989.

Any damned fool can write a plan. It's the execution that gets you screwed up.
General James F. Hollingsworth, quoted in John D. Lock, *To Fight With Intrepidity*, 1998.

The plan was smooth on paper, only they forgot about the ravines.
Russian military proverb.

POLITICS AND THE MILITARY

Whatever talents I may possess (and they are but limited) are military talents. My education and training are military. I think the military and the civil talents are distinct, if not different, and full duty in either sphere is about as much as one man can qualify himself to perform. I shall not do the people the injustice to accept high civil office with whose questions it has not been my business to become familiar.
General Robert E. Lee's conversation with Senator B.H. Hill, recounted in a post-war speech, quoted in William Jones, *Personal Reminiscences, Anecdotes, and Letters of Gen. Robert E. Lee*, 1874.

Avoid all political entanglements. I have always thought the most slavish life any man could lead was that of a politician. Besides I do not believe any man can be successful as a soldier whilst he has an anchor ahead for other advancement. I know of no circumstances likely to arise which could induce me to accept of any political office whatever. My only desire will be, as it has always been, to whip out the rebellion in the shortest way possible and to retain as high a position in the Army afterwards as the Administration then in power may think me suitable for.
General Ulysses S. Grant, 16 February 1864, letter to Commander Daniel Ammen, *Papers of USG*, X, 1982, pp. 132–133.

The Army will hear nothing of politics from me, and in return I expect to hear nothing of politics from the Army.
Herbert Asquith, speech at Ladybank, England, 4 April 1914.

War cannot for a single minute be separated from politics.
Mao Tse-tung, 1938, lecture.

My life is a mixture of politics and war. The latter is bad enough – but I've been trained for it! The former is straight and unadulterated venom! But I have to devote lots of my time, and much more of my good disposition, to it.
General of the Army Dwight D. Eisenhower, 27 September 1943, letter to his wife.

There was a constant infiltration of enemy forces from the Laotian border, who were protected by the political restrictions from Washington that

POWER

prevented us from knocking out their reinforcement bases or supply lines. We would constantly be pushing them back, but they had a damned sanctuary across the border guaranteed by American politicians. So we had to take back the same areas over and over again. I don't think there was a guy, from private to general, who didn't have a stench in his mouth from the politics of the war.
    Major-General James Day, 'Bushido Code', Al Santoli, ed., *Leading the Way: How Vietnam Veterans Rebuilt the U.S. Military*, 1993.

In a normal state, you cannot drive the army into politics with a stick. But our army [Russian] has been systematically politicized. Recall the slogans of 1993, not very long ago – the Army is outside politics. But who decided the outcome of the political confrontation in the fall of that very year at the White House? The Army. Who turned out to be the country's main political figure? The commander of a tank regiment. It was he who presented his 125-mm political arguments and decided the outcome of the campaign. One can only guess what would have happened had he presented his arguments on the other side.
    General Aleksandr I. Lebed, speaking of the participation of Russian Army elements in the suppression of the October 1993 coup in Moscow, *Literaturnaya Gazeta* (Moscow), 17 May 1995.

POWER

The right use of the sword is that it should subdue the barbarians while lying gleaming in its scabbard. If it leaves its sheath it cannot be said to be used rightly. Similarly the right use of military power is that it should conquer the enemy while concealed in the breast. To take the field with an army is to be found wanting in the real knowledge of it. Those who hold the office of Shogun are to be particularly clear on this point.
    Tokugawa Ieyasu (1543–1616), quoted in A.L. Sadler, *The Maker of Modern Japan: The Life of Tokugawa Ieyasu*, 1937.

Power is my mistress. I have worked too hard in conquering her to allow anyone to take her from me or even to covet her.
    Napoleon, in conversation with Pierre-Louis Roederer, 1804, *Journal du comte P.-L. Roederer*, 1909.

There are no manifestos like cannon and musketry.
    The Duke of Wellington (1769–1852).

The only prize much cared for by the powerful is power. The prize of the general is not a bigger tent, but command.
    Justice Oliver Wendell Holmes, Jr, 15 February 1913, speech to Harvard Law School Association of New York.

Power without responsibility: the prerogative of the harlot thoughout the ages.
    Rudyard Kipling, *Kipling Journal*, December 1971.

All peoples, whether they be orientals or occidentals, are far more impressed by an exhibition of power than by the perusual of diplomatic notes.
    Major-General John A. Lejeune, *The Reminiscences of a Marine*, 1929.

Political power emanates from the barrel of a gun.
    Mao Tse-tung, *On Guerrilla Warfare*, 1938.

Power for the sake of lording it over fellow creatures or adding to personal pomp, is rightly judged base. But power in a national crisis, when a man believes he knows what orders should be given, is a blessing.
    Winston S. Churchill, *The Second World War: Their Finest Hour*, 1.1, 1959.

PRAISE

Praise from a friend, or censure from a foe,

PRAISE

Are lost on hearers that our merits know.
Homer, *The Iliad*, c. 800 BC, tr. alexander Pope, 1743.

For men can endure to hear others praised only so long as they can severally persuade themselves of their ability to equal the actions recounted; when this point is passed, envy comes and with it incredulity.
Pericles, quoted in Thucydides, *History of the Peloponnesian War*, 2.120, c. 404 BC, tr. Richard Crowley, 1910.

When a general gives a public speech he ought also to say something in praise of the enemy. This will convince our men, even when you are praising others, that you will never deprive us of the praise we might receive from others and adorn them with our honors.
The Emperor Maurice, *The Strategikon*, c. AD 600 (*Maurice's Strategikon*, tr. George Dennis, 1984).

The desire to imitate brave actions will be aroused by praise. And these trifles will diffuse a spirit of emulation among troops which affects both officers and soldiers and in time will make them invincible.
Field Marshal Maurice Comte de Saxe, *My Reveries*, 1732, tr. Phillips, 1940.

The general can even discuss the war with some of his corps commanders who are most intelligent, and permit them to express their sentiments freely in conversation. If you find some good among what they say, you should not remark about it then, but make use of it. When this has been done, you should speak about it in the presence of many others, it was so-and-so who had this idea; praise him for it. This modesty will gain the general the friendship of thinking men, and he will more easily find persons who will speak their sentiments sincerely to him.
Frederick the Great, *Instructions to His Generals*, 1747, tr. Phillips, 1940.

Praise from enemies is suspicious; it cannot flatter an honorable man unless it is given after the cessation of hostilities.
Napoleon, *The Military Maxims of Napoleon*, tr. Burnod, 1827.

Never in the field of human conflict was so much owed by so many to so few.
Sir Winston Churchill, 20 August 1940, praise to the RAF Fighter Command for its heroism and in the Battle of Britain.

And don't forget a good word for the cooks.
Lieutenant-General Sir Leslie Morshead (1889–1959), at a commanders' conference after the Battle of El Alamein, quoted in Horner, *The Commanders*, 1984.

Humility must always be the portion of any man who receives acclaim earned in the blood of his followers and the sacrifices of his friends.
General of the Army Dwight D. Eisenhower, 12 June 1945, address in London.

All a soldier desires to drive him forward is recognition and appreciation of his work.
General George S. Patton, Jr (1885–1947).

You must get around and show interest in what your subordinates are doing even if you don't know much about the technique of their work. And when you are making these visits, try to pass out praise then due, as well as corrections and criticisms.
General of the army Omar N. Bradley, 16 May 1967, speech to the US Army Command and General Staff College.

A general must never be chary in alloting praise where it is due. People

like to be praised when they have done well. In this connection Sir Winston Churchill once told me of the reply made by the Duke of Wellington, in his last years, when a friend asked him: 'If you had your life over again, is there any way in which you could have done better?' The old Duke replied: 'Yes, I should have given more praise.'
Field Marshal Viscount Montgomery of Alamein, *A History of Warfare*, 1968.

And, you led the fighting into Baghdad the day the statue of the dictator was pulled down.
President George W. Bush, 15 September 2003, address to the 3rd Infantry Division, Ft. Stewart, Georgia, upon its return from Iraq, *AUSA News Release*

PRAYERS AND HYMNS

It is God that girdeth me with strength, and maketh my way perfect.
He maketh my feet like hinds' feet, and setteth me upon my high places.
He teacheth my hands to war, so that a bow of steel is broken by my own arms.
Thou hast also given me the shield of thy salvation: and thy right hand hath holden me up, and thy gentleness hath made me great.
Thou hast enlarged my steps under me, that my feet did not slip.
I have pursued mine enemies, and overtaken: neither did I turn again till they were consumed.
I have wounded them that they were not able to rise: they are fallen under my feet.
For thou hast girded me with strength unto the battle: thou hast subdued under me those that rose up against me.
Thou hast also given me the necks of mine enemies; that I might destroy them that hate me.
They cried, but there was none to save them: even unto the LORD, but he answered them not.
King David (d. 962 BC), Psalms 18:32–41.

Dear Lord, I pray Thee to suffer me not to see Thy Holy City, since I cannot deliver it from the hands of Thy enemies.
King Richard I, 'The Lion Heart', King of England, 1192, on approaching Jerusalem, the object of his Crusade, which he was never to conquer.

O Lord God, when Thou givest to thy servants to endeavour any great matter, grant us also to know that it is not the beginning, but the continuing of the same until it is thoroughly finished which yieldeth the true glory.
Sir Francis Drake (c. 1540–1596).

Whom do you fear, seeing that God the Father is with you?
We fear nothing.
Whom do you fear, seeing that God the Son is with you?
We fear nothing.
Whom do you fear, seeing that God the Holy Spirit is with you?
We fear nothing.
The Gaelic Blessing of 1589, quoted in Admiral Sir J.F. 'Sandy' Woodward, *One Hundred Days: The Memoirs of the Falklands Battle Group Commander*, 1992, p. 43.

Oh Lord! Thou knowest how busy I must be this day: if I forget Thee, do not Thou forget me. March on, boys!
Sir Jacob Hill, 23 October 1642, before the Battle of Edgehill in the English Civil War.

For what we are about to receive may the Lord make us truly thankful.
British sergeant of the First Battalion of Foot Guards (now Grenadier Guards) awaiting the volley of the French at Fontenoy, 11 May 1745, quoted in John Manchip White, *Marshal of France*, 1962, p. 159.

Oh God, let me not be disgraced in my old days. Or if Thou wilt not help me, do not help these scoundrels; but leave us to try it ourselves.

PRAYERS AND HYMNS

Leopold I of Anhalt-Dessau, 'The Old Dessauer', 14 December 1745, before the Battle of Kesselsdorf.

Down, down with the French! ...is my constant prayer.
    Admiral Viscount Nelson, December 1798, quoted in Robert Southey, *Life of Nelson*, 1813.

May the Great God whom I worship grant to my Country, and for the benefit of Europe in general, a great and glorious victory; and may no misconduct in any way tarnish it; and may humanity after Victory be the predominant feature in the British Fleet. For myself individually, I commit my life to Him who made me, and may his blessing light upon my endeavours for serving my Country faithfully. To Him I resign myself and the just cause which is entrusted to me to defend. Amen. Amen. Amen.
    Admiral Viscount Nelson, 21 October 1805, diary prayer before the Battle of Trafalgar.

O Lord, if Thou wilt not be for us today, we ask that Thou be not against us. Just leave it between the French and ourselves.
    Lieutenant-General Sir Alan Campbell of Erracht, prayer before battle, c. 1809.

Hymn Before Action, 1896
    The earth is full of anger
    The seas are dark with wrath,
    The nations in their harness
    Go up against our path:
    Ere yet we loose the legions
    Ere yet we draw the blade,
    Jehovah of the Thunders,
    Lord God of Battles, aid!
    Rudyard Kipling.

Please God – let there be victory, before the Americans arrive.
    Field Marshal Douglas Earl Haig, 1917 diary entry – attributed.

High Flight
Oh! I have slipped the surly bonds of Earth
And danced the skies on laughter-silvered wings;
Sunward I've climbed, and joined the tumbling mirth
of sun-split clouds, – and done a hundred things
You have not dreamed of – wheeled and soared and swung
High in the sunlit silence. Hov'ring there,
I've chased the shouting wind along, and flung
My eager craft through footless halls of air...

Up, up the long delirious, burning blue
I've topped the wind-swept heights with easy grace
Where never lark nor ever eagle flew –
And, while with silent lifting mind I've trod
The high untrespassed sanctity of space,
Put out my hand, and touched the face of God.
    Flying Officer John G. Magee, Royal Canadian Air Force, 3 September 1941. An American volunteer in the RCAF, Magee flew Spitfires in Britain and was killed on a routine training mission on 11 December 1941. He had just sent the sonnet on the back of a letter, saying, 'I am enclosing a verse I wrote the other day. It started at 30,000 feet, and was finished soon after I landed.'

Finally, knowing the vanities of man's effort and the confusion of his purpose, let us pray that God may accept our services and direct our endeavours, so that when we shall have done all we shall see the fruits of our labours and be satisfied.
    Major-General Orde Wingate, February 1943, *Order of the Day*, Imphal.

Almighty and most merciful Father, we

PREPAREDNESS

humbly beseech Thee, of Thy great goodness, to restrain these immoderate rains with which we have had to contend. Grant us fair weather for Battle. Graciously harken to us as soldiers who call upon Thee that, armed with Thy power, we may advance from victory to victory, and crush the oppression and wickedness of our enemies, and establish Thy justice among men and nations. Amen.

Colonel James H. O'Neil, 24 December 1944, the famous weather prayer that General George S. Patton, Jr, ordered him to write as Chaplain of the Third Army during the Battle of the Bulge, quoted in Ladislas Farago, *Patton, Ordeal and Triumph*, 1964.

PREPAREDNESS

Now, it happened through this god, the lord of gods, that I was prepared and armed to trap them like wild fowl. He furnished my strength and caused my plans to prosper. I went forth, directing these marvelous things. I equipped my frontier in Zahi, prepared before them. The chiefs, the captains of infantry, the nobles, I caused to equip the harbor-mouths, like a strong wall, with warships, galleys and barges... They were manned completely from bow to stern with valiant warriors bearing their arms, soldiers of all the choicest of Egypt, being like lions roaring upon the mountain tops. The charioteers were warriors... and all good officers... ready of hand. Their horses were quivering in their every limb, ready to crush the countries under their feet. I was valiant Montu, stationed before them, that they might behold the hand-to-hand fighting of my own arms. I, King Rameses III, was made a far-striding hero, conscious of his might, valiant to lead his army in the day of battle.

Rameses III, Pharaoh of Egypt, c. 1190 BC, of the Northern War against Peoples of the Sea, quoted in James Breasted, *Ancient Records of Egypt*, 1906.

If we undertake this war without preparation, we should by hastening its commencement only delay its conclusion.

Archidamus II, King of Sparta, 432 BC, speech on the prospects of war with Athens, quoted in Thucydides, *History of the Peloponnesian War*, 1.84, c. 404 BC, tr. Richard Crawley, 1910.

He, therefore, who aspires to peace should prepare for war.

Flavius Vegetius Renatus, *Military Institutions of the Romans*, c. AD 378, tr. Clark, 1776.

Sages are very careful not to forget about danger when secure, nor to forget about chaos in times of order. Even when there is peace in the land, it will not do to abandon the military altogether. If you lack foresight, you will be defenseless. It is necessary to develop cultured qualities internally while organizing military preparedness externally. Be considerate and gentle with foreigners, beware of the unexpected. Routine military exercises in each of the four seasons is the way to show that the nation is not oblivious to warfare. Not forgetting about warfare means teaching the people not to give up the practice of martial arts.

The rule is 'Even if the land is at peace, to forget about warfare leads to collapse.'

Liu Ji (1310–1375), *Lessons of War* (*Mastering the Art of War*, tr. Thomas Cleary, 1989).

Happy is that city which in time of peace thinks of war.

Inscription in the Armoury of Venice.

After a fatal procrastination, not only of vigorous measures but of preparation for such, we took a step as decisive as the passage of the Rubicon, and now find ourselves plunged at once in a most serious war without a single requisition, gunpowder excepted, for carrying it on.

PREPAREDNESS

General Sir John Burgoyne, letter from Boston, April 1775, after the Battles of Lexington and Concord.

There is nothing so likely to produce peace as to be well prepared to meet an enemy.
General George Washington, 29 January 1780, letter to Elbridge Gerry.

That people will continue the longest in the enjoyment of peace who timely prepare to vindicate themselves and manifest a determination to protect themselves whenever they are wronged.
Tecumseh, late 1811, to a great Indian council of the southern tribes, urging alliance against the expanding Americans.

DEFENSIVE POLICY of the U.S. – Does not alter PACIFIC POLICY of U.S. Be as pacific as you please but do not let the other fellow catch you unprepared.
Lieutenant-General Robert L. Bullard, 1911, journal entry.

Last month I paid a visit to Berlin; it was a waste of time because when I came back I found that Whitehall knows all about Berlin. The Germans are a peaceful and hard-working nation, full of culture and human kindness. They have an army which consists of about a million of soldiers in the ranks and about four million trained men in civil employment, mostly waiters in Piccadilly. The object of the German army is to maintain the peace of Europe and every year they maintain peace a little more by adding to their Army. We main peace by cutting our Army down – and our way is better because it is cheaper.
Field Marshal Sir Henry Wilson visiting Germany before 1914, quoted in C.R. Ballard, *Kitchener*, 1930, p. 205.

Preparedness does not mean militarism or an aggressive military spirit; it means simply the application to the military questions of the day of something of the experience and lessons of the past as well as those of the present. A man armed against thieves is not prone to become a thief unless he is one at heart. A nation can be strong without being immoral or a bully.
Major-General Leonard Wood, *Our Military History: Its Facts and Fallacies*, 1916.

Naturally enough, military problems are anathema to the men in power. Nobody, whether a spendthrift or a miser likes getting bills. Though, in the last analysis, armaments are the consequences of policy, governments are afraid to impose them until immediate danger makes them obviously essential. On the contrary, when a long period of peace is promised, they are the first to put ships out of commission and disband regiments.
General Charles de Gaulle, *The Edge of the Sword*, 1932.

To maintain in peace a needlessly elaborate military establishment entails economic waste. But there can be no compromise with minimum requirements. In war there is no intermediate measure of success. Second best is to be defeated, and military defeat carries with it national disaster – political, economic, social, and spiritual disaster.
General of the Army Douglas MacArthur, *Annual Report of the Chief of Staff, June 30, 1935.*

Want of foresight, unwillingness to act when action would be simple and effective, lack of clear thinking, confusion of counsel until the emergency comes, until self-preservation strikes its jarring gong – these are the features which constitute the endless repetition of history.
Sir Winston S. Churchill, speech in the House of Commons, 2 May 1935, *Winston*

S. Churchill: *His Complete Speeches, 1897–1963*, VI, 1974, p. 5592.

We must be prepared... [It] is good to be patient, it is good to be circumspect, it is good to be peace-loving – but it is not enough. We must be strong, we must be self-reliant.
Sir Winston S. Churchill (1874–1965).

Without preparedness superiority is not real superiority and there can be no initiative either. Having grasped this point, a force which is inferior but prepared can often defeat a superior enemy by surprise attack.
Mao Tse-tung, *On Protracted War*, May 1938.

Virtuous motives, trammeled by inertia and timidity, are no match for armed and resolute wickedness. A sincere love of peace is not excuse for muddling hundreds of millions of humble folk into total war. The cheers of weak, well-meaning assemblies soon cease to echo, and their votes soon cease to count. Doom marches on.
Sir Winston S. Churchill, *The Gathering Storm*, 1948.

No foreign policy can have validity if there is no adequate force behind it and no national readiness to make the necessary sacrifices to produce that force.
Sir Winston S. Churchill, *The Gathering Storm*, 1948.

We sleep safely in our beds because rough men stand ready in the night to visit violence on those who would harm us.
George Orwell (1903–1950), quoted in George Will, 'Rough and Ready', *The Washington Post*, 22 November 1998.

Generals have often been reproached with preparing for the last war instead of for the next – an easy gibe when their fellow countrymen and their political leaders, too frequently, have prepared for no war at all. Preparation for war is an expensive, burdensome business, yet there is one important part of it that costs little – study. However changed and strange the new conditions of war may be, not only generals, but politicians and ordinary citizens, may find there is much to be learned from the past that can be applied to the future and, in their search for it, that some campaigns have more than others foreshadowed the coming modern war. I believe that ours in Burma was one of these.
Field Marshal Viscount Slim of Burma, *Defeat Into Victory*, 1956.

It is customary in the democratic countries to deplore expenditures on armaments as conflicting with the requirements of social services. There is a tendency to forget that the most important social service a government can do for its people is to keep them alive and free.
Air Marshal Sir John Slessor (1897–1979).

THE PRESS

Newspapers should be limited to advertising.
Napoleon (1769–1821), quoted *In The Words of Napoleon*, p. 9, tr. Daniel Savage Gray, 1977.

Four hostile newspapers were more to be feared than a thousand bayonets.
Napoleon, quoted in Liddell Hart, *The Sword and the Pen*, 1976.

You see me master of France; well, I would not undertake to govern her for three months with liberty of the press.
Napoleon, to Prince Metternich, quoted in Louis de Bourrienne, *The Memoirs of Napoleon Bonaparte*, I, 1855.

Possible? Is anything possible? Read the newspapers.
The Duke of Wellington (1769–1821), quoted in William Fraser, *Words of Wellington*, 1889.

THE PRESS

I was honoured by a good deal of abuse from some at home for telling the truth. But I could not tell lies or make things pleasant... The only thing the partisans of misrule could allege was that I did not 'make things pleasant' for the authorities and that, amid the filth and starvation, and deadly stagnation of the camp, I did not go about 'babbling of green fields' of present abundances and of prospect of victory.
> William Howard Russell, the first of the great war correspondents, of his reporting during the Crimean War, *Crimea Despatches, 1854–1855*, 1966.

So you think the papers ought to say more about your husband! My brigade is not a brigade of newspaper correspondents.
> Lieutenant-General Thomas J. 'Stonewall' Jackson, 5 August 1861, letter to his wife after the First Battle of Manassas.

I am sorry, as you say, that the movements of the armies cannot keep pace with the expectations of the editors of papers. I know they can regulate matters satisfactorily to themselves on paper. I wish they could do so in the field. No one wishes them more success than I do & would be happy to see them have full swing. Genl Floyd has the benefit of three editors on his staff. I hope something will be done to please them.
> General Robert E. Lee, letter to his wife, 7 October 1861, quoted in *Wartime Papers of Robert E. Lee*, 1961.

Why, sir, in the beginning we appointed all our worst generals to command the armies, and all our best generals to edit the newspapers. As you know, I have planned some campaigns and quite a number of battles. I have given the work all the care and thought I could, and sometimes, when my plans were completed, as far as I could see, they seemed to be perfect. But when I have fought them through, I have discovered defects and occasionally wondered why I did not see some of the defects in advance. When it was all over, I found by reading a newspaper that these best editor generals saw all the defects plainly from the start. Unfortunately, they did not communicate their knowledge to me until it was too late... I have no ambition but to serve the Confederacy, and do all I can to win our independence. I am willing to serve in any capacity to which the authorities assign me. I have done the best I could in the field, and have not succeeded as I could wish. I am willing to yield my place to these best generals, and I will do my best for the cause editing a newspaper.
> General Robert E. Lee's conversation with Senator B.H. Hill, as recounted in a post-war speech, quoted in J. William Jones, *Personal Reminiscences, Anecdotes, and Letters of Gen. Robert E. Lee*, 1874.

Every country is finally answerable for the wanton mischief done by its newspapers, and the reckoning is liable to be presented some day in the shape of a final decision from some other country.
> Prince Otto von Bismarck, speech to the Reichstag, 6 February 1888, quoted in Lewis Copeland, ed., *The World's Great Speeches*, 1952.

The printing press is the greatest weapon in the armoury of the modern commander.
> Colonel T.E. Lawrence (1888–1935).

The essence of successful warfare is secrecy. The essence of successful journalism is publicity.
> *British Regulations for War Correspondents*, 1958.

In general, journalism appears to nurture the pontifical judgment. I was on occasion reminded of General Eisenhower's remark to a publisher who had told him at length what was wrong with the conduct of World War II. 'I

thought it was only in the world's oldest profession,' General Eisenhower said, 'that amateurs think they can do better than professionals.'
>General William Westmoreland, A Soldier Reports, 1976.

The Press, as far as I was concerned, could do something very difficult to itself.
>Sandy Woodward, comment on the press version of his first briefing on 26 April 1982, One Hundred Days: The Memoirs of the Falklands Battle Group Commander, 1982, p. 111.

Wars by definition require a hardness of heart that looks terrible on television. Ulysses Grant would have lost his job in a week if he had had to discuss his methods (industrial warfare: the grinder) with Deborah Norville.
>Lance Morrow, 'A New Test of Resolve', Time, 3 September 1990.

Military leaders should plan adequately for dealing with the media. The objective is to ensure military operations are put in the proper context for the American public and audiences around the world. Commanders or senior military officials can no longer get away with a 'no comment' answer regarding American troop use in trouble spots worldwide. Ignoring the media does not make them go away – it just forces them to contact alternative sources for their stories.

Refusing to talk to the media also guarantees the military's perspective will not be heard. Thus, it is to the military's benefit to cooperate with the media because 'We don't win unless CNN says we win.'
>General H. Hugh Shelton, Military Review, November–December 1995.

[T]hey've [the press] got the attention span of gnats.
>Secretary of Defense Donald Rumsfeld, 31 October 2001, quoted in Bob Woodward, 'Doubts and Debate Before Victory over the Taliban,' Washington Post, 18 November 2002, p. A01.

They [television reports] are no more than snapshots of a particular time and a particular place. They tell you very little if anything at all of the progress of the campaign at a strategic level.
>General Sir Michael Jackson, Chief of the General Staff, commenting on the nature of television war reporting in Iraq, quoted by Jo Dillon, The Independent, 30 March 2003.

THE PRINCIPLES OF WAR

The art of war owns certain elements and fixed principles. We must acquire that theory, and lodge it in our heads – otherwise we will never get very far.
>Frederick the Great, quoted in Duffy, The Military Life of Frederick the Great, 1985.

1. Nothing but the offensive. 2. Speed in marches, swiftness. 3. Methodism is not needed. 4. Full authority to the commanding general. 5. Attack the enemy and hit him in the field. 6. Lose no time in sieges... take fortresses chiefly by assaults or open force... 7. Never divide forces for security of different points. If the enemy went around them, so much the better: he approaches in order to be defeated... Never occupy yourself with vain maneuvers.
>Marshal Aleksandr V. Suvorov, his plan for the Italian Campaign (1799), quoted in Voyennaya istoriya (Moscow), 1971.

It is a principle of warfare, that when it is possible to make use of thunderbolts, they should be preferred to cannon.
>Napoleon, dictation on St Helena, Correspondance de Napoléon Ier, publiée par ordre de l'Empereur Napoléon III, XXXI, p. 429, 1858–1870, quoted in Christopher Herold, ed., The Mind of Napoleon, 1955.

It is true that Jomini always argues for fixed principles. Genius works by

THE PRINCIPLES OF WAR

inspiration. What is good in certain circumstances may be bad in others; but one ought to consider principles as an axis which holds certain relations to a curve. It may be good to recognize that on this or that occasion one has swerved from fixed principles of war.
Napoleon, quoted in Gaspard Gourgaud, Talks of Napoleon at St. Helena, 1816.

Tactics can be learned from treatises, somewhat like geometry, and so can the various military evolutions or the science of the engineers and the gunner; but knowledge of the grand principles of warfare can be acquired only through the study of military history and of the battles of the great captains and through experience. There are no precise, determinate rules: everything depends on the character that nature has bestowed on the general, on his qualities and defects, on the troops, on the range of the weapons, on the season of the year, and on a thousand circumstances which are never twice the same.
Napoleon (1769–1821), dictation on St. Helena, Correspondance de Napoléon Ier, publiée par ordre de l'Empereur Napoléon III, XXXI, p. 365, 1858–1870, quoted in Christopher Herold, ed., The Mind of Napoleon, 1955.

Principles of military art shine in history like the Sun on the horizon, so much the worse for blind men incapable of seeing them.
Napoleon, quoted in The Principles of Combined Arms Battle (Moscow), 1992.

Get your principles straight; the rest is a matter of detail.
Napoleon (1769–1821).

These three things you must always keep in mind: concentration of strength, activity, and a firm resolve to perish gloriously. They are the three principles of the military art which have disposed luck in my favor in all my operations.
Napoleon, 1804, letter to General Lauriston, quoted in Christopher Herold, ed., The Mind of Napoleon, 1955.

The principles of war are the same as those of a siege. Fire must be concentrated at one point, and as soon as the breach is made, the equilibrium is broken and the rest is nothing.
Napoleon, Maxims of War, 1831.

Only those general principles and attitudes that result from clear and deep understanding can provide a *comprehensive* guide to action. It is to these that opinions on specific problems should be anchored. The difficulty is to hold fast to these results of contemplation in the torrent of events and new opinions. Often there is a gap between principles and actual events that cannot always be bridged by a succession of logical deductions. Then a measure of self-confidence is needed, and a degree of skepticism is also salutary. Frequently nothing short of an imperative principle will suffice, which is not part of the immediate thought-process, but dominates it: that principle is in all doubtful cases *to stick to one's first impression and to refuse to change unless forced to do so by a clear conviction.* A strong faith in the overriding truth of tested principles is needed; the *vividness* of transient impressions must not make us forget that such truth as they contain is of a lesser stamp. By giving preference, in case of doubt, to our earlier convictions, by holding to them stubbornly, our actions acquire that quality of steadiness and consistency which is termed strength of character.
Major-General Carl von Clausewitz, On War, 1.3, 1832, tr. Michael Howard and Peter Paret, 1976.

One great principle underlies all the operations of war – a principle which must be followed in all good

THE PRINCIPLES OF WAR

combinations. It is embraced in the following maxims:
1. To throw by strategic movements the mass of an army, successively, upon the decisive points of a theater of war, and also upon the communications of the enemy as much as possible without compromising one's own.
2. To maneuver to engage fractions of the hostile army with the bulk of one's forces.
3. On the battlefield, to throw the mass of the forces upon the decisive point, or upon the portion of the hostile line which it is of the first importance to overthrow.
4. To so arrange that these masses shall not only be thrown upon the decisive point, but that they shall engage at the proper times and with ample energy.

Lieutenant-General Antoine-Henri Baron de Jomini, *Summary of the Art of War*, 1838, tr. Mendell and Craighill, 1862.

War acknowledges principles, and even rules, but these are not so much fetters, or bars, which compel its movement aright, as guides which warn us when it is going wrong.

Rear Admiral Alfred Thayer Mahan (1840–1914).

(1) Cut line of communication; (2) cause enemy to form front to flank; (3) operate on internal lines; (4) separate bodies of enemy and fight in detail; and (5) direct attack.

General George S. Patton, Jr, quoted in Carlo d'Este, *Patton: A Genius for War*, 1995.

The fundamental principles of war are neither very numerous nor in themselves very abstruse, but the application of them is difficult and cannot be made subject to rules. The correct application of principles to circumstances is the outcome of sound military knowledge, built up by study and practice until it has become instinct.

British Army *Field Service Regulations 1909*, quoted in Fuller, *The Foundations of the Science of War*, 1926.

There are eight principles of war, and they constitute the laws of every scientifically fought boxing match as of every battle. These principles are:
1st Principle. – The Principle of the objective
2nd Principle. – The Principle of the offensive
3rd Principle. – The Principle of security
4th Principle. – The Principle of concentration
5th Principle. – The Principle of economy of force
6th Principle. – The Principle of movement
7th Principle. – The Principle of surprise
8th Principle. – The Principle of co-operation.

No one of the above eight principles is of greater value than the other. No plan of action can be considered in harmony unless all are in harmony, and none can be considered in harmony unless weighed against the conditions which govern their application. Seldom can a perfect plan be arrived at because the fog of war seldom, if ever, rises. It is, however, an undoubted fact that the general who places his trust in the principles of war, and who trusts in them the more strongly the fog of war thickens, almost inevitably beats the general who does not.

Major-General J.F.C. Fuller, *The Reformation of War*, 1923. Fuller was the first to define the principles of war as a coherent, interactive set of concepts; these principles, subsequently modified, are the basis of the Principles of War used by the US and British Armies.

I would give you a word of warning on the so-called principles of war, as laid down in *Field Service Regulations*. For heaven's sake, don't treat those as holy

THE PRINCIPLES OF WAR

writ, like the Ten Commandments, to be learned by heart, and as having by their repetition some magic, like the incantations of savage priests. They are merely a set of common-sense maxims, like 'cut your coat according to your cloth,' 'a rolling stone gathers no moss,' 'honesty is the best policy,' and so forth... Clausewitz has a different set, so has Foch, so have other military writers. They are all simply common sense, and are instinctive to the properly trained soldier.
    Field Marshal Viscount Wavell of Cyrenaica, c. 1930, lecture to officers at Aldershot.

The military student does not seek to learn from history the minutia of method and technique. In every age these are decisively influenced by the characteristics of weapons currently available and by means at hand for maneuvering, supplying and controlling combat forces. But research does bring to light those fundamental principles, and their combinations and applications, which, in the past, have been productive of success. These principles know no limitation of time. Consequently, the army extends its analytical interest to the dust-buried accounts of wars long past as well as to those still reeking with the scent of battle.
    General of the Army Douglas MacArthur, *Annual Report of the Chief of Staff for Fiscal Year Ending June 30, 1935.*

Respect for the opinion of this or that great soldier must never be allowed to go so far that nobody dares to discuss it. A sure sense of reality must be aroused. Given a well-founded knowledge of basic principles, any man of reasonably cool and logical mind can work out most of the details for himself, provided he is not inhibited in his thinking.
    Field Marshal Erwin Rommel, *The Rommel Papers*, 1953.

Against the principle of surprise – continuous activity by the various intelligence agencies. Against the principle of maintenance-of-aim – tactical diversionary attacks and strategical, psychological and political offensives. Against the principle of economy-of-force – attacks against lines of communications and stores in the rear, thereby pinning down the enemy's forces and dispersing them. Against the principle of coordination – strike against the channels of administration. Against the principle of concentration – diversionary attacks and air activity to split up the enemy's forces. Against the principles of security – sum total of the above activities and those that follow. Against the principle of offensive-spirit – offensive spirit. Against the principle of mobility – destruction of the lines of communication.
    General Yigael Yadin, *Israeli Forces Journal*, 9/1949.

The principles of war, not merely one principle, can be condensed into a single word – 'concentration'. But for truth this needs to be amplified as the 'concentration of strength against weakness'. And for any real value it needs to be explained that the concentration of strength against weakness depends on the dispersion of your opponent's strength, which in turn is produced by a distribution of your own that gives the appearance, and partial effect of dispersion. Your dispersion, his dispersion, your concentration – such is the sequence, and each is a sequel. True concentration is the fruit of calculated dispersion.
    Captain Sir Basil Liddell Hart, *Strategy*, 1954.

In most military text books, more particularly *Field Service Regulations*, a list of the principles of war will be found... There are several versions of these so-called principles, but they are no more than pegs on which to hang our tactical thoughts. There is nothing irrevocable about them; sometimes they may be discarded with impunity; but as a study of military history will show, they should only be discarded after deep consideration. They are very important guides rather than principles, and in the writer's opinion the simplest and most useful are derived from the seven tactical elements... aim or object, security, mobility, offensive power, economy of force, concentration of force, and surprise. Further, they are as applicable to strategy (operations in plan) as to tactics (operations in action), two terms which should never be separated by a bulkhead, because their components flow into each other and together constitute the art of war.

Major-General J.F.C. Fuller, *The Generalship of Alexander the Great*, 1958.

War remains an art and, like all arts, whatever its variation, will have its enduring principles. Many men, skilled either with sword or pen, and sometimes with both, have tried to expound those principles. I heard them once from a soldier of experience for whom I had a deep and well-founded respect. Many years ago, as a cadet hoping someday to be an officer, I was poring over the 'Principles of War,' listed in the old Field Service Regulations, when the sergeant-major came upon me. He surveyed me with kindly amusement. 'Don't bother your head about all them things, me lad,' he said. 'There's only one principle of war, and that's this. Hit the other fellow as quick as you can and as hard as you can, where it hurts him the most, when he ain't lookin'!' As a recruit, I earned that great man's reproof often enough; now, as an old soldier, I would hope to receive his commendation. I think I might, for we of the Fourteenth Army held to his 'Principle of War.'

Field Marshal Viscount Slim of Burma, *Defeat Into Victory*, 1956.

PRISONERS OF WAR (PoWS)

No prisoner pleads well his case,
Framing it sorrowfully. But,
To comfort his distress, he can make a song.
My friends are many, though their gifts are few.
Shame upon them, if my ransom is not paid
These two winters in prison.

My friends and barons know well –
English, Norman, Poitevin and Gascon all –
That no friend of mine so low
Would I leave in prison for want of cash.
I mean no reproach
But still am I held captive.

Richard I *Coeur de Lion*, 1192, tr. Caitlin Matthews, song written while a prisoner of Leopold of Austria after having been captured on his return from Crusade.

I enjoy too much pleasure in softening the Hardships of Captivity to with-hold any comfort from Prisoners; and beg you to do me the Justice to conclude, that no Requisition of this Nature, that should be made, will ever be denied.

General George Washington, 13 May 1777, to Leopold Phillip von Heister, *The Writings of George Washington*, VIII, p. 58, John C. Fitzgerald, ed., 1933.

When the war is concluded, I am definitely of the opinion that all animosity should be forgotten, and that all prisoners should be released.

The Duke of Wellington, 1804, letter to E.S. Warring.

Maxim 109. Prisoners of war do not belong to the power for which they have fought; they all are under the safeguard of honor and generosity of the nation that has disarmed them.

Napoleon, *The Military Maxims of Napoleon*, tr. Burnod, 1827.

Losses incurred during the battle consist

PRISONERS OF WAR

mostly of dead and wounded; after the battle, they are usually greater in terms of captured guns and prisoners. While the former are shared more or less evenly by winner and loser, the latter are not. For that reason they are usually only found on one side, or at any rate in significant numbers on one side.

That is why guns and prisoners have always counted as the real trophies of victory: they are also its measure, for they are the tangible evidence of its scale. They are a better index to the degree of superior morale than any other factors, even when one relates them to the casualty figures…

Major-General Carl von Clausewitz, *On War*, 4.4, 1832, tr. Michael Howard and Peter Paret, 1976.

An officer congratulating the General upon the great number of his prisoners, said jocularly, that they surrendered too easily, for the Confederacy would be embarrassed with their maintainence. He answered, smiling, 'It's cheaper to feed them than to fight them.'

Lieutenant-General Thomas J. 'Stonewall' Jackson, during the Seven Days' Battles, 30 June 1862, quoted in R.L. Dabney, *Life of Lieut.-Gen. Thomas J. Jackson*, II, 1866.

Prisoner of war guard companies, or an equivalent organization, should be as far forward as possible in action to take over prisoners of war, because troops heated with battle are not safe custodians. Any attempt to rob or loot prisoners of war by escorts must be strictly dealt with.

General George S. Patton, Jr, *War As I Knew It*, 1947.

A prisoner of war is a man who tries to kill you and fails, and then asks you not to kill him.

Sir Winston S. Churchill, 1952, quoted in *The Observer* (London).

I saw a group of soldiers clustered around a prisoner; as I drove up, one of them was hitting the Egyptian. I court-martialed the soldier on the spot, sentencing him to thirty-five days in the stockade. It was the kind of thing that brought my anger to the boiling point. In battle you fight and you have to kill. That's the nature of it. But once a man is your prisoner you never touch him.

General Ariel Sharon, *Warrior*, 1989.

I made sure that we treated prisoners humanely. Sadism is bad for morale, and, in the long run, humane treatment yields more information than brutality.

Major-General John K. Singlaub, of his command of the Studies and Observations Group (unconventional warfare) during the opening stages of the U.S. deployment to Vietnam, *Hazardous Duty*, 1991.

There was only one moment at which Ahmad showed any emotion. He presented an accounting of coalition prisoners of war held by Iraq. 'We have forty-one in all,' he said. I made notes as he read them off:

17 Americans
2 Italians
12 British
1 Kuwaiti
9 Saudis

This left a number of people unaccounted for, and I quickly brought out our list of MIAs, but he stopped me. 'And we would like to have the numbers of the POWs on our side as well.'

'As of last night, sixty thousand,' I replied. 'Or sixty thousand plus, because it is difficult to count them completely.' His face went completely pale: he had had no concept of the magnitude of their defeat.

When the Iraqis treated for peace at Safwan, 31 March 1991, General H. Norman Schwarzkopf, *It Doesn't Take a Hero*, 1992.

Keep this saying before your eyes: 'It is not fitting for a Prophet that he should have prisoners of war until he hath

thoroughly subdued the land.'
'Therefore, when ye meet the unbelievers (in fight), smite at their necks.'
  Osama bin Ladin, statement, 12 February 2003, *BBC News World Edition*.

THE PROFESSION OF ARMS

Upon being greeted by two friars with the words, 'God give you peace,' Sir John is reputed to have replied, 'God take away your alms. For as you live by charity, so do I by war, and to me it is as genuine a vocation as yours.
  Sir John Hawkwood (1320–1394).

War is my homeland,
My armour is my house,
And fighting is my life.
  German soldiers' saying, 16th century, quoted in Langer, *The Thirty Years War*, 1980.

Generally speaking, the Way of the warrior is resolute acceptance of death. Although not only warriors but priests, women, peasants and lowlier folk have been known to die readily in the cause of duty or out of shame, this is a different thing. The warrior is different in that studying the Way of strategy is based on overcoming men.
  Miyamoto Musashi, *The Book of the Five Rings*, 1645, tr. Victor Harris, 1984.

The Profession of a Souldier is allowed to be lawful by the Word of God; and so Famous and Honourable amongst Men, that Emperours and Kings do account it a Great Honour to be of the Profession, and to have Experience in it.
  General George Monck, *Observations Upon Military and Political Matters*, 1671.

War is a trade for the ignorant and a science for the expert.
  Chevalier Jean Charles Folard (1669–1752).

The man who devotes himself to war should regard it as a religious order into which he enters. He should have nothing, know no other home than his troop, and should hold himself honored in his profession.
  Field Marshal Maurice Comte de Saxe, *My Reveries*, 1732, tr. Phillips, 1940.

A sense of vocation is the greatest virtue of the military man.
  Field Marshal Prince Aleksandr V. Suvorov (1729–1800), quoted in *Morskoi Sbornik*, 5/1981.

From the nature of our profession we hold life by a more precarious tenure than many others, but when we fall, we trust it is to benefit our Country. So fell your Son by a cannon-ball under my immediate command at the Siege of Bastia. I had taken him on shore with me, from his abilities and attention to his duty.
  Admiral Viscount Nelson, 1794, letter to the father of Seaman Thomas Davies of the *Agamemnon*, who died at his side, quoted in David Walder, *Nelson*, 1978.

There are workers in buildings, in colours, in forms and in words – myself, I'm a worker in battles. It's my calling. – At thirty-five I've already made eighteen of them, called Victories. – I have to be paid for my work. And it's not expensive to pay with a throne. – Besides, I shall always go on working. You'll see plenty more...
  Napoleon, 1804, conversation with Pope Pius VII at Fontainebleau, quoted by Alfred Comte de Vigny, *Servitude and Grandeur of Arms*, 1857.

In any case, who is asking anything of you? Who has asked you to feed me? If you stopped your provisions and I were hungry, these brave soldiers would take compassion on me. I could go to the mess of their grenadiers, and I am sure they would not deny the first, the oldest soldier of Europe.
  Napoleon, 18 August 1816, to Admiral Cockburn on St Helena, referring to the

THE PROFESSION OF ARMS

men of the 58th Foot, the garrison of the island, cited in R.M. Johnston, ed., *The Corsican*, 1910.

I like only those who make war.
Napoleon, 1816, on St Helena, quoted in Christopher Herold, ed., *The Mind of Napoleon*, 1955.

No matter how clearly we see the citizen and the soldier in the same man, how strongly we conceive of war as the business of the entire nation, opposed diametrically to the pattern set by the *condottieri* of former times, the business of war will always remain individual and abstract. Consequently for as long as they practice this activity, soldiers will think of themselves as members of a kind of guild, in whose regulations, laws, and customs the spirit of war is given pride of place. And that does seem to be the case. No matter how much one may be inclined to take the most sophisticated view of war, it would be a serious mistake to underrate professional pride (*esprit de corps*) as something that may and must be present in an army to greater or lesser degree. Professional pride is the bond between the various natural forces that activate the military virtues; in the context of this professional pride they crystallize more readily.
Major-General Carl von Clausewitz, *On War*, 3.5, tr. Michael Howard and Peter Paret, 1976.

Arms is a profession that, if its principles are adhered to for success, requires an officer do what he fears may be wrong, and yet, according to military experience, must be done, if success is to be attained.
Lieutenant-General Thomas J. 'Stonewall' Jackson, 11 April 1862, letter to his wife.

I wish you to understand that the day is at hand when you will be called upon to meet the enemy in the worst form of our profession... Hot and cold shot will, no doubt, be freely dealt to us, and there must be stout hearts and quick hands to extinguish the one and stop the holes in the other.
Admiral David G. Farragut, general order before the attack past the forts guarding New Orleans on 24 April 1862, quoted in Loyall Farragut, *The Life of David Glasgow Farragut*, 1879.

In no event will there be money in it; but there may always be honor and quietness of mind and worthy occupation – which are better guarantees of happiness.
Rear Admiral Alfred Thayer Mahan, *The Navy as a Career*, 1895.

I do not regard and have never regarded permanent soldiering as an attractive proposition for any man who has some other profession at his command... if a man could command an income no larger in private practice than he could in military employment, I would recommend to him to stick to private practice every time. There is something about permanent military occupation which seems to confine a man's scope and limit his opportunities, and after he has had a few years under the circumscribed conditions of official routine, he generally finds himself wholly out of touch with civil occupation.
General Sir John Monash, 6 January 1911, quoted in Horner, *The Commanders*, 1984.

There is scope and occupation in the army for the highest mental as well as the highest physical efficiency. Whereas delight in the trade of arms, inherited and indulged from generation to generation, drives one man to the army, another may join from the desire to devote his ability and knowledge to the most immediate form of national service; one is attracted by a profession which gives pride of place to manly strength and personal value, and other by the prospect of activity in the open, another because he is interested in the machinery of war, and yet another because he loves horses.

αὐτὸς γὰρ ἐφέλκεται ἄνδρα σίδηρος
autos gar ephelketai andra sideros
Iron attracts man. [*Odyssey* 16.294]
  Colonel-General Hans von Seekt, *Thoughts of a Soldier*, 1930.

Men who adopt the profession of arms submit, of their own free will, to a law of perpetual constraint. Of their own accord they reject the right to live where they choose, to say what they think, to dress as they like. From the moment they become soldiers it needs but an order to settle them in this place, to move them to that, to separate them from their families and dislocate their normal lives. On the word of command they must rise, march, run, endure bad weather, go without sleep or food, be isolated in some distant post, work till they drop. They have ceased to be the masters of their fate. If they drop in their tracks, if their ashes are scattered to the four winds, that is all part and parcel of their job.
  General Charles de Gaulle, *The Edge of the Sword*, 1932.

Our calling is most ancient, and like all other old things it has amassed through the ages certain customs and traditions which decorate it and ennoble it, which render beautiful the otherwise prosaic occupation of being professional men-at-arms: killers.
  General George S. Patton, Jr, quoted in Fitton, ed., *Leadership*, 1990.

The point I wish to make here, and to repeat it for emphasis, is that the professional military man has three primary responsibilities:
  First, to give his honest, fearless, objective, professional military opinion of what he needs to do the job the Nation gives him.

Second, if what he is given is less than the minimum he regards as essential, to give his superiors an honest, fearless, objective opinion of the consequences.

Third, and finally, he has the duty whatever the final decision, to do the utmost with whatever is furnished.
  Generacl Matthew B. Ridgway, address to the Army Staff as Chief of Staff, 1954, quoted in Russel B. Reynolds, *The Officer's Guide, 1970–1971 Edition*, 1970.

War is the professional soldier's time of opportunity.
  Captain Sir Basil Liddell Hart, *Defence of the West*, 1956.

I find I have liked all of the soldiers of different races who have fought with me and most of those who have fought against me. This is not strange, for there is a freemasonry among fighting soldiers that helps them to understand one another even if they are enemies.
  Field Marshal Viscount Slim of Burma, *Unofficial History*, 1959.

The soldier, be he friend or foe, is charged with the protection of the weak and unarmed. It is the very essence and reason of his being. When he violates this sacred trust, he not only profanes his entire cult but threatens the fabric of international society.
  General of the Army Douglas MacArthur, quoted in William Manchester, *American Caesar: Douglas MacArthur, 1880–1964*, 1978.

The function of the profession of arms is the ordered application of force in the resolution of a social problem. Harold Lasswell describes it as the management of violence, which is rather less precise. The bearing of arms among men for the purpose of fighting other men is found as far back as we can see. It has become at some times and in some places a calling resembling the priesthood in its dedication. It has never

THE PROFESSION OF ARMS

ceased to display a strong element of the vocational.

It has also become a profession, not only in the wider sense of what is professed, but in the narrower of an occupation with a distinguishable corpus of specific technical knowledge and doctrine, a more or less exclusive group coherence, a complex of institutions peculiar to itself, an educational pattern adapted to its own needs, a career structure of its own and a distinct place in the society which has brought it forth. In all these respects it has strong points of resemblance to medicine and law, as well as holy orders.
General Sir John Hackett, *The Profession of Arms*, 1963.

There is discipline. There is drill... When you are relying on your mates and they are relying on you, there's no room for slackness or sloppiness. If you're not prepared to accept the rules, you're better off where you are.
British Army recruiting poster, 1976.

It is well to remind those who quote the injunction in the Sermon on the Mount to 'turn the other cheek also' to him who 'shall smite thee on the right cheek', that, in the version in Saint Matthew's Gospel, this is preceded by the words, 'I say unto you that ye resist not evil'. The sad fact of life is that, if evil is not resisted, it will prevail. That is the justification for the use of force to deter, and if necessary, defeat those who turn to it to further their own ends, the justification for maintaining in the service of the community and the state, forces who are trained, skilled and well-equipped to meet that challenge when and wherever it arises. Their profession is an honourable one.
Field Marshal Lord Carver, *Warfare Since 1945*, 1980.

The man-at-arms serves under an unlimited liability... As John Ruskin said, the professional soldier has to offer himself to be slain.
General Sir John Hackett, quoted in 'Are Soldiers Civilized', *Parameters*, Summer 1984.

The other distinct thing is that military virtues, as Arnold Toynbee calld them – fortitude, personal loyalty, consistency and compassion – make any group of men a better group to be in than one in which those qualities are absent.

The difference between the profession of arms and other professions then is that these qualities are not simply agreeable ones to have, but are functional necessities.
General Sir John Hackett, quoted in 'Are Soldiers Civilized?', *Parameters*, Summer 1984.

Most of us were fascinated by the intellectual challenge [US Army Command and General Staff College]. This certainly didn't mean we were eager for war. There aren't many professional soldiers who actually find combat pleasurable. Critics of the military – especially liberal academics and journalists – mistakenly believe that soldiers enjoy war. That's as logical as assuming surgeons enjoy cancer. We study the profession of arms not to wage war, but to defend the values of our civilization. This is an uncomplicated patriotic notion that is hard for sophisticated civilians to swallow. But our dedication to professional excellence does not mean career soldiers are especially cruel, or that we value human life less than our civilian counterparts. In fact, there is abundant proof to the contrary to be found in the thousands of citations for combat gallantry in actions to save lives.
Major-General John K. Singlaub, *Hazardous Duty*, 1991.

This is a very busy and demanding time for the army and many of you will have

PROMOTION

had repeated operational tours in recent years as well as other duties that have taken you away from your homes and families. We are an army for use and it is right that we should be engaged.

> General Sir Michael Jackson, Chief of the General Staff, quoted in Trevor Royle, 'Soldiers... who needs them?' *Sunday Herald*.

PROMOTION

I sailed with King Amenhotep I triumphant, when he ascended the river to Kush in order extend the borders of Egypt... I was at the head of our army; I fought incredibly; his majesty beheld my bravery. I brought off two hands, and took them to his majesty... I brought his majesty in two days to Egypt from the upper well; he [the king] presented me with gold. Then I brought away two female slaves, in addition to those which I had taken to his majesty. One [the king] appointed me 'Warrior of the Ruler.'

> Ahmose son of Abana, Egyptian naval officer during the reign of Pharaoh Amenhotep I (1557–1501 BC), quoted in James Breasted, *Ancient Records of Egypt*, 1906.

Hang him, or else make him a Marshal. I can't stand half-measures.

> Marshal of France Maurice de Saxe, to Louis XV who had just disparaged the future Marshal Ulrich-Frederik-Waldemar von Löwendahl for the brutal taking of Bergen, 1747, quoted in John Manchip White, *Marshal of France*, 1962, p. 228.

Mad, is he? Than I hope he will *bite* some of my other generals.

> George II, comment to the Duke of Newcastle who complained that General James Wolfe was mad, quoted in Henry Beckles Wilson, *The Life and Letters of James Wolfe*, 1909, 17.

When I find any officer that answers me with firmness, intelligence, and clearness, I set him down in my list for making use of his service on proper occasions.

> Frederick the Great (1712–1786), quoted in Douglas Sladen, ed., 'Morning the Fourth,' *Confessions of Frederick the Great*, 1915.

I assembled the officers and men; I asked them who had done well; and I promoted those who could read and write.

> Napoleon, 23 November 1806, having reviewed the Fourth Corps following the Prussian campaign, quoted in R.M. Johnston, ed., *The Corsican*, 1910.

If the skill of a general is one of the surest elements of victory, it will readily be seen that the judicious selection of generals is one of the most delicate points in the science of government and one of the most essential parts of the military policy of a state. Unfortunately, this choice is influenced by so many petty passions that chance, rank, age, favor, party spirit, or jealousy will have as much to do with it as the public interest and justice.

> Lieutenant-General Antoine-Henri Baron de Jomini, *Summary of the Art of War*, 1838, tr. Mendell and Craighill, 1862.

I cannot afford to lose this man. He fights.

> Abraham Lincoln, refusing to sack Grant after accusations that he had been drunk at the battle of Shiloh, April 1862, quoted in Dixon Wecter, *The Hero in America*, 1941, 12.2.

In my opinion, an officer's pay, when he reaches the highest grade, is given him as a reward for long and faithful services. But the ignorant, as well as the wise and efficient, have risen to the command of squadrons and stations, according to their position on the Register; and when it was attempted to change the order of things, the law was ignored by the country and the President... I have always been an advocate of promotion or distinction for professional services

PROMOTION

of a chivalrous character, and would never object to the promotion of any brother officer, who had the good fortune to be at the right place at the right time, and did an act by which he risked his life and reputation in the performance of his duty.
>Admiral David G. Farragut, quoted in Loyall Farragut, *The Life of David Glasgow Farragut*, 1879.

Differences of opinion must exist between the best of friends as to policies in war, and of judgment as to men's fitness. The officer who has the command, however, should be allowed to judge the fitness of the officers under him, unless he is very manifestly wrong.
>General of the Army Ulysses S. Grant, *Personal Memoirs*, II, 1885.

We are now at war, fighting for our lives, and we cannot afford to confine Army appointments to persons who have excited no hostile comment in their career.
>Sir Winston S. Churchill, 19 October 1940, note to the Chief of the Imperial General Staff.

In war, you must either trust your general or sack him.
>Field Marshal John Dill, letter to Gen. Archibald Percival whom he sacked in 1941, quoted in Ronald Lewis, *The Chief*, 5, 1980.

More and more does the 'System' tend to promote to *control*, men who have shown themselves efficient *cogs* in the machine... There are few commanders in our higher commands. And even these, since their chins usually outweigh their foreheads are themselves outweighed by the majority – of commanders who are essentially staff officers.
>Captain Sir Basil Liddell Hart, *Thoughts on War*, 1944.

The flaw in all these conscientious attempts to improve the system of promotion is that they tend inevitably to stress the formal record at the expense of the human element and that they aim at an improved flow of promotion when nothing less than an eruption will avail. In fact, the effort to make the Army 'fairer' as a profession is contrary to the very nature of its object – war, which is essentially 'unfair'. The professional attitude was epitomized in a recent article by Admiral Harper, who scathingly criticized Mr. Churchill's selection of an officer for a certain command on the ground that Mr. Churchill had said that his choice was based on personal impressions and conversations rather than on official records. Yet outside the closely regulated professions what employer choosing a man for an important post would not weigh personality more than record? Admittedly, the value of a choice based on personal knowledge depends on the man who makes it, and even at the best it is highly unfair to those who fail through no fault of their own to catch the selector's eye. But it is the only system, or lack of a system, under which genius can be brought rapidly to the top. The lesson of military history is that genius and not mere competence decides the fate of warring nations. It was the shrewd dictum of one who, whatever his defects, was assuredly a genius – Lord Fisher – that 'favouritism is the secret of efficiency'. In our zeal for equalitarian fairness we have almost ruled out 'favouritism', and the stalemate of 1914–1918 is a testimonial to our good intentions – which 'paved the path to hell'.
>Captain Sir Basil Liddell Hart, *Thoughts on War*, 1944.

My only fear is that the extra stripe is going to interfere with my drinking arm.
>Admiral William 'Bull' Halsey, upon the

award of his fifth star and promotion to Fleet Admiral, quoted in E.B. Potter, *Bull Halsey*, 1985.

If evolution were destined to favour the rise of those who, in the tragic hours when the storm sweeps away conventions and habits, are the only ones to remain on their feet and to be, therefore, necessary, would not that be all to the good?
General Charles de Gaulle, *War Memoirs: The Call to Honour 1940–1942*, 1955.

My own feeling now, after having been through two world wars, is that an extensive use of weedkiller is needed in the senior ranks after a war; this will enable the first class younger officers who have emerged during the war to be moved up.
Field Marshal Viscount Montgomery of Alamein, *Memoirs of Field Marshal Montgomery*, 1958.

By good fortune in the game of military snakes and ladders, I found myself a general.
Field Marshal Viscount Slim of Burma, *Unofficial History*, 1959.

We make generals today on the basis of their ability to write a damned letter. Those kinds of men can't get us ready for war.
Lieutenant-General Lewis B. 'Chesty' Puller, quoted in Burke Davis, *Marine*, 1962.

This is a long tough road we have to travel. The men that can do things are going to be sought out just as surely as the sun rises in the morning. Fake reputations, habits of glib and clever speech, and glittering surface performance are going to be discovered and kicked overboard. Solid, sound leadership, with inexhaustible nervous energy to spur on the efforts of lesser men, and ironclad determination to face discouragement, risk, and increasing work without flinching, will always characterize the man who has a sure-enough, bang-up fighting unit. Added to this he must have a darned strong tinge of imagination – I am continuously astonished by the utter lack of imaginative thinking among so many of our people that have reputations for being really good officers. Finally, the man has to be able to forget himself and his personal fortunes.
General of the Army Dwight D. Eisenhower, *At Ease: Stories I Tell My Friends*, 1967.

PROMPTNESS

You will have seen by what I have had occasion to delineate concerning war that promptness contributes a great deal to success in marches and even more in battles. That is why our army is drilled in such a fashion that it acts faster than others. Drill is the basis of these maneuvers which enable us to form in the twinkling of an eye.
Frederick the Great, *Instructions to His Generals*, 1747, tr. Phillips, 1940.

I have always been a quarter of an hour before my time, and it has made a man of me.
Admiral Viscount Nelson (1758–1805).

In war, procrastination is a crime, and promptness is a handmaiden of victory.
Major-General John A. Lejeune, *The Reminiscences of a Marine*, 1929.

PROPAGANDA

His victories were published in all lands, to cause that every land together may see, to cause the glory of his conquests to appear; King Merneptah, the Bull, lord of strength, who slays his foes, beautiful upon the field of victory... He has penetrated the land of Temeh in his lifetime, and put eternal fear in the heart of Meshwesh. He has turned back Libya, who invaded Egypt, and great

PROPAGANDA

fear of Egypt is in their hearts.
> Merneptah, Pharaoh of Egypt, c. 1220 BC, celebrating his victory over the Libyans, quoted in James Breasted, *Records of Ancient Egypt*, 1906.

We are here to guide public opinion, not to discuss it.
> Napoleon, 1804, meeting of the *Conseil d'État*, quoted in Christopher Herold, ed., *The Mind of Napoleon*, 1955.

Barere still believes that the masses must be stirred. On the contrary, they must be guided without noticing it.
> Napoleon, 1804, letter to Fouché, quoted in Christopher Herold, ed., *The Mind of Napoleon*, 1955.

This won't do; it will drive the people in England mad. Write me down a victory.
> The Duke of Wellington, upon the news of the heavy British casualties suffered in the Battle of Albuera on 16 May 1811, which was described as a defeat, cited in Arthur Bryant, *The Great Duke*, 1972.

We were hypnotized by the enemy propaganda as a rabbit is by a snake. It was exceptionally clever and conceived on a great scale. It worked by strong mass-suggestion, kept in the closest touch with the military situation, and was unscrupulous as to the means it used... The Army found no ally in a strong propaganda directed from home. While her Army was victorious on the field of battle, Germany failed in the fight against the *moral* of the enemy peoples.
> General Erich Ludendorff, *My War Memories, 1914–1918*, 1919.

Propaganda, as inverted patriotism, draws nourishment from the sins of the enemy. If there are no sins, invent them! The aim is to make the enemy appear so great a monster that he forfeits the rights of a human being. He cannot bring a libel action, so there is no need to stick at trifles.
> General Sir Ian Hamilton, *The Soul and Body of an Army*, 1921.

It was more subtle than tactics, and better worth doing, because it dealt with uncontrollables, with subjects incapable of direct command. It considered the capacity for mood of our men, their complexities and mutability, and the cultivation of whatever in them promised to profit our intention. We had to arrange their minds in order of battle just as carefully and as formally as other officers would arrange their bodies. And not only our men's minds, though naturally they came first. We must also arrange the minds of the enemy, so far as we could reach them; then those other minds of the nation supporting us behind the firing line, since more than half the battle passed there in the back; then the minds of the enemy nation waiting the verdict; and of the neutrals looking on; circle beyond circle.
> Colonel T.E. Lawrence, *The Seven Pillars of Wisdom*, 1935.

During the great war a battle of propaganda was waged by all belligerents, though at its beginning few were prepared to wage it. Our object was to prove that the Germans were 'dirty dogs,' and that it was they who had started the war. I do not suggest that our contentions were wrong, but I cannot help feeling that when the Germans retaliated the means we employed to protect our national character were not of the best. In place of maintaining our reputation for fair play we hired a pack of journalists to defend us. These people, who had spent their lives in raking filth out of the law courts, went to mud with the alacrity of eels, and, though they undoubtedly succeeded in blackening the German nation, we ourselves became somewhat piebald in these gutter attacks.

PUBLIC OPINION

All they need do really is quietly let people know the truth. There is no need to bang the big drum. Official reports should stick to the absolute truth – once you start lying, the war's as good as lost. Information Division's outlook is all wrong. All this talk of guiding public opinion and maintaining the national morale is so much empty puff.
>Admiral Isoroku Yamamoto, 1942, quoted in Agawa, *The Reluctant Admiral*, 1979.

If democracy and Parliamentary institutions are to triumph in this war, it is absolutely necessary that Governments resting upon them shall be able to act and dare, that the servants of the Crown shall not be harassed by nagging and snarling, that enemy propaganda shall not be fed needlessly out of our own hands and our reputation disparaged and undermined throughout the world.
>Winston S. Churchill, in the House of Commons, 2 July 1943, quoted in 'Finest Hour', *Journal of the Churchill Centre & Society*, Number 121, Winter 03–04.

The political skill of the Communist government of North Vietnam in portraying their cause in terms that would elicit the support of liberal Americans must be judged one of the most successful in history. They oppressed the Catholic Church and somehow secured the support of a large number of American Roman Catholic clergy. They savagely suppressed any movement that seemed to diverge from the official party line and were supported by large numbers of academics, students, and clergy in the United States and around the world. They were good soldiers but they were world-class propagandists. They won in the media that which they could not win on the battlefield.
>Lieutenant-General William J. McCaffrey, 4 August 1988, letter, quoted in Lee and Cragg, *Inside the VC and the NVA*, 1992.

It's not about propaganda or spin, it is about human decency.
>General Sir Michael Jackson, Chief of the General Staff, telling journalists to think about the effects of what they wrote and broadcast on the bereaved families, quoted by Jo Dillon, *The Independent*, 30 March 2003.

PUBLIC OPINION

A popular outcry will drown the voice of military experience.
>Rear Admiral Alfred Thayer Mahan, *Naval Strategy*, 1911.

Nothing is more dangerous in wartime than to live in the temperamental atmosphere of a Gallup Poll, always feeling one's pulse and taking one's temperature. I see that a speaker at the week-end said that this was a time when leaders should keep their ears to the ground. All I can say is that the British nation will find it very hard to look up to leaders who are detected in that somewhat ungainly posture.
>Sir Winston S. Churchill, speech in the House of Commons, 30 September 1941, *Winston S. Churchill: His Complete Speeches, 1897–1963*, VI, 1974, p. 6495.

We are told that we must not alarm the easygoing voter and public. How thin and paltry these arguments will sound if we are caught a year or two hence, fat, opulent and defenceless.
>Sir Winston S. Churchill, quoted in James Hume, *The Wit & Wisdom of Winston S. Churchill*, 1995.

PUNISHMENTS/ AND REWARDS

The general creates awesomeness by executing the great and becomes enlightened by rewarding the small. Prohibitions are made effective and laws

>Major-General J.F.C. Fuller, *The Foundations of the Science of War*, 1926.

PUNISHMENTS AND REWARDS

implemented by careful scrutiny in the use of punishments. Therefore if by executing one man the entire army will quake, kill him. If by rewarding one man the masses will be pleased, reward him. In executing, value the great; in rewarding, value the small. When you kill the powerful and the honored, this is punishment that reaches the pinnacle. When rewards extend down to the cowherds, grooms, and stablemen, these are rewards penetrating downward to the lowest. When punishments reach the pinnacle and rewards penetrate to the lowest, then your awesomeness has been effected.

   The T'ai Kung, c. 1100 BC (*The Seven Military Classics of Ancient China*, tr. Ralph Sawyer, 1993).

Bestow rewards without regard to rule, issue orders without regard to previous arrangements and you will be able to handle a whole army.

   Sun Tzu, *The Art of War*, c. 500 BC, tr. Giles, 1910.

And so by means of such incentives even those who stay at home feel the impulse to emulate such achievements in the field no less than those who are present and see and hear what takes place... When we consider this [the Roman] people's almost obsessive concern with military rewards and punishments, and the immense importance which they attach to both, it is not surprising that they emerge with brilliant success from every war in which they engage.

   Polybius, *Histories*, 6.39, c. 125 BC (*The Rise of the Roman Empire*, tr. Ian Scott-Kilvert, 1987).

A policy of rewards and penalties means rewarding the good and penalizing wrongdoers. Rewarding the good is to promote achievement; penalizing wrongdoers is to prevent treachery.

   It is imperative that rewards and punishments be fair and impartial. When they know rewards are to be given, courageous warriors know what they are dying for; when they know penalties are to be applied, villains know what to fear.

   Therefore, reward should not be given without reason, and penalties should not be applied arbitrarily. If rewards are given for no reason, those who have worked hard in public service will be resentful; if penalties are applied arbitrarily, upright people will be bitter.

   Zhuge Liang (AD 180–234), *The Way of the General* (*Mastering the Art of War*, tr. Thomas Cleary, 1989).

Punishment, and fear thereof, are necessary to keep soldiers in order in quarters; but in the field they are more influenced by hope and rewards.

   Flavius Vegetius Renatus, *Military Institutions of the Romans*, c. AD 378, tr. Clark, 1776.

He ought not to be easy with those who have committed offenses out of cowardice or carelessness in the hope of being regarded as a good leader, for a good leader does not encourage cowardice or laziness. On the other hand, he ought not to punish hastily and without a full investigation just to show he can act firmly. The first leads to contempt and disobedience; the other to well-deserved hatred with all its consequences. Both of these are extremes. The better course is to join fear with justice, that is, impose a fitting punishment upon offenders after proof of guilt. This, for reasonable men, is not punishment, but correction, and aids in maintaining order and discipline.

   The Emperor Maurice, *The Strategikon*, c. AD 600 (*Maurice's Strategikon*, tr. George Dennis, 1984).

Paradise is under the shadow of our swords. Forward!

PUNISHMENTS AND REWARDS

For soldiers to strive to scale high walls in spite of deep moats and showers of arrows and rocks, or for soldiers to plunge eagerly into the fray of battle, they must be induced by serious rewards; then they will prevail over any enemy.

The rule is 'Where there are serious rewards, there will be valiant men.'
Liu Ji (1310–1375), *Lessons of War* (*Mastering the Art of War*, tr. Thomas Cleary, 1989).

O sons, brothers, and nephews who are here in the presence of the majesty of our ruler Motecuhzoma, I, Tlacaelel, wish to give more courage to those who are of a strong heart and embolden those who are weak so that the prize for their deeds will encourage them. I wish to compare the strong and the weak and mention their rewards or their retribution. When you go to the marketplace and see a precious ear ornament or nose pendant, or when you see splendid and beautiful feathers or a rich gilded shield, or weapons done in featherwork, do you not covet them, do you not pay the price that is asked? Know now that the king, who is here before you, has willed that labrets, golden garlands, many-colored feathers, ear ornaments, armbands, shields, weapons, insignia, rich mantles, and breechcloths are no longer to be purchased in the market by brave men. Now the sovereign will deliver them as payment and prize for heroic feats, for memorable deeds. Each one of you, when he goes into battle, to perform great feats, must keep in mind that, while he is carrying out these heroic acts in war, he has in reality gone to the marketplace where he will find all these priceless things, for upon returning from the war he will receive them according to his merits and so that he can display these as proof of his worth.

Omar I ibn al-Khattab, Caliph, AD 637, at the Battle of Quadisiya.

Tlacaelel, c. 1450, the promulgation of the Aztec code of the warrior, quoted in Diego Durán, *The History of the Indies of New Spain*, 1581, tr. Doris Heyden, 1994.

If a man who serves indolently and a man who serves well are treated in the same way, the man who serves well may begin to wonder why he does so.
Asakura Toshikage (1428–1481), *The 17 Articles of Asakura Toshikage* (*Ideals of the Samurai*, tr. William Wilson, 1982).

The consciousness of having attempted faithfully to discharge my duty, and the approbation of my Country will be a sufficient recompense for my Services.
General George Washington, 18 March 1783, to the President of Congress, *The Writings of George Washington*, XXVI, p. 232, John C. Fitzgerald, ed., 1938.

I intend to make the generals and officers who have served me well so rich that there can be no excuse for their dishonouring the most noble of professions by their greed, while drawing upon themselves the contempt of the soldiers.
Napoleon, 15 November 1805, after the victory at Ulm, quoted in R.M. Johnston, ed., *The Corsican*, 1910.

...Then there is this further observation (which I entreat the board to bear in mind), that the regularity, and order, and discipline of the corps, is not merely a public affair; it is not only that the regiment may be fit to do its service for the public, but I say that it is a positive breach of faith with the good man, if discipline should not be enforced. I will suppose that there are 100 men in a company, 80 of whom behave exceedingly well, and submit to all your regulations (and I apprehend that will be found to be pretty nearly the usual proportion); there will be 80 of them who will never commit a fault of any description; they lead a quiet life in the barracks, and do every thing that you

PURSUIT

require from them; but there are 20 that will not do so. What happens? These 20 are constantly disturbing the peace and the comfort of the 80, and there is perpetual riot and disorder going on in the barracks besides. These men are perpetually in a state of punishment; they are either *billed-up*, or they are in confinement in the guard house, or they are in confinement in the gaol, or in some way or other. The consequence is, that these 80 are obliged to do their duty for them; so that, by not enforcing your own orders, and not punishing when those orders are not obeyed, you are doing the grossest injustice to those who do obey your orders.
> The Duke of Wellington, evidence given before the Royal Commission for Inquiring into Military Punishments, n.d., *The Dispatches of Field Marshal the Duke of Wellington, During His Various Campaigns in India, Denmark, Portugal, Spain, the Low Countries, and France*, VIII, 1834–38, pp. 351–352.

Measured military punishment, together with a short and clear explanation of the offense, touches the ambitious soldier more than brutality which drives him to despair.
> General William T. Sherman, *Memoirs*, II, 1875.

The great merit in giving rewards is to give them promptly; it can not be considered generous when they are doled out too carefully.
> Admiral David D. Porter, *Incidents and Anecdotes of the Civil War*, 1885.

PURSUIT

A pursuit gives even cowards confidence.
> Xenophon, 400 BC, speech to the Greek Army at Calpe on the Black Sea, *Anabasis*, c. 360 BC.

No, you must realize what has been accomplished in this battle. In five days you will feast victoriously in the Capitol. Follow me. I will go ahead with the cavalry and they will learn of your arrival before they know you are coming.
> Maharbal Barca, the commander of the Carthaginian cavalry after the battle of Cannae, when his brother, Hannibal, preferred to rest the army rather than pursue, 2 August 216 BC, quoted in Livy, *Annals of the Roman People*, c. 25 BC–AD 17 (*Livy: A History of Rome*, 22, tr. Moses Hadas and Joe Poe, 1962).

It is a great feat to steer a policy to a successful conclusion or to overcome one's enemies in a campaign, but it requires a great deal more skill and caution to make good use of such triumphs. Thus we find that those who have won victories are far more numerous than those who have used them well...
> Polybius, *Histories*, 10.36, c. 125 BC (*The Rise of the Roman Empire*, tr. Ian Scott-Kilvert, 1979).

The words of the proverb: 'A bridge of gold should be made for the enemy,' is followed religiously. This is false. On the contrary, the pursuit should be pushed to the limit. And the retreat which had appeared such a satisfactory solution will be turned into a route [*sic*]. A detachment of ten thousand men can destroy an army of one hundred thousand in flight. Nothing inspires so much terror or occasions so much damage, for everything is lost. Substantial efforts are required to restore the defeated army, and in addition you are rid of the enemy for a good time. But many generals do not worry about finishing the war so soon.
> Field Marshal Maurice Comte de Saxe, *My Reveries*, 1732, tr. Phillips, 1940.

Only pursuit destroys a running enemy.
> Field Marshal Prince Aleksandr V. Suvorov (1729–1800).

Follow a defeated enemy ruthlessly and put the fear of Shaka into him.

QUAGMIRE

Shaka Zulu (c. 1787–1828), military maxim, quoted in E.A. Ritter, *Shaka Zulu*, 1955.

Maxim 83. a general-in-chief should never allow any rest either to the conqueror or to the conquered.
Napoleon, *The Military Maxims of Napoleon*, tr. Burnod, 1827.

The importance of the victory is chiefly determined by the vigor with which the immediate pursuit is carried out. In other words, pursuit makes up the second act of victory and in many cases is more important than the first. Strategy at this point draws near to tactics in order to receive the completed assignment from it; and its first exercise of authority is to demand that the victory should really be complete.
Major-General Carl von Clausewitz, *On War*, 4.12, 1832, tr. Michael Howard and Peter Paret, 1976.

...To follow up the success we gain with the utmost energy. The pursuit of the enemy when defeated is the only means of gathering up the fruits of victory.
Major-General Carl von Clausewitz, *On War*, 1832, tr. Michael Howard and Peter Paret, 1976.

When you strike the enemy and overcome him, never give up the pursuit as long as your men have strength to follow; for an enemy routed, if hotly pursued, becomes panic-stricken, and can be destroyed by half their number.
Lieutenant-General Thomas J. 'Stonewall' Jackson (1824–1863).

Do not fail in that event to make the most of the opportunity by the most vigorous attack possible, as it may save us what we have most reason to apprehend – a slow pursuit, in which he gains strength as we lose it.
General of the Army William T. Sherman, 1864, quoted in Liddell Hart, *Sherman*, 1930.

Strenuous, unrelaxing pursuit is therefore as imperative after a battle as is courage during it.
Rear Admiral Alfred Thayer Mahan, *Naval Strategy*, 1911.

In pursuit you must always stretch possibilities to the limit. Troops having beaten the enemy will want to rest. They must be given as objectives, not those that you think they will reach, but the farthest they could possibly reach.
Field Marshal Viscount Allenby of Meggido, 1917, Order to XXI Corps, Philistia.

Pursuit will hold the first place in your thoughts. It is at the moment when the victor is most exhausted that the greatest forfeit can be extracted from the vanquished.
Sir Winston S. Churchill, 13 December 1940, to Lord Wavell after his victory over the Italians in the Libyan desert.

While coolness in disaster is the supreme proof of a commander's courage, energy in pursuit is the surest test of his strength of will.
Field Marshal Viscount Wavell of Cyrenaica, unpublished *Recollections*, 1946.

QUAGMIRE

I don't do quagmires.
Secretary of Defense Donald Rumsfeld, 24 July 2003, press conference.

QUALITY

At my first going out into this engagement I saw our men beaten at every hand... 'Your troops,' said I, 'are most of them old decayed serving-men, and tapsters, and such like fellows; and,' said I, 'their troops are gentlemen's sons, younger sons and persons of quality: do you think that the spirits of such base mean fellows will ever be able to encounter gentlemen, that have honor and courage and resolution in them?
...You must get men of spirit: and take it not ill what I say, – I know you will

QUOTATIONS AND MAXIMS

not, – of a spirit that is likely to go on as far as gentlemen will go: or else you will be beaten still.'
   Oliver Cromwell, October 1642, to John Hampden after the Battle of Edgehill.

It is not the big armies that win battles; it is the good ones.
   Field Marshal Maurice Comte de Saxe, *My Reveries*, 1732, tr. Phillips, 1940.

More is not better, better is better.
   General Gordon R. Sullivan, quoted in *Military Review*, July–August 1996.

QUARTER

Glasdale, Glasdale, yield to the King of Heaven. You have called me 'whore': I pity your soul and the souls of your men.
   Joan of Arc, 7 May 1429, to the English commander in the assault on the bridge at Orléans, *Joan of Arc in her Own Words*, 1996.

We must spare our enemies or it will be our loss, since they and all that belong to them must soon be ours.
   Anonymous Inca prince.

Magdeburg Quarter!
   For years after the massacre of the city of Magdeburg in May 1631, Imperialist soldiers attempting to surrender would be met with this Protestant response as they were shot down, quoted in C.V. Wedgwood, *The Thirty Years War*, 1938.

Tarleton Quarter!
   Battle cry of the American troops at the Battle of Cowpens, 17 January 1781, during the Revolution, referring to the merciless conduct of the British commander, Lieutenant-Colonel Banastre Tarleton, who had offered no quarter to the Americans in previous engagements.

QUOTATIONS AND MAXIMS

I hate quotations. Tell me what you know.
   Ralph Waldo Emerson, *Journals*, 1849.

It is a good thing for an uneducated man to read books of quotations.
   Sir Winston S. Churchill, *My Early Life*, 1930.

Catchwords and trite phrases are not the same thing as quotations, although not unrelated; for quotations also tend to have ridiculous and dangerous associations. At the same time it is undoubtedly convenient to find that someone else has already expressed the same thought in a happy and generally accepted form; not to mention the fact that literary people are agreeably surprised or impressed when they find a soldier occasionally quoting Goethe or even Greek, suggesting an intellectual capacity in excess of that required for reading the drill book. That is why I sometimes make quotations myself.
   Colonel-General Hans von Seekt, *Thoughts of a Soldier*, 1930.

Professional soldiers are sentimental men, for all the harsh realities of their calling. In their wallets they carry bits of philosophy, fragments of poetry, quotations from the Scriptures which in time of stress and danger speak to them with great meaning.
   General Matthew B. Ridgway, *My Battles in War and Peace*, 1956.

Nothing so comforts the military mind as the maxim of a great but dead general.
   Barbara W. Tuchman, *The Guns of August*, 1962.

R

RANK

The rank of Officers, which to me, Sir, is much dearer than the Pay.
  General George Washington, 29 May 1754, to Robert Dinwiddie, *Writings*, I, 1931, p. 83.

Do not hesitate to give commands to officers in whom you repose confidence without regard to claims of others on account of rank.
  General of the Army Ulysses S. Grant, August 1864, letter to Major-General Philip Sheridan, *Papers of USG*, XI, 1982, pp. 379–380.

I think of rank but of trivial importance so that it is sufficient for the individual to exercise command.
  General Robert E. Lee (1807–1870).

After the close of the war I was once conversing with Gen. Grant when he remarked playfully that in all the changes in rank which occurred to officers in our service he thought the one which was felt to be the most important was when one was promoted from Brevet 2nd Lieutenant to 2nd Lieutenant! (The General himself had served for more than two years a Brevet 2nd Lieutenant.) I admitted that that was certainly an important event in one's history, but according to my observation there was one change which beat it all to pieces. 'What is that?' he asked. I replied 'when one drops from a Major General to a Captain.'
  Major-General John Gibbon, referring to the loss of wartime rank in the post-war army, *Personal Reminiscences of the Civil War*, 1928.

The barrier of rank is the highest of all barriers in the way of access to the truth.
  Captain Sir Basil Liddell Hart, *Thoughts on War*, 1944.

The badge of rank which an officer wears on his coat is really a symbol of servitude to his men.
  General Maxwell D. Taylor, *The Field Artillery Journal*, January/February 1947.

One of the problems inherent in possessing rank and owning the responsibility of command is that people don't think you're quite human. You can't get drunk even every now and then. (But there have never been too many teetotalers among the military, if we leave out Stonewall Jackson and – I suppose – Sir Galahad.)
  General Curtis LeMay, *Mission With LeMay*, 1965.

RASHNESS

I may be accused of rashness but not of sluggishness.
  Napoleon, 6 May 1796, *Correspondance de Napoléon Ier, publiée par ordre de l'Empereur Napoléon III*, No. 337, Vol. I, 1858–1870.

Rashness succeeds often, still more often fails.
  Napoleon (1769–1821).

Rashness in war is prudence.
  Admiral Sir John Fisher, *Memories*, 1919.

READINESS

The art of war teaches us to rely not on the likelihood of the enemy's not coming, but on our own readiness to

RECONCILIATION

receive him; not on the chance of his not attacking but rather on the fact that we have made our positions unassailable.
  Sun Tzu, *The Art of War*, 8, c. 500 BC, tr. Giles, 1910.

A general who desires peace must be prepared for war, for the barbarians become very nervous when they face an adversary all set to fight.
  The Emperor Maurice, *The Strategikon*, c. AD 600 (*Maurice's Strategikon*, tr. George Dennis, 1984).

Maxim 7. An army should be ready every day, every night, and at all times of the day and night, to oppose all the resistance of which it is capable. With this view, the soldier should be invariably complete in arms and ammunition; the infantry should never be without its artillery, its cavalry, and its generals.
  Napoleon, *The Military Maxims of Napoleon*, tr. George D'Aguilar, 1831 (David Chandler, 1987).

The Gods who give the crown of victory to those who, by their training in peaceful times, are already victorious before they fight, refuse it to those who, satisfied with one victory, rest contentedly in peace. As the old saying goes, 'After a victory, tighten the strings of your helmet.'
  Admiral Marquis Heihachiro Togo, 21 December 1905, farewell to the Japanese Fleet after the Battle of Tsushima, quoted in Ogasawara, *Life of Admiral Togo*, 1934.

It is too late to learn the technique of warfare when military operations are already in progress, especially when the enemy is an expert in it.
  General Aleksei A. Brusilov, *A Soldier's Notebook*, 1931.

I am not a bit anxious about my battles. If I am anxious I don't fight them. I wait until I am ready.
  Field Marshal Viscount Montgomery of Alamein (1887–1976) – attributed.

Being ready is not what matters. What matters is winning after you get there.
  Lieutenant-General Victor H. 'The Brute' Krulak, *Life Magazine*, 30 April 1965.

As you know, you have to go to war with the Army you have, not the Army you want.
  Secretary of Defense Donald Rumsfeld, 9 December 2004, at Camp Buehring, Kuwait, in response to a soldier's question of why they had to fabricate their own armour for vehicles, quoted in 'Troops put thorny questions to Rumsfeld,' *CNN.com*.

RECONCILIATION

Let us try in this way, if we can, to win back public opinion and gain a lasting victory. For all too many others have incurred hatred through their cruelty and failed to maintain their victory for long, with the exception of Sulla, and I do not wish to emulate him. Let this be a new way of gaining victory; let us secure ourselves through mercy and magnanimity!
  Julius Caesar, early March 45 BC, letter after the Roman Civil War.

It is the prime consideration of a general to secure victory – with its achievement unavoidable wrongs can later be remedied by kindness.
  Gonzalo de Corboba, 'El Gran Capitan', (1453–1515).

The war will be eternal if nobody is forgiven... I am decidedly of opinion that all animosity should be forgotten.
  The Duke of Wellington, 12 March 1804 and 22 January 1804, in India, cited in Arthur Bryant, *The Great Duke*, 1972.

I want no one punished. Treat them liberally all around. We want those people to return to their allegiance to the Union and submit to the laws.
  President Abraham Lincoln, 28 March

RECONNAISSSANCE

1865, instructions to Generals Grant and Sherman and to Admiral Porter just before Lee's surrender.

RECONNAISSANCE

Agitate the enemy and ascertain the pattern of his movement. Determine his dispositions and so ascertain the field of battle. Probe him and learn where his strength is abundant and where deficient.
> Sun Tzu, *The Art of War*, 6, c. 500 BC, tr. Giles, 1910.

Single men in the night will be more likely to ascertain facts than the best glasses in the day.
> General George Washington, 10 July 1779, letter to General Anthony Wayne, *Writings*, XV, 1936, p. 397.

...Skilfully reconnoitering defiles and fords, providing himself with trusty guides, interrogating the village priest and the chief of relays, quickly establishing relations with the inhabitants, seeking out spies, seizing letters.
> Napoleon, on the duties of a chief of reconnaissance, *Maxims of War*, 1831.

Bear in mind that your telegrams may make the whole Army strike tents, and night or day, rain or shine, take up the line of march. Endeavor, therefore, to secure accurate information... Above all, vigilance! vigilance! vigilance!
> Major J.E.B. Stuart, 4 April 1864, letter to John R. Chambliss, quoted in Douglas Southall Freeman, *Lee's Lieutenants: Gettysburg to Appomattox*, 1944.

Time spent on reconnaissance is seldom wasted.
> British Army *Field Service Regulations*, 1912.

Only the enemy in front, every other bugger behind.
> Unofficial motto of the British Infantry Reconnaissance Corps, WWII.

You can never do too much reconnaissance.
> General George S. Patton, Jr, *War As I Knew It*, 1947.

Junior officers of reconnaissance units must be very inquisitive. Their reports must be accurate and factual. Negative information is as important as positive information.
> General George S. Patton, Jr, *War As I Knew It*, 1947.

Do your own reconnaissance. See for yourself and then get down to the job without delay.
> Field Marshal Erich von Manstein (1887–1973).

RECRUITS/RECRUITMENT

An army raised without proper regard to the choice of its recruits was never yet made good by the length of time.
> Flavius Vegetius Renatus, *Military Institutions of the Romans*, c. AD 378, tr. Clark, 1776.

Men just dragged from the tender Scenes of domestick life; unaccustomed to the din of Arms; totally unacquainted with every kind of Military skill ...when opposed to Troops regularly train'd, disciplined, and appointed, superior in knowledge, and superior in Arms, makes them timid and ready to fly from their own shadows.
> General George Washington, 24 September 1776, letter to the President of Congress, *Writings*, VI, 1932, p. 110.

I will receive 200 able-bodied men if they will present themselves at my headquarters by the first of June with a good horse and gun. I wish none but those who desire to be actively engaged. My headquarters for the present is at Corinth, Miss. Come on, boys, if you want a heap of fun and to kill some Yankees.
> N.B. Forrest

REFLECTION/MEDITATION

Colonel, Commanding
Forrest's Regiment
> Lieutenant-General Nathan Bedford Forrest, April 1862, recruiting advertisement in the *Appeal*, run while he recuperated from wounds suffered at the Battle of Shiloh, quoted in Robert Henry, *First With the Most Forrest*, 1944.

The country lad is generally the most humble, obedient, and easily governed; but the city-born recruit is, as a rule, the smartest, tidiest, and most easily trained.
> John Menzies, *Reminiscences of an Old Soldier*, 1883.

For the Red Gods call us out and we must go!
> Rudyard Kipling, 'The Feet of the Young Men'.

The director of the CIA in my time was Army Gen. Beetle Smith, whom I had met on some occasion during the war in Africa. One day as he entered the meeting, late, he announced that he had just left the Oval Office. He said the President (Truman) asked him why the Army was rejecting so many recruits for lack of physical fitness for military duty. 'Why,' said the President, 'when I entered the Army one doctor would look up your rear while another looked down your throat. If they couldn't see each other, they'd sign you up.'
> General Hamilton H. Howze, *A Cavalryman's Story*, 1996.

REFLECTION/MEDITATION

The wise must contemplate the intermixture of gain and loss. If they discern advantage [in difficult situations], their efforts can be trusted. If they discern harm [in prospective advantage], difficulties can be resolved.
> Sun Tzu, *The Art of War*, 8, *c.* 500 BC, tr. Ralph Sawyer, 1994.

He ponders the dangers inherent in the advantages, and the advantages inherent in the dangers.
> The 'Martial' Emperor Ts'ao Ts'ao (AD 155–220), commentary to *Sun Tzu*, 8, *c.* 500 BC, tr. Samuel Griffith, 1963.

The general should not go to sleep before reflecting on what he should have done that he might have neglected and on what he has to do the next day.
> The Emperor Maurice, *The Strategikon*, *c.* AD 600 (*Maurice's Strategikon*, tr. George Dennis, 1984).

What is the good of experience if you do not reflect.
> Frederick the Great (1712–1786).

The winters wear away, so do our years, and so does life itself; and it matters little where a man passes his days and what station he fills, or whether he be great or considerable, but it imports him something to look to his manner of life. This day I am twenty-five years of age, and all that time is as nothing... But it is worth a moment's consideration that one may be called away on a sudden, unguarded and unprepared; and the oftener these thoughts are entertained the less will be the dread or fear of death. You will judge by this sort of discourse that it is the dead of night, when all is quiet and at rest, and one of those intervals when men think of what they really are, and what they really should be; how much is expected and how little performed... The little time taken for meditation is the best employed in all our lives; for, if the uncertainty of our state and being is then brought before us, and that compared with our course of conduct, who is there that won't immediately discover the inconsistency of all his behaviour and the vanity of all his pursuits.
> Major-General Sir James Wolfe, quoted in Liddell Hart, *Great Captains Unveiled*, 1927.

THE REGIMENT

If I appear to be always ready to reply to everything, it is because, before undertaking anything, I have meditated for a long time – I have foreseen what might happen. It is not a spirit which suddenly reveals to me what I have to say or do in a circumstance unexpected by others: it is reflexion, meditation.
Napoleon, quoted in Barrett Parker, ed., *Famous British Generals*, 1951.

Until tomorrow; night brings counsel.
Napoleon, his method of dealing with the matters of the greatest importance, compared to the quick dispatch of the great mass of his daily business, quoted in Chandler, *The Campaigns of Napoleon*, 1966.

Generalship, at least in my case, came not by instinct, unsought, but by understanding, hard study and brain concentration. Had it come easily to me, I should not have done it as well.
Colonel T.E. Lawrence, 1932, letter to Liddell Hart.

An acquaintance with the 'friction' of military movements and the actual conditions of the battlefield is invaluable. But the duration of experience matters less than the capacity for reflection. And the measure of this, in turn, counts for more than the extent of study. Given some acquaintance with actual war, and time for study, the mastery of strategy and tactics is likely to be in proportion to the capacity for reflection, analysis, and originality of thought.
Captain Sir Basil Liddell Hart, *Thoughts on War*, 1944.

THE REGIMENT

After the train had been captured by 150 Boers, the last four men, though completely surrounded, and with no cover continued to fire until three were killed, the fourth wounded. On the Boers asking the survivor the reason why they had not surrendered, he replied, 'Why, man, we are the Gordon Highlanders.'
Field Marshal Viscount Kitchener, 10 August 1901, telegram from Pretoria to King Edward VII during the Boer War.

The noble courage that has its origins in love of country and sense of duty is not confined to the well-born; it is to be equally found in the uneducated private soldier. What can be finer than his love of regiment, his devotion to its reputation, and his determination to protect its honour! To him 'The Regiment' is mother, sister and mistress. That its fame may live and flourish he is prepared to risk all and to die without a murmur. What early cause calls forth greater enthusiasm? It is a high, an admirable phase of patriotism, for, to the soldier, his regiment is his country.
Field Marshal Viscount Wolseley, *The Story of a Soldier's Life*, 1903.

Heads up the Warwicks! Show the _____ yer cap-badges!
Field Marshal Viscount Slim of Burma, quoting the 'Incorrigible Rogue', a soldier in his company during the Mesopotamian campaign, 1916, *Unofficial History*, 1959 (see Preface).

Never forget: the Regiment is the foundation of everything.
Field Marshal Viscount Wavell of Cyrenaica (1883–1950), quoted in Fergusson, *The Trumpet in the Hall*, 1970.

Though I shall maintain that good soldiers are not bred from bad stock, I do not doubt that many unpromising specimens were transformed by training; in particular by that part of training which consists in inculcating esprit de corps. I remember men recruited at the street corner by starvation who came to act on the principle that if the Regiment lived it did not matter if they died, though they did not

REGULATIONS

put it that way. This was their source of strength, their abiding faith, it was the last of all the creeds that in historical times have steeled men against death.
    Lord Moran, *The Anatomy of Courage*, 1945.

The art of command is the art of dealing with human nature. The soldier is governed through his heart and not through his head. The complete acceptance in the soldiers' training of that creed with all its implications is the particular contribution England has made to the business of making soldiers. It began, I think, with Moore – it was the foundation of the discipline of his camp at Shorncliffe. The men's response to kindness led them to accept the religion of the Regiment – that only the Regiment matters; a faith which made our professional army before the last war [1914–1918], in German words, 'a perfect thing apart'.
    Lord Moran, *The Anatomy of Courage*, 1945.

They will remember for a little while in England. The soldier does have his day. I want to remind you this afternoon that it is not enough to remember now. We've got to show what we think of their sacrifice in the way we conduct ourselves in the days ahead.
    The Chaplain of the First Battalion, Gloucestershire Regiment (The Glosters), April 1951, memorial service in Korea for the battalion, which had been overrun after a heroic defence.

There is a great source of strength in the regimental system itself. This sets up in the group a continuing focus of affection, trust and loyalty. It uses insignia, totems and a good deal of almost mystical paraphernalia to increase the binding grip of the whole upon its members. Some day I want to turn the ethnologists on to study of the regimental system as a means of strengthening group resistance to stress. Their advice may well be that instead of a quaint and decorative traditional survival, we have in the British regimental system a military instrument of deadly efficiency.
    General Sir John Hackett, *The Profession of Arms*, 1983.

REGULARS

Regular Troops alone are equal to the exigencies of modern war, as well as for defence as offence, and whenever a substitute is attempted it must prove illusory and ruinous.
    General George Washington, 15 September 1780, to the President of Congress, *Writings*, XX, 1937, p. 49.

When defending itself against another country, a nation never lacks men, but too often, *soldiers*.
    Napoleon, *Political Aphorisms*, 1848.

I longed for more Regular troops with which to rebuild and expand the Army. Wars are not won by heroic militia.
    Sir Winston S. Churchill, *Their Finest Hour*, 1949.

REGULATIONS

Nobody in the British Army ever reads a regulation or an order as if it were to be a guide for his conduct, or in any other manner than as an amusing novel.
    The Duke of Wellington (1769–1852).

Regulations are all very well for drill but in the hour of danger they are no more use… You must learn to think.
    Marshal of France Ferdinand Foch (1851–1929).

Method, I explained, was laid down in the Training Manuals, which were written not for sages, but for normal men, many of whom are fools. Though they must be followed, 'do not imagine for a moment,' I said, 'that they have been written to exonerate you from thinking.'
    Major-General J.F.C. Fuller, *Memoirs of an*

REINFORCE SUCCESS

Unconventional Soldier, 1936.

Fudge the regulations. They're sometimes made to be broken for the good of the whole.
> General of the Army Douglas MacArthur, quoted in William Ganoe, *MacArthur Close-Up*, 1955, p. 28.

You do not rise by the regulations, but in spite of them. Therefore in all matters of active service the subaltern must never take 'No' for an answer. He should get to the front at all costs.
> Sir Winston S. Churchill, *Ian Hamilton's March*.

One of the first regulations might be to think.
> Major-General Orlando Ward, under the pseudonym 'MacKenzie Hill', *The Field Artillery Journal*, c. 1938.

REINFORCE SUCCESS

The moment it becomes certain that an assault cannot succeed, suspend the offensive; but when one does succeed, push it vigorously, and if necessary pile in troops at the successful point from wherever they can be taken.
> General of the Army Ulysses S. Grant, 3 June 1864, at the Battle of Cold Harbor, quoted in Horace Porter, *Campaigning With Grant*, 1906.

RELIEF

I must remove this man, yes. Disgrace him I will not, lest in doing so I paralyze the initiative of another man.
> General Robert E. Lee, quoted in Douglas Southall Freeman, *Douglas Southall Freeman on Leadership*, 1993.

The service cannot afford to keep a man who does not succeed.
> Lieutenant-General Thomas J. 'Stonewall' Jackson, quoted in Henderson, *Stonewall Jackson*, 1898.

Jackson: Colonel, why do you not get your brigade together, keep it together, and move on?
Colonel: It is impossible, General, I can't do it.
Jackson: Do not say it is impossible! Turn your command over to the next officer. If he cannot do it, I will find someone who can, if I have to take him from the ranks!
> Lieutenant-General Thomas J. 'Stonewall' Jackson, 2 June 1862, quoted in Henry Kyd Douglas, *I Rode With Stonewall*, 1940.

There were those who had been relieved from their commands who came in to tell me of the injustice which had been done them. It was hard to talk to men of this class, because in most instances I was convinced by what they told me that their relief was justified. To discuss with an old friend the smash-up of his career is tragic and depressing at best, and more practically when he feels that he has been treated unfairly and an honorable record forever besmirched.
> General of the Army George C. Marshall, *Memoirs of My Service in the World War*, written between 1923 and 1926, published in 1976.

In war you must either trust your general or sack him.
> Field Marshal Sir John Dill, 1941, letter to Field Marshal Wavell.

There were instances in Europe where I relieved commanders for their failure to move fast enough. And it is possible that some were the victims of circumstance. For how can the blame for failure be laid fairly on a single man when there are in reality so many factors that can effect the outcome of any battle? Yet each commander must always assume total responsibility for every individual in his command. If his battalion or regimental commanders fail him in the attack, then he must relieve them or be relieved himself. Many a division commander has failed not

RELIGION

because he lacked the capacity for command but only because he declined to be hard enough on his subordinates.

In the last analysis, however, the issue of relief resolves itself into one of mutual confidence.
> General of the Army Omar N. Bradley, *A Soldier's Story*, 1951.

RELIGION

If thou wishest the gods to be propitious to thee, thou must honour the gods.
> Xenophon, *Memorabilia*, c. 360 BC.

Before getting into danger, the general should worship God. When he goes into danger, then, he can with confidence pray to God as a friend.
> The Emperor Maurice, *The Strategikon*, c. AD 600 (*Maurice's Strategikon*, tr. George Dennis, 1984).

I think that when God grants me victory over the rest of Palestine I shall divide my territories, make a will stating my wishes, then set sail on this sea for their far-off lands and pursue the Franks there, so as to free the earth of anyone who does not believe in God, or die in the attempt.
> Saladin (1138–1193), quoted by Baha Ad-Din in Gabarieli, *Arab Historians of the Crusades*, 1957.

He who is steadfast unto death shall be saved and they who suffer in a just cause theirs is the kingdom of heaven.
> Edward, 'The Black Prince', 19 September 1356, address to his knights before the Battle of Poitiers, quoted in Froissart, *Chronicles*, 1523–1525.

It will give me great pleasure to fight for my God against your gods, who are a mere nothing.
> Hernan Cortes, 1521, to the Aztec priesthood, quoted in *Five Letters*, 1522–1525.

You may win salvation under my command, but hardly riches.
> King Gustavus II Adolphus (1594–1632).

I can say this of Naseby that when I saw the enemy drawn up and march in gallant order towards us, and we a company of poor ignorant men, to seek how to order our battle, the General having commanded me to order all the horse, I could not, riding alone about my business, but smile out to God in praises in assurance of victory, because He would, by things that are not, bring to naught things that are. Of which I had great assurance – and God did it.
> The Lord Protector Oliver Cromwell, after the Battle of Naseby, 14 June 1644, quoted in Churchill, *A History of the English Speaking Peoples: The New World*, 1956.

Truly I think he that prays and preaches best will fight best.
> The Lord Protector Oliver Cromwell, 25 December 1650, letter to Colonel Francis Hacker, quoted in Ashely, ed., *Cromwell*, 1969.

Put your trust in God, but be sure to see that your powder is dry.
> The Lord Protector Oliver Cromwell (1599–1658), to his troops when they were about to cross a river.

When I lay me down to sleep, I recommend myself to the care of Almighty God; when I awake I give myself up to His direction. Amidst all the evils that threaten me, I will look up to Him for help, and question not but that He will either avert them or turn them to my advantage. Though I know neither the time nor the place of my death, I am not at all solicitous about it, because I am sure that He knows them both, and that He will not fail to support and comfort me.
> Admiral Viscount Nelson, 1793, journal entry, quoted in David Howarth, *Nelson: The Immortal Memory*, 1988.

Now I can do no more. We must trust

RELIGION

to the Great Disposer of all events, and the justice of our cause. I thank God for this great opportunity of doing my duty.

> Admiral Viscount Nelson, 21 October 1805, moments after his last signal to the fleet at Trafalgar, 'England expects every man to do his duty', quoted in Robert Southey, *Life of Nelson*, 1813.

The soldiers of the army have permission to go to mass, so far as this: they are forbidden to go into the churches during the performance of divine service, unless they go to assist in the performance of the service. I could not do more, for in point of fact soldiers cannot by law attend the celebration of mass excepting in Ireland. The thing now stands exactly as it ought; any man may go to mass who chooses, and nobody makes any inquiry about it. The consequence is, that nobody goes to mass, and although we have whole regiments of Irishmen, and of course Roman Catholics, I have not seen one soldier perform any one act of religious worship in these Catholic countries, excepting making the sign of the cross to induce the people of the country to give them wine.

> The Duke of Wellington, 1809, *The Dispatches of Field Marshal the Duke of Wellington, During His Various Campaigns in India, Denmark, Portugal, Spain, the Low Countries, and France*, V, pp. 134–135, 1834–1838.

I never had so grand and awful an idea of the resurrection as ...[when] I saw ...more than five hundred Britons emerging from the heaps of their dead comrades, all over the plain rising up, and ...coming forward ...as prisoners.

> Major-General Andrew Jackson, of the survivors of the British attack on his defences during the Battle of New Orleans, 8 January 1815, quoted in James Parton, *A Life of Andrew Jackson*, II, 1859–1860.

Everything proclaims the existence of a God, that cannot be questioned; but all our religions are evidently the work of men. Why are there so many? – Why has ours not always existed? – Why does it consider itself exclusively the right one? – What becomes in this case of all the virtuous men who have gone before us? – Why do these religions revile, oppose, exterminate one another? – Why has this been the case ever and every where? – Because men are ever men; because priests have ever and every where introduced fraud and falsehood. However, as soon as I had power, I immediately re-established religion. I made it the ground-work and foundation upon which I built. I considered it as the support of sound principles and good morality, both in doctrine and in practice. Besides, such is the restlessness of man, that his mind requires that something undefined and marvellous which religion offers; and it is better for him to find it there, than to seek it of Cagliostro, of Mademoiselle Lenormand, or of the fortune-tellers and imposters.

> Napoleon, 1 June 1816, conversation on St Helena, quoted in Emmanuel Las Casas, *The Life, Exile, and Conversations of Napoleon Bonaparte*, II, 1890.

Look at a man in his full vigor, look at those columns ready to throw themselves into the battlefield: the drum beats the charge; now they move; the grapeshot mows them down. There is no question of priests or confession.

> Napoleon, 1819, conversation on St Helena, quoted in F. Antommarchi, *Mémoires*, I, 1825, and in Christopher Herold, ed., *The Mind of Napoleon*, 1955.

Although I have served in my profession in several countries, and among foreigners, some of whom professed various forms of the Christian religion, while others did not profess it at all; I never was in one in which it was not the

REMINISCENCES

bounden duty of the soldier to pay proper deference and respect to whatever happened to be the religious institutions or ceremonies of the place.
   The Duke of Wellington, 8 April 1829, speech to the House of Lords.

Hell, is that all? Detail forty men at once for baptism!
   Major-General Daniel E. Sickles, 1862, when his brigade chaplain expressed concern that other brigades were converting more men than he was, quoted in *The Journals of Joseph Hopkins Twichell*, item dated 16 December 1905.

Captain, my religious belief teaches me to feel as safe in battle as in bed. God has fixed the time for my death. I do not concern myself about that, but to be always ready, no matter when it may overtake me... That is the way all men should live, and then all would be equally brave.
   Lieutenant-General Thomas J. 'Stonewall' Jackson, to Brigadier-General John Imboden, quoted in *Battles and Leaders of the Civil War, Being for the Most Part Contributions by Union and Confederate Officers*, I, Johnston and Buel, ed., 1887.

In one thing we took 'the touch of elbows.' It was not uncommon incident that from close opposing bivouacs and across hushed breastworks at evening voices of prayer from over the way would stir our hearts, and floating songs of love and praise be caught up and broadened into a mighty and thrilling chorus by our men softening down in cadences like enfolding wings. Such moments were surely a 'Truce of God.'
   Major-General Joshua L. Chamberlain, *The Passing of the Armies*, 1915.

God disposes. This ought to satisfy us.
   General Robert E. Lee, a note found in his desk at Washington College after his death, quoted in A.L. Long, *Memoirs of Robert E. Lee*, 1886.

I pray daily to do my duty, retain my self-confidence, and accomplish my destiny. No one can live under the aweful responsibility I have without Divine help. Frequently I feel that I don't rate it.
   General George S. Patton, Jr, 19 June 1943, diary entry.

A chaplain visits our company. In a tired voice, he prays for the strength of our arms and for the souls of the men who are to die. We do not consider his denomination. Helmets come off. Catholics, Jews, and Protestants bow their heads and finger their weapons. It is the front-line religion: God and the Garand.
   Lieutenant Audie Murphy, *To Hell and Back*, 1949.

Finally, I do not believe that today a commander can inspire great armies, or single units, or even individual men, and lead them to achieve great victories, unless he has a proper sense of religious truth. He must always keep his finger on the spiritual pulse of his armies; he must be sure that the spiritual purpose which inspires them is right and true, and is clearly expounded to one and all. Unless he does this, he can expect no lasting success. For all leadership, I believe, is based on the spiritual quality, the power to inspire others to follow; this spiritual quality may be for good, or evil. In many cases in the past this quality has been devoted towards personal ends, and was partly or wholly evil; whenever this was so, in the end it failed. Leadership which is evil, while it may temporarily succeed, always carries within itself the seeds of its own destruction.
   Field Marshal Viscount Montgomery of Alamein, *The Memoirs of Field Marshal Montgomery*, 1958.

REMINISCENCES

Chief of the sailors, Ahmose, son of

REMINISCENCES

Ebana, triumphant; he says: 'I will tell you, O all ye people; I will cause you to know the honors which came to me. I was presented with gold seven times in the presence of the whole land; male and female slaves likewise. I was endowed with many fields.' The fame of one valiant in achievements shall not perish in this land forever.

> Ahmose, son of Abana, Egyptian naval officer in the reign of Pharaohs Ahmose, Amenhotep I and Thutmose I (1580–1501 BC), quoted in James Breasted, *Ancient Records of Egypt*, 1906.

That was once a man.

> The Duke of Marlborough, 1716, gazing up in the twilight of his life at his magnificent portrait by Kneller showing him in his glory as one of the great captains of history, quoted in S.I. Reid, *John and Sarah Marlborough*, 1914, p. 413.

I should have given Suchet an army-corps under my command, I should have sent Davout a month earlier to organize my army, and appointed Clauzel Minister of War. Or I ought to have given Soult the command of the Guard. He did not wish me to employ Ney. I should have spent the night of the 15th in Fleurus, and given Grouchy's command to Suchet, and given the former the command of all the cavalry, as I had not got Murat. The soldiers did not know each other well enough to possess the proper *esprit de corps*. The cavalry were better than the infantry. It is a pity that I did not fall at Waterloo, for that would have been a fine ending. My situation is frightful! I am like a dead man, yet full of life!

> Napoleon (1769–1821), of the Waterloo campaign, quoted in F.M. Kircheisen, *The Memoirs of Napoleon I*, 1929.

To be at the head of a strong column of troops, in the execution of some task that requires brain, is the highest pleasure of war – a grim one and terrible, but which leaves on the mind and memory the strongest mark; to detect the weak point of an enemy's line; to break through with vehemence and thus lead to victory; or to discover some key-point and hold it with tenacity; or to do some other distinct act which is afterward recognized as the real cause of success. These all become matters that are never forgotten.

> General of the Army William T. Sherman, *The Memoirs of General W.T. Sherman*, II, 1875.

To conclude, I would again put on record the deep love and gratitude that I have always felt for all the troops who were in my charge. At a word from me, they faced wounds, agony, and death for Russia's sake. And all in vain! May I be pardoned, for the guilt is not mine; and I could not foresee the future.

> General Aleksei A. Brusilov, *A Soldier's Notebook*, 1931.

It was seventeen years ago – those days of old have vanished, tone and tint; they have gone glimmering through the dreams of things that were. Their memory is a land where flowers of wondrous beauty and varied colors spring, watered by tears and coaxed and caressed into fuller bloom by the smiles of yesterday. Refrains no longer rise and fall from that land of used-to-be. We listen vainly, but with thirsty ear, for the witching melodies of days that are gone. Ghosts in olive drab and sky blue and German gray pass before our eyes; voices that have stolen away in the echoes from the battlefields no more ring out. The faint, far whisper of forgotten songs no longer floats through the air. Youth, strength, aspirations, struggles, triumphs, despairs, wide winds sweeping, beacons flashing across uncharted depths, movements, vividness, radiance, shadows, faint bugles sounding reveille, far drums beating the long roll, the crash of guns, the rattle of

REORGANISATION

musketry – the still white crosses! And tonight we are met to remember.
> General of the Army Douglas MacArthur, 14 July 1935, address to the veterans of the 42nd 'Rainbow' Division, *A Soldier Speaks*, 1965.

This I have always felt; as a youth and now as a man well in middle age: That truth is courage intellectualized. Thus the idea of the great man is the human coping stone of my philosophy as it was of the philosophy of that great man – Thomas Carlyle. Therefore in my study of war I have always put the great man first. As my system was founded on the organization of the human body, of necessity it follows that to breathe life into it, those who can do so must be men who at least aspire towards greatness.
> Major-General J.F.C. Fuller, *Memoirs of an Unconventional Soldier*, 1936.

At a difficult time a prince of my royal family once sent me a small portrait of Frederick the Great on which he had inscribed these words that the great king addressed to his friend, the Marquis d'Argens, when his own defeat seemed imminent. 'Nothing can alter my inner soul: I shall pursue my own straight course and shall do what I believe to be right and honorable.' The little picture I have lost, but the king's words remain engraved on my memory and are for me a model. If, despite everything, I could not prevent the defeat of my country, I must ask my readers to believe that this was not for lack of a will to do so.
> Colonel-General Heinz Guderian, *Panzer Leader*, 1953.

Well, I'd like to do it all over again. The whole thing.

And more than that – more than anything – I'd like to see once again the face of every Marine I've ever served with.
> General Lewis 'Chesty' Puller, 26 June 1960, to his wife when she asked him what he would like to do after his retirement, quoted in Burke Davis, *Marine*, 1962.

With every passing year we draw further and further away from those years of war. A new generation of people has grown up. For them war only means our reminiscences of it. And the numbers of us who took part in those historic events are dwindling fast. But I am convinced that time cannot cause the greatness to fade. Those were extraordinarily difficult, but also truly glorious years. Once a person has undergone great trials and come through victorious, then throughout his life he draws strength from this victory.
> Marshal of the Soviet Union Georgi K. Zhukov, *Reminiscences and Reflections*, 1974.

I would like to be rated not by what I wrote but by how I lived, for I wrote to live and not the other way around. Having tried in this accounting to bestride the various periods in my journeying as if they were so many stages along a road, largely with the hope of explaining fortune that I hardly understand myself, I find on looking back that there is little to lament and much to cheer. Good companions were always alongside to help me through trial; I needed them as they in turn needed me. My yesterdays were the best the country ever knew, or so I believe, for in living them to the hilt I helped make them.

I think I will leave it at that.
> Brigadier-General S.L.A. Marshall, *Bringing Up the Rear*, 1979.

REORGANISATION

Any military reorganization should conform to certain set principles: –

(1) Power must go with responsibility.
(2) The average human brain finds its effective scope in handling from three to six other brains.

REPLACEMENTS

If a man divides the whole of his work into two branches and delegates his responsibility, freely and properly, to two experienced heads of branches he will not have enough to do. The occasions when they would have to refer to him would be too few to keep him fully occupied. If he delegates to three heads he will be kept fairly busy whilst six heads of branches will give most bosses a ten hours' day. Those data are the results of centuries of the experiences of soldiers, which are greater, where organization is in question, than those of politicians, business men or any other class of men by just so much as an Army in the field is a bigger concern than a general election, the Bank of England, the Standard Oil Company, the Steel Trusts, the Railway Combines of America, or any other part of politics or business. Of all the ways wasted there is none so vicious as that of your clever politician trying to run a business concern without having any notion of self-organization. One of them who took over Munitions for a time had so little idea of organizing his own energy that he nearly died of overwork through holding up the work of others; by delegating responsibility coupled with direct access to himself to seventeen sub-chiefs!

Now, it will be understood why a battalion has four companies (and not seventeen); why a brigade has three or four battalions (and not seventeen).

General Sir Ian Hamilton, The Soul and Body of an Army, 1921.

The hypothesis of every reorganization of an army is, first of all, peace on the outer borders. That means several years of external peace and a state of political calm... Before these conditions are achieved a successful military reorganization cannot be accomplished. Success cannot be attained while you are in a continual state of war.

Colonel-General Hans von Seekt, while a military advisor to Chiang Kai-shek during the 1930s, Denkschrift für Marschall Chiang Kai-shek, n.d.

REPLACEMENTS

Those people think they are indispensable; they don't know I have a hundred division commanders who can take their place.

Napoleon, repeated remark about his marshals, quoted in Christopher Herold, ed., The Mind of Napoleon, 1955.

It looks hard to put those brigades, now numbering less than 800 men, into battle. They feel discouraged, whereas if we could have a steady influx of recruits the living would soon forget the dead.

General William T. Sherman, letter to Major-General Halleck, 4 September 1864, O.R., Series I, Vol. XXXVIII, Part 5, pp. 793–794.

The fighting troops are not being replaced effectively, although masses of drafts are sent to the technical and administrative services, who were originally on the most lavish scale, and who have since hardly suffered at all by the fire of the enemy. The first duty of the war office is to keep up the rifle infantry strength.

Sir Winston S. Churchill, 3 May 1943, note to the Chief of the Imperial General Staff.

Replacements are like spare parts – supplies. They must be asked for in time by the front line, and the need for them must be anticipated in the rear.

General George S. Patton, Jr, War As I Knew It, 1947.

REPORTS

I never suffer reports, unsupported by proofs, to have weight in my Mind.

General George Washington, 15 July 1781, to Richard Henry Lee, Writings, XXII, 1937, p. 382.

REPUTATION

All reports should be written clearly, precisely, as far as possible avoiding any inaccuracy, length or beauty of expression, in order not to cloud the thought.
> Field Marshal Prince Aleksandr V. Suvorov, 1799, during his campaign in Northern Italy.

Reports are not self executive.
> Florence Nightingale, 1857, marginal comment on a document.

The officers say I don't believe anything. I certainly believe very little that comes in the shape of reports. They keep everybody stirred up. I mean to be whipped or to whip my enemy, and not to be scared to death.
> Admiral David G. Farragut, quoted in Alfred Thayer Mahan, *Admiral Farragut*, 1892.

As it was, nothing came of all the loss and effort, except a report which I sent over to the British headquarters in Palestine for the Staff's consumption. It was mainly written for effect, full of quaint smiles and mock simplicities; and made them think me a modest amateur, doing his best after the great models; not a clown, leering after them where they with Foch, bandmaster, at their head went drumming down the old road of effusion of blood into the house of Clausewitz. Like the battle, it was a nearly-proof parody of regulation use. Headquarters loved it, and innocently, to crown the jest, offered me a decoration on the strength of it. We should have more bright breasts in the Army if each man was able without witnesses, to write his own despatch.
> Colonel T.E. Lawrence, *The Seven Pillars of Wisdom*, 1926.

In war nothing is ever as bad, or as good, as it is reported to Higher Headquarters. Any reports which emanate from a unit after dark – that is, where the knowledge has been obtained after dark – should be viewed with skepticism by the next higher unit. Reports by wounded men are always exaggerated and favor the enemy.
> General George S. Patton, Jr, *War As I Knew It*, 1947.

In a battle nothing is ever as good or as bad as the first reports of excited men would have it.
> Field Marshal Viscount Slim of Burma, *Unofficial History*, 1959.

Bad news is not like fine wine – it does not improve with age.
> General Creighton Abrams, quoted in *Parameters*, Summer 1987.

REPUTATION

Odysseus probed his own great fighting
   heart:
'O dear god, what becomes of
   Odysseus now?
A disgraceful thing if I should break
   and run,
fearing their main force – but it's far
   worse
if I'm taken all alone. Look, Zeus just
   drove
the rest of my comrades off in panic
   flight.
But debate, my friend, why thrash
   things out?
Cowards, I know, would quit the
   fighting now
but the man who wants to make his
   mark in war
must stand his ground and brace for all
   he's worth –
suffer his wounds or wound his man to
   death.
> Homer, *The Iliad*, 11.483–487, c. 800 BC, tr. Robert Fagles, 1990.

At last is *Hector* stretch'd upon the
   plain,
Who fear'd no vengeance for Patrocles
   slain:
Then Prince! you should have fear'd,

REPUTATION

what now you feel;
Achilles absent was *Achilles* still.
> Homer, *The Iliad*, 22.415–419, c. 800 BC, tr. Alexander Pope, 1743; Achilles taunting the dying Hector that he had forgotten there was deadly substance behind Achilles' reputation.

Questioned as to how he had gained his great reputation, he said: 'By having despised death.'
> Plutarch, quoting Agesilaus, Eurypontid King of Sparta (400–360 BC), *The Lives*, c. AD 100 (*Plutarch on Sparta*, tr. Richard Talbert, 1988).

What of the two men in command? You have Alexander, they – Darius!
> Alexander the Great, 333 BC, address to his army before the Battle of Issus, quoted in Arrian, *The Campaigns of Alexander*, 2.7, c. AD 150, tr. Aubrey de Sélincourt.

But how many ships do you reckon my presence is worth?
> King Antigonus II Gonatas (c. 319–239 BC).

But what most commonly keeps an army united, is the reputation of the general, that is, of his courage and good conduct; without these, neither high birth nor any sort of authority is sufficient.
> Niccolò Machiavelli, *The Art of War*, 1521.

I have tamed men of iron in my day – shall I not crush men of butter?
> Alvarez de Toledo, Duke of Alva (1507–1582), of his attempt to pacify the Netherlands.

It is more the nature of men to be less interested in things which relate to others than about those in which they themselves are concerned. The reputation of an organization becomes personal just as soon as it is an honor to belong to it.
> Field Marshal Maurice Comte de Saxe, *My Reveries*, 1732, tr. Phillips, 1940.

Substantial service is what constitutes lasting reputation; and your reports this campaign are the best panegyric that can be given of your actions.
> Major-General Nathaniel Greene, 27 January 1782, to Major-General (then Lieutenant-Colonel) Henry 'Light Horse Harry' Lee, quoted in *Henry Lee, Memoirs of the War in the Southern Department*, 1869.

He who fears to lose his reputation is sure to lose it.
> Napoleon, 1797, letter, quoted in Christopher Herold, ed., *The Mind of Napoleon*, 1955.

A great reputation is a great noise; the more there is made, the farther off it is heard. Laws, institutions, monuments, nations, all fall; but the noise continues and resounds in after ages.
> Napoleon, quoted in Louis de Bourrienne, *Memoirs of Napoleon Bonaparte*, I, 1855.

My power proceeds from my reputation, and my reputation from the victories I have won. My power would fall if I were not to support it with more glory and more victories. Conquest has made me what I am; only conquest can maintain me.
> Napoleon, 30 December 1802, quoted in R.M. Johnston, ed., *The Corsican*, 1910.

Had I succeeded, I should have died with the reputation of the greatest man that ever lived.
> Napoleon (1769–1821), quoted in Barry O'Meara, *Napoleon in Exile*, 1822.

Many faults, no doubt, will be found in my career; but Arcole, Rivoli, the Pyramids, Marengo, Austerlitz, Jena, Friedland – these are granite: the tooth of envy is powerless here.
> Napoleon, 1815, conversation, quoted in Christopher Herold, ed., *The Mind of Napoleon*, 1955.

Go back, give my compliments to them

RESERVES

and tell the Stonewall Brigade to maintain her reputation!
> Lieutenant-General Thomas J. 'Stonewall' Jackson, 30 August 1862, on the second day of the Battle of Second Bull Run, when the Stonewall Brigade was hard-pressed and appealed to Jackson for reinforcements, quoted in James Robertson, *Stonewall Jackson*, 1997.

What can I say to you as I go away? When the heart is full it is not easy to speak. But I would say this to you:
'YOU have made this Army what it is. YOU have made its name a household word all over the world. Therefore, YOU must uphold its good name and its traditions.'
> Field Marshal Viscount Montgomery of Alamein (then General), August 1943, to the Eighth Army, 'Personal Message from the Commander – to be read to all the troops'.

Don't forget, you don't know I'm here at all. No word of that fact is to be mentioned in any letter. The world is not supposed to know what the hell they did with me. I'm not supposed to be commanding this army. I'm not even supposed to be in England. Let the first bastards to find out be the goddam Germans. Some day I want them to raise up their hind legs and howl: 'Jesus Christ, it's that goddam Third Army and that sonofabitch Patton again!'
> General George S. Patton, Jr, part of his famous pre-battle speech to units of the Third Army before D-Day, 1944, quoted in Carlo d'Este, *Patton: A Genius For War*, 1995.

RESERVES

When one has a good reserve, one does not fear one's enemies.
> Richard I, 'The Lion Heart', King of England, 1194, after his victory at the Battle of Freteval, quoted in *Histoire de Guillaume le Maréchal*, c. 1220.

Providence is always on the side of the last reserve.
> Napoleon (1769–1821), *Sayings of Napoleon*.

Fatigue the opponent, if possible, with few forces and conserve a decisive mass for the critical moment. Once this decisive mass has been thrown in, it must be used with the greatest audacity.
> Major-General Carl von Clausewitz, *Principles of War*, 1812.

The great secret of battle is to have reserve. I always had.
> The Duke of Wellington (1769–1852).

Maxim 96. A general who retains fresh troops for the day after a battle is almost always beaten. He should, if helpful, throw in his last man, because on the day after a complete success there are no more obstacles in front of him; prestige alone will insure new triumphs to the conqueror.
> Napoleon, *The Military Maxims of Napoleon*, tr. Burnod, 1827.

In the battle line, tactics merely consist in overcoming hostile resistance by a slow and progressive wear of the enemy's resources; for that purpose, the fight is kept up everywhere. It must be supported, and such is the use made of the reserves. They must become warehouses into which one dips to replace the wear and tear as it occurs. Art consists in still having a reserve when the opponent no longer has one, so as to have the last word in a struggle in which wearing-down is the only argument employed.
> Marshal of France Ferdinand Foch, *Principles of War*, 1913.

I have no more reserves. The only men I have left are the sentries at my gates. I will take them with me to where the line is broken, and the last of the English will die fighting.

RESISTANCE TO REFORM

Field Marshal Sir John French, 1914, message to General Foch during the Battle of Ypres, quoted in Spears, *Liaison 1914, A Narration of the Great Retreat*.

There is always the possibility of accident, of some flaw in materials, present in the general's mind: and the reserve is unconsciously held to meet it.
Colonel T.E. Lawrence (1888–1935).

It is in the use and withholding of their reserves that the great Commanders have generally excelled. After all, when once the last reserve has been thrown in, the Commander's part is played… The event must be left to pluck and to the fighting troops.
Sir Winston S. Churchill, *Painting as a Pastime*, 1932.

To fight without a reserve is similar to playing cards without capital – sheer gambling. To trust to the cast of dice is not generalship.
Major-General J.F.C. Fuller, *Memoirs of an Unconventional Soldier*, 1936.

A study of Napoleon's tactics will show us that the first step he took in battle was not to break the enemy's front, and then when his forces were disorganized risk being hit by the enemy's reserves; but instead to draw the enemy's reserves into the fire fight, and directly they were drawn in to break through them or envelop them. Once this was done, security was gained; consequently, a pursuit could be carried out, a pursuit being more often than not initiated by troops disorganized by victory against troops disorganized by defeat.
Major-General J.F.C. Fuller, *Memoirs of an Unconventional Soldier*, 1936.

RESISTANCE TO REFORM

I have not written this book for military monks, but for civilians, who pay for their alchemy and mysteries. In war there is nothing mysterious, for it is the most common-sense of all the sciences… If it possess a mystery, then that mystery is unprogressiveness, or it is a mystery that, in a profession which may, at any moment, demand the risk of danger and death, men are found willing to base their work on the campaigns of Waterloo and Sedan when the only possible war which confronts them is the next war.
Major-General J.F.C. Fuller, *The Reformation of War*, 1923.

It [the Army] has been in the family for three hundred years, and he is naturally very loath to part with it and inhabit some horrible ferro-concrete house. He cannot afford to modernize it, and, to make both ends meet, he shuts up room after room, and so 'economizes' his reduced income and hopes for better times. He cannot tear himself away from its memories and traditions and family ghosts, and so dry-rot creeps through its foundations and the rain percolates through its roof.
Major-General J.F.C. Fuller, *The Reformation of War*, 1923.

There are two kinds of treason. The first is constitutional and means betraying one's own country. The second means action whereby a party 'betrays their trust.' The Army and Navy are treasonable under that head for not giving proper improvements to the air service. Of course, I refer to the system and not to any individual.
Brigadier-General William 'Billy' Mitchell, 1925, at his court-martial.

The problem which faces the reformer of armies in peace might be likened to that of an architect called on to alter and modernize an old-fashioned house without increasing its size, with the whole family still living in it (often grumbling at the architect's improvements, since an extra bathroom can only be added at the expense of

RESOLUTION/RESOLVE

someone's dressing room) and under the strictest financial limitations.
  Field Marshal Viscount Wavell of Cyrenaica, 'The Army and the Prophets', *Royal United Services Institute (RUSI)*, May 1930.

An army is an institution not merely conservative but retrogressive by nature. It has such natural resistance to progress that it is always insured against the danger of being pushed too fast. Far worse and more certain, as history abundantly testifies, is the danger of it slipping backward. Like a man pushing a barrow up a hill, if the soldier ceases to push, the military machine will run back and crush him. To be deemed a revolutionary in the army is merely an indication of vitality, the pulse-beat which shows that the mind is still alive. When a soldier ceases to be a revolutionary it is a sure sign he has become a mummy. (Jan. 1931.)
  Captain Sir Basil Liddell Hart, *Thoughts on War*, 1944.

RESOLUTION/RESOLVE

I have not the particular shining bauble or feather in my cap for crowds to gaze at or kneel to, but I have power and resolution for foes to tremble at.
  Oliver Cromwell (1599–1658).

The true prudence of a general consists of energetic resolve.
  Napoleon (1769–1821).

Is it not the manner in which the leaders carry out the task of command, of impressing their resolution in the hearts of others, that makes them warriors, far more than all other aptitudes or faculties which theory may expect of them?
  General Gerhardt von Scharnhorst, at the time of Blücher's appointment to command the army of Silesia in 1813, quoted in Ferdinand Foch, *Principles of War*, 1913.

Hardship, blood, and death create enthusiasts and martyrs and give birth to bold and desperate resolutions.
  Napoleon (1769–1821), on St. Helena, quoted in Christopher Herold, ed., *The Mind of Napoleon*, 1955.

Great extremities require extraordinary resolution. The more obstinate the resistance of an army, the greater the chances of success. How many seeming impossibilities have been accomplished by men whose only resolve was death!
  Napoleon, *Maxims of War*, 1831.

The President of the United States ordered me to break through the Japanese lines and proceed from Corregidor to Australia for the purpose, as I understand it, of organizing the American offensive against Japan, a primary object of which is the relief of the Philippines. I came through, and I shall return.
  General of the Army Douglas MacArthur, 17 March 1942, upon his arrival in Australia, *Reminiscences*, 1964.

Since the war began, tens of thousands of officers and men of matchless loyalty and courage have done battle at the risk of their lives, and have died to become guardian gods of our land.
  Ah, how can I ever enter the imperial presence again? With what words can I possibly report to the parents and brothers of my dead comrades?
  The body is frail, yet with a mind firm with unshakable resolve I will drive deep into the enemy's positions and let him see the blood of a Japanese man.
  Wait but a while, young men! – one last battle, fought gallantly to the death, and I will be joining you!
  Admiral Isoroku Yamamoto, late September 1942, a poem found in his desk before the fatal flight during which he was shot down by US aircraft, quoted in Agawa, *Reluctant Admiral*, 1979.

RESPONSIBILITY/ACCOUNTABILITY

I have returned. By the grace of God, our forces stand again on Philippine soil.
> General of the Army Douglas MacArthur, 20 October 1944, landing on Leyte.

Never take 'No' for an answer. Never submit to failure.
> Sir Winston S. Churchill (1874–1965).

We have no wish to shed blood, but we shall not acquiesce in an act of unprovoked aggression – undertaken, presumably, in the false belief that we lacked the courage and the will to respond. Let the world be under no illusion. These people are British and we mean to defend them. We are in earnest, and no one should doubt our resolve.
> Prime Minister Margaret Thatcher, 7 April 1982, in the House of Commons during the Falklands War, quoted in *The Falklands Campaign: A Digest of Debates in the House of Commons*, 1982, p. 69.

This is not time to go wobbly.
> Prime Minister Margaret Thatcher, 26 August 1990, to President George Bush as she sought to strengthen his resolve to take military action against Saddam Hussein's invasion of Kuwait.

I will not forget the wound to our country and those who inflicted it. I will not yield, I will not rest, I will not relent in waging this struggle for freedom and security for the American people.

The course of this conflict is not known, yet its outcome is certain. Freedom and fear, justice and cruelty, have always been at war, and we know that God is not neutral between them. Fellow citizens, we'll meet violence with patient justice, assured of the righteousness of our cause and confident of the victories to come.

In all that lies before us, may God grant us wisdom and may He watch over the United States of America.
> President George W. Bush, address to a joint session of Congress, Thursday, 20 September 2001, *The Washington Post*, 21 September 2001, p. A24.

That lesson has had to be relearned throughout the ages: the lesson that weakness is provocative, that a refusal to confront gathering dangers can increase, not reduce, future peril and that victory ultimately comes only to those who are purposeful and steadfast.
> Secretary of Defense Donald Rumsfeld, 18 October 2004, *The Guardian*

RESPONSIBILITY/ACCOUNTABILITY

Inspired with Nestor's voice and sent by Zeus,
the dream cried out, 'Still asleep, Agamemnon?
The son of Atreus, the skilled breaker of horses?
How can you sleep all night, a man weighed down with duties?
Your armies turning over their lives to your command –
responsibilities so heavy.
> Homer, *The Iliad*, 2.25–29, c. 800 BC, tr. Robert Fagles, 1990.

Responsibility is the test of a man's courage.
> Admiral Lord St Vincent (1735–1833).

This was all my fault, General Pickett. This has been my fight and blame is mine. Your men did all men could do. The fault is entirely my own.
> General Robert E. Lee, 3 July 1863, at the Battle of Gettysburg, after Pickett's Charge when Lee rode out to meet the remnants of the attacking force, quoted in Stackpole, *They Met at Gettysburg*, 1956.

...Responsibility, the best of educators.
> Rear-Admiral Alfred Thayer Mahan, *The Life of Nelson*, 1897.

I don't know who won the Battle of the

RETIREMENT

Marne, but if it had been lost, I know who would have lost it.
> Marshal of France J.C. Joffre, 1919, to the Briey Parliamentary Commission.

Leave that to me. I am the only one who must apologize to His Majesty.
> Admiral Isoroku Yamamoto, 4 June 1942, when the question was raised as to how the Japanese Fleet could apologize to the Emperor for the defeat it had just suffered at Midway, quoted in John Deane Potter, *Yamamoto*, 1967.

In forty hours I shall be in battle, with little information and on the spur of the moment will have to make most momentous decisions, but I believe that one's spirit enlarges with responsibility and that, with God's help, I shall make them and make them right. It seems that my whole life has been pointed to this moment. When this job is done, I presume I will be pointed to the next step in the ladder of destiny. If I do my full duty, the rest will take care of itself.
> General George S. Patton, Jr, 6 November 1942, *War As I Knew It*, 1947.

A General Officer who will invariably assume the responsibility for failure, whether he deserves it or not, and invariably give the credit for success to others, whether they deserve it or not, will achieve outstanding success.
> General George S. Patton, Jr, *War As I Knew It*, 1947.

Hindenburg once said that he was occasionally held responsible for victories, but always for defeats.
> Field Marshal Albert Kesselring, *Memoirs of Field Marshal Kesselring*, 1953.

The price of greatness is responsibility.
> Sir Winston S. Churchill (1874–1965).

Any clever person can make plans for winning a war if he has no responsibility for carrying them out.
> Sir Winston S. Churchill (1874–1965).

RETIREMENT

He says, 'I followed the Kings of Upper and Lower Egypt, the gods; I was with their majesties when they went to the South and North country, in every place where they went; from King Ahmose I, triumphant, King Amenhotep I, triumphant, King Thutmose I, triumphant, King Thutmose II, triumphant, until this Good God, King Thutmose III who is given life forever.

I attained a good old age, having had a life of royal favor, having had honor under their majesties and the love of me having been in court.
> Ahmose-Pen-Nekhbet, Egyptian general of the 18th Dynasty, c. 1580–1500 BC, quoted in James Breasted, *Ancient Records of Egypt*, 1906.

I am just beginning to experience that ease, and freedom from public cares which, however desirable, takes some time to realize; for strange as it may tell, it is nevertheless true, that it was not 'till lately I could get the better of my usual custom of ruminating as soon as I waked in the Morning, on the business of the ensuing day; and of my surprise, after having revolved many things in my mind, to find that I was no longer a public Man, or had any thing to do with public transactions.
> General George Washington, 20 February 1784, *Writings*, XXVII, 1938, p. 340.

The time factor ...rules the profession of arms. There is perhaps none where the dicta of the man in office are accepted with such an uncritical deference, or where the termination of an active career brings a quicker descent into careless disregard. Little wonder that many are so affected by the sudden transition as to cling pathetically to the trimmings of the past.
> Captain Sir Basil Liddell Hart, *Thoughts on War*, 1944.

RETREAT

I just hope he never retires. He'll have to run the house then, and I'll have to get out.
>Mrs Victor H. Krulak, wife of 'The Brute', *Life Magazine*, 30 November 1965.

On Friday, August 30, I put on my battle fatigues and reported to the personnel office at Central Command. A young female soldier handed me my discharge form and said, 'Sir, this is your DD-214. We recommend you put this form into your safe deposit box because this is the only real proof that you were ever in the service.'
>General H. Norman Schwarzkopf, *It Doesn't Take a Hero*, 1992.

RETREAT

No shame in running.
Fleeing disaster, even in pitch darkness.
Better to flee from death than feel its grip.
>Homer, *The Iliad*, 14.96–98, c. 800 BC, tr. Robert Fagles, 1990; Agamemnon, having lost his nerve as the Trojans threatened the Greek camp and ships.

I do not flee, but I give ground like the rams, in order that I might make my charge the mightier.
>King Philip II of Macedon (382–336 BC), quoted in Alfred S. Bradford, *Philip II of Macedon: A Life from the Ancient Sources*, 1992.

There is no retreat, or, to put it more mildly, retirement, which does not bring an infinity of woes to those who make it, to wit: shame, hunger, loss of friends, goods and arms, and death, which is the worst of them, but not the last for infamy endures for ever.
>Hernan Cortes, 1520, address to his army, quoted in de Gomara, *Istoria de la Conquista de Mexico*, 1552, tr. Simpson.

To withdraw is not to run away, and to stay is no wise action, when there's more reason to fear than to hope.
>Miguel de Cervantes, *Don Quixote*, 1605–1615, tr. Peter Moteux and John Ozell.

You have acquired as much honour covering so great a retreat as if you had gained the battle.
>The Earl of Crawford to the Household Cavalry at the battle of Fontenoy, 1745, quoted in 'Battle of Fontenoy,' http://www.war-art.com/battle_of_fontenoy.htm.

If you have been obliged to withdraw repeatedly before the enemy, the troops get frightened. The same thing happens if you have suffered some check; even when a good opportunity presents itself to you, it must be allowed to escape as you see any wavering among the troops.
>Frederick the Great, *Instructions to His Generals*, 1747, tr. Phillips, 1940.

Never sound the retreat. Never. Warn the men that if they hear it, it is only a ruse on the part of the enemy.
>Field Marshal Prince Aleksandr V. Suvorov (1729–1800), quoted in Blanch, *The Sabres of Paradise*, 1960.

There are few generals that have run oftener, or more lustily than I have done. But I have taken care not to run too far, and commonly have run as fast forward as backward, to convince the Enemy that we were like a Crab, that could run either way.
>Major-General Nathaniel Greene, 18 July 1781, to Major-General Henry Knox.

If we are defeated, we can think about retreating then, and in any case, I shall be dead, so why should I worry.
>Marshal of France Pierre Augereau, Duc de Castiglione, 1796, advice to Napoleon who considered retreating before the Battle of Castiglione, quoted in Delderfield, *Napoleon's Marshals*, 1962.

People who think of retreating before a battle has been fought ought to have stayed home.

RETREAT

Marshal of France Michel Ney, Duc d'Elchingen, Prince de La Moskova, 1805, to Baron Jomini, quoted in Raymond Horricks, *Marshal Ney, The Romance and the Real*, 1982.

Honorable retreats are no ways inferior to brave charges, as having less fortune, more of discipline, and as much valour.
Lieutenant-General Sir William Napier, *A History of the War in the Peninsula*, 1828–1840

Maxim 6... However skilful the manoeuvres, a retreat will always weaken the *moral* [morale] of an army, because in losing the chances of success, these last are transferred to the enemy. Besides, retreats cost always more men and *matériel* than the most bloody engagements, with this difference, that in a battle the enemy's loss is nearly equal to your own, whereas in a retreat the loss is on your side only.
Napoleon, *The Military Maxims of Napoleon*, tr. George D'Aguilar, 1831 (David Chandler, ed., 1987).

To know when to retreat and to dare to do it.
The Duke of Wellington, when asked for his opinion on the best test of greatness in a general, quoted in William Fraser, *Words on Wellington*, 1889.

When a battle is lost, the strength of the army is broken – its moral even more than its physical strength. A second battle without the help of new and favorable factors would mean outright defeat, perhaps even absolute destruction. That is a military axiom. It is in the nature of things that a retreat should be continued until the balance of power is reestablished – whether by means of reinforcements or the cover of strong fortresses or major natural obstacles or the overextention of the enemy.
Major-General Carl von Clausewitz, *On War*, 4.13, 1832, tr. Michael Howard and Peter Paret, 1976.

The officers should feel ...that firmness amid reverses is more honorable than enthusiasm in success, – since courage alone is necessary to storm a position, while it requires heroism to make a difficult retreat before a victorious and enterprising enemy always opposing to him a firm and unbroken front. A fine retreat should meet with a reward equal to that given for a great victory.
Lieutenant-General Antoine-Henri Baron de Jomini, *Summary of the Art of War*, 1838, tr. Mendell and Craighill, 1862.

My wounded are behind me and I will never pass them alive.
Major-General Zachary Taylor, 1847, upon receiving the order to retire at the Battle of Buena Vista.

Retreat, hell! We just got over here.
Major-General Wendell Neville, of the Marine Brigade, 3 May 1918, at the Second Battle of the Marne, when his Fifth Marines were moving up to the front and a French officer urged him to retreat, quoted in Lawrence Stallings, *The Doughboys*, 1963.

It will be readily understood that troops which have once begun to retreat lose heart, discipline slackens, and it is hard to say when or how they will stop and what their conditions will be when they do.
General Aleksei A. Brusilov, *A Soldier's Notebook*, 1931.

There had been last-minute reinforcements, a battalion of U.S. Army troops which fought its way through the enemy with heavy losses. Its colonel reported to Puller [USMC] for orders.
'Take your position along those hills and have your men dig in.'
'Yes, sir. Now, where is my line of retreat?'

REVENGE

Puller's voice became low and hard: 'I'm glad you asked me that. Now I know where you stand. Wait one minute.' He took a field telephone and called his tank commander. The Army officer listened to the Marine order:

'I've got a new outfit,' Puller said. He gave its position in detail. 'If they start to pull back from that line, even one foot, I want you to open fire on them.' He hung up the telephone and turned to the Army officer:

'Does that answer your question?'

Lieutenant-General Lewis 'Chesty' Puller, 1950, at the Chosin Reservoir, quoted in Burke Davis, *Marine*, 1962.

The soldier who has been forced to retreat through no fault of his own loses confidence in the higher command; because he has withdrawn already from several positions in succession he tends to look upon retreat as an undesirable but natural outcome of a battle.

Field Marshal Earl Alexander of Tunis, *The Alexander Memoirs 1940–1945*, 1961.

REVENGE

The officers and soldiers of the army must recollect that their nations are at war with France solely because the ruler of the French nation will not allow them to be at peace, and is desirous of forcing them to submit to his yoke; and they must not forget that the worst of the evils suffered by the enemy in his profligate invasion of Spain and Portugal have been occasioned by the irregularities of the soldiers, and their cruelties authorised and encouraged by their chiefs toward the unfortunate and peaceful inhabitants of the country.

To revenge this conduct on the peaceable inhabitants of France would be unmanly and unworthy of the nations to whom the Commander of the Forces would now address himself, and at all events would be the occasion of similar and worse evils to the army at large than those which the enemy's armies have suffered in the Peninsula, and would eventually prove highly injurious to the public interests.

The Duke of Wellington, 23 June 1813, at Salvatierra, upon the invasion of France, *Selections from the General Orders of Field Marshal the Duke of Wellington During His Various Commands 1799–1818*, 1847.

We must take revenge for the many sorrows inflicted upon the nations, and for so much arrogance. If we do not, then we are miserable wretches indeed, and deserve to be shocked out of our lazy peace every two years and threatened with the scourge of slavery... We must return the visits of the French to our cities, by visiting them in theirs. Until we do, our revenge and triumph will be incomplete. If the Silesian Army reaches Paris first, I shall at once have the bridges of Austerlitz and Jena blown up, as well as the Arc de Triomphe.

Field Marshal August Graf von Gneisenau, at the beginning of the 1814 campaign, quoted in David Hamilton-Williams, *The Fall of Napoleon*, 1994.

This was a hard sight. Our men got very much exasperated & one day when I brought up the rear, I [had] some sad work in protecting helpless women and children from outrage, when the rebels had been firing from their houses on us, and the men were bent on revenge. I invariably gave them my protection which any man of honor will give any woman as long as she is a woman. But I have no doubt they were all 'burnt out' before the whole army got by. It was a sad business. I am willing to fight men in arms but not *babes in arms*.

Major-General Joshua L. Chamberlain, 14 December 1864, letter to his sister, describing the attitude of his troops after Union stragglers had had their throats cut by the Confederates.

When we arrived there, 90 percent of Afghanistan's residents welcomed us

THE REVOLUTION IN MILITARY AFFAIRS

but when we left, 100 percent of them hated us. It is not difficult to understand why. When Afghan fighters fired on a Russian convoy from a village, for example, a Soviet commander used to simply raze that village to the ground. At first it was just a village, then 10 villages, then 100 villages, until half of the country was destroyed, and then there was a nationwide war, when local men were deeply upset by the fact that their children and wives were being killed and their houses being destroyed and that they were deprived of means of survival. The same thing happened in Chechnya. When nine settlements were razed to the ground, the most uncontrollable and wild unit of fighters was formed from the residents of Bamut which had been destroyed. They had nothing to lose.
> General Aleksandr I. Lebed, *Argumenty I Fakty* (Moscow), 22 October 1996.

Revenge is a dish that should be eaten cold.
> English proverb.

REVOLUTION

In time of revolution, with perseverance and courage, a soldier should think nothing impossible.
> Napoleon, *Political Aphorisms*, 1848.

I always had a horror of revolutionising any country for a political object. I always said, if they rise of themselves, well and good, but do not stir them up; it is a fearful responsibility.
> The Duke of Wellington, 1808, upon the suggestion of raising insurrections in the Spanish colonies in the Americas, quoted in The Earl of Stanhope, *Notes of Conversations with the Duke of Wellington*, 1888.

He who serves a revolution ploughs the sea.
> General Simon Bolivar, December 1830, quoted in Salvador de Madariaga, *Bolivar*, 1852.

Be not deceived. Revolutions do not go backward.
> Abraham Lincoln, 19 May 1856, speech.

Every great revolution brings ruin to the old army.
> Leon Trotsky, 1921, to Professor Milyukov.

Revolution is not a dinner party, not an essay, or a painting, nor a piece of embroidery; it cannot be advanced softly, gradually, carefully, considerately, respectfully, politely, plainly and modestly.
> Mao Tse-tung, quoted in *Time*, 18 December 1950.

The fundamental principle of revolutionary war: strike to win, strike only when success is certain; if not, then, don't strike.
> General Vo Nguyen Giap, *People's War – People's Army*, 1961.

THE REVOLUTION IN MILITARY AFFAIRS (RMA)

Gentlemen: I hear many say, 'What need so much ado and great charge in caliver, musket, pike, and corselet? Our ancestors won many battles with bows, black bills, and jacks.' But what think you of that?
Captain: Sir, then was then, and now is now. The wars are much altered since the fiery weapons first came up.
> Robert Barret, *The Theory and Practice of Modern Wars*, 1598, quoted in John Smythe, *Certain Discourses Military*, 1966.

But since the discovery of gunpowder has changed the art of war, the whole system has, in consequence, been changed. Strength of body, the first quality among the heros of antiquity, is at present of no significance. Stratagem vanquishes strength, and art overcomes courage. The understanding of the general has more influence on the fortunate or unfortunate consequences

RISK

of the campaign than the prowess of the combatants. Prudence prepares and traces the route that valor must pursue; boldness directs the execution, and ability, not good fortune, wins the applause of the well informed.
> Frederick the Great (1712–1786), 'Reflections on Charles XII', *Posthumus Works of Frederick the Great*, tr. Holcroft, 1789, quoted in Jay Luvaas, ed., *Frederick the Great on the Art of War*, 1966.

Whereas we had available for immediate purposes one hundred and forty-nine first class warships, we have now two, these being the *Warrior* and her sister *Ironside*. There is not now a ship in the English navy apart from these two that it would not be madness to trust to an engagement with that little [American] *Monitor*.
> *The Times* (London), 1862, quoted in John Taylor Wood, 'The First Fight of Iron-Clads, *Battles and Leaders of the Civil War, Being for the Most Part Contributions by Union and Confederate Officers*, II, Johnston and Buel, ed., 1888.

Cambrai had become the Valmy of a new epoch in war, the epoch of the mechanical engineer.
> Major-General J.F.C. Fuller, 1917, writing of the effect of the introduction of tanks at the Battle of Cambrai (20 Nov–6 Dec 1917), quoted in Robert Asprey, *The German High Command in War*, 1991, p. 346.

The distinguishing feature of the development of the means of armed combat, under present-day conditions is the appearance of qualitatively new types of weapons and military equipment and their rapid mass introduction in the armed forces, which sharply increased in fighting capabilities of the latter and led to a fundamental break in the organization forms of the armed forces and the means for carrying out military operations in every scale. In military strategy, in military art, in military affairs as a whole, a revolution has taken place.
> Marshal of the Soviet Union V.D. Sokolovskiy, quoted in Harriet Fast Scott, ed., *Soviet Military Strategy* (third edition), 1968.

In the images of falling statues, we have witnessed the arrival of a new era. For a hundred years of war, culminating in the nuclear age, military technology was designed and deployed to inflict casualties on an ever-growing scale. In defeating Nazi Germany and Imperial Japan, Allied forces destroyed entire cities, while enemy leaders who stared the conflict were safe until the final days. Military power was used to end a regime by breaking a nation.

Today, we have the greater power to free a nation by breaking a dangerous and aggressive regime. With new tactics and precision weapons, we can achieve military objectives without directing violence against civilians. No device of man can remove the tragedy from war. Yet it is a great advance when the guilty have far more to fear from war than the innocent.
> President George W. Bush, 1 May 2003, speech aboard the *USS Abraham Lincoln*, announcing the end of the military phase of the conquest of Iraq.

RISK

Risks should not be taken without necessity or real hope of gain. To do so is the same as fishing with gold as bait.
> The Emperor Maurice, *The Strategikon*, c. AD 600 (*Maurice's Strategikon*, tr. George Dennis, 1984).

We must risk something for God!
> Hernan Cortes, 1519, when he smashed the idol of the God of the Smoking Mirror (Huitzilopochtli) in the Great Teocalli (Templo Mayor) in Tenochtitlan.

You say that I run too much risk. I don't do it because I want to but because I am obliged to. If I don't go into danger, nobody else will. They are

RISK

all volunteers; I can't force them.
King Henry IV (Henry of Navarre) of France (1553–1610).

It is true that I must run great risk; no gallant action was ever accomplished without danger.
Admiral John Paul Jones, 1778, letter to the American Commissioners in Paris.

The rules of conduct, the maxims of action, and the tactical instincts that serve to gain small victories may always be expanded into the winning of great ones with suitable opportunity; because in human affairs the sources of success are ever to be found in the fountains of quick resolve and swift stroke; and it seems to be a law inflexible and inexorable that he who will not risk cannot win.
Admiral John Paul Jones, 1791, letter to Vice-Admiral Kersaint.

If the art of war consisted merely in not taking risks glory would be at the mercy of very mediocre talent.
Napoleon (1769–1821).

In all great action there is risk.
The Duke of Wellington, September 1803, in India to his brother, shortly before the Battle of Assaye, quoted in Arthur Bryant, The Great Duke, 1972.

Every naval expedition we have attempted since I have been at the head of the government has failed, because the admirals see double and have picked up the idea, I don't know where, that you can make war without running risks.
Napoleon, 12 September 1804, to his naval minister, Decrès, quoted in R.M. Johnston, ed., The Corsican, 1910.

I was aware that I was risking infinitely too much, but something must be risked for the honor of the Service.
General Sir John Moore, December 1808, letter from Spain.
First reckon, then risk.
Field Marshal Helmuth Graf von Moltke (1800–1891).

Calculated risk guided by skill is the right way to interpret the motto, 'l'audace, toujours l'audace'. (Sept. 1925.)
Captain Sir Basil Liddell Hart, Thoughts on War, 1944.

Men have bad luck and their luck may change. But anyhow you will not get generals to run risks unless they feel they have behind them a strong government.
Sir Winston S. Churchill (1874–1965).

Dearest Lu,
    We've been attacking since the 31st with dazzling success. There'll be consternation amongst our masters in Tripoli and Rome, perhaps in Berlin too. I took the risk against all orders and instructions because the opportunity seemed favourable. No doubt it will all be pronounced good later and they'll all say they'd have done exactly the same in my place. We've already reached our first objective, which we weren't supposed to get to until the end of May. The British are falling over each other to get away. Our casualties small. Booty can't yet be estimated. You will understand that I can't sleep for happiness.
Field Marshal Erwin Rommel, 3 April 1941, letter to his wife, The Rommel Papers, 1953.

The habit of gambling contrary to reasonable calculations is a military vice which, as the pages of history reveal, has ruined more armies than any other cause.
Captain Sir Basil Liddell Hart, Thoughts on War, 1944.

Take calculated risks. That is quite different from being rash.
General George S. Patton, Jr, 6 June 1944, letter to his son, The Patton Papers, II, 1974.

ROUT

In making war there must be three parts risk and seven parts security, in addition to one's own subjective effort.
> General Lin Piao, 1946, quoted in Dennis and Ching Ping Bloodworth, *The Chinese Machiavelli*, 1976.

War is risk. Either its ends permit of honest differences of opinion about what should best be done, or operations long since would have become an exact science and general staff work would be as routine as logarithms.
> Brigadier-General S.L.A. Marshall, *Men Against Fire*, 1947.

In enterprises in which one risks everything, there usually comes a moment when the person responsible feels that fate is being determined. By a strange convergence the thousand trials in the midst of which he is struggling seem suddenly to blossom into a decisive episode. If it turns out well, fortune will be in his hands.
> General Charles de Gaulle, *War Memoirs: The Call to Honor 1940–1942*, 1955.

When you stop a dictator there are always risks. But there are greater risks in not stopping a dictator.
> Prime Minister Margaret Thatcher, 5 April 1982, interview.

ROUT

Surely if there is one military maxim of universal value, it is to press hard on a rout.
> Colonel T.E. Lawrence, quoted in Liddell Hart, *Colonel Lawrence: The Man Behind the Legend*, 1935.

Marshal Timoshenko told me that the army had been so utterly routed by the enemy that the only way to rally the troops was to set up mobile field kitchens and hope that the soldiers would return when they got hungry.
> Nikita Khrushchev, of the Soviet disaster at Kiev in 1941, quoted in Charles Messenger, *The Blitzkrieg Story*, 1976.

RUSE

'Behold ye, the great might of my father, Amon-Re. The countries which came from their isles in the midst of the sea, they advanced on Egypt, their hearts relying upon their arms. The net was made ready for them, to ensnare them. Entering stealthily into the harbor-mouth, they fell into it. Caught in their place, they were dispatched, and their bodies stripped. I showed you my might which was in that which my majesty wrought while I was alone. My arrow struck, and none escaped my arms nor my hand. I flourished like a hawk among fowl; my talons descended upon their heads. Amon-Re was upon my right and upon my left, his might and power were in my limbs, a tumult for you; commanding for me that my counsels and my designs should come to pass...
> Rameses III, Pharaoh of Egypt, c. 1190 BC, the defeat of the Sea Peoples in their descent upon Egypt, quoted in James Breasted, *Ancient Records of Egypt*, 1906.

Ruses are of great usefulness. They are detours which often lead more surely to the objective than the wide road which goes straight ahead...You outwit the enemy to force him to fight, or to prevent him from it.
> Frederick the Great, *Instructions to His Generals*, 1747, tr. Phillips, 1940.

We are bred up to feel it a disgrace even to succeed by falsehood; the word spy conveys something as repulsive as slave; we will keep hammering along with the conviction that honesty is the best policy, and that truth always wins in the long run. These pretty little sentiments do well for a child's copy book, but a man who acts on them had better sheathe his sword forever.
> Field Marshal Viscount Wolseley, *A Soldier's Pocket-Book*, 1869.

S

SACRIFICE

So fight by the ships, all together. and
that comrade
who meets his death and destiny,
speared or stabbed,
Let him die! He dies fighting for
fatherland –
no dishonor there!
He'll leave behind him wife and sons
unscathed,
His house and estates unharmed – once
these Argives sail for home,
the fatherland they love.
> Homer, *The Iliad*, 15.574–580, c. 800 BC, tr. Robert Fagles, 1990; Hector, rallying the Trojans and urging them to storm the Greek camp.

When the ephors said: 'Haven't you decided to take any action beyond blocking the passes against the Persians?', 'In theory, no,' he said, 'but in fact I plan to die for the Greeks.'
> Leonidas, King of Sparta, 480 BC, reflecting on the Delphic oracle that victory required the death of a Spartan king, quoted in Plutarch, *The Lives*, c. AD 100 (*Plutarch on Sparta*, tr. Richard Talbert, 1988).

God forbid that I should live an Emperor without an Empire! As my city falls, I will fall with it! Whosoever wishes to escape, let him save himself if he can; and whoever is ready to face death, let him follow me!
> Emperor Constantine XI, May 1453, at the fall of Constantinople, quoted in Mijatovich, *Constantine: The Last Emperor of the Greeks*, 1892.

We swore obedience when we joined the Order. We swore on the vows of chivalry that our lives would be sacrificed for the faith whenever, and wherever, the call might come. Our brethren in St. Elmo must now make that sacrifice.
> Jean de la Valette, Grand Master of the Order of the Knights of St John, 20 June 1565, letter to the Viceroy of Sicily, on the fate of the Order's defenders of the fort of St Elmo during the Turkish siege of Malta in 1565.

Who lives for honour must know how to die for the universal good.
> Gustavus II Adolphus, 30 April 1632, to his commanders, who commented on how close he had come to death in the day's fighting along the River Lech, quoted in Theodore Ayrault Dodge, *Gustavus Adolphus*, 1895.

Perish yourself but rescue your comrade!
> Field Marshal Prince Aleksandr V. Suvorov (1729–1800).

I have never had much value for the public spirit of any man who does not sacrifice his private views and convenience when it is necessary.
> The Duke of Wellington, 25 March 1801, to his brother Henry Wellesley, cited in Arthur Bryant, *The Great Duke*, 1972.

Never, I believe, have soldiers shown such great self-sacrifice as mine have for me! In spite of of all my defeats, no soldier has ever cursed me, not even when dying. Never have troops served a man more faithfully than they have

SACRIFICE

served me! To the last drop of blood that flowed from their veins, they called: 'Long live the Emperor!'
> Napoleon, quoted in F.M. Kircheisen, *Memoirs of Napoleon I*, 1929.

There is no more gallant deed recorded in history. I ordered these men in there because I saw I must gain five minutes in time. Reinforcements were coming on the run, but I knew that before they could reach the threatened point, the Confederates would seize the position. I would have ordered that regiment in if I had known every man would have been killed. It had to be done. I was glad to find such a gallant body of men at hand willing to make the terrible sacrifice that the occasion demanded.
> Major-General Winfield Scott Hancock, of his order throwing the First Minnesota into the path of two oncoming Confederate brigades at the crisis of the battle at Gettysburg on 2 July 1863.

I, Nogi Maresuke, commander-in-chief of the Third Imperial Army before Port Arthur, celebrate with sake and many offerings a fete in honor of you... I wish to tell you that your noble sacrifice has not been in vain, for the enemy's fleet has been destroyed, and Port Arthur has at last surrendered. I, Nogi Maresuke, took oath with you to conquer or seek oblivion in death. I have survived to receive the Imperial thanks, but I will not monopolize the glory. With you, Spirits of the Dead, who achieved this great result, I desire to share my triumph.
> General Maresuke Nogi, 1905, quoted in Dennis and Peggy Warner, *The Tide at Sunrise*, 1974.

But war is a ruthless taskmaster, demanding success regardless of confusion, shortness of time, and paucity of tools. Exact justice for the individual and a careful consideration of his rights is quite impossible. One man sacrifices his life on the battlefield and another sacrifices his reputation elsewhere, both in the same cause. The hurly-burly of the conflict does not permit commanders to draw fine distinctions; to succeed, they must demand results, close their ears to excuses, and drive subordinates beyond what would ordinarily be considered the limit of human capacity. Wars are won by the side that accomplishes the impossible.
> General of the Army George C. Marshall, *Memoirs of My Services in the World War* (written between 1923 and 1926), 1976

The solemn characteristic feature of the soldier's profession is the readiness to die in discharge of duty. Other professions, too, may require the risk of life in discharge of duty, every man may be faced, outside his profession, with the necessity for the last great sacrifice, as an ethical duty; but in no other profession do killing and its corollary, readiness to die, form the essence of professional duty. If the true art of war lies in destroying the enemy, then its exponent must also be prepared to be destroyed himself. This conception of the soldier's function justifies us in speaking of soldiering as something unique. It is the responsibility for life and death which gives the soldier his special character, his gravity, and self-consciousness – not only the responsibility for his own life, which may be sacrificed, not light-heartedly but from a feeling of duty, but the simultaneous responsibility for the lives of comrades and, in the end, for the life of the enemy, whose death is not an act of independent free will on the part of the killer, but an acknowledgement of professional duty. The feeling of responsibility for oneself and others is one of the most vital characteristics of the soldier's life. Responsibility towards oneself demands the most exacting inward and outward training for the military profession, so that the final

sacrifice may not be made in vain.
> Colonel-General Hans von Seekt, *Thoughts of a Soldier*, 1930.

The soldier, above all other men, is required to perform the highest act of religious teaching – sacrifice. In battle and in the face of danger and death he discloses those divine attributes which his Maker gave when He created man in his own image. No physical courage and no brute instincts can take the place of the divine annunciation and spiritual uplift which will alone sustain him. However horrible the incidents of war may be, the soldier who is called upon to offer and to give his life for his country is the noblest development of mankind.
> General of the Army Douglas MacArthur, 14 July 1935, address to the veterans of the Rainbow (42nd) Division, *A Soldier Speaks*, 1965.

I would say to the House, as I said to those who have joined this Government: 'I have nothing to offer but blood, toil, tears, and sweat.'

We have before us an ordeal of the most grievous kind. We have before us many, many long months of struggle and of suffering. You ask, What is our policy? I will say: 'It is to wage war, by sea, land, and air, with all our might and with all the strength that God can give us: to wage war against a monstrous tyranny, never surpassed in the dark, lamentable catalogue of human crime. That is our policy.' You ask, What is our aim? I can answer in one word: Victory – victory at all costs, victory in spite of all terror, victory however long and hard the road may be; for without victory there is no survival. Let that be realized; no survival for the British Empire; no survival for all the British Empire has stood for; no survival for the urge and impulse of the ages, that mankind will move forward towards its goal. But I take up my task with buoyancy and hope. I feel sure that our cause will not be suffered to fail among men. At this time I feel entitled to claim the aid of all, and I say, 'Come, then, let us go forward together with our united strength.'
> Sir Winston S. Churchill, 13 May 1940, first speech to the House of Commons as Prime Minister, *Blood, Sweat, and Tears*, 1941.

Out of the depths of sorrow and sacrifice will be born again the glory of mankind.
> Sir Winston S. Churchill (1874–1965).

I consider it no sacrifice to die for my country. In my mind we came here to thank God that men like these have lived rather than to regret that they died.
> General George S. Patton, Jr, 11 November 1943, speech at an Allied cemetery in Italy, quoted in Semmes, *Portrait of Patton*, 1955.

Refreshingly
After the violent storm
The moon rose radiant.
> Vice-Admiral Takijiro Ohnishi, Commander, Kamikaze Special Attack Force – his final haiku before committing suicide, 15 August 1945, quoted in Inoguchi and Nakajima, *The Divine Wind*, 1955.

SAFETY

For they had learned that true safety was to be found in long previous training, and not in eloquent exhortations uttered when they were going into action.
> Thucydides, *The History of the Peloponnesian War*, c. 404 BC.

The desire for safety stands against every great and noble enterprise.
> Cornelius Tacitus, *Annals*, c. AD 116.

Every attempt to make war easy and safe will result in humiliation and

SAILORS/SEAMEN

disaster.
> General of the Army William T. Sherman, *The Memoirs of General W.T. Sherman*, II, 1875.

Self-preservation is the keystone in the arch of war, because it is the keystone in that greater arch called life. No normal man wishes to be killed in battle, though he may long to die in battle rather than to die in his bed. He does not wish to do so, because there is not virtue in merely dying, for virtue is to be sought in living and living rightly.
> Major-General J.F.C. Fuller, *Generalship: Its Diseases and Their Cure*, 1933.

Discipline apart, the soldier's chief cares are, first, his personal comfort, i.e., regular rations, proper clothing, good billets, and proper hospital arrangements (square meals and a square deal in fact); and secondly, his personal safety, i.e., that he shall be put into a fight with as good a chance for victory and survival.
> Field Marshal Viscount Wavell of Cyrenaica (1883–1950).

'Safety first' is the road to ruin in war.
> Sir Winston S. Churchill, 3 November 1940, telegram to Anthony Eden.

Success cannot be guaranteed. There are no safe battles.
> Sir Winston S. Churchill (1874–1965).

SAILORS/SEAMEN

It must be kept in mind that seamanship, just like anything else, is a matter requiring skill, and will not admit of being taken up occasionally as an occupation for times of leisure; on the contrary, it is so exacting as to leave leisure for nothing else.
> Pericles, 432 BC, arguing for war with Sparta, quoted in Thucydides, *History of the Peloponnesian War*, 1.142, c. 404 BC, tr. Richard Crawley, 1910.

My seamen are now what British seamen ought to be... almost invincible; they really mind shot no more than peas.
> Admiral Viscount Nelson, February 1794, off Bastia in Corsica, quoted in David Howarth, *Nelson: The Immortal Memory*, 1988.

The sailor in a squadron fights only once in every campaign; the soldier fights every day. The sailor, whatever may be the fatigues and dangers on the sea, undergoes fewer of these than the soldier. He never suffers from hunger or thirst; he has always with him his quarters, his kitchen, his hospital and his pharmacy.
> Napoleon, *Maxims of War*, 1831.

SALUTE

The salute is the mutual greeting of respect and loyalty between members of a fighting organization.
> General George S. Patton, Jr, May 1941, address to the officers of the Second Armored Division, *The Patton Papers*, II, 1974.

One morning long ago, as a brand new second lieutenant, I was walking on to parade. A private soldier passed me and saluted; I acknowledged his salute with an airy wave of the hand. Suddenly, behind me, a voice rasped out my name. I spun round. There was my Colonel, for whom I had a most wholesome respect, and with him the Regimental Sergeant Major, of whom, if truth must be told, I stood in some awe. 'I see,' said the Colonel, 'you don't know how to return a salute. Sergent Major, plant your staff in the ground and let Mr. Slim practise saluting it until he does know how to return a salute!' So to and fro I marched in sight of the whole battalion, saluting the Sergeant Major's cane. I could cheerfully have murdered the Colonel, the Sergeant Major and, more than cheerfully, my fellow

THE SCHOOL SOLUTION

subalterns grinning at me. At the end of ten minutes the Colonel called me up to him. All he said was, 'Now remember, discipline begins with the officers!'
> Field Marshal Viscount Slim of Burma, *Courage and Other Broadcasts*, 1957.

A salute from an unwilling man is as meaningless as the moving of a leaf on a tree; it is a sign only that the subject has been caught by a gust of wind. But a salute from the man who takes pride in the gesture because he feels privileged to wear the uniform of the United States, having found military service good, is the epitome of military virtue.
> Brigadier-General S.L.A. Marshall, *The Officer As Leader*, 1950.

SANG FROID

Chatfield, there seems to be something wrong with our bloody ships today.
> Admiral Sir David Beatty, 31 May 1916, observing how British battle cruisers blew up after direct hits at the battle of Jutland, quoted in Winston Churchill, *The World Crisis*, 1927, chapter 41.

Just hold on a little longer, boys – we're sucking them into forty-millimeter range!
> An unknown gunnery officer aboard the USS *Fanshaw Bay*, 25 October 1944, as the Japanese main battle fleet raced to close with the six slow and lightly armed escort carriers of Task Force Taffy 3 at the Battle of Samar, quoted in James D. Hornfischer, *The Stand of the Tin Can Sailors*, 2004, p. 235.

THE SCHOOL SOLUTION

In this art as in poetry and eloquence, there are many who can trace the rules by which a poem or an oration should be composed, and even compose, according to the exactest rule; but for want of that enthusiastic and divine fire, their productions are languid and insipid; so in our profession, many are to be found who know every precept of it by heart; but alas! when called upon to apply them, are immediately at a stand. They then recall their rules and want to make everything, the rivers, the woods, ravines, mountains, etc. subservient to them; whereas their precepts should, on the contrary, be subject to these, who are the only rules, the only guide we ought to follow; whatever manoeuvre is not formed on them is absurd and ridiculous.

These form the great book of war; and he who cannot read it, must be forever content with the title of a brave soldier and never aspire to that of great general.
> Major-General Henry Lloyd, *History of the Late War in Germany, 1766–1782*.

One need only be on one's guard against the bottomless pit of systematic rules.
> Field Marshal Prince Aleksandr V. Suvorov (1729–1800), quoted in Blease, *Suvorof*, 1920.

There are no precise or determined rules; everything depends on the character that nature has given to the general, on his qualities, his shortcomings, on the nature of the troops, on the range of the firearms, on the season and on a thousand other circumstances which are never the same.
> Napoleon (1769–1821), 'Notes sur l'art de la guerre', *Correspondance de Napoléon Ier, publiée par ordre de l'Empereur Napoléon III*, Vol. XXXI, 1858–1870.

Maxim 42. There is no authority in war without exception…
> Napoleon, *The Military Maxims of Napoleon*, tr. George D'Aguilar, 1831 (David Chandler, ed., 1987).

Unhappy the general that comes upon the field of battle with a system.

SCORCHED EARTH

Napoleon, *Maxims*, 1804–1815.

Pity the warrior who is contended to crawl about in the beggardom of rules! ...What genius does must be the best of rules, and theory cannot do better than show how and why it is so.
  Major-General Carl von Clausewitz, *On War*, 2.2, 1832, tr. J.J. Graham, 1908.

If men make war in slavish obedience to rules, they will fail.
  General of the army Ulysses S. Grant, *Personal Memoirs*, I, 1885.

Whenever I met one of them fellers that fit by note, I generally whipped h–ll out him before he got his tune pitched.
  Lieutenant-General Nathan Bedford Forrest, quoted in *As They Saw Forrest*, 1956.

No servitude is more hopeless than that of unintelligent submission to an idea formally correct, yet incomplete. It has all the vicious misleading of a half-truth unqualified by appreciation of modifying conditions; and so seamen who disdained theories, and hugged the belief in themselves as 'practical' became doctrinaires in the worst sense.
  Rear-Admiral Alfred Thayer Mahan, *Types of Naval Officers*, 1901.

When confronted with a situation, do not try to recall examples given in any particular book on the subject; do not try to remember what your instructor said... do not try to carry in your minds patterns of particular exercises or battles, thinking they will fit new cases, because no two sets of circumstances are alike.
  General of the Armies John J. Pershing, 1918, address to the First Infantry Division, quoted in Vandiver, *Black Jack*, 1977.

The principles which regulate the use of all available means in war – the economic employment by a commander of the forces at his disposal, the building up of a concentration (and, consequently, advances by phases or forward bounds), surprise for the enemy, security for oneself – are of value only (how often has this been said before!) insofar as they are adapted to the circumstances of the given situation. There is nothing specifically military about this generalization; it is as true for politics and industry as it is for armies in the field.
  General Charles de Gaulle, *The Edge of the Sword*, 1932.

Official manuals, by the nature of their compilation, are merely registers of prevailing practice, not the log-books of a scientific study of war.
  Captain Sir Basil Liddell Hart, *Thoughts on War*, 1944.

There is no approved solution to any tactical situation.
  General George S. Patton, Jr, *War As I Knew It*, 1947.

Normally, there is no ideal solution to military problems; every course has its advantages and disadvantages. One must select that which seems best from the most varied aspects and then pursue it resolutely and accept the consequences. Any compromise is bad.
  Field Marshal Erwin Rommel, *The Rommel Papers*, 1953.

SCORCHED EARTH

In case of a forced retreat of Red Army units, all rolling stock must be evacuated; to the enemy must not be left a single engine, a single railway car, not a single pound of grain or gallon of fuel... In occupied regions conditions must be made unbearable for the enemy and all his accomplices. They must be hounded and annihilated at every step and all their measures frustrated.
  Joseph V. Stalin, 3 July 1941, address to the Soviet people.

SEA POWER

SEA POWER

Not wholly can Pallas win the heart of
Olympian Zeus,
Though she prays him with many
prayers and all her subtlety;
Yet will I speak to you this other word,
as firm as adamant:
Though all else shall be taken within
the bound of Cecrops
and the fold of the holy mountain of
Cithaeron,
Yet Zeus the all-seeing grants to
Athene's prayer
That the wooden wall only shall not
fall, but help you and your children.
But await not the host of horse and
foot coming from Asia,
Nor be still, but turn your back and
withdraw from the foe.
Truly a day will come when you will
meet him face to face.
Divine Salamis, you will bring death to
women's sons
When the corn is scattered, or the
harvest gathered in.

Delphic Oracle, June 480 BC, the famous oracle who predicted that the Athenian Navy would be the decisive factor in the defence of Greece against the Persian invasion, quoted in Peter Green, *The Greco-Persian Wars*, 1996.

He who commands the sea has command of everything.

Themistocles (c. 528–462 BC) quoted in Cicero, *Epistolae ad Atticum*.

Minos is the first to whom tradition ascribes the possession of a navy. He made himself master of a great part of what is now termed the Hellenic Sea; he conquered the Cyclades, and was the first colonizer of most of them, expelling the Carians and appointing his own sons to govern them. Lastly, it was he who, from a natural desire to protect his growing revenues, sought, as far as he was able, to clear the sea of pirates.

Thucydides, *The Peloponnesian War*, I.24, c. 404 BC, tr. Benajmin Jowett, 1881.

For our naval skill is of more use to us for service on land, than their military skill for service at sea. Familiarity with the sea they will not find an easy acquisition. If you who have been practising at it ever since the Median invasion have not yet brought it to perfection, is there any chance of anything considerable being effected by an agricultural, unseafaring population, who will besides be prevented from practising by the constant presence of strong squadrons of observation from Athens? With a small squadron they might hazard an engagement, encouraging their ignorance by numbers; but the restraint of a strong force will prevent their moving, and through want of practice they will grow more clumsy, and consequently more timid. It must be kept in mind that seamanship, just like everything else, is a matter of art, and will not admit of being taken up occasionally as an occupation for times of leisure; on the contrary, it is so exacting as to leave leisure for nothing else.

Pericles, 432/431 BC, address to the Athenians on the virtues of sea power, quoted in Thucydides, *The History of the Peloponnesian War*, 1.142, c. 404 BC, tr. Richard Crawley, 1910.

He who rules on the sea will very shortly rule on the land also.

Khayer ad-Din, 'Barbarossa', (d. 1546).

This much is certain; that he that commands the sea is at great liberty, and may take as much and as little of the war as he will. Whereas these, that be strongest by land, are many times nevertheless in great straits.

Francis Bacon, *On the True Greatness of Kingdoms and Estates*, 1597.

Whosoever commands the sea commands the trade; whosoever commands

SEA POWER

the trade of the world commands the riches of the world, and consequently the world itself.
> Sir Walter Raleigh, *Historie of the Worlde*, 1616.

A man of war is the best ambassador.
> Oliver Cromwell (1599–1658).

When the army is landed, the business is half done.
> Major-General Sir James Wolfe, 1758, quoted in *Army Magazine*, February 2001, p. 24.

Without a respectable Navy – alas, America!
> Admiral John Paul Jones, 17 October 1776, letter to Robert Morris.

Without a decisive naval force we can do nothing definitive, and with it everything honorable and glorious.
> General George Washington, 20 December 1780, letter to Benjamin Franklin.

I do not say the Frenchman will not come; I only say he will not come by sea.
> Admiral Lord St Vincent, 1803, as First Lord of the Admiralty to the British Cabinet on the possibility of a French invasion, quoted in G.J. Marcus, *The Age of Nelson*, 1971.

Wherever wood can swim, there I am sure to find this flag of England.
> Napoleon, July 1815, at Rochefort upon his surrender to the Royal Navy.

My arm was strong enough, it is true, to stop with a single check, all the horses of the continent. But I could not bridle the English fleets: and there lay all the mischief. Had not people the sense enough to see this?
> Napoleon, at St Helena 1816, quoted in Emmanuel Las Cases, *The Life, Exile, and Conversations of the Emperor Napoleon by the Count de Las Cases*, IV/7, 1835.

Nor must Uncle Sam's Web-feet be forgotten. At all the watery margins they have been present. Not only on the deep sea, the broad bay, and the rapid river, but also up the narrow muddy bayou, and wherever the ground was a little damp, they have been, and made their track.
> President Abraham Lincoln, 26 August 1863, letter to James C. Conkling.

It is not the taking of individual ships or convoys, be they few or many, that strikes down the money power of a nation; it is the possession of that overbearing power on the sea which drives the enemy's flag from it, or allows it to appear only as the fugitive; and by controlling the great common, closes the highway by which commerce moves to and from the enemy's shores. This overbearing power can only be exercised by great navies.
> Rear-Admiral Alfred Thayer Mahan, *The Influence of Sea Power Upon History*, 1890.

The world has never seen a more impressive demonstration of the influence of sea power upon history. Those far distant, storm-beaten ships, upon which the Grand Army never looked, stood between it and the dominion of the world.
> Rear-Admiral Alfred Thayer Mahan, referring to the Royal Navy, *The Influence of Sea Power Upon the French Revolution and Empire*, II, 1892.

The whole principle of naval fighting is to be free to go anywhere with every damned thing the Navy possesses.
> Admiral Sir John Fisher, 1914.

When we speak of command of the seas, it does not mean command of every part of the sea at the same moment, or at every moment. It only means that we can make our will prevail ultimately in any part of the seas which may be selected for operations, and thus

SECRECY

indirectly make our will prevail in every part of the seas.
> Sir Winston S. Churchill, 11 October 1940, to the House of Commons.

If we lose the war at sea, we lose the war.
> Admiral of the Fleet Sir Dudley Pound, First Sea Lord, 5 March 1942, as the depredations of the German submarines were decimating the British and American merchant marines, quoted in Correlli Barnett, *Engage the Enemy More Closely*, 1991.

History shows that those states which do not have naval forces at their disposal have not been able to hold the status of a great power for very long.
> Admiral of the Fleet of the Soviet Union Sergei G. Gorshkov, *Red Sea Rising*, 1974.

SECRECY

O divine art of subtlety and secrecy! Through you we learn to be invisible, through you inaudible; and hence hold the enemy's fate in our hands.
> Sun Tzu, *The Art of War*, 6, c. 500 BC, tr. Giles, 1910.

To keep your actions and your plans secret always has been a very good thing. For that reason Metellus, when he was with armies in Spain, replied to one who asked him what he was going to do the next day, that if his shirt knew he would burn it. Marcus Crassus said to one who asked him when he was going to move the army: 'Do you believe that you will be the only one not to hear the trumpet?'
> Niccolò Machiavelli, *The Art of War*, 1521.

Kalckstein: Your Majesty, am I right in thinking there is going to be a war?
Frederick: Who can tell!
Kalckstein: The movement seems to be directed on Silesia.
Frederick: Can you keep a secret? (Taking him by the hand.)
Kalckstein: Oh yes, Your Majesty.
Frederick: Well, so can I!
> Frederick the Great, 1740, during his preparation for the seizure of Silesia, quoted in Christopher Duffy, *The Military Life of Frederick the Great*, 1985.

The necessity of procuring good Intelligence is apparent and need not be further urged – All that remains for me to add, is that you keep the whole matter as secret as possible. For upon Secrecy, Success depends in most Enterprizes of the kind, and for want of it, they are generally defeated, however well planned + promising a favourable issue.
> General George Washington, 26 July 1777, letter to Colonel Elias Dayton.

Take nobody into your confidence, not even your chief of staff.
> Napoleon, 26 May 1812, to his brother Jérôme, *Correspondance de Napoléon Ier, publiée par ordre de l'Empereur Napoléon III*, No. 18727, XXIII, 1858–1870.

Frederick the Great was right when he said that if his night-cap knew what was in his head he would throw it into the fire.
> Lieutenant-General Antoine-Henri Baron de Jomini, *Summary of the Art of War*, 1838, tr. Mendell and Craighill, 1862.

Lincoln adopted a set form to meet one question and, according to the *Chicago Journal*, the dialogue ran:
Visitor: When will the army move?
Lincoln: Ask General Grant.
Visitor: General Grant will not tell me.
Lincoln: Neither will he tell me.
> President Abraham Lincoln, quoted in Carl Sandburg, *Abraham Lincoln*, II, 1964.

No serving soldier can tell his fellow-countryman anything about an Army which is not (1) quite commonplace; (2) an expression of the views of the Authorities at the moment.

SECURITY

General Sir Ian Hamilton, *The Soul and Body of an Army*, 1921.

War and truth have a fundamental incompatibility. The devotion to secrecy in the interests of the military machine largely explains why, throughout history, its operations commonly appear in retrospect the most uncertain and least efficient of human activities.

Captain Sir Basil Liddell Hart, *Thoughts on War*, 1944.

SECURITY
Principle of War

In making tactical dispositions, the highest pitch you can attain is to conceal them; conceal your dispositions and you will be safe from the prying of the subtlest of spies, from the machinations of the wisest brains. How victory may be produced for them out of the enemy's own tactics – that is what the multitude cannot comprehend.

All men can see these tactics whereby I conquer, but what none can see is the strategy out of which victory is evolved.

Sun Tzu, *The Art of War*, c. 500 BC, tr. Giles, 1910.

A general who has been defeated in a pitched battle, although skill and conduct have the greatest share in beating him, may in his defense throw the blame to fortune. But if he has suffered himself to be surprised or drawn into the snares of his enemy, he has no excuse for his fault, because he might have avoided such a misfortune by taking proper precautions and employing spies on whose intelligence he could depend.

Flavius Vegetius Renatus, *Military Institutions of the Romans*, c. AD 378, tr. Clark, 1776.

It was not my wish to disclose to all what I am thinking. For talk carried about through a camp cannot keep secrets, for it advances little by little until it is carried out even to the enemy.

Count Belisarius, AD 541, to his officers when they questioned his judgement during the Persian Wars, quoted in Procopius, *History of the Wars*, 2.18, c. AD 560, tr. H.B. Dewing.

Even in friendly territory a fortified camp should be set up; a general should never have to say: 'I did not expect it.'

The Emperor Maurice, *The Strategikon*, c. AD 600 (*Maurice's Strategikon*, tr. George Dennis, 1984).

No enterprise is more likely to succeed than one concealed from the enemy until it is ripe for execution.

Niccolò Machiavelli, *The Art of War*, 1521.

Do not neglect the principles of foresight and know that often, puffed up with success, armies have lost the fruit of their heroism through a feeling of false security.

Frederick the Great, *Instructions to His Generals*, 1747, tr. Phillips, 1940.

The Principle of Security. The objective in battle being to destroy or paralyse the enemy's fighting strength, consequently the side which can best secure itself against the action of its antagonist will stand the best chance of winning, for by saving its men and weapons, its organization and moral [morale], it will augment its offensive power. Security is, therefore, a shield and not a lethal weapon, consequently the defensive is not the strongest form of war, but merely a prelude to the accomplishment of the objective – the defeat of the enemy by means of the offensive invigorated by defensive measures. The offensive being essential to success, it stands to reason that security without reference to the offensive is no security at all, but merely delayed suicide.

Major-General J.F.C. Fuller, *The Reformation of War*, 1923.

SELF-CONFIDENCE

As danger and the fear of danger are the chief moral obstacles of the battlefield, it follows that the imbuing of troops with a sense of security is one of the chief duties of a commander; for, if weapons be of equal power, battles are won by a superiority of nerve rather than by a superiority of numbers. This sense of security, though it may be supplemented by earth works or mechanical contrivances, is chiefly based on the feeling of moral ascendence due to fighting efficiency and confidence in command. Thus, a man who is a skilled marksman will experience a greater sense of security when lying in the open than an indifferent rifleman in a trench.

Given the skilled soldier, the moral ascendency resulting from his efficiency will rapidly evaporate unless it be skilfully directed and employed. As in all undertakings – civil or military, ultimately we come back to the impulse of the moment, to the brains which control impulse and to each individual nerve which runs through the military body. To give skilled troops to an unskilled leader is tantamount to throwing snow on hot bricks. Skill in command is, therefore, the foundation of security, for a clumsy craftsman will soon take the edge off his tools.
  Major-General J.F.C. Fuller, *The Foundations of the Science of War*, 1926.

In Wartime, truth is so precious that she should always be attended with a bodyguard of lies.
  Winston S. Churchill, 30 November 1943, remark to Joseph Stalin at the Teheran Conference, *The Second World War: Closing the Ring*, 24, 1951.

A balance must be struck between the needs of free speech in a democracy – we need to communicate information, particularly in the House – and the operational requirements to which I have jut referred. It is a balance that is sometimes extremely difficult to strike. Understandably, Right Hon. and Hon. Members want as much information as they can get, but it is my responsibility to ensure that I do not make information available that would in any way compromise the operational effectiveness of the armed forces.
  Secretary of State for Defence Geoff Hoon, answering questions in the House of Commons, 25 October 2004.

SELF-CONFIDENCE

Often there is a gap between principles and actual events that cannot always be bridged by a succession of logical deductions. Then a measure of self-confidence is needed.
  Major-General Carl von Clausewitz, *On War*, 1, tr. Michael Howard and Peter Paret, 1976.

Never take counsel of your fears.
  Lieutenant-General Thomas J. 'Stonewall' Jackson, 18 June 1862, to Major Hotchkiss.

Wilson, I am a damned sight smarter than Grant. I know a great deal more about war, military history, strategy, and grand tactics than he does; I know more about organization, supply, and administration, and about everything else than he does. But I tell you where he beats me, and where he beats the world. He don't care a damn for what the enemy does out of his sight, but it scares me like hell… I am more nervous than he is. I am more likely to change my orders, or to countermarch my command than he is. He uses such information as he has, according to his best judgment. He issues his orders and does his level best to carry them out without much reference to what is going on about him.
  General of the Army William T. Sherman, quoted in Harry T. Williams, *McClellan, Sherman and Grant*, 1962.

SELF-CONTROL

The most vital quality a soldier can possess is self-confidence, utter, complete and bumptious.
> General George S. Patton, Jr, 6 June 1944, letter to his son, *The Patton Papers*, II, 1974.

One of the most interesting points to my mind about all this business of making war is the way that people try and shake your confidence in what you are doing, and suggest that your plan is not good, and that you ought to do this, or that. If I had done all that was suggested I would still be back in the Alamein area.
> Field Marshal Viscount Montgomery of Alamein, quoted in Nigel Hamilton, *Master of the Battlefield*, 1983.

Don't ever, ever second-guess what you are doing. You are doing a wonderful job. Get your heads up and it will turn out just fine
> General Tommy Franks to his staff during Operation Iraqi Freedom, quoted by Joseph L. Galloway, 'Gen. Franks tells how Iraq war plan came together,' Knight Ridder Newspapers, 19 June 2003.

I don't do helmets,
> Colonel Mike Riddell-Webster, Commander, Black Watch, famous for walking wearing his red hackle through Az Zubayr in Iraq as soon as it was taken, quoted by David Moore, *Daily Telegraph*, 28 April 2003.

SELF-CONTROL

Self-control is the chief element in self-respect, and self-respect is the chief element in courage.
> Thucydides, *The History of the Peloponnesian War*, c. 404 BC.

Let the army see that you are not unduly elated over successes nor utterly cast down by failures.
> The Emperor Maurice, *The Strategikon*, c. AD 600 (*Maurice's Strategikon*, tr. George Dennis, 1984).

There is nothing more base than for a man to lose his temper too often. No matter how angry one becomes, his first thought should be to pacify his mind and come to a clear understanding of the situation at hand. Then, if he is in the right, to become angry is correct.
> Shiba Yoshimasa (1350–1410), *The Chikubasho*, 1380, (*The Ideals of the Samurai*, tr. William Wilson, 1982).

But it might be closer to the truth to assume that the faculty known as self-control – the gift of keeping calm even under the greatest stress – is rooted in temperament. It is itself an emotion which serves to balance the passionate feelings in strong characters without destroying them, and it is this balance alone that assures the dominance of the intellect...
> Major-General Carl von Clausewitz, *On War*, 1.3, 1832, tr. Michael Howard and Peter Paret, 1976.

General Meade was an officer of great merit with drawbacks to his usefulness which were beyond his control... He made it unpleasant at times, even in battle, for those around him to approach him with information.
> General of the Army Ulysses S. Grant, *Personal Memoirs*, II, 1885.

If there are persons with steel nerves on a patrol boat, then their night attacks will be completely successful. Persons with great self-control can achieve miracles, while weak will of the executors and a lack of tenacity to a significant degree diminish the result.
> Admiral Stepan O. Makarov (1849–1904), quoted in Ablamonov, *Admiral*.

An officer should never speak ironically or sarcastically to an enlisted man, since the latter doesn't have a fair chance to answer back. The use of profanity and epithets comes under the same heading. The best argument for a man keeping

SIMPLICITY

his temper is that nobody else wants it; and when he voluntarily throws it away, he loses a main prop to his own position.

>Brigadier-General S.L.A. Marshall, *The Armed Forces Officer*, 1950.

Explosions of temper do not necessarily ruin a general's reputation or influence with his troops; it is almost expected of them ('the privileged irascibility of senior officers', someone has written), and it is not always resented, sometimes even admired, except by those immediately concerned. But sarcasm is always resented and seldom forgiven. In the Peninsula the bitter sarcastic tongue of Craufurd, the brilliant but erratic leader of the Light Division, was much more wounding and feared than the more violent outbursts of Picton, a rough, hot-tempered man.

>Field Marshal Viscount Wavell of Cyrenaica, *Soldiers and Soldiering*, 1953.

SEX

Fighting men are 'frolicking' men, and I'm not going to do a damned thing about it. If you Frenchmen can't keep your daughters out of trouble, don't count on the armed forces of the United States to do it for you.

>Fleet Admiral William 'Bull' Halsey, to the French authorities of New Caledonia, 1943, who complained of the number of pregnancies among their daughters caused by American servicemen stationed on the island, quoted in E.B. Potter, *Bull Halsey*, 1985, p. 206.

Once before I had passed an evening in the ADSEC billet at the Hotel Harscamp in Namur while motoring back to Luxembourg from First Army. I recall the name of that hotel clearly. It was dusk and we had stopped in town to inquire of a GI the way to the Harscamp.

'Whores' camp?' At first he looked puzzled but then he brightened with the thought, 'say – d'ya mean they really got one here?'

>General of the Army Omar N. Bradley, *A Soldier's Story*, 1951.

SIC TRANSIT GLORIA MUNDI

The eagle has ceased to scream, but the parrots will now begin to chatter. The war of the giants is over and the pigmies will now start to squabble.

>Sir Winston S. Churchill, 7 May 1945, comment upon the announcement of VE Day, quoted in Kay Halle, *Irrepressible Churchill*, 1966.

Thus, then, on the night of the tenth of May, at the outset of this mighty battle, I acquired the chief power in the State, which henceforth I wielded in ever-growing measure for five years and three months of world war, at the end of which time, all our enemies having surrendered unconditionally or being about to do so, I was immediately dismissed by the British electorate from all further conduct of their affairs.

>Sir Winston S. Churchill, *The Gathering Storm*, 1948.

Seven months ago, I could give a single command and 541,000 people would immediately obey. Today, I can't get a plumber to come to my house.

>General H. Norman Schwarzkopf, c. April 1992.

When a soldier's tour of duty is over, it's over.

>General Wesley Clark, July 1999, upon the news that, despite his victory over Serbia, he would be dismissed early from his position as NATO Commander.

SIMPLICITY
Principle of War

The art of war is a simple art; everything is in the performance. There is nothing vague in it; everything in it is common

SKILL

sense; ideology does not enter in.
>Napoleon, dictation on St Helena, quoted in Christopher Herold, ed., *The Mind of Napoleon*, 1955.

The art of war is like everything else that is beautiful and simple. The simplest moves are the best. If MacDonald, instead of doing whatever he did, had asked a peasant for the way to Genoa, the peasant would have answered, 'Through Bobbio' – and that would have been a superb move.
>Napoleon, 1818, conversation, quoted in Christopher Herold, ed., *The Mind of Napoleon*, 1955.

In war so much is always unknown that it frequently happens that even the simplest actions rapidly become exceedingly complex. As from the simple to the complex is the rule in war, therefore the simpler, more direct and clearer the beginning the less likely is action to get out of hand.
>Major General J.F.C. Fuller, quoted in Waldemar Erfurth, *Surprise*, 1943, notes, p. 158.

In war only what is simple can succeed.
>Field Marshal Paul von Hindenburg (1847–1934).

Spartan simplicity must be observed. Nothing will be done merely because it contributes to beauty, convenience, comfort, or prestige.
>Circular from the Office of the Chief Signal Officer, US Army, 29 May 1945.

The KISS principle: KEEP IT SIMPLE, STUPID!
>Anglo-American military maxim.

SKILL

Force has no place where there is need of skill
>Herodotus (c. 484–420 BC), *The Histories of Herodotus*.

Certes the Frenchmen and Rutters, deriding our new archery… will not let in open skirmish, if any leisure serve, to turn up their tails and cry, 'Shoot, English!' and all because our strong shooting is decayed and laid in bed. But if some of our Englishmen now lived that served Edward the Third in his wars with France, the breech of such a varlet should have been nailed to his bum with an arrow and another feathered in his bowels before he should have turned about to see who shot the first.
>William Harrison, quoted in Holinshed, *Chronicles*, 1807.

Mr. Amery, author of *The Times* history of the War, probed a weak spot in the prevailing European theory by arguing that superior skill now counted more than superior numbers, and that its proportionate value would increase with material progress. The same note was struck by General Baden-Powell, who urged that the way to develop it was to give officers responsibility when young – he was left to find his channel for proving this in the Boy Scout movement, and not in the Army.
>Captain Sir Basil Liddell Hart, *A History of the World War 1914–1918*, 1935.

Very few people will use skill if brute force will do the trick. The worst thing for a good general is to have superior numbers.
>Colonel T.E. Lawrence, 31 March 1929, letter to B.H. Liddell Hart, *T.E. Lawrence to his Biographers*, 1963.

SOLDIERS

Just think of how the soldier is treated. While still a child he is shut up in the barracks. During his training he is always being knocked about. If he makes the least mistake he is beaten, a burning blow on his body, another on his eye, perhaps his head is laid open with a wound. He is battered and

SOLDIERS

bruised with flogging. On the march he has to carry bread and water like the load of an ass; the joints of his back are bowed; they hang heavy loads round his neck like that of an ass...
> Anonymous Egyptian scribe expressing the usual distaste of the pen for the sword, quoted in Cotrell, *The Warrior Pharaohs*.

You are the disaster.
Would to god you commanded another army,
a ragtag crew of cowards, instead of ruling us,
the men whom Zeus decrees, from youth to old age,
must wind down our brutal wars to the bitter end
until we drop and die, down to the last man.
> Homer, *The Iliad*, 14.103–107, c. 800 BC, tr. Robert Fagles, 1990; Odysseus to Agamemnon, who panicked and urged the Greeks to flee from the danger of the Trojan assault on their camp.

If, however, you are indulgent, but unable to make your authority felt; kind-hearted but unable to enforce your commands; and incapable, moreover, of quelling disorder, then your soldiers must be likened to spoiled children; they are useless for any practical purpose.
> Sun Tzu, *The Art of War*, 10, c. 500 BC, tr. Giles, 1910.

Remember that zeal, honor, and obedience mark the good soldier.
> Brasidas, 422 BC, speech to the Spartans and their allies before the Battle of Amphipolis, quoted in Thucydides, *History of the Peloponnesian War*, 5.9, c. 404 BC, tr. Richard Crawley, 1910.

Many think the genius of the soldier wants subtlety, because military law, which is summary and blunt, and apt to appeal to the sword, finds no exercise for the refinements of the forum.
> Cornelius Tacitus, *Agricola*, AD 98 (*The Complete Works of Tacitus*, tr. Alfred Church and William Brodribb, 1942); to this preface, he contrasted the character of his father-in-law, Cnaeus Julius Agricola, later governor of Britain.

For who ought to be more faithful than a man that is entrusted with the safety of his country, and has sworn to defend it to the last drop of his blood? Who ought to be fonder of peace than those that suffer by nothing but war? Who are under greater obligations to worship God than Soldiers, who are daily exposed to innumerable dangers, and have most occasion for his protection?
> Niccolò Machiavelli, *The Art of War*, 1521.

The distinctive mark of the soldier is that all his desires are despotic; that of the civilian is that he submits everything to discussion, to truth, to reason.
> Napoleon, 4 May 1802, quoted in R.M. Johnston, ed., *The Corsican*, 1910.

Soldiers generally win battles; generals generally get the credit for them.
> Napoleon, 1815, to General Gaspard Gourgaud on St Helena.

I hate that cheering. If once you allow soldiers to express an opinion, they may on some other occasion hiss instead of cheer.
> The Duke of Wellington, 1815, to Lady Shelley in Paris, *The Diary of Frances Lady Shelley*, II, 1912.

Sir, when soldiers have been christened by the fire of the battle-field, they have all one rank in my eyes.
> Napoleon, 16 May 1816, to Major-General Sir Hudson Lowe, the Governor of St Helena and his jailer, who had suggested that officers of his staff rather than the officer established at Napoleon's residence, accompany him on his rides; quoted in Emmanuel Las Casas, *The Life, Exile, and*

SOLDIERS

Conversations of Emperor Napoleon Bonaparte, II, 1891.

Mme. de Montholon: Which were the best troops?
Napoleon: Those that win battles, madam. And they are fickle, they must be taken on their day, like you ladies. The best troops have been, the Carthaginians under Hannibal, the Romans under the Scipios, the Macedonians under Alexander, the Prussians under Frederick. Some day my army of Italy and that of Austerlitz may be equalled, but, surely never surpassed.
> Napoleon, 28 August 1816, quoted in R.M. Johnston, ed., *The Corsican*, 1910.

Troops are made to let themselves be killed.
> Napoleon, 1817, conversation on St Helena, cited in Christopher Herold, ed., *The Mind of Napoleon*, 1955.

Maxim 58. The first qualification of a soldier is fortitude under fatigue and privation. Courage is only the second; hardship, poverty, and want are the best school of the soldier.
> Napoleon, *The Military Maxims of Napoleon*, tr. George D'Aguilar, 1831 (David Chandler, ed., 1987).

I have never been able to join in the popular cry about the recklessness, sensuality, helplessness of the soldier. On the contrary I should say (& no woman perhaps has ever seen more of the manufacturing & agricultural classes of England than I have – before I came out here) that I have never seen so teachable & helpful a class as the army generally.
Give them an opportunity promptly & securely to send money home – & they will use it.
Give them a School & lecture & they will come to it.
Give them a book & a game & a Magic Lanthorn & they will leave off drinking.
Give them suffering & they will bear it.
Give them work & they will do it.
I had rather have to do with the Army generally than with any other class I have ever attempted to serve.
And I speak with intimate experience of 18 months which I have had since I 'joined the Army' – no woman (or man either) having seen them under such conditions.
And when I compare them with the Medical Staff Corps, the Land Transport Corps, the Army Works Corps, I am struck with the soldier's superiority as a moral & even as an intellectual being.
If Officers would but think thus of their men, how much might not be done for them.
> Florence Nightingale, 6 March 1856, Scutari, The Barracks Hospital, letter to Lieutenant-Colonel Lefroy, quoted in Goldie, ed., *'I have done my duty': Florence Nightingale in the Crimean War 1854–1856*, 1987.

Boys are soldiers in their hearts already – and where the harm? Soldiers are not pugnacious. Paul bade Timothy to be a good soldier. Christ commended the centurion. Milton urged teachers to fit their pupils for all the offices of war. The very thought of danger and self-sacrifice are inspirations.
> General Sir Ian Hamilton, *The Soul and Body of an Army*, 1921.

As in the shades of a November evening I, for the first time, led a platoon of Grenadiers across the sopping fields ... the conviction came into my mind with absolute assurance that the simple soldiers and their regimental officers, armed with their cause, would by their virtues in the end retrieve the mistakes and ignorance of Staffs and Cabinets, or Admirals, Generals, and politicians – including, no doubt, many of my own. But, alas, at what a needless cost!
> Sir Winston S. Churchill, *The World Crisis*, 1923.

SOLDIERS

What are the qualities of a good soldier, by the development of which we make the man war-worthy – fit for any war? ...The following four – in whatever order you place them – pretty well cover the field: discipline, physical fitness, technical skill in the use of weapons, battle-craft.
> Field Marshal Viscount Wavell of Cyrenaica, 15 February 1933, lecture at the Royal United Service Institution (RUSI).

The soldier is the Army. No army is better than its soldiers. The soldier is also a citizen. In fact, the highest obligation and privilege of citizenship is that of bearing arms for one's country. Hence it is a proud privilege to be a soldier – a good soldier. Anyone, in any walk of life, who is content with mediocrity is untrue to himself and to American tradition. To be a good soldier a man must have discipline, self-respect, pride in his unit and in his country, a high sense of duty and obligation to his comrades and to his superiors, and self-confidence born of demonstrated ability.
> General George S. Patton, Jr, *War As I Knew It*, 1947.

The soldier is a man; he expects to be treated as an adult, not a schoolboy. He has rights; they must be made known to him and thereafter respected. He has ambition; it must be stirred. He has a belief in fair play; it must be honored. He has a need of comradeship; it must be supplied. He has imagination; it must be stimulated. He has a sense of personal dignity; it must be sustained. He has pride; it can be satisfied and made the bedrock of character once he has been assured that he is playing a useful and respected role. To give a man this is the acme of inspired leadership. He has become loyal because loyalty has been given to him.
> General of the Army George C. Marshall (1880–1959).

I relapse into silence and go into the tent. For a long time afterwards the man Steen is the subject of my thoughts. Now I understand him better than I did. I realize how much virile strength and strength-giving understanding can be passed from one man to another in a quiet talk at the front. It is not the soldier's way to be communicative. He expresses himself very differently from a civilian. His talk is every bit as uncivilian and tongue-tied as it is popularly represented. And because war jerks a man out of all pretence and hypocrisy, the things a warrior says, even if they only take the form of an oath or a primitive sentimentality, are integrally sincere and genuine, and therefore finer stuff than all the glib rhetoric of the civilian world.
> Colonel Hans Ulrich Rudel, *Stuka Pilot*, 1958.

I have written much of generals and staff officers... the war in Burma was a soldier's war. There comes a moment in every battle against a stubborn enemy when the result hangs in the balance. Then the general, however skilful and farsighted he may have been, must hand over to his soldiers, to the men in the ranks and to their regimental officers and leave them to complete what he has begun.
> Field Marshal Viscount Slim of Burma, *Defeat Into Victory*, 1956.

By people I do not mean 'personnel.' I do not mean 'end Strength.' I mean living, breathing, serving, human beings. They have needs and interests and desires. They have spirit and will, and strengths and abilities. They have weaknesses and faults; and they have names. They are the heart of our preparedness... and this preparedness – as a nation and as an Army – depends upon the spirit of our soldiers. It is the spirit that gives the Army ... life.

SOLDIERS AND POLITICIANS

Without it we cannot succeed.
  General Creighton Abrams (1914–1974), quoted in *Parameters*, 6/1981.

If in danger or in doubt, get the bloody brew-can out.
  British Army saying, quoted in Julian Thompson, *No Picnic: 3 Commando Brigade in the South Atlantic*: 1982, p. 89.

SOLDIERS AND POLITICIANS

It is easy enough to say to a general, go to Italy, win battles, and sign peace at Vienna. But the doing of it is not so easy. I have never paid the least attention to the plans sent to me by the Directoire. Only fools could take stock in such rubbish.
  Napoleon, 21 March 1797, in R.M. Johnston, ed., *The Corsican*, 1910.

I understand why Bonaparte is accused; it's for concluding peace. But I warn you, I speak in the name of 80,000 men; the time when cowardly lawyers and low chatterers could send soldiers to the guillotine has passed, and if you drive them to it, the soldiers of Italy will march to the Clichy gate with their general: but, if they do, look out for yourselves!
  Napoleon, 30 June 1797, to the Directoire, quoted in R.M. Johnston, *The Corsican*, 1910.

If ever I have the luck to set foot in France again, the reign of chatter is over.
  Napoleon, 11 August 1799, in Cairo, quoted in R.M. Johnston, ed., *The Corsican*, 1910.

I had rather talk to soldiers than to lawyers. Those ... made me nervous. I am not accustomed to assemblies; it may come in time.
  Napoleon, 10 November 1799, after a stormy session with the Council of Ancients, Paris, quoted in R.M. Johnston, ed., *The Corsican*, 1910.

I do not desire to place myself in the most perilous of all positions: – a fire upon my rear, from Washington, and the fire, in front, from the Mexicans.
  Lieutenant-General Winfield Scott, 21 May 1846, letter to the Secretary of War.

I start early in the morning... Those hounds in Washington are after me again.
  Major-General George B. McClellan, *McClellan's Own Story*, 1887.

If they [politicians] would only follow the example of their ancestors, enter a herd of swine, run down some steep bank and drown themselves in the sea, there would be some hope of saving the country.
  Major-General Henry W. Halleck, *The War of the Rebellion: A Compilation of the Official Records of the Union and Confederate Armies*, Vol. XXII, Part 1, p. 793.

I am a soldier. It is my duty to obey orders. It is enough to turn one's hair grey to spend one day in the Congress. The members are patriotic and earnest, but they will neither take the responsibility of action nor will they clothe me with the authority to act for them.
  General Robert E. Lee, quoted in John B. Gordon, *Reminiscences*, 1903.

So long as we can gain success, the interference of politicians in military matters can be resisted; but on the first disaster they press upon us like a pack of hungry wolves.
  Major-General Henry W. Halleck, letter to Schofield, 7 July 1863, John M. Schofield Manuscripts (LoC).

Keep your hands off the regiment, ye iconoclastic civilian officials who meddle and muddle in Army matters. Clever politicians you may be, but you are not soldiers and you do not understand them; they are not pawns on

SPEECHES AND THE SPOKEN WORD

a chessboard. Leave the management of our fighting men to soldiers of experience in our British Army of old renown, and do not parody us by appearing in public decked for the nonce in a soldier's khaki coat. You might as well put your arm in a sling, or tie your head up in the bandage of some poor maimed soldier, to whom, when wounded and unable to earn a livelihood, your regulations allow a pension of sixpence a day!

    Field Marshal Viscount Wolseley, *The Story of a Soldier's Life*, 1903.

All the reproaches that have been levelled against the leaders of the armed forces by their countrymen and by the international courts have failed to take into consideration one very simple fact: that policy is not laid down by soldiers, but by politicians. This has always been the case and is so today. When war starts, the soldiers can only act according to the political and military situation as it then exists. Unfortunately it is not the habit of politicians to appear in the conspicuous places when the bullets begin to fly. They prefer to remain in some safe retreat and to let the soldiers carry out 'the continuation of politics by other means.'

    Colonel-General Heinz Guderian, *Panzer Leader*, 1953.

The soldier always knows that everything he does... will be scrutinized by two classes of critics – by the government that employs him and by the enemies of that Government.

    Field Marshal Viscount Slim of Burma, *Unofficial History*, 1959.

SPEECHES AND THE SPOKEN WORD

Student: Do you mean that in addition to his other duties a cavalry leader must take care to be a good speaker?
Socrates: Did you suppose that commander of cavalry should be mum?
Did you never reflect that all the best we learned according to custom – the learning, I mean, that teaches us how to live – we learned by means of words, and that every other good lesson to be learned is learned by means of words; that the best teachers rely most on the spoken word and those with the deepest knowledge of the greatest subjects are the best talkers?

    Socrates (470–399 BC), quoted by Xenophon, *Memorabilia*, 3.11–12, c. 360 BC, tr. E.C. Marchant, 1923.

Great eloquence, like fire, grows with its material; it becomes fiercer with movement, and brighter as it burns.

    Cornelius Tacitus, *Oratory*, 36, c. AD 75 (*The Complete Works of Tacitus*, tr. Alfred Church and William Brodribb, 1942).

The general who possesses some skill in public speaking is able, as in the past, to rouse the weak-hearted to battle and restore courage to a defeated army.

    The Emperor Maurice, *The Strategikon*, c. AD 600 (*Maurice's Strategikon*, tr. George Dennis, 1984).

You know what words can do to soldiers.

    Napoleon, 12 March 1800, to General Brunne, *Correspondance de Napoléon Ier, publiée par ordre de l'Empereur Napoléon III*, VI, p. 178, quoted in Christopher Herold, ed., *The Mind of Napoleon*, 1955.

Maxim 61. It is not speeches at the moment of battle that render soldiers brave. The veteran scarcely listens to them, and the recruit forgets them at the first discharge. If discourses and harangues are useful, it is during the campaign; to do away unfavourable impressions, to correct false reports, to keep alive proper spirit in the camp, and to furnish materials and amusement for the bivouac. All printed orders should keep in view these objects.

SPEED

Napoleon, *The Military Maxims of Napoleon*, tr. George D'Aguilar, 1831 (David Chandler, ed., 1987).

Action employs men's fervour, but words arouse it.
> General Charles de Gaulle, quoted in F.M. Richardson, *Fighting Spirit*, 1978.

Old man, when you have something to say to officers or men, make it snappy. The fewer words, the better. They won't believe you if you shoot bull. When you face ranks of men and try that, you can hear 'em sigh in despair when you open your mouth, if they sense you're a phoney. They can usually look at you and tell. Maybe it doesn't sound like it, but that's an important thing in a Marine's career.
> Lieutenant-General Lewis 'Chesty' Puller, quoted in Burke Davis, *Marine*, 1962.

Men who can command words to serve their thoughts and feelings are well on their way to commanding men to serve their purposes.
> Brigadier-General S.L.A. Marshall, *The Armed Forces Officer*, 1950.

We found that the best way to convince men that what they were doing was worthwhile was to tell them yourself. The spoken word, delivered in person – not over the wireless – is the greatest instrument. An occasional talk by the man who holds the responsibility for the show counts a lot. It doesn't need an orator. Any man who holds control over others should be able to talk to them, provided he has two qualifications. First, that he is clear in his own mind about what he wants to put over, and secondly that he believes it himself. That last is important.
> Field Marshal Viscount Slim of Burma, *Courage and Other Broadcasts*, 1957.

The troops must be brought to a state of wild enthusiasm before the operation begins. They must have that offensive eagerness and that infectious optimism which comes from physical well-being. They must enter the fight with the light of battle in their eyes and definitely wanting to kill the enemy. In achieving this end, it is the spoken word which counts, from the commander to his troops; plain speech is far more effective than any written word.
> Field Marshal Viscount Montgomery of Alamein, *The Memoirs of Field Marshal Montgomery*, 1958.

Churchill mobilized the English language and sent it into battle.
> President John F. Kennedy, quoted in *The Gazette*, Library of Congress, Vol 15. No 8, 27 February 2004, p. 9.

SPEED

Rapidity is the essence of war. Take advantage of the enemy's unreadiness, make your way by unexpected routes, and attack unguarded spots.
> Sun Tzu, *The Art of War*, 11, c. 500 BC, tr. Giles, 1910.

If you move rapidly, above all if the work of your scouts, your intelligence and your couriers is reliable, you can be certain of defeating a battalion with a detachment, an army with a battalion.
> Anonymous Byzantine general, 10th-century treatise on tactics, quoted in Guerdan, *Byzantium, Its Triumphs and Tragedy*.

In military practice one must plan quickly and carry on without delay, so as to give the enemy no time to collect himself.
> Field Marshal Prince Aleksandr V. Suvorov (1729–1800), quoted in W. Lyon Blease, *Suvorof*, 1920.

Swiftness and impact are the soul of genuine warfare.
> Field Marshal Prince Aleksandr V. Suvorov (1729–1800), quoted in V.Ye.. Savkin, *Basic Principles*, 1972.

SPIRIT

Just as lightning has already struck when the flash is seen, so when the enemy discovers the head of the army, the whole should be there, and leave him no time to counteract its dispositions.
    Field Marshal Francois Comte de Guibert (1744–1790).

It is very advantageous to rush unexpectedly on an enemy who has erred, to attack him suddenly and come down on him with thunder before he sees the lightning.
    Napoleon (1769–1821).

In order to smash, it is necessary to act suddenly.
    Napoleon (1769–1821).

The third rule is never to waste time. Unless important advantages are to be gained from hesitation, it is necessary to set to work at once. By this speed a hundred enemy measures are nipped in the bud, and public opinion is won most rapidly.
    Major-General Carl von Clausewitz, *Principles of War*, 1812, tr. Gatzke, 1942.

Speed is the essential requisite for a first-class ship of war – but essential only to go into action, not out of it.
    Rear-Admiral John A. Dahlgren (1809–1870).

The true speed of war is not headlong precipitancy, but the unremitting energy which wastes no time.
    Rear-Admiral Alfred Thayer Mahan, *Lessons of the War With Spain*, 1899.

Speed of mind to seize an opening and exploit an unforeseen advantage is the product not only of the commander's training and natural ability, but also of the promptness of his information – through personal touch with the progress of the battle, and the instant rendering of reports. (Oct. 1925.)
    Captain Sir Basil Liddell Hart, *Thoughts on War*, 1944.

One of the first lessons I had drawn from my experience of motorized warfare was that speed of manoeuvre in operations and quick reaction in command are decisive. Troops must be able to carry out operations at top speed and in complete coordination. To be satisfied with norms is fatal. One must constantly demand and strive for maximum performance, for the side which makes the greater effort is the faster – and the faster wins the battle.
    Field Marshal Erwin Rommel, *The Rommel Papers*, 1953.

Except for Napoleon, probably no other general appreciated as fully as Alexander the value of mobility in war. From the opening of his career until its close, speed dominated all his movements, and the result was that, by increasing the time at his disposal, in any given period he could proportionately accomplish more than his opponent.
    Major-General J.F.C. Fuller, *The Generalship of Alexander the Great*, 1960.

There is only one sound way to conduct war as I read history: Deploy to the war zone as quickly as possible sufficient forces to end it at the earliest moment. Anything less is a gift to the other side.
    Brigadier-General S.L.A. Marshall, 'Thoughts of Vietnam', in Thompson, *The Lessons of Vietnam*, 1977.

Lightning speed, more lightning speed; boldness, more boldness.
    Motto of the North Vietnamese General Staff, quoted in Colonel Harry Summers, *On Strategy*, 1981.

Speed is life.
    United States Marine Corps maxim.

SPIRIT

Now as for the host of the Vandals, let no one of you consider them. For not

by numbers of men nor by measure of body, but by valour of soul, is war want to be decided.
> Count Belisarius, December AD 533, address to his army before the Battle of Tricamarum in North Africa, in which he destroyed the last hope of the Vandals, quoted in Procopius, *History of the Wars*, 4.1, c. AD 560, tr. H.B. Dewing.

The spirit of the commander is naturally communicated to the troops, and there is an ancient saying that that it is better to have an army of deer commanded by a lion than an army of lions commanded by a deer.
> The Emperor Maurice, *The Strategikon*, c. AD 600 (*Maurice's Strategikon*, tr. George Dennis, 1984).

Do not compare your physical forces with those of the enemy's, for the spirit should not be compared with matter. You are men, they are beasts; you are free, they are slaves. Fight and you shall conquer. God grants victory to constancy.
> General Simon Bolivar, 1814, quoted in Salvador de Madariaga, *Bolivar*, 1952.

A warlike spirit, which alone can create and civilize a state, is absolutely essential to national defense and to national perpetuity... The more warlike the spirit of the people, the less need for a large standing army... Every male brought into existence should be taught from infancy that the military service of the Republic carries with it honor and distinction, and his very life should be permeated with the ideal that even death itself may become a boon when a man dies that a nation may live and fulfill its destiny.
> General of the Army Douglas MacArthur, Infantry Journal, 3/1927.

There is no substitute for the spiritual, in war. Miracles must be wrought if victories are to be won, and to work miracles men's hearts must needs be afire with self-sacrificing love for each other, for their units, for their division, and for their country.
> Major-General John A. Lejeune, *The Reminiscences of a Marine*, 1929.

As long as you are generous and true, and also fierce, you cannot hurt the world or even seriously distress her. She was made to be wooed and won by youth. She has lived and thrived only by repeated subjugations.
> Sir Winston S. Churchill, *My Early Life*, 1930.

It is the cold glitter in the attacker's eye not the point of the questing bayonet that breaks the line. It is the fierce determination of the drive to close with the enemy not the mechanical perfection of the tank that conquers the trench. It is the cataclysmic ecstasy of conflict in the flier not the perfection of his machine gun that drops the enemy in flaming ruin.
> General George S. Patton, Jr (1885–1947).

To be among them made every day good. Theirs was a beaten, battered division, almost drained of flesh and blood, its rifle companies cut to fifteen or twenty men per unit. They understood the enormity of their misfortune, yet the few who survived glorified it by their spirit.
> Brigadier-General S.L.A. Marshall, of the men of the Second Infantry Division in 1951, *Bringing Up the Rear*, 1979.

SPIT AND POLISH

Bad luck to this marching,
Pipe-claying and starching;
How neat one must be to be killed by the French!
> Soldiers' song from the Peninsular War, 1811, Mickey Free's lyrical complaint to the 14th Light Dragoons, quoted in C. Lever, *Charles O'Malley*, 1905.

THE STAFF

The easiest and quickest path into the esteem of traditional military authorities is by the appeal to the eye rather than to the mind. The 'polish and pipeclay' school is not yet extinct, and it is easier for the mediocre intelligence to become an authority on buttons than on tactics. (March 1925.)
 Captain Sir Basil Liddell Hart, *Thoughts on War*, 1944.

The way of endurance lay primarily in deadening reflection with action; but where action was restricted, trivialities came to the rescue. Here rests the defence not only for gossip and 'camp rumours' but for much seemingly futile 'spit and polish'. In the training camp its abuse was a hindrance to the development of military intelligence, and thus a handicap on war efficiency, yet in the rest camp it had a healing virtue that has often been misunderstood, not least by its advocates.
 Captain Sir Basil Liddell Hart, *The Memoirs of Captain Liddell Hart*, 1965.

When spit-and-polish are laid on so heavily that they become onerous, and the ranks cannot see any legitimate connection between the requirements and the development of an attitude which will serve as clear fighting purpose, it is to be questioned that the exactions serve any good object whatever.
 Brigadier-General S.L.A. Marshall, *The Armed Forces Officer*, 1950.

SPORT

Upon the fields of friendly strife
Are sown the seeds
That, upon other fields, on other days
Will bear the fruits of victory.
 General of the Army Douglas MacArthur, *Reminiscences*, 1964, p. 82; lines written by MacArthur while Superintendent at West Point (1919–1922), which he ordered to be carved over the entrance to the gymnasium.

Nothing more quickly than competitive athletics brings out the qualities of leadership, quickness of decision, promptness of action, mental and muscular co-ordination, aggressiveness, and courage. And nothing so firmly establishes that indefinable spirit of group interest and pride which we know as morale.
 General of the Army Douglas MacArthur, c. 1919–1922, while Superintendent at West Point, quoted in Francis Miller, *General Douglas MacArthur*, 1942.

In sport, in courage, and in the sight of Heaven, all men must meet on equal terms.
 Sir Winston S. Churchill (1874–1965).

THE STAFF

Large staffs – small victories.
 Field Marshal Prince Aleksandr V. Suvorov (1729–1800).

Every staff officer must be considered as acting under the direct orders and superintendence of the superior officer for whose assistance he is employed, and who is responsible for his acts. To consider the relative situation of the general officer and the staff officer in any other light, would tend to alter the nature of the Service, and, in fact, might give the command of the troops to a subaltern staff officer instead of to their general officer.
 The Duke of Wellington, cited in Sir Charles Oman, *Wellington's Army*, 1913.

A good staff has the advantage of being more lasting than the genius of a single man.
 Lieutenant-General Antoine-Henri Baron de Jomini, *Summary of the Art of War*, 1838, tr. Mendell and Craighill, 1862.

I do not want to make an appointment on my staff except of such as are early risers.
 Lieutenant-General Thomas J. 'Stonewall'

THE STAFF

Jackson, 1862, letter to his wife.

The more intimately it [the staff] comes into contact with the troops, the more useful and valuable it becomes. The almost entire separation of the staff from the line, as now practiced by us, and hitherto by the French, has proved mischievous, and the great retinues of staff-officers with which some of our earlier generals began the war were simply ridiculous.
General of the Army William T. Sherman, *The Memoirs of General W.T. Sherman*, 1875.

Great captains have no need for counsel. They study the questions which arise, and decide them, and their entourage has only to execute their decisions. But such generals are stars of the first magnitude who scarcely appear once in a century. In the great majority of cases, the leader of the army cannot do without advice. This advice may be the outcome of the deliberations of a small number of qualified men. But within this small number, one and only one opinion must prevail. The organization of the military hierarchy must ensure subordination even in thought and give the right and duty of presenting a single opinion for the examination of the general-in-chief to one man and only one.
Field Marshal Helmuth Graf von Moltke, 1862, letter.

The typical staff officer is the man past middle life, spare, unwrinkled, intelligent, cold, passive, noncommittal; with eyes like a codfish, polite in contact but at the same time unresponsive, cool, calm and as damnably composed as a concrete post or a plaster-of-paris cast; a human petrification with a heart of feldspar and without charm or the friendly germ; minus bowels, passion or a sense of humor. Happily they never reproduce and all of them finally go to hell.

General Freiherr Kurt von Hammerstein Equord (1878–1943).

Without a staff, an army could not peel a potato.
Lieutenant-General Hunter Liggett (1857–1935).

There is a type of staff officer who seems to think that it is more important to draft immaculate orders than to get out a reasonably well-worded order in time for action to be taken before the situation changes or the opportunity passes. (June 1933.)
Captain Sir Basil Liddell Hart, *Thoughts on War*, 1944.

The staff becomes an all-controlling bureaucracy, a paper octopus squirting ink and wriggling its tentacles into every corner. Unless pruned with an axe it will grow like a fakir's mango tree, and the more it grows the more it overshadows the general. It creates work, it creates offices, and, above all, it creates rear-spirit. No sooner is a war declared than the general-in-chief (and many a subordinate general also) finds himself a Gulliver in Lilliput, tied down to his office stool by innumerable threads woven out of the brains of his staff and superior staffs.
Major-General J.F.C. Fuller, *Generalship: Its Diseases and Their Cure*, 1933.

A yes man on a staff is a menace to a commander. One with the courage of his convictions is an asset.
Major-General Orlando Ward, c. 1934.

The military staff must be adequately composed: it must contain the best brains in the fields of land, air, and sea warfare, propaganda war, technology, economics, politics and also those who know the people's life.
General Erich Ludendorff, *Total War*, 1935.

Them potted palms up at division, they

THE STAFF

get promoted because they can write a good letter.

General Lewis 'Chesty' Puller, 1942, when a battalion commander on Guadalcanal, quoted in Merrill B. Twining, *No Bended Knee*, 1994.

Well, could they die a nobler death?

General of the Army Douglas MacArthur, 1944, response to a visitor from the War Department who suggested that MacArthur seemed to be working his staff to death, quoted in Geoffrey Parret, *Old Soldiers Never Die: The Life of Douglas MacArthur*, 1996, p. 362.

One of the inevitable problems with the Marine Corps or any other military service is that staff officers take over the minute a war is ended. The combat people run things when the chips are down and the country's life is at stake but when the guns stop, nobody's got a use for a combat man.

The staff officers are like rats; they stream out of hiding and take over. It's true. Just watch what happens to paperwork. God, in peacetime they put out enough to sink a small-size nation into the sea – and when war breaks, most of it just naturally stops. That's the way they do everything. There must always be staff people, of course, or we'd never get anything done – but if we don't stop this empire building of the staff, somebody's going to come along and lick us one of these days. We'll be so knotted in red tape that we can't move.

Lieutenant-General Lewis 'Chesty' Puller, June 1946, letter to his wife, quoted in Burke Davis, *Marine*, 1962.

There is certainly no black magic about becoming a successful Army staff officer in the field. Youth and brains are the important things. With a comparatively small amount of training and experience most staff appointments can be filled by non-regulars. In 21st Army Group a number of important posts were held by young ex-civilians. One could not have wished for better men. Montgomery backed youth and the 'clever chap,' and this policy paid him enormously.

Major-General Sir Francis de Guingand, *Operation Victory*, 1947.

I now want you to consider the general in relation to his troops. I will begin with a few words about his staff, who are the means by which he controls and directs his army. I will give you two simple rules which every general should observe: first, never to try to do his own staff work; and, secondly, never to let his staff get between him and his troops. What a staff appreciates is that it should receive clear and definite instructions, and then be left to work out the details without interference. What troops and subordinate commanders appreciate is that a general should be constantly in personal contact with them, and should not see everything simply through the eyes of his staff. The less time a general spends in his office and the more with his troops the better.

Field Marshal Viscount Wavell of Cyrenaica, *Soldiers and Soldiering*, 1953.

I was evacuated to hospital in England and for some months I took no further part in the war. I had time for reflection in hospital and came to the conclusion that the old adage was probably correct: the pen was mightier than the sword. I joined the staff.

Field Marshal Viscount Montgomery of Alamein, after being wounded in 1914, *The Memoirs of Field Marshal Montgomery*, 1958.

My war experience led me to believe that the staff must be the servants of the troops, and that a good staff officer must serve his commander and the troops but himself be anonymous.

STRATEGY

The qualities of a leader are not limited to commanders. The requirements for leadership are just as essential in the staff officer, and in some respects more exacting, since he does not have that ultimate authority which can be used when necessary and must rely even more than his commander on his own strength of character, his tact and persuasion in carrying our his duties.
General Matthew B. Ridgway, 'Leadership', *Military Review*, 10/1966.

Field Marshal Viscount Montgomery of Alamein, *The Memoirs of Field Marshal Montgomery*, 1958.

It's like dropping a pound of mercury on the floor. It scatters all over but pretty soon you find it all together over in the corner of the room.
General Creighton Abrams (1914–1974), speaking of the difficulty in reducing the Army Staff, quoted in *Military Review*, 4/1985.

The staff is there because it serves a vital purpose. Recognition is rare, medals are few, but the successful staff officer can at least take pride in having served as midwife at the birth of great events.
General Merrill B. Twining, *No Bended Knee*, 1994.

It's my 'porcupine' theory of command. No commander can know everything. He must rely on deputies, competent in the narrow areas assigned them. His responsibility is to make sure none of them tugs the blanket to one side of the bed. A deputy who answers 'Yes, sir' to every stupid thing his commander says can get his boss into serious trouble. He must have the courage to take a stand and be able to defend it.
Lieutenant-General Aleksandr Lebed, *Time Magazine*, 27 February 1995.

STRATEGY

Attack by stratagem. Sun Tzu said: In the practical art of war, the best thing of all is to take the enemy's country whole and intact; to shatter and destroy it is not so good. So, too, it is better to capture an army entire than to destroy it, to capture a regiment, detachment or a company entire than to destroy them.
Hence to fight and conquer in all your battles is not supreme excellence; supreme excellence consists of breaking the enemy's resistance without fighting.
Sun Tzu, *The Art of War*, 3, c. 500 BC, tr. Giles, 1910.

I know well that war is a great evil and the worst of all evils. But since our enemies clearly look upon the shedding of blood as one of their basic duties and the height of virtue, and since each one must stand up for his country and his own people with word, pen, and deed, we have decided to write about strategy. But putting it into practice we shall be able not only to resist our enemies but even to conquer them... Strategy is the means by which a commander may defend his own lands and defeat his enemies. The general is the one who practices strategy.
Anonymous Byzantine general, *On Strategy*, c. AD 527–565, quoted in George Dennis, *Three Byzantine Military Treatises*, 1985.

The most complete and happy victory is this: to compel one's enemy to give up his purpose, while suffering no harm oneself.
Count Belisarius (c. AD 505–565).

Strategy is the craft of the warrior.
Miyamoto Musashi, *A Book of Five Rings*, 1645, tr. Victor Harris, 1984.

Strategy is based on the forces you have, on the strength of the enemy, on the situation of the country where you want to carry the war, and on the actual political condition of Europe.

STRATEGY

Frederick the Great (1712–1786), 'Pensées et règles générales pour la guerre', *Oeuvres de Frederic le Grand*, XXVIII, 1846–1856.

Strategy is the science of making use of space and time. I am more jealous of the latter than of the former. We can always recover lost ground, but never lost time.
  Field Marshal August Graf von Gneisenau (1761–1831), quoted in Hermann Foertsch, *The Art of Modern War*, 1940.

...The art of employment of battles as a means to gain the object of the war.
  Major-General Carl von Clausewitz, *On War*, 1832, tr. Michael Howard and Peter Paret, 1976.

Strategy is a system of makeshifts. It is more than a science, it is the application of science to practical affairs; it is carrying through an originally conceived plan under a constantly shifting set of circumstances. It is the art of acting under the pressure of the most difficult kind of conditions. Strategy is the application of common sense in the work of leading an army; its teachings hardly go beyond the first requirement of common sense; its value lies entirely in its concrete application. It is a matter of understanding correctly at every moment a constantly changing situation, and then doing the simplest and most natural thing with energy and determination. This is what makes war an art, an art that is served by many sciences. Like every art, war cannot be learned rationally, but only by experience. In war, as in art, there can be no set standards nor can code of rules take the place of brains.
  Field Marshal Helmuth Graf von Moltke, *On Strategy*, 1871.

A clever military leader will succeed in many cases in choosing defensive positions of such an offensive nature from the strategic point of view that the enemy is compelled to attack us in them.
  Field Marshal Helmuth Graf von Moltke (1800–1891).

The main thing in true strategy is simply this: first deal as hard blows at the enemy's soldiers as possible, and then cause so much suffering to the inhabitants of a country that they will long for peace and press their Government to make it. Nothing should be left to the people but eyes to lament the war.
  General of the Army Phillip H. Sheridan, quoted in Forbes, *Memories and Studies in War and Peace*, 1895.

As in a building, which, however fair and beautiful the superstructure, is radically marred and imperfect if the foundation be insecure – so, if the strategy be wrong, the skill of the general on the battlefield, the valor of the soldier, the brilliancy of the victory, however otherwise decisive, fail of their effect.
  Rear-Admiral Alfred Thayer Mahan, *Naval Administration and Warfare*, 1908.

Principles of strategy never transcend common sense.
  Motto of the highest school of the German Army before 1914, quoted in Leon Trotksy, *Military Writings*, 1967.

Where the strategist is empowered to seek a military decision, his responsibility is to seek it under the most advantageous circumstances in order to produce the most profitable results. Hence his true aim is not so much to seek battle, as to seek a strategic situation so advantageous that if it does not of itself produce the decision, its continuation by a battle is sure to achieve this. In other words, dislocation is the aim of strategy; its sequel may be either the enemy's dissolution or his disruption in battle.

STRENGTH

Dissolution may involve some partial measure of fighting, but this has not the character of a battle. (Oct. 1928.)
>Captain Sir Basil Liddell Hart, *Thoughts on War*, 1944.

Kill Japs, kill Japs, and keep on killing Japs.
>Admiral of the Fleet William 'Bull' Halsey, strategy for winning the war, November 1942, quoted in E.B. Potter, *Bull Halsey*, 1985.

Strategy is the determination of the direction of the main blow – The plan of strategy is the plan of the organization of the decisive blow in the direction in which the blow can most quickly give the maximum results.
  In other words, to define the direction of the main blow means to predetermine the nature of operations in the whole period of war, to determine nine-tenths of the fate of the entire war. In this is the main task of strategy.
>Joseph Stalin, quoted in R.L. Garthoff, *Soviet Military Doctrine*, 1953.

The aim of strategy is to clinch a political argument by means of force instead of words. Normally this is accomplished by battle, the true object of which is not physical destruction, but mental submission on the part of the enemy.
>Major-General J.F.C. Fuller, *Armoured Warfare*, 1943.

The history of strategy is, fundamentally, a record of the application and evolution of the indirect approach.
>Captain Sir Basil Liddell Hart, *Strategy*, 1954.

…The art of distributing and applying the military means to fulfill the ends of policy.
>Captain Sir Basil Liddell Hart, *Strategy*, 1954.

The laws of strategy are objective and apply impartially to both hostile sides.
>Marshal of the Soviet Union V.D. Sokolovksiy, *Soviet Military Strategy* (Third edition), 1968.

Our strategy to go after this army is very, very simple. First we're going to cut it off, and then we're going to kill it.
>General Colin Powell, The Washington Post, 23 January 1991, speaking of the Iraqi Army just before the beginning of Operation Desert Storm.

STRENGTH

All of us right behind you, hearts intent on battle.
Nor do I think you'll find us short on courage,
long as our strength will last. Past his strength
no man can go, though he's set on mortal combat.
>Homer, *The Iliad*, 13.908–912, c. 800 BC, tr. Robert Fagles, 1990; Prince Paris to his brother, Hector.

Peace is our aim and strength the only way of getting it.
>Sir Winston S. Churchill (1874–1965).

If we are strong, our character will speak for itself. If we are weak, words will be of no help.
>President John F. Kennedy, undelivered address, Dallas, 22 November 1963.

With regard to resignation – No. Now is the time for strength and resolution.
>Prime Minister Margaret Thatcher, 6 April 1982, during the Falklands War when MP Bob Cryer in the House of Commons suggested she resign because her government was caught unprepared by the Argentine attack, quoted in *The Falklands Campaign: A Digest of Debates in the House of Commons*, 1982, p. 22.

STRIKE WEAKNESS

Military tactics are like unto water; for

water in its natural course runs away from high places and hastens downwards. So in war, the way to avoid what is strong is to strike what is weak.
    Sun Tzu, *The Art of War*, 6, c. 500 BC, tr. Giles, 1910.

Another sound principle is always to aim for the weak point of the enemy's position, even if it is a long way off. After all, pushing yourself to make a great effort is less risky than engaging a superior force.
    Xenophon, *How To Be A Good Cavalry Commander*, c. 360 BC (*Hiero the Tyrant and Other Treatises*, tr. Robin Waterfield, 1997).

To make sure you attack where there is no defense is what is crucial to warriorship...
    Sun Bin, *The Lost Art of War*, c. 350 BC, tr. Thomas Cleary.

Go into emptiness, strike voids, bypass what he defends, hit him where he does not expect you.
    The 'Martial' Emperor Ts'ao Ts'ao (AD 155–220), commentary to Sun Tzu, *The Art of War*, 6, c. 500 BC, tr. Samuel Griffith, 1963.

President of the Directoire: 'You often defeated a stronger foe with fewer forces.'
Bonaparte: 'But in this case, too, the lesser forces still suffered defeat at the hands of the larger forces. With an enemy army superior in numbers against me, I dashed like lightning to its flank, smashed it, took advantage of the enemy's disarray, and again rushed with all my forces to other points. Thus, I inflicted defeat piecemeal, and the victory which I won was, as you see, nothing more than the victory of the stronger over the weaker.'
    Napoleon, quoted in Colonel V.Ye. Savkin, *Operational Art and Tactics*, 1972.

Everybody has a weak spot, and the first thing I try to do is to find out where it is, and pitch into it with the biggest shell or shot that I have, and repeat that dose until it operates.
    Admiral David G. Farragut, quoted in Alfred Thayer Mahan, *Admiral Farragut*.

We won the victory where it was easy to win and not over something that was hard to conquer... each division of the combined squadrons did its work well and not more. There was nothing remarkable in our bravery. We regard ourselves as a fit example of good fight...
    Admiral Marquis Heihachiro Togo, remarking on the Battle of Tsushima (27–28 May 1905), quoted in Dennis and Peggy Warner, *The Tide at Sunrise*, 1974.

Have been giving everyone a simplified directive of war. Use steamroller strategy; that is, make up your mind on course and direction of action, and stick to it. But in tactics, do not steamroller. Attack weakness. Hold them by the nose and kick them in the pants.
    General George S. Patton, Jr, 6 November 1942, *War As I Knew It*, 1947.

SUBORDINATES

Encourage and listen well to the words of your subordinates. It is well known that gold lies hidden underground.
    Nabeshima Naoshige (1538–1618), quoted in William Wilson, *Ideals of the Samurai*, 1982.

Today I decided to stay in... There is nothing that I can do at the front except bother people, as they are all doing a swell job, and I believe advancing very rapidly. One of the hardest things that I have to do to – and I presume any General has to do – is not to interfere with the next echelon of command when the show is going all right.
    General George S. Patton, Jr, 18 November 1944, letter to his sister Nita, *The Patton Papers, II*, 1974.

SUCCESS

There have been great and distinguished leaders in our military Services at all levels who had no particular gifts for administration and little for organizing the detail of decisive action either within battle or without. They excelled because of a superior ability to make use of the brains and command the loyalty of well-chosen subordinates. Their particular function was to judge the goal according to their resources and audacity, and then to hold the team steady until the goal was gained. So doing, they complemented the power of the faithful lieutenants who might have put them in the shade in any IQ test.
Brigadier-General S.L.A. Marshall, *The Armed Forces Officer*, 1950.

Of course a commander must know in what way to give verbal orders to his subordinates. No two will be the same; each will require different treatment. Some will react differently from the others; some will be happy with a general directive whilst others will like more detail. Eventually a mutual confidence on the subject will grow up between the commander and his subordinates; once this has been achieved there will never be any more difficulties or misunderstandings.
Field Marshal Viscount Montgomery of Alamein, *The Memoirs of Field Marshal Montgomery*, 1958.

Closely akin to the relationship with staff officers is keeping in close personal touch with your principal subordinate commanders – in the division, with your brigade and separate battalion commanders; in the corps with your division commanders, their chiefs of staff, and as many of the commanders of attached corps units as you can; and in the army, with corps and division commanders and their chiefs of staff. There is always time for these visits; administrative work can be done at night. By day you belong with your troops.

General Matthew B. Ridgway, 'Leadership', *Military Review*, 10/1966.

SUCCESS

'Tis man's to fight, but Heaven's to give success.
Homer, *The Iliad*, 6.427, c. 800 BC, tr. Alexander Pope, 1743.

Success is a matter of planning and it is only careless people who find that Heaven will not help their mortal designs.
Themistocles, 480 BC, the year of the Battle of Salamis, quoted in Herodotus (c. 484–420 BC), *The Struggle for Greece*, tr. Cavandar.

Whosoever desires constant success must change his conduct with the times.
Niccolò Machiavelli, *Discourses*, 1517.

Although I cannot insure success, I will endeavour to deserve it.
Admiral John Paul Jones, 1780, letter.

I am sure you will deserve success. To mortals is not given the power of commanding it.
Admiral Lord St Vincent, 14 July 1797, letter to Nelson before sailing against Tenerife.

They won't draw me from my cautious system. I'll fight them only where I am pretty sure of success.
The Duke of Wellington, autumn 1810, while watching Marshal Masséna's army starving outside his lines at Torres Vedres, *Supplementary Despatches, Correspondence, and Memoranda of Field Marshal Arthur Duke of Wellington, K.G., 1794–1818*, VI, 1858–1872, p. 612.

...In war many roads lead to success, and that they do not all involve the opponent's defeat. They range from the destruction of the enemy's forces, the conquest of his territory, to a temporary occupation or invasion, to projects

with an immediate purpose, and finally passively awaiting the enemy's attacks. Any one of these may be used to overcome the enemy's will: the choice depends on the circumstances.
> Major-General Carl von Clausewitz, *On War*, 1.2, 1832, tr. Michael Howard and Peter Paret, 1976.

Napoleon was right in many things, and especially in one thing – he forgave everything but want of success.
> The Duke of Wellington, April 1838, quoted in Andrew Rogers, *Napoleon and Wellington*, 2001.

Success in war, like charity in religion, covers a multitude of sins.
> General Sir William Napier (1785–1860).

Obedience is the principle, but man stands above the principle... who is right in battle is decided in most cases by success...
> Field Marshal Helmuth Graf von Moltke (1800–1891).

My success had been so uninterrupted that the men thought that victory was chained to my standard. Men who go into a fight under the influence of such feelings are next to invincible, and generally are the victors before it begins.
> Colonel John S. Mosby, *Mosby's War Reminiscences*, 1887.

Too much success is not wholly desirable; an occasional beating is good for men – and nations.
> Rear-Admiral Alfred Thayer Mahan, *The Life of Nelson*, 1897.

A man who says that his success is due to himself is a fool. Success is one-third ability and two-thirds good fortune – or perhaps more exactly, one-third ability, one third luck (or providence), and one-third the power to 'stick it', to keep on pegging away until fortune turns, as it usually does sooner or later. (Oct. 1932.)

> Captain Sir Basil Liddell Hart, *Thoughts on War*, 1944.

Success is how high you bounce when you hit bottom.
> General George S. Patton (1889–1945), quoted in *Reader's Digest*, April 1995, p. 177.

No one can guarantee success in war, but only deserve it.
> Sir Winston S. Churchill, *Their Finest Hour*, 1948.

What is success? I think it is a mixture of having a flair for the thing that you are doing; knowing that it is not enough, that you have got to have hard work and a certain sense of purpose.
> Prime Minister Margaret Thatcher, *Parade*, 13 July 1986.

There are no secrets to success; don't waste time looking for them... Success is the result of perfection, hard work, learning from failure, loyalty to those for whom you work and persistence. You must be ready for opportunity when it comes.
> General Colin Powell, quoted in *The Washington Post*, 15 January 1989.

SUICIDE

And the battle went sore against Saul, and the archers hit him; and he was sore wounded of the archers. Then said Saul unto his armourbearer Draw thy sword, and thrust me through therewith; lest these uncircumcised come and thrust me through, and abuse me. But his armourbearer would not; for he was sore afraid. Therefore Saul took a sword, and fell upon it.
> Samuel I, 31.3–4; of the death of King Saul fighting the Philistines at the Battle of Mount Gilboa c. 1006 BC.

SUPERIORITY

Yes, we must escape, but this time with our hands, not our feet.
> Marcus Brutus, shortly after his defeat at the Second Battle of Philippi, 26 November 42 BC, just before he fell upon his sword, quoted in Plutarch, *The Lives*, c. AD 100 (*Makers of Rome*, tr. Ian Scott-Kilvert, 1965).

You must not pity me in this last turn of fate. You should rather be happy in the remembrance of our love and in the recollection that of all men I was once the most powerful and now at the end have fallen not dishonourably, a Roman by a Roman vanquished.
> Mark Antony, 30 BC, to Cleopatra as he lay dying after falling upon his own sword following his flight from the destruction of the fleet at the Battle of Actium.

I feel that I am reserved for one end or other.
> Robert Lord Clive (1725–1774), comment when his pistol failed to go off twice when attempting to commit suicide.

There is as much courage in supporting with firmness the afflictions of the soul as there is in standing steady under the grape of a battery of guns. To give one's self up to grief without resistance, to kill one's self to escape it, is to abandon the battlefield defeated.
> Napoleon, 14 May 1802, at St Cloud, quoted in R.M. Johnston, ed., *The Corsican*, 1910.

To kill myself would be a gambler's death. I am condemned to live. Besides, only the dead do not return.
> Napoleon, 11 April 1814, after his abdication, *Correspondance de Napoléon Ier, publiée par ordre de l'Empereur Napoléon III*, Vol. XXXI, pp. 485–486, 1858–1870.

When the nerves break down, there is nothing left but to admit that one can't handle the situation and to shoot oneself.
> Adolf Hitler, February 1943, upon hearing of the surrender of Field Marshal Paulus at Stalingrad, quoted in Clark, *Barbarossa*, 1965.

SUPERIORITY

By discovering the enemy's dispositions and remaining invisible ourselves, we can keep our forces concentrated while the enemy must be divided.

We can form a single united body, while the enemy must split up into fractions. Hence there will be a whole pitted against separate parts of a whole, which means that we shall be many to the enemy's few.

And if we are thus able to attack an inferior force with a superior one, our opponents will be in dire straits.
> Sun Tzu, *The Art of War*, 6, c. 500 BC, tr. Giles, 1910.

Relative superiority, that is, the skillful concentration of superior strength at the decisive point, is much more frequently based on the correct appraisal of this decisive point, on suitable planning from the start; which leads to appropriate disposition of forces, and on the resolution needed to sacrifice nonessentials for the sake of essentials – that is, the courage to retain the major part of one's forces united. This is particularly characteristic of Frederick the Great and Bonaparte.
> Major-General Carl von Clausewitz, *On War*, 3.8, 1832, tr. Michael Howard and Peter Paret, 1976.

The fundamental object in all military combinations is to gain local superiority by concentration.
> Rear-Admiral Alfred Thayer Mahan, *Naval Strategy*, 1911.

One is always liable to be smashed by superior force, but one should never be caught unprepared to do one's best.
> Field Marshal Viscount Allenby of Megiddo, 1900, letter to his wife in the Boer War.

SURPRISE

Armies and nations are mainly composed of normal men, not of abnormal heroes, and once these realize the permanent superiority of the enemy they will surrender to *force majeure*.

Captain Sir Basil Liddell Hart, *Thoughts on War*, 1944.

SURPRISE
Principle of War

Behold, the wretched, vanquished chief of Kheta [Hittite king] together with numerous allied countries, were stationed in battle array, concealed on the northwest of the city of Kadesh, while his majesty [Ramses II] with his bodyguard, and the division of Amon was marching behind him. The division of Re crossed over the river-bed on the south side of the town of Shabtuna, at the distance of [app. 1.4 miles] from the division of Amon; the division of Ptah was on the south of the city of Aranami; and the division of Sutekh was marching on upon the road. His majesty had formed the first rank of all the leaders of his army...

They [Hittites] came forth from the southern side of Kadesh, and they cut through the division of Re in its middle, while they were marching without knowing and without being drawn up for battle. The infantry and the chariotry of his majesty retreated before them. Now, his majesty had halted on the north of the city of Kadesh, on the western side of the Orontes. Then came one to tell his majesty.

His majesty shone like his father Montu, when he took the adornments of war; as he seized his coat of mail, he was like Baal in his hour. The great span [chariot] which bore his majesty called: 'Victory in Thebes,' from the great stables of Ramses II, was in the midst of the leaders. His majesty halted the rout; by himself and none other with him. When his majesty went to look behind him, he found 2,500 chariotry surrounding him...

The Poem of Pentaur, Rameses II suffering the first great reported surprise attack in history at the Battle of Kadesh, 1294 BC (Rameses would have also suffered the first recorded great defeat had he not by personal example fought his way out and rallied his army to pull off at least a draw), quoted in James Breasted, *Ancient Records of Egypt*, 1906.

The execution of a military surprise is always dangerous, and the general who is never taken off his guard himself, and never loses an opportunity of striking at an unguarded foe, will be most likely to succeed in war.

Thucydides, *History of the Peloponnesian War*, c. 404 BC.

Among the enemy, when they saw him advancing so unexpectedly, there was a total lack of steadiness. Some were running to take up their positions, others forming into line, others bridling their horses, others putting on their breastplates. The general impression was one of people expecting to suffer rather than cause damage.

Xenophon, of the onset of the Theban Army led by Epaminondas at the Battle of Mantinea in 362 BC, *Hellenica*, c. 360 BC (*A History of My Times*, tr. Rex Warner, 1966).

Sameness is inadequate to attain victory; therefore difference is used, for surprise. Therefore stillness is surprise to the mobile, relaxation is surprise to the weary, fullness is surprise to the hungry, orderliness is surprise to the unruly, many are a surprise to the few. When the initiative is direct, holding back is surprise; when a surprise attack is launched without retaliation, that is victory. Those who have an abundance of surprises excel in gaining victories.

Sun Bin, *The Lost Art of War*, c. 350 BC, tr. Thomas Cleary, 1966

Even brave men are dismayed by

SURPRISE

sudden perils.
Cornelius Tacitus, *Annals*, 15.59, c. AD 115–117 (*The Complete Works of Tacitus*, tr. Alfred Church and William Brodribb, 1942).

Novelty and surprise throw an enemy into consternation but common incidents have no effect.
Flavius Vegetius Renatus, *Military Institutions of the Romans*, c. AD 378, tr. Clark, 1776.

The general must make it one of his highest priorities and concerns to launch secret and unexpected attacks upon the enemy whenever possible. If he is successful in this sort of operation, with only a small group of men, he will put large numbers of the enemy to flight.
Emperor Nikephoros II Phokas, *Skirmishing*, c. AD 969 (*Three Byzantine Military Treatises*, tr. George Dennis, 1985).

Everything which the enemy least expects will succeed the best. If he relies for security on a chain of mountains that he believes impracticable and you pass these by roads unknown to him, he is confused to start with, and if you press him he will not have time to recover from his consternation. In the same way, if he places himself behind a river to defend the crossing and you find some ford above or below by which to cross, this surprise will derange and confuse him.
Frederick the Great, *Instructions to His Generals*, 1747, tr. Phillips, 1940.

Any officer or non-commissioned officer who shall suffer himself to be surprised... must not expect to be forgiven.
Major-General Sir James Wolfe, 1759, General Orders during the Quebec Expedition.

Who can surprise well must conquer.
Admiral John Paul Jones, 10 February 1778, letter to Benjamin Franklin, quoted in Lorenz, *John Paul Jones*, 1943.

To astonish is to vanquish.
Field Marshal Prince Aleksandr V. Suvorov (1729–1800), quoted by Chuikov, in Leites, *The Soviet Style in War*, 1982.

The enemy reckons we're sixty miles away... Suddenly we're on him like a cloudburst. His head whirls. Attack! Cut down, stab, chase, don't let him get away.
Field Marshal Aleksandr V. Suvorov (1729–1800), 1799, quoted in Philip Longworth, *The Art of Victory*, 1966.

...Nothing so pregnant with dangerous consequences, or so disgraceful to an Officer in arms, as a surprise.
Admiral Lord St Vincent, 1798, memorandum to the Fleet off Cadiz.

It is a principle of warfare, that when it is possible to make use of thunderbolts, they should be preferred to cannon.
Napoleon (1769–1821), dictation on St Helena, quoted in Christopher Herold, ed., *The Mind of Napoleon*, 1955.

Maxim 95. War is composed of nothing but surprises. While a general should adhere to general principles, he should never lose the opportunity to profit by these surprises. It is the essence of genius. In war there is only one favorable moment. Genius seizes it.
Napoleon, *The Military Maxims of Napoleon*, tr. Burnod, 1827.

To be defeated is pardonable; to be surprised – never!
Napoleon, *Maxims of War*, 1831.

For the side that can benefit from the psychological effects of surprise, the worse the situation is, the better it may turn out, while the enemy finds himself incapable of making coherent decisions.

SURPRISE

This holds true not only for senior commanders, but for everyone involved; for one peculiar feature of surprise is that it loosens the bonds of cohesion, and individual action can easily become significant.
   Major-General Carl von Clausewitz, On War, 3.9, 1832, tr. Michael Howard and Peter Paret, 1976.

The way to destroy the enemy's morale, to show him that his cause is lost, is therefore surprise in every sense of the word, bringing into the struggle something 'unexpected and terrible,' which therefore has a great effect. It deprives the enemy of the power to reflect, and consequently to discuss.
   Under various forms we always find the same principle of surprise, seeking to produce on the enemy the same moral result, terror; creating in him, by the sudden appearance of something unexpected and overwhelming, the feeling of impotence, the assurance that he cannot win; that is, that he is beaten.
   Marshal of France Ferdinand Foch, Principles of War, 1913.

Inaction leads to surprise, and surprise to defeat, which is after all only a form of surprise.
   Marshal of France Ferdinand Foch, Precepts and Judgments, 1919.

Surprise – the pith and marrow of war!
   Admiral Sir John Fisher, Memories, 1919.

An enemy may be surprised, which implies that he is thrown off balance. This is the best method of defeating him, for it is so economical, one man taking on to himself the strength of many. Surprise may be considered under two main headings: surprise effected by doing something that the enemy does not expect, and surprise effected by doing something that the enemy cannot counter. The first may be denoted as moral surprise, the second as material.
   Major-General J.F.C. Fuller, The Reformation of War, 1923.

The subject of surprise is an immense one, and one which influences all forms and modes of war. It is one which is nearly always lost sight of during peacetime, because danger and fear are more often than not abstract quantities; but in wartime they manifest, and with them manifests surprise – the demoralizing principle.
   Major-General J.F.C. Fuller, The Foundations of the Science of War, 1926.

Of all keys to success in war 'unexpectedness' is the most important. By it a commander, whether of an army or a platoon, can often unlock gates which are impregnable to sheer force. (July 1929.)
   Captain Sir Basil Liddell Hart, Thoughts on War, 1944.

Movement generates surprise, and surprise gives impetus to movement.
   Captain Sir Basil Liddell Hart, 'Strategy', Encyclopaedia Britannica (1929 edition).

Tactics must change if surprise is to be obtained, for surprise does not merely mean hitting the enemy at some unexpected point, but also hitting him in an unexpected way.
   Major-General J.F.C. Fuller, Memoirs of an Unconventional Soldier, 1936.

A military man can scarcely pride himself on having 'smitten a sleeping enemy'; it is more a matter of shame, simply, for the one smitten.
   Admiral Isoroku Yamamoto, 9 January 1942, referring to Pearl Harbor, quoted in Agawa, The Reluctant Admiral, 1979.

We have inflicted a complete surprise on the enemy. All our columns are inserted in the enemy's guts.
   Major-General Orde Wingate, 11 March

SURRENDER

1944, Order of the Day to the Third Indian Division.

...To do something which the enemy cannot prevent, and to do something which he does not suspect. The first action may be compared to surprising a man with his eyes open, in the other, the man has his eyes shut.
Major-General J.F.C. Fuller, *The Generalship of Alexander the Great*, 1958.

Since time immemorial there have been lively, self-confident commanders who have exploited the principle of surprise – the means whereby inferior forces may snatch victory, and turn downright impossible conditions to their own advantage. The effects on the morale of both parties are immense – but his every element of incalculability proves a deterrent to the ponderous spirits, and this is probably why they are so reluctant to embrace new weapons, even when the inadequacy of the old ones is all too clear.
General Heinz Guderian, *Achtung – Panzer!*, 1992 (first published in German in 1937).

Surprises are a commonplace in war – and reconsidered opinions too.
General Robert L. Eichelberger (1886–1961).

Most opponents are at their best if they are allowed to dictate the battle; they are not so good when they are thrown off balance by manoeuvre and are forced to react to your own movements and thrusts. Surprise is essential. Strategical surprise may often be difficult, if not impossible, to obtain; but tactical surprise is always possible and must be given an essential place in training.
Field Marshal Viscount Montgomery of Alamein, *A History of Warfare*, 1968.

A force within striking distance of an enemy must be suitably disposed with regard to its battle positions, being ready at all times to fight quickly if surprised.
Field Marshal Viscount Montgomery of Alamein, *A History of Warfare*, 1968.

SURRENDER

Defeat is one thing; disgrace is another.
Winston S. Churchill, on the surrender of the British garrison at Tobruk in June 1942, *The Second World War: The Hinge of Fate*, 1.22, 1950.

Sir, –
I have the mortification to inform your excellency, that I have been forced to give up the posts of York and Gloucester, and to surrender the troops under my command, by capitulation, on the 19th instant, as prisoners of war, to the combined forces of America and France.
General Charles Marquis Cornwallis, 17 October 1781, to General Sir Henry Clinton, announcing the surrender of his army at Yorktown, quoted in Henry Lee, *War in the Southern Department*, 1869.

Maxim 69. There is but one honourable mode of becoming prisoner of war. That is, by being taken separately; by which is meant, being cut off entirely, and when we can no longer make use of our arms. In this case there can be no conditions, for honour can impose none. We yield to an irresistable necessity.
Napoleon, *The Military Maxims of Napoleon*, tr. George D'Aguilar, 1831 (David Chandler, ed., 1987).

Your Royal Highness: victimized by the factions which divide my country, and by the hostility of the European powers, I have ended my political career; and I come, as Themistocles did, to claim a seat by the hearth of the British people. I put myself under the protection of British law – a protection which I claim from your Royal Highness, as the strongest, the stub-

SURRENDER

bornest, and most generous of my foes.
  Napoleon, 14 July 1815, letter to the Prince Regent of England from Ile d'Aix after his defeat at Waterloo, *Correspondance de Napoléon Ier, publiée par ordre de l'Empereur Napoléon III*, XXVIII, 22066, 1858–1870.

Sir: yours of this date proposing armistice and appointment of commissioners to settle terms of capitulation, is just received. No terms except an unconditional and immediate surrender can be accepted. I propose to move immediately upon your works.
  General of the Army Ulysses S. Grant, 16 November 1862, to the Confederate commander of Fort Donelson, *Personal Memoirs*, I, 1885.

Then there is nothing left for me but to go and see General Grant, and I would rather die a thousand deaths.
  General Robert E. Lee, 9 April 1865.

After four years of arduous service, marked by unsurpassed, courage and fortitude, the Army of Northern Virginia has been compelled to yield to overwhelming numbers and resources. I need not tell the survivors of so many hard-fought battles, who have remained steadfast to the last, that I have consented to this result from no distrust of them; but, feeling that valor and devotion could accomplish nothing that could compensate for the loss that would have attended the continuation of the contest, I have determined to avoid the useless sacrifice of those whose past services have endeared them to their countrymen.

  With an increasing admiration of your constancy and devotion to your country, and a grateful remembrance of your kind and generous consideration of myself, I bid you an affectionate farewell.
  General Robert E. Lee, farewell message to the Army of Northern Virginia, 10 April 1865, after the surrender at Appomattox.

Tell General Howard I know his heart. What he told me before I have in my heart. I am tired of fighting. Our chiefs are killed. Looking Glass is dead. Too-hul-hule-sute is dead. The old men are all dead. It is the young men who say yes or no. He who once led them is dead. It is cold and we have no blankets.

  The little children are freezing to death. My people, some of them, have run away to the hills, and have no blankets, no food; no one knows where they are – perhaps freezing to death. I want to have time to look for my children and see how many of them I can find. Maybe I shall find them among the dead. Hear me, my chiefs. I am tired; my heart is sick and sad. From where the sun now stands I will fight no more forever.
  Chief Joseph of the Nez Perce, upon his surrender to a representative of General O.O. Howard, 5 October 1877, quoted in Mathieson, *The Nez Perce Indian War*, 1964.

I would fight without a break. I would fight in front of Amiens. I would fight in Amiens. I would fight behind Amiens. I would never surrender.
  Marshal of France Ferdinand Foch, 26 March 1918, to Sir Douglas Haig during the German Friedenstürm offensive.

Nations which went down fighting rose again, but those which tamely surrendered were finished.
  Sir Winston S. Churchill (1874–1965).

It is with a broken heart and head bowed in sadness but not in shame, that I report to Your Excellency that I must go today to arrange terms for the surrender of the fortified islands of Manila Bay.

  There is a limit of human endurance, and that limit has long since been past. Without prospect of relief I feel it is my duty to my country and to my

SURRENDER

gallant troops to end this useless effusion of blood and human sacrifice. If you agree, Mr. President, please say to the nation that my troops and I have accomplished all that is humanly possible and that we have upheld the best tradition of the United States and its Army. May God bless and preserve you and guide you and the nation in the effort to ultimate victory.

With profound regret and with continued pride in my gallant troops, I go to meet the Japanese commander. Goodbye, Mr. President.
> General Jonathan M. Wainwright's message to President Roosevelt, 6 May 1942, as he was about to surrender Corrigedor to the Japanese under General Homma, quoted in Duane Schultz, *Hero of Bataan*, 1981.

Let no man surrender so long as he is unwounded and can fight.
> Field Marshal Viscount Montgomery of Alamein, 30 October 1942, message to the Eighth Army on the eve of the Battle of Alamein.

Surrender is forbidden. Sixth Army will hold their positions to the last man and the last round, and by their heroic endurance will make an unforgettable contribution to the establishment of a defensive front and the salvation of the western world.
> Adolf Hitler, 23 February 1943, to Field Marshal Paulus, commander of the German Sixth Army surrounded in Stalingrad.

They [the German delegation sent by Admiral Doenitz to open negotiations for surrender] were brought to my caravan site and were drawn up under the Union Jack, which was flying proudly in the breeze. I kept them waiting for a few minutes and then came out of my caravan and walked towards them. They all saluted, under the Union Jack. It was a great moment; I knew the Germans had come to surrender and that the war was over.
> Field Marshal Viscount Montgomery of Alamein, 5 May 1945, taking the surrender of the Germans in northern Germany, *The Memoirs of Field Marshal Montgomery*, 1958.

To receive the unconditional surrender of half a million enemy soldiers, sailors and airmen must be an event which happens to few people in the world. I was very conscious that this was the greatest day of my life.
> Admiral Earl Mountbatten, 12 September 1945, upon the Japanese surrender in Singapore, *Personal Diary of Admiral the Lord Louis Mountbatten*, 1988.

There is almost always a way out, even of an apparently hopeless position, if the leader makes up his mind to face the risks.
> General Paul von Lettow-Vorbeck, *East African Campaigns*, 1957.

I was convinced that an effective way really to impress on the Japanese that they had been beaten in the field was to insist on this ceremonial surrender of swords. No Japanese soldier, who had seen his general march up and hand over his sword, would ever doubt that the Invincible Army was invincible no longer. We did not want a repetition of the German First War legend of an unconquered army. With this in mind, I was dismayed to be told that General MacArthur in his over-all instruction for the surrender had decided that the 'archaic' ceremony of the surrender of swords was not to be enforced. I am afraid I disregarded his wishes. In Southwest Asia all Japanese officers surrendered their swords to British officers of similar or higher rank; the enemy divisional and army commanders handed theirs in before large parades of their already disarmed troops. Field-Marshal Terauchi's sword is in Admiral Mountbatten's hands; General Kimura's

THE SWORD AND THE PEN

is on my mantelpiece, where I always intended that one day it should be.
> Field Marshal Viscount Slim of Burma, *Defeat Into Victory*, 1956.

THE SWORD AND THE PEN

It is said that the warrior's is the twofold Way of pen and sword, and he should have a taste for both Ways.
> Miyamoto Musashi, *A Book of Five Rings*, 1645, tr. Victor Harris, 1984.

I would rather have written that poem, gentlemen, than take Quebec tomorrow.
> Major-General Sir James Wolfe, 12 September 1759, referring to Grey's 'Elegy' on the day before his death in victory at the Battle of Quebec.

It is not the business of naval officers to write books.
> Rear-Admiral F.M. Ramsay, 1893, endorsement on the unfavourable fitness report made on Alfred Thayer Mahan.

It is often said that a man who writes well cannot be a good soldier; most of the great commanders, from King David, Xenophon, and Caesar to Wellington, not only wrote well, but extremely well.
> Field Marshal Viscount Wolseley, *The Story of a Soldier's Life*, II, 1903.

Our regulations governing the publication by soldiers of their views on military matters are veritable Lettres de Cachet, consigning the intellects of our Service to the Bastille of ignorance.
> General Sir Ian Hamilton, *The Soul and Body of an Army*, 1921.

The written essay or appreciation is a good test of character and grasp of the principles of war. It is apt to explode the legend of the 'strong, silent man' – who is usually silent because his mind is so hazy that he fears to commit himself to the risk of logical argument. The man who writes gives proof that at any rate he possesses some knowledge, whereas it is quite a possibility that the mind of the inarticulate one may be a military vacuum. Further, arguments that would pass muster in verbal conferences are easily seen to be lacking in logic when written, for cold print is a merciless exposer of mental fog. (March 1923.)
> Captain Sir Basil Liddell Hart, *Thoughts on War*, 1944.

There is one quality above all which seems to me essential for a good commander, the ability to express himself clearly, confidently, and concisely, in speech and on paper... My experience of getting on for fifty years' service has shown me that it is a rare quality amongst Army Officers, to which not nearly attention is paid in their education. It is one which can be acquired, but seldom is, because it is seldom taught.
> Field Marshal Viscount Wavell of Cyrenaica, *Soldiers and Soldiering*, 1953.

The notion that the pen is mightier than the sword is a fantasy. Try waving a book at the man who comes after you with a machete or a gun. Yet the pen can inform the sword.
> Lieutenant-Colonel Ralph Peters, *Fighting for the Future*, 1999.

T

TACTICS

To ensure that your whole host may withstand the brunt of the enemy's attack and remain unshaken – this is effected by maneuvers direct and indirect.

That the impact of your army may be like a grindstone dashed against an egg – that is effected by science of weak and strong points.

In all fighting, the direct methods may be used for joining battle, but indirect methods will be needed in order to secure victory.

Indirect tactics, efficiently applied, are inexhaustible as Heaven and Earth, unending as the flow of rivers and streams; like the sun and moon, they end but to begin anew; like the four seasons, they pass but to return once more.

There are not more than five musical notes, yet the combinations of these five give rise to more melodies than can ever be heard. There are not more than five primary colors, yet in combination they produce more hues than can ever be seen. There are not more than five cardinal tastes, yet combinations of them yield more flavours than can ever be tasted.

In battle, there are not more than two methods of attack – the direct and indirect; yet these two in combination give rise to an endless series of maneuvers. The direct and indirect lead on to each other in turn. It is like moving in a circle – you never come to an end. Who can exhaust the possibilities of their combinations?

Sun Tzu, *The Art of War*, 5, c. 500 BC, tr. Giles, 1910.

In tactics every engagement, great or small, is a defensive one if we leave the initiative to the enemy.

Major-General Carl von Clausewitz, *On War*, 1832, tr. Michael Howard and Peter Paret, 1976.

Tactics is an art based on the knowledge of how to make men fight with maximum energy against fear, a maximum which organization alone can give.

Colonel Charles Ardant du Picq, *Battle Studies*, 1880, tr. Greely, 1957.

I perceived at once, however, that Hardee's tactics – a mere translation from the French with Hardee's name attached – was nothing more than common sense... I found no trouble in giving commands that would take my regiment where I wanted it to go... I do not believe that the officers of the regiment ever discovered that I had never studied the tactics that I used.

General of the army Ulysses S. Grant, *Personal Memoirs*, 1885.

In war we resemble a man endeavoring to seek an enemy in the dark, and the principles which govern our action will be similar to those which he would naturally adopt. The man stretches out one arm in order to grope for his enemy (Discover). On touching his adversary he feels his way to the latter's throat (Reconnoiter). as soon as he has reached it, he seizes him by the collar or throat so that his antagonist cannot wriggle away or strike back at him effectively (Fix). Then with his other fist he strikes his enemy, who is unable to

TACTICS

avoid the blow, a decisive knock-out blow (Decisive attack). Before the enemy can recover, he follows up his advantage by taking steps to render him finally powerless (Exploit).

>Captain Sir Basil Liddell Hart, 'The Essential Principles of War...', *United Services Magazine*, 4/1920.

Nine-tenths of tactics are certain, and taught in books; but the irrational tenth is like the kingfisher flashing across the pool and that is the test of generals. It can only be ensured by instinct, sharpened by thought practising the stroke so often that at the crisis it is as natural as a reflex.

>Colonel T.E. Lawrence, 'The Science of Guerrilla Warfare', *Encyclopaedia Britannica*, 1929.

Tactics are the cutting edge of strategy, the edge which chisels out the plan into an action; consequently the sharper this edge is the clearer will be the result.

>Major-General J.F.C. Fuller, *Grant and Lee, A Study in Personality and Generalship*, 1933.

There is only one tactical principle which is not subject to change. It is: to use the means at hand to inflict the maximum amount of wounds, death and destruction on the enemy in the minimum of time.

>General George S. Patton, Jr, *War As I Knew It*, 1947.

It seems also that he who devises or develops a new system of tactics deserves special advancement on the military honor role of fame. All tactics since the earliest days have been based on evaluating an equation in which x=mobility, y=armour, and z=hitting power. Once a satisfactory solution has been found and a formula evolved, it tends to remain static until some thinking soldier (or possibly civilian) recognizes that the values of x,y,z have been changed by the progress of inventions since the last formula was accepted and that a new formula and new systems of tactics are required.

>Field Marshal Viscount Wavell of Cyrenaica, *Soldiers and Soldiering*, 1953.

On the opening page of his great work *On War*, Clausewitz makes a very simple yet profound remark. It is, that 'War is nothing but a duel on an extensive scale', and he likens it to a struggle between two wrestlers; between two pugilists would be a more apt comparison. If so, then the primary elements of tactics are to be seen in their simplest form in a fight between two unarmed men. They are: to think, to guard, to move and to hit.

Before a bout opens, each man must consider how best to knock out his adversary, and though as the fight proceeds he may be compelled to modify his means, he must never abandon his aim. At the start he must assume a defensive attitude until he has measured up his opponent. Next, he must move under cover of his defence towards him, and lastly by foot-play, and still under cover of his defence, he must assume the offensive and attempt to knock him out. In military terms, the four primary tactical elements are: the aim or object, security, mobility and offensive power.

If two pugilists are skilled in their art, they will recognize the value of three accentuating elements. They will economize their physical force, so as not to exhaust themselves prematurely; they will concentrate their blows against the decisive point selected, the left or right of their opponents jaw, or his solar plexus, and throughout will attempt to surprise him – that is, take him off-guard, or do something which he does not expect, or cannot guard against. In military terms these accentuating elements are: economy of force, concentration of force, and surprise.

TACTICS AND STRATEGEY

TACTICS AND STRATEGY

All men can see these tactics whereby I conquer, but what none can see is the strategy out of which victory evolved.
Sun Tzu, *The Art of War*, c. 500 BC.

Sir, my strategy is one against ten, my tactics are ten against one.
The Duke of Wellington (1769–1852).

The conduct of war... consists in the planning and conduct of fighting... It consists of a greater or lesser number of single acts each complete in itself, which... are called 'engagements' and which form new entities. This gives rise to the completely different activity of planning and executing these engagements themselves, and of coordinating each of them with the others in order to further the object of the war. One has been called tactics, and the other strategy.
  According to our classification, then, tactics teaches the use of armed forces in the engagement; strategy, the use of engagements for the object of the war.
Major-General Carl von Clausewitz, *On War*, 2.1, 1832, tr. Michael Howard and Peter Paret, 1976.

Strategy is the art of making war upon the map, and comprehends the whole theater of operations. Grand tactics is the art of posting troops upon the battlefield according to the characteristics of the ground, of bringing them into action, and of fighting upon the ground, in contradiction to planning upon a map. Logistics comprises the means and arrangements which work out the plans of strategy and tactics. Strategy decides where to act; logistics brings the troops to this point; grand tactics decides the manner of execution and the employment of the troops.
Major-General J.F.C. Fuller, *The Generalship of Alexander the Great*, 1958.

Lieutenant-General Antoine-Henri Baron de Jomini, *Summary of the Art of War*, 1838, tr. Mendell and Craighill, 1862.

Strategy furnishes tactics with the opportunity to strike and with the prospect of success, through its conduct of the armies and of their concentration on the field of battle. On the other hand, however, it accepts the results of every single engagement, and builds upon them. Strategy retires when a tactical victory is in the making, in order to exploit the newly created situation.
Field Marshal Helmuth Graf von Moltke, quoted in Hermann Foertsch, *The Art of Modern War*, 1940.

A tactical success is only really decisive, if it is gained at the strategically correct spot.
Field Marshal Helmuth Graf von Moltke, quoted in Hermann Foertsch, *The Art of Modern War*, 1940.

...Contact (a word which perhaps better than any other indicates the dividing line between strategy and tactics)...
Rear-Admiral Alfred Thayer Mahan, *The Influence of Sea Power Upon History*, 1890.

The greater the certainty with which the success of the main force attack can be predicted in advance, the easier the task set that main force will be. One must not count on the heroism of the troops. The strategy must ensure that the tactical task is a readily feasible one.
Marshal of the Soviet Union Mikhail N. Tukhachevskiy, 1924, quoted in Richard Simpkin, *Deep Battle*, 1987.

As regards the relation of strategy to tactics, while in execution the borderline is often shadowy, and it is difficult to decide exactly where a strategical movement ends and a tactical movement begins, yet in conception the two are distinct. Tactics lies in, and fills the province of, fighting. Strategy not only

stops on the frontier, but has for its purpose the reduction of fighting to the slenderest proportions. (Oct. 1928.)
> Captain Sir Basil Liddell Hart, *Thoughts on War*, 1944.

In peace we concentrate so much on tactics that we are apt to forget that it is merely the handmaiden of strategy. (Sept. 1930.)
> Captain Sir Basil Liddell Hart, *Thoughts on War*, 1944.

So far as I can see strategy is eternal, and the same and true: but tactics are the ever-changing languages through which it speaks. A general can learn as much from Belisarius as from Haig – but not a soldier. Soldiers have to know their means.
> Colonel T.E. Lawrence, Whit Monday 1933, quoted in *T.E. Lawrence to His Biographers*, 1963.

Tactics is the art of handling troops on the battlefield, strategy is the art of bringing forces to the battlefield in a favourable position.
> Field Marshal Viscount Wavell of Cyrenaica, quoted in David Chandler, *The Atlas of Military Strategy*, 1980.

I rate the skilful tactician above the skilful strategist, especially him who plays the bad cards well.
> Field Marshal Viscount Wavell of Cyrenaica, *Soldiers and Soldiering*, 1953.

…Definition of strategy as – 'the art of distributing and applying military means to fulfill the ends of policy'. For strategy is concerned not merely with the movement of forces – as its role is often defined – but with the effect. When the application of the military instrument merges into actual fighting, the dispositions for and control of such direct action are termed 'tactics'. The two categories, although convenient for discussion, can never be truly divided into separate compartments because each not only influences but merges into the other.
> Captain Sir Basil Liddell Hart, *Strategy*, 1954.

In strategy… calculation is simpler and a closer approximation to truth possible than in tactics. For in war the chief incalculable is the human will, which manifests itself in resistance, which in turn lies in the province of tactics. Strategy has not to overcome resistance, except from nature. Its purpose is to diminish the possibility of resistance, and it seeks to fulfil this purpose by exploiting the elements of movement and surprise.
> Captain Sir Basil Liddell Hart, *Strategy*, 1954.

TALENT FOR WAR

The best general is not the man of noble family, but the man who can take pride in his own deeds.
> The Emperor Maurice, *The Strategikon*, c. AD 600 (*Maurice's Strategikon*, tr. George Dennis, 1984).

Unless a man is born with a talent for war, he will never be other than a mediocre general. It is the same with all talents; in painting, or in music, or in poetry, talent must be inherent for excellence. all sublime arts are alike in this respect. that is why we see so few outstanding men in science. Centuries pass without producing one. application rectifies ideas but does not furnish a soul for that is the work of nature.
> Field Marshal Maurice Comte de Saxe, *My Reveries*, 1732, tr. Phillips, 1940.

It is bad to lack good fortune, but it is a misfortune to lack talent… The fortune of war is on the side of the soldier of talent.
> Field Marshal Aleksandr V. Suvorov (1729–1800), quoted in K. Ossipov, *Suvorov*, 1945.

TANKS

War is a serious sport, in which one can endanger his reputation and his country: a rational man must feel and know whether or not he is cut out for this profession.

Napoleon, to Prince Eugene, 30 April 1809, *Correspondance de Napoléon Ier, publiée par ordre de l'Empereur Napoléon III*, No. 15144, Vol. XVIII, 1858–1870, p. 525.

No written examination is a test of character, personality or leadership. Examinations are not a true test of ability and are not test at all of character or leadership. Some shine at examinations, others fail to disclose their knowledge of the subject. Brains of themselves are only of academic value. I have seen so many scholars fail in life through lack of application, character and personality... Entrance examinations into the Army used to be particularly harmful. How can written examinations be a test of good officership?... Academic knowledge without character and personality is useless. The boy with average intelligence, with drive, character and practical ability – I knew so many of them who have had brilliant careers, leaving scholars behind. Scholars seldom enter the Army, preferring professions or politics. Successful soldiers have not been scholars, the one exception being Lord Wavell. I can think of no soldier who has reached the top of the ladder in recent years whose main assets have been scholarly ability. Our leading soldiers in both World Wars reached the top through character, personality, determination and leadership. All the very necessary and arduous brainwork is done by scholarly staff officers.

Colonel Richard Meinertzhagen, *Army Diary 1899–1926*, 1960.

Everyone is born for something, whether it is to be a soldier or a journalist. I was born to be a soldier. Even in my horoscope, which I do not believe in. I am an aries born in the year of the tiger – therefore, a warrior twice over. I spent 26 years in the army and it felt as if it was a day. It was not heavy work. I was never afraid. I did not mind the mud, or the cold, or the blood.

Lieutenant-General Aleksandr I. Lebed, *El Pais* (Madrid), 15 March 1998.

TANKS

I managed to get astride one of the German trenches... and opened fire with Hotchkiss machine-guns. There were some Germans in the dug-outs and I shall never forget the look on their faces when they emerged...

Captain H.W. Mortimore, 15 September 1916, commander of the first tank in action, quoted in Trevor Wilson, *The Myriad Faces of War*, 1986.

Employment of tanks in mass is our greatest enemy.

General Erich Ludendorff, quoted in Richard Simpkin, *Race to the Swift*, 1985.

That the application of petrol to land warfare will prove as great a step in tactics as that of steam in naval warfare.

That the characteristics of security, offensive power and mobility, which the tank combines in a higher degree than any single other weapon, are those which are fundamental to success in war.

That the tank enables the main advantages of sea warfare, unrestricted movement, to be to a great extent superimposed on that of land warfare...

That the application of machinery to land warfare is as great a saver of man-power as its application to manufacture. That is: the tank does not create another man-power problem, but is a solution to existing problems.

Major-General J.F.C. Fuller, memorandum of 1918, *Memoirs of an Unconventional Soldier*, 1936.

But it is not, of course, enough that the

TANKS

Tank offers protection to those who fight in it. A trench or a hole in the ground will do the same. But the Tank is essentially a mobile weapon of offense. It is the weapon for the nation which does not fight willingly, but when it fights, fights to win, and to win quickly with as little bloodshed as possible. It is the weapon for men who, if they must fight, like to fight like intelligent beings still subjecting the material world to their will, and who are most unwillingly reduced to the roles of mere marching automata, bearers of burdens and diggers of the soil, roles from which the patient German did not seem adverse.
    Major Clough Williams-Ellis, *The Tank Corps*, 1919.

The tank marks as great a revolution in land warfare as an armored steamship would have marked had it appeared amongst the toilsome triremes of Actium.
    General Sir Ian Hamilton, *The Soul and Body of an Army*, 1921.

'It was Mr. Churchill who, as First Lord of the Admiralty, gave the first order for eighteen tanks, or 'landships' as they were then called, on March 26, 1915. He did not inform either the War Office or the Treasury – an almost unprecedented and certainly unconstitutional reticence, dictated by fear that conventional minds might stifle a great idea.'
    Sir Winston S. Churchill, quoted in Colin Coote, ed., *Maxims and Reflections*, 1949.

Once appreciate that tanks are not an extra arm or mere aid to infantry but the modern form of heavy cavalry and their true military use is obvious – to be concentrated and used in as large masses as possible for a decisive blow against the Achilles' heel of the enemy army, the communications and command centres which form its nerve system.
    Captain Sir Basil Liddell Hart, writing in the *Westminster Gazette*, 1925.

...The internal-combustion engine which is ready to carry whatever one wants, wherever it is needed, at all speeds and distances... the internal-combustion engine which, if it is armoured, possesses such a fire power and shock power that the rhythm of the battle corresponds to that of its movements.
    General Charles de Gaulle, *The Army of the Future*, 1934.

On the other side of the coin, a group of comrades looked not a little askance at the introduction into the army of tanks in mass. They were scared of rain, snow, autumn, spring and the like. Some comrades said that, because of our roads and climatic conditions, tanks would only be able to operate 'for something like six weeks in the year.' However the tank's talent for lively action soon stood the misgivings of these timid theorists on their heads. Tanks perform perfectly well in summer, winter, spring, and autumn.
    Marshal of the Soviet Union Mikhail N. Tukhachevskiy, Red Army Field Service Regulations (1936), quoted in Richard Simpkin, *Deep Battle*, 1987.

Until our critics can produce some new and better method of making a successful land attack other than self-massacre, we shall continue to maintain our belief that tanks – properly employed, needless to say – are today the best means available for a land attack.
    Colonel-General Heinz Guderian, *Achtung! Panzer!*, 1937.

(The tank) is therefore the weapon of potentially decisive attack. Mobility and firepower will only be exploited to the full if the attack achieves deep penetration and the armoured force, having broken out, can go over to the pursuit

TANKS

...The higher the concentration of tanks, the faster, greater and more sweeping will be the success – and the smaller our own losses... Tanks must attack with surprise, and as far as possible where the enemy is known or presumed to be weak... The tank needs supporting arms which complement it and can go everywhere with it... Even in defence, the tank must be employed offensively. Concentration is even more important here, so that the enemy's superiority can be offset at least in one spot.
>   Colonel-General Heinz Guderian, *Panzer Marsch!*, 1937, quoted in Richard Simpkin, *The Race to the Swift*, 1985.

The shorter the battle, the fewer men will be killed and hence the greater their self-confidence and enthusiasm. To produce a short battle, tanks must advance rapidly but not hastily... Mobile forces should be used in large groups and [be] vigorously led. They must attempt the impossible and dare the unknown.
>   General George S. Patton, Jr, 1939, after reading an article by Heinz Guderian on tanks, *The Patton Papers, II*, 1974.

Although the tactical consequences of motorisation and armour had been pre-eminently demonstrated by British military critics, the responsible British leaders had not taken the risk either of using this hitherto untried system as a foundation for peacetime training, or of applying it in war.
>   Field Marshal Erwin Rommel, *The Rommel Papers*, 1953.

The officers of a panzer division must learn to think and act independently within the framework of the general plan and not wait until they receive orders.
>   Field Marshal Erwin Rommel, *The Rommel Papers*, 1953.

It was principally the books and articles of the Englishmen, Fuller, Liddell Hart, and Martel, that excited my interest and gave me food for thought. These far-sighted soldiers were even then trying to make of the tank something more than just an infantry support weapon. They envisaged it in relationship to the growing motorization of our age, and thus they became the pioneers of a new type of warfare on the largest scale.

I learned from them the concentration of armor, as employed in the battle of Cambrai. Further, it was Liddell Hart who emphasized the use of armored forces for long-range strokes, operations against the opposing army's communications, and also proposed a type of armored division combining panzer and panzer-infantry units. Deeply impressed by these ideas I tried to develop them in a sense practicable for our own army. So I owe many suggestions of our further development to Captain Liddell Hart.
>   Colonel-General Heinz Guderian, *Panzer Leader*, 1953.

I was brought up in the Israeli Army which no doubt, was very much influenced by your unorthodoxic [sic] school of thought, and of course I am strongly in favor of your ideas.
>   General Ariel Sharon, letter to Liddell Hart.

While we are emphasizing making ourselves more deployable, this recent action in Iraq shows that when we have to shift to high-intensity combat our heavy equipment really does make a difference. When an Abrams tank turns up on a corner, everybody takes notice.
>   Major-General Buford 'Buff' Blount III, Commander, 3rd Infantry Division, quoted in the *Atlanta Journal Constitution*, 7 September 2003, '1000 Brits Could Die,' Mirror.co.uk

Armor – The Combat Arm of Decision.

Motto of the US Army Armor Branch.

Fear naught.
Motto of the British Royal Tank Corps.

TEAM/TEAMWORK

Nestor, the mission stirs my fighting blood.
I'll slip right into enemy lines at once –
these Trojans, camped at our flank.
If another comrade would escort me, though,
there'd be more comfort in it, confidence too.
When two work side-by-side, one or the other spots the opening first if a kill's at hand.
When one looks out for himself, alert but alone,
his reach is shorter – his sly moves miss the mark.
> Homer, *The Iliad*, 10.258–266, c. 800 BC, tr. Robert Fagles, 1990; King Diomedes of Argos, asking for a companion when he volunteered to reconnoitre the Trojan camp.

Four brave men who do not know each other will not dare to attack a lion. Four less brave, but knowing each other well, sure of their reliability and consequently of their mutual aid, will attack resolutely.
> Colonel Charles Ardant du Picq, *Battle Studies*, 1880, tr. Greely, 1957.

But after all, no leadership or expert naval direction could be successful unless it were supported by the whole body of officers and men of the Navy. It is upon these faithful, trusty servants in the great ships and cruisers that the burden falls directly day after day.
> Sir Winston S. Churchill, speech in the House of Commons, 18 December 1939, *Blood, Sweat, and Tears*, 1941.

An Army is a team. It eats, sleeps, lives and fights as a team. All this stuff you've been hearing about individuality is a bunch of crap. The billious bastard who wrote that kind of stuff for the *Saturday Evening Post* didn't know any more about real battle than he did about fucking…
…Every man, every department, every unit, is important to the vast scheme of things. The ordnance men are needed to supply the guns, the QM to bring up the food and the clothes for us, for where we are going, there isn't a hell of a lot for us to steal. Every man in the mess hall, even the one who heats the water to keep us from getting diarrhea, has a job to do. Even the chaplain is important, for if we get killed and he isn't there to bury us, we would all go to hell.
> General George S. Patton, Jr, speech to the Third Army, June 1944, 'A General Talks to His Army', 1944.

TEMPO

Defeat the enemy with cold steel, bayonets, swords, and pikes… Don't slow down during an attack. When the enemy is broken, shattered, then pursue him at once, and don't give him time either to collect or re-form. If he surrenders, spare him; only order him to throw down his arms. During the attack call on the enemy to surrender… Spare nothing, don't think of your labours; pursue the enemy night and day, so long as anything is left to be destroyed.
> Field Marshal Prince Aleksandr V. Suvorov, *The Science of Victory*, 1796.

Keep going all the time and always forward.
> Major-General John A. Lejeune, personal motto, *The Public Ledger* (Philadelphia), 1 August 1920.

A commander must accustom his staff to a high tempo from the outset, and continuously keep them up to it. If he once allows himself to be satisfied with norms, or anything less than an all-out

TENACITY

effort, he gives up the race from the starting post, and will sooner or later be taught a bitter lesson.
  Field Marshal Erwin Rommel, *The Rommel Papers*, 1953.

It is physically impossible to sustain a high tempo of activity indefinitely, although clearly there will be times when it is advantageous to push men and equipment to the limit. Thus, the tempo of war will fluctuate – from periods of intense activity to periods in which activity is limited to information gathering, replenishment, or redeployment. Darkness and weather can influence the tempo of war but need not halt it. A competitive rhythm will develop between the opposing wills, with each belligerent trying to influence and exploit tempo and the continuous flow of events to suit his purposes.
  General Alfred M. Gray and Major John Schmitt, *Warfighting*, US Marine Corps, 1989.

It is complex. But all wars are complex. We perhaps have been lulled into something of a false sense of security over the last few years that somehow the application of military force can act rather like a magic wand, without penalty. It is not always like that. Armies cannot move forever... believe it or not, soldiers need a bit of sleep from time to time. It is a pause while people get themselves sorted out for what comes next.
  General Sir Michael Jackson, Chief of the General Staff, 30 March 2003, referring to the operational pause by the Coalition ground forces in the drive on Baghdad, quoted in '1,000 Brits Could Die,' Mirror.co.uk.

TENACITY

If the military leader is filled with high ambition and if he pursues his aims with audacity and strength of will, he will reach them in spite of all obstacles.
  Major-General Carl von Clausewitz, *Principles of War*, 1812, tr. Gatzke, 1942.

The enemy hold our front in very strong force, and evince a strong determination to interpose between us and Richmond to the last. I shall take no backward steps...
  General of the Army Ulysses S. Grant, letter to Major-General Halleck during the Battle of Spotsylvania, 10 May 1864, quoted in Horace Porter, *Campaigning With Grant*, 1897.

I have seen your despatch expressing your unwillingness to break your hold where you are. Neither am I willing. Hold on with a bulldog grip, and chew and choke as much as possible.
  President Abraham Lincoln, 17 August 1864, dispatch to General Grant on the James River, supporting Grant's refusal of General Halleck's suggestion that he detach troops to help quell expected disturbances over another draft call, quoted in Horace Porter, *Campaigning With Grant*, 1897.

Tenacity of purpose and untiring energy in execution can repair a first mistake and baffle deeply laid plans.
  Rear-Admiral Alfred Thayer Mahan (1840–1914).

Victory will come to the side that outlasts the other.
  Marshal of France Ferdinand Foch, 7 September 1914, order during the First Battle of the Marne.

I knew my ground, my material and my allies. If I met fifty checks I could yet see a fifty-first way to my object.
  Colonel T.E. Lawrence (1888–1935).

Once an action is begun, it should be carried through regardless of difficulties. If the going is hard, one should think of the enemy's losses and his fear, and so carry on to victory. A commander should not ignore the

TERRAIN

possibility of help coming to the enemy, nor should he be overly worried about it. An action begun and broken off when tenaciousness might have secured victory only breaks down morale. To persist and win through reveals a commander's strong will to win. On the other hand, when an action has been ordered, or even initiated, and circumstances so change the situation as to make it hopeless, a reversal of the order reveals a commander's versatility.
    General Lin Piao, 1946, quoted in Ebon, *Lin Piao*, 1970.

Positions are seldom lost because they have been destroyed, but almost invariably because the leader has decided in his own mind that the position cannot be held.
    General Alexander A. Vandegrift (1887–1973), quoted in *Warfighting*, Fleet Marine Force Manual (FMFM) 1, 1990.

TERRAIN

The natural formation of the country is the soldier's best ally.
    Sun Tzu, *The Art of War*, 10, c. 500 BC, tr. Giles, 1910.

The nature of the ground is often of more consequence than courage.
    Flavius Vegetius Renatus, *The Military Institutions of the Romans*, c. AD 378, tr. Clark, 1776.

A general should possess a perfect knowledge of the localities where he is carrying on a war.
    Niccolò Machiavelli, *Discourses*, 1517.

Knowledge of the country is to a general what a rifle is to an infantryman and what the rules of arithmetic are to a geometrician. If he does not know the country he will do nothing but make gross mistakes. Without this knowledge his projects, be they otherwise admirable, become ridiculous and often impracticable. Therefore study the country where you are going to act!
    Frederick the Great, *Instructions to His Generals*, 1747, tr. Phillips, 1940.

As many different terrains as there are in existence, just so many different battles will there be.
    Frederick the Great (1712–1786), quoted in Hugo von Freytag-Loringhoven, *Generalship in the Great War*, 1920.

Where a deer can cross, a soldier can cross.
    Field Marshal Prince Aleksandr V. Suvorov (1729–1800), quoted in K. Ossipov, *Suvorov*, 1945.

Maxim 104. An army can march anywhere and at any time of the year, wherever two men can place their feet.
    Napoleon, *The Military Maxims of Napoleon*, tr. Burnod, 1827.

Terrain difficulties can be overcome only by a special faculty we sometimes call 'a sense of locality [ortsinn].' It is an ability to form an accurate mental picture of the country, and readily orient one's self with it. This sense of locality is largely imagination, coming in part from actual visual impressions, and part from a mental process which uses knowledge gained from study and experience to fill in missing points and make a complete whole of the fragments determined by physical sight. The power of imagination then brings this whole picture before the minds eye, makes it into a picture, or a mental map, and maintains it permanently present and unbroken.
    Major-General Carl von Clausewitz, *On War*, 1.3, 1832, tr. Michael Howard and Peter Paret, 1976.

Every day I feel more and more in need of an atlas, as the knowledge of the geography in its minutest details is essential to a true military education.
    General of the Army William T. Sherman,

quoted in Chandler, ed., *Atlas of Military Strategy*, 1980.

Natural hazards, however formidable are inherently less dangerous and uncertain than fighting hazards.
Captain Sir Basil Liddell Hart, *Strategy*, 1954.

TERRORISM

Terrorism can be commendable, and it can be reprehensible. The terrorism we practice is of the commendable kind for it is directed at the tyrants and the aggressors and the enemies of Allah.
Osama bin Ladin to journalist John Miller, 1998.

It is the duty of Muslims to prepare as much force as possible to terrorize the enemies of God.
Osama bin Ladin, 1998, in statement titled 'The Nuclear Bomb of Islam'.

There are two types of terror, good and bad. What we are practising is good terror. We will not stop killing them and whoever supports them.
Osama bin Ladin, undated video, reported on 11 November 2001 by the *Daily Telegraph*.

This terrorism is a divine commandment. Allah has said: 'Make ready for them whatever you can of armed strength and of mounted pickets at the frontier, whereby you may daunt the enemy of Allah and your enemy and others beyond them whom you know not' [Koran 8:60]. Striking horror, panic, and fear in the hearts of the enemies of Allah is a divine commandment, and the Muslim has in this matter two choices: Either he believes in these verses, which are clear, or he denies these verses, and [becomes] an infidel. The Muslim has no other option.
Abu Hafs the Mauritanian, senior al-Qaida leader fighting in Chechnya, interviewed in *Al Jazeera*, 30 November 2001.

The purpose of terrorism lies not just in the violent act itself. It is in producing terror. It sets out to inflame, to divide, to produce consequences which they then use to justify further terror.
Prime Minister Tony Blair, speech to the House of Commons, 18 March 2003, his final justification for war with Iraq.

THEORY AND PRACTICE

Every art has its rules and maxims; they must be studied. Theory facilitates practice. The lifetime of one man is not sufficiently long to enable him to acquire perfect knowledge and experience; theory helps to supplement it; it provides youth with early experience and makes him skilful also through the mistakes of others. In the profession of war the rules of the art are never transgressed without punishment from the enemy.
Frederick the Great, Oeuvres, 29. 58–59, quoted in Azer Gat, *The Origins of Military Thought*, 1989.

Pity the theory that conflicts with reason! No amount of humility can gloss over the contradiction; indeed, the greater the humility, the sooner it will be driven off the field of real life by ridicule and contempt.
Major-General Carl von Clausewitz, *On War*, 2.2, 1832, tr. Michael Howard and Peter Paret, 1976.

Theory becomes infinitely more difficult as soon as it touches the realm of moral values.
Major-General Carl von Clausewitz, *On War*, 2.2, 1832, tr. Michael Howard and Peter Paret, 1976.

Theory exists so that one need not start afresh each time sorting out the material and plowing through it, but will find it ready to hand and in good order. It is meant to educate the mind of the future commander or, more accurately, to guide him in his self-education, not to accompany him to the battlefield.

THOSE LEFT BEHIND

Major-General Carl von Clausewitz, *On War*, 2.2, 1832, tr. Michael Howard and Peter Paret, 1976.

One of the surest ways of forming good combinations in war would be to order movements only after obtaining perfect information of the enemy's proceedings. In fact, how can any man say what he should do himself, if he is ignorant of what his adversary is about? Even as it is unquestionably of the highest importance to gain this information, so it is a thing of the utmost difficulty, not to say impossibility. This is one of the chief causes of the great difference between the theory and the practice of war.
 Lieutenant-General Antoine-Henri Baron de Jomini, *Summary of the Art of War*, 1838, tr. Mendell and Craighill, 1862.

If theory went wrong, it is due to the fact that very few theorists had seen war...
 General Mikhail I. Dragomirov (1830–1905), quoted in Ferdinand Foch, *Principles of War*, 1913.

For us theory is the basis of confidence in the actions being taken...
 V.I. Lenin (1870–1924).

To those comrades who wish to build in military affairs by means of the Marxist method I recommend that they review our field statutes in this light and indicate just what changes – from the standpoint of Marxism – should be introduced into the rules for gathering intelligence, for securing one's lines, for artillery preparation, or for attack. I should very much like to hear at least a single new word in this sphere arrived at through the Marxist method – not just 'an opinion or so' but something new and practical.
 Leon Trotksy, 8 May 1922, opening remarks to the Military Scientific Society, *Military Writings*, 1969.

Due to the facility with which men, material, and even situations may be created on paper, many writers produce theses on war as admirable as they are impractical. Due also to our school experiences where such intangible factors as morale, training, discipline, fatigue, equipment, and supply cannot be considered, we adopt the simple course of first omitting reference to these factors, and later of assuming them all satisfactory...
 Duped by the historians who explain defeat and enhance victory by assuring us that both result from the use of PERFECT armies, the fruit of super-thinking, we never stop to consider the inaptness of the word perfect as a definition for armies.
 General George S. Patton, Jr, January 1928, *The Patton Papers, I*, 1972.

I have nothing to say against theoretical training, and certainly nothing against practical training. Whoever would become master of his craft must have served as apprentice and journeyman; only a genius can bridge gaps in this sequence of instruction. Every man of action is an artist, and he must know the material with which, in which, and against which he works before he begins his task.
 Colonel-General Hans von Seekt, *Thoughts of a Soldier*, 1930.

THOSE LEFT BEHIND

So his whole head was dragged down in
 the dust.
And now his mother began to tear her
 hair...
she flung her shining veil to the ground
 and raised
a high, shattering scream, looking down
 at her son.
Pitifully his loving father groaned and
 round the king
his people cried with grief and wailing
 seized the city –
for all the world as if all Troy were
 torched and smoldering

down from the looming brows of the
  citadel to her roots.
Priam's people could hardly hold the
  old man back,
frantic, mad to go rushing out the
  Dardan Gates.
He begged them all, groveling in the
  filth,
crying out to them, calling each man by
  name,
'Let go, my friends! Much as you care
  for me,
let me hurry out of the city, make my
  way,
all on my own, to Achaea's waiting
  ships!
I must implore that terrible, violent
  man...
Perhaps – who knows? – he may respect
  my age,
may pity an old man. He has a father
  too,
as old as I am – Peleus sired him once,
Peleus reared him to be the scourge of
  Troy
but most of all to me – he made my life
  a hell.
So many sons he slaughtered, just
  coming into bloom...
and stabbing grief for him will take me
  down to Death –
my Hector – would to god he had
  perished in my arms!
Then his mother who bore him – oh so
  doomed,
she and I could glut ourselves with grief.'
  Homer, *The Iliad*, 22.477–503, *c.* 800 BC, tr.
  Robert Fagles, 1990; the grief of King
  Priam of Troy, having just witnessed the
  death of his son Hector at the hands of
  Achilles beneath the walls of the city.

Those who are not here are eating earth that we might live. Are we to forget the sorrowing mothers who bore them, and let their younger brothers and sisters go in want, because they gave their lives for our Zululand? Have we not a saying that a grieving mother's heart is soothed by a stomach full of meat? Well then, let us give to the bereaved the reward which the departed warriors would have received had they lived, and let us give a double measure, with both hands, to take the taste of bitter aloes from the mouths of the sorrowing ones.
  Shaka Zulu, 1819, after the Second
  Ndwandwe War, quoted in E.A. Ritter,
  Shaka Zulu, 1955.

July 14, 1861
Camp Clark, Washington
My very dear Sarah:
  The indications are very strong that we shall move in a few days – perhaps tomorrow, and lest I should not be able to write again, I feel impelled to write a few lines that may fall under your eye when I shall be no more...
  I have no misgivings about, or lack of confidence in the cause in which I am engaged, and my courage does not halt or falter. I know how American Civilization now leans on the triumph of the Government, and how great a debt we owe to those who went before us through the blood and suffering of the Revolution. And I am willing – perfectly willing – to lay down all my joys in this life, to help maintain this Government, and to pay the debt...
  Sarah my love for you is deathless, it seems to bind me with mighty cables that nothing but Omnipotence could break; and yet my love of Country comes over me like a strong wind and bears me unresistably with all these chains to the battle field.
  The memories of the blissful moments I have enjoyed with you come crowding over me, and I feel most deeply grateful to God and to you that I have enjoyed them so long. And how hard it is for me to give them up and burn to ashes the hopes of future years, when, God willing, we might still have lived and loved together, and see our boys grown up to honourable manhood, around us. If I do not return, my dear Sarah, never forget how much I

TIME/TIMING

loved you, nor when my last breath escapes me on the battle field, it will whisper your name. Forgive me my faults, and the many pains I have caused you. How thoughtless, how foolish I have often times been!

But, O Sarah! if the dead can come back to this earth and flit unseen around those they love, I shall always be with you; in the brightest day and the darkest night... always, always; and when the soft breeze fans your cheek, it shall be my breath, or the cool air your throbbing temple, it shall be my spirit passing by. Sarah do not mourn me dead; think I am gone and wait for me, for we shall meet again....

> Major Sullivan Ballou, Second Rhode Island Volunteers, letter to his wife in Smithfield, written a week before his death at the Battle of First Bull Run, quoted in Ken Burns, *The Civil War*, 1990.

Wait for Me
Wait for me, and I'll return.
Only wait very hard . . .
Wait, when you are filled with sorrow
As you watch the yellow rain,
Wait, when the wind sweeps the snowdrifts,
Wait, in the sweltering heat,
Wait, when others have stopped waiting,
Forgetting their yesterdays.
Wait, even when from afar no letters come to you,
Wait, even when others are tired of waiting,
Wait, even when my mother and son think I am no more,
And when friends sit around the fire drinking to my memory,
Wait, and do not hurry to drink to my memory too;
Wait for I'll return,
Defying every death,
And let those who do not wait
Say I was lucky.
They will never understand
That in the midst of death,
You with your waiting saved me.
Only you and I will know how I survived, –
It's because you waited,
As no one else did.

> Konstantin Simonov, c. 1943, the most famous Russian poem of WWII.

I don't think there's ever been... even in the history of warfare, been a successful count of the dead... The people who will know [the count] best, unfortunately, are the families that won't see their loved ones come home.

> General H. Norman Schwarzkopf, 27 February 1991, briefing to the press corps, Saudi Arabia.

They are the gold star children, war's innocent victims, and their pain shimmers across the years pure and undimmed. They pass through life with an empty room in their hearts where a father was supposed to live and laugh and love.

All their lives they listen for the footstep that will never fall, and long to know what might have been.

> Lieutenant-General Harold G. Moore, *We Were Soldiers Once... and Young*, 1992.

TIME/TIMING

It is better to be at the right place with ten men than absent with ten thousand.

> Tamerlane (AD 1336–1405), quoted in Harold Lamb, *Tamerlane, The Earth Shaker*, 1928.

Disintegration happens to everything. When a house crumbles, a person crumbles, or an adversary crumbles, they fall apart by getting out of rhythm with the times.

In the art of war on a large scale, it is also essential to find the rhythm of opponents as they come apart, and pursue them so as not to let openings slip by. If you miss the timing of vulnerable moments, there is the likelihood of counterattack.

441

TIME/TIMING

In the individual art of war it also happens that an adversary will get out of rhythm in combat and start to fall apart. If you let such a chance get by you, the adversary will recover and thwart you. It is essential to follow up firmly on any loss of poise on the part of an adversary, to prevent the opponent from recovering.
Miyamoto Musashi, *Book of the Five Spheres* (*The Japanese Art of War*, tr. Thomas Cleary, 1991).

The moment provides the victory. One moment decides the outcome of a battle and one – the success of a campaign. I do not operate in hours, but rather in minutes.
Field Marshal Prince Aleksandr V. Suvorov (1729–1800), quoted in *Voyenny Vestnik*, 11/1986.

Admiral, we have no time to lose. Fortune gave me three days only.
Napoleon, 1 July 1798, to Admiral Brueys at Alexandria when the admiral protested at Napoleon's command to land his army immediately; Nelson's fleet had only departed three days before, quoted in S.J. Watson, *By Command of the Emperor*, 1957.

What a pity it is that I cannot move on for want of grain! My troops are in high health, order and spirits, but the unfortunate defect of arrangement... previous to my arrival has ruined everything... If I had been able to reach the river one day sooner I should have been across before it filled... How true it is that in military operations, time is everything.
The Duke of Wellington, 30 June 1800, to Lieutenant-Colonel Close, after failing by one day to ford a river before the monsoon broke, *The Dispatches of Field Marshal the Duke of Wellington, During His Various Campaigns in India, Denmark, Portugal, Spain, the Low Countries, and France*, Vol. II, 1834–1838.

Go, sir, gallop, and don't forget that the world was made in six days. You can ask me for anything you like, except time.
Napoleon, 11 March 1803, to a courier to the Tsar, quoted in R.M. Johnston, ed., *The Corsican*, 1910.

Time is everything; five minutes make the difference between victory and defeat.
Admiral Viscount Nelson (1758–1805).

The loss of time is irretrievable in war; the excuses that are advanced are always bad ones, for operations go wrong only through delays.
Napoleon, 20 March 1806, to Joseph Bonaparte, *Correspondance de Napoléon Ier, publiée par ordre de l'Empereur Napoléon III*, No. 9997, Vol. XII, 1858–1870.

It may be that in the future I may lose a battle, but I shall never lose a minute.
Napoleon, quoted in Liddell Hart, *Strategy*, 1954.

Strategy is the art of making use of time and space. I am less chary of the latter than of the former; space we can recover, time never.
Napoleon, *Correspondance de Napoléon Ier, publiée par ordre de l'Empereur Napoléon III*, No. 140707, Vol. XVIII, 1858–1870, p. 218.

Force and time in this kind of operation amount to almost the same thing, and each can to a very large extent be expressed in terms of each other.
   A week lost was about the same as a division. Three divisions in February could have occupied the Gallipoli Peninsula with little fighting. Five could have captured it after March 18. Seven were insignificant at the end of April, but nine just might have done it. Eleven might have sufficed at the beginning of July. Fourteen were to prove insufficient on Aug. 7.
Sir Winston S. Churchill, of the Gallipoli campaign in 1915.

TIMIDITY

Our cards were speed and time, not hitting power, and these gave us strategical rather than tactical strength. Range is more to strategy than force.
    Colonel T.E. Lawrence, 'The Science of Guerrilla Warfare', *Encyclopaedia Britannica*, 1929.

Can you fill 'the unforgiving minute?'
    Rudyard Kipling 'If…' (a question General George Patton frequently quoted to himself and others) quoted in Eric Larrabee, *Commander in Chief: Franklin Delano Roosevelt, His Lieutenants, and their War*, 1987, p. 234.

One of the deeper truths of war has been aptly expressed by Lawrence – 'timing in war depends on the enemy as much as upon yourself.' Timetables too often enable the enemy to turn the tables on you. (Aug. 1933.)
    Captain Sir Basil Liddell Hart, *Thoughts on War*, 1944.

The history of failure in war can almost be summed up in two words: too late. Too late in comprehending the deadly purpose of a potential enemy; too late in realizing the mortal danger; too late in preparedness; too late in uniting all possible forces for resistance; too late in standing with one's friends. Victory in war results from no mysterious alchemy of wizardry but entirely upon the concentration of superior force at the critical points of combat. To face an adversary in detail has been the prayer of every conqueror in history.
    General of the Army Douglas MacArthur, 16 September 1940, statement for the Committee to Defend America by Aiding the Allies, *A Soldier Speaks*, 1965.

Death in battle is a function of time. The longer the troops remain under fire, the more men get killed. Therefore, everything must be done to speed up movement.
    General George S. Patton, Jr, April 1943, *The Patton Papers, II*, 1974.

Time is neutral; but it can be made the ally of those who will seize it and use it to the full.
    Sir Winston S. Churchill (1874–1965).

A good solution applied with vigor now is better than a perfect solution ten minutes later.
    General George S. Patton, Jr, *War As I Knew It*, 1947.

A battalion this week can often be more effective than a division next month.
    British Army motto.

TIMIDITY

A general good at commanding troops is like one sitting in a leaking boat or lying under a burning roof. For there is not time for the wise to offer counsel nor the brave to be angry. all must come to grips with the enemy. and therefore it is said that of all the dangers in employing troops, timidity is the greatest and that the calamities which overtake an army arise from hesitation.
    Wu Ch'i (430–381 BC), *Art of War*, 3.2, in appendix I, Sun Tzu, *The Art of War*, c. 500 BC, tr. Samuel Griffith, 1963.

It is essential to be cautious and take your time in making plans, and once you come to a decision to carry it out right away without any hesitation or timidity. Timidity, after all is not caution, but the invention of wickedness.
    The Emperor Maurice, *The Strategikon*, c. AD 600 (*Maurice's Strategikon*, tr. George Dennis, 1984).

He either fears his fate too much,
    or his deserts are small,
That puts it not unto the touch,
    to win or lose it all.
    James Graham, Marquis of Montrose (1612–1650).

TOMMY ATKINS

'Help, danger,' and other figments of the imagination are all right for old women, who are afraid to get off the stove because they may break their legs, and for lazy, luxurious people, and blockheads – for miserable self-protection, which in the end, whether good or bad, in fact, always passes for bravery with the story tellers.
   Field Marshal Prince Aleksandr V. Suvorov (1729–1800), quoted in W. Lyon Blease, *Suvorof*, 1920.

The torment of precautions often exceeds the dangers to be avoided. It is sometimes better to abandon one's self to destiny.
   Napoleon (1769–1821).

Given the same amount of intelligence, timidity will do a thousand times more damage in war than audacity.
   Major-General Carl von Clausewitz, *On War*, 3.6, 1832, tr. Michael Howard and Peter Paret, 1976.

When told by Hamilton of [Douglas] Haig's appointment to take charge of a column, he remarked, 'Haig will do nothing! He's quite all right, but he's too_____cautious: he will be so fixed on not giving the Boers a chance he'll never give himself one. If I were to go to him one evening and offer to land him at daybreak next morning within galloping distance of 1,000 sleeping Boers, I know exactly what he'd do: he'd insist on sending out someone else to make sure the Boers were really there – to make sure no reinforcements were coming up to them, and to make so dead sure in fact, then when he did get there not a single d_____d Boer would be within ten miles of him.
   Colonel Aubrey Woolls-Sampson, comment during the Boer War (1899–1902) on Douglas Haig, who was later to command the British Expeditionary Force (BEF) in WWI, quoted in Liddell Hart, *Through the Fog of War*, 1938.

TOMMY ATKINS

Our God and Souldiers we alike adore;
Ev'n at the Brink of danger; not before:
After deliverance, both alike required;
Our God's forgotten, and our Souldiers slighted.
   Francis Quarles, 'Of Common Devotion', *The Complete Works in Prose and Verse of Francis Quarles*, II, 2, Alexander B. Grosart, ed., 1880.

We are like cloaks – one thinks of us only when it rains.
   Marshal of France Maurice de Saxe, quoted in Samuel Bent, *Familiar Short Sayings of Great Men*, 1887, p. 475.

Thus terminated the war, and with it all remembrance of the veteran's services.
   General Sir William Napier, *History of the War in the Peninsula*, 1828–1840.

For it's Tommy this, an' Tommy that,
   an' 'Chuck him out, the brute!'
But it's 'Saviour of 'is country' when
   the guns begin to shoot;
An' it's Tommy this, an' tommy that,
   an' anything you please;
An' Tommy ain't a bloomin' fool – you
   bet that Tommy sees!
   Rudyard Kipling, 'Tommy Atkins'.

In war time no soldier is free to say what he thinks; after a war no one cares what a soldier thinks.
   General Sir Ian Hamilton, *Ian Hamilton's Dispatches From the Dardenelles*, 1917.

I had finished riding up the New York streets on the back seat of a Cadillac convertible like many before, waving, nodding, and watching torn pieces of paper come down like snow. The police were busy holding back a mob of well-wishers. A middle-aged man with a thin face and graying at the temples broke through the line and grabbed me by the arm. A policeman grabbed him and started to put him behind the line, but I said: 'Wait a minute. I think he wants to

TRADITION

tell me something.'

He did: 'Enjoy it today, my boy, because they won't give you a job cleaning up the streets tomorrow.'
>Colonel Gregory 'Pappy' Boyington, *Baa Baa Black Sheep*, 1958.

TRADITION

...None have inherited a past more crowded with great deeds; and many are heartened by such a heritage and encouraged to care for virtue and prove their gallantry.
>Socrates (470–399 BC), of the Athenians, quoted by Xenophon, *Memorabilia*, 5.3, c. 360 BC, tr. E.C. Marchant, 1923.

Nothing is so disgraceful as slavishness to custom; this is both a result of ignorance and a proof of it.
>Field Marshal Maurice Comte de Saxe, *My Reveries*, 1732, 1732, tr. Phillips, 1940.

Every trifle, every tag or ribbon that tradition may have associated with the former glories of a regiment should be retained, so long as its retention does not interfere with efficiency.
>Colonel Clifford Walton, *History of the British Standing Army, 1660–1700*, 1894.

The value of tradition to the social body is immense. The veneration for practices, or for authority, consecrated by long acceptance, has a reserve of strength which cannot be obtained by any novel device.
>Rear-Admiral Alfred Thayer Mahan, 'The Military Rule of Obedience', *National Review*, 3/1902.

The regular officer has the traditions of forty generations of serving soldiers behind him, and to him the old weapons are the most honoured.
>Colonel T.E. Lawrence (1888–1935).

As descendents of the men who gained such splendid victories in so many battles from 1702 onwards we are simply unable to be cowardly. We've got to win our battles, whatever the cost, so that people will say 'They were worthy descendents of the 32nd' and that's saying a hell of a lot.
>Unidentified British colonel of the regiment, quoted in F.M. Richardson, *Fighting Spirit*, 1978.

You will hardly fade away until the sun fades out of the sky and the earth sinks into the universal blackness. For already you form part of that great tradition of the Dardenelles which began with Hector and Achilles. In another few thousand years the two stories will have blended into one, and whether when 'the iron roaring went up to the vault of heaven through the unharvested sky,' as Homer tells us, it was the spear of Achilles or whether it was a 100-lb shell from Asiatic Annie won't make much odds to the Almighty.
>General Sir Ian Hamilton, in a preface addressed to his Gallipoli soldiers, quoted in Alan Moorehead, *Gallipoli*, 1956.

It takes the Navy three years to build a ship. It would take three hundred to rebuild a tradition.
>Admiral Sir Andrew Browne Cunningham, May 1941, rejecting his staff's recommendation to abandon British ground forces on Crete in order to protect the Royal Navy's ships.

The spirit of discipline, as distinct from its outward and visible guises, is the result of association with martial traditions and their living embodiment.
>Captain Sir Basil Liddell Hart, *Thoughts on War*, 1944.

Fortune is rightly malignant to those who break with the traditions and customs of the past.
>Sir Winston S. Churchill, 23 April 1943, note to the Foreign Secretary.

However good and well trained a man

may be as an individual, he is not a good soldier till he has become absorbed into the corporate life of his unit and has been entirely imbued with its traditions.
> Field Marshal Viscount Wavell of Cyrenaica, *Soldiers and Soldiering*, 1953.

However praiseworthy it may be to uphold tradition in the field of soldierly ethics, it is to be resisted in the field of military command.
> Field Marshal Erwin Rommel, *The Rommel Papers*, 1953.

There can be no doubt that old traditions are in theory of great value to any army. The characters of the more eminent General Staff officers of the past should have provided a younger generation with good models without, at the same time, hindering or perhaps even preventing contemporary development. But in practice tradition is not always regarded as simply supplying ideals of behavior, but rather as a source of practical example, as though an imitation of what was done before could reproduce identical results despite the fact that meanwhile circumstances and methods have completely altered. Hardly any mature institution can avoid this fallacious aspect of tradition.
> Colonel-General Heinz Guderian, *Panzer Leader*, 1953.

Some people scoff at tradition. They are right if tradition is taken to mean that you must never do something for the first time; but how wrong they are if you regard tradition as a standard of conduct, handed down to you, below which you must never fall. Then tradition, instead of being a pair of handcuffs to fetter you, will be a handrail to steady and guide you in steep places. One of those traditions, perhaps the greatest of them, is leadership, for the be-all and the end-all of an officer is to be a leader.
> Field Marshal Viscount Slim of Burma, *Courage and Other Broadcasts*, 1957.

Whatever else may be said about the traditions of the Royal Navy, their appropriateness to today and their value, there is one at least that I hold fundamental to all the rest. I call it the 'Jervis bay Syndrome'. This refers to the armed merchant cruiser HMS *Jervis Bay* which had formerly been a 14,000-ton passenger liner, built in 1922, and called to duty in the Second World War with seven old six-inch guns mounted on her deck. She was assigned to convoy protection in the North Atlantic and placed under the command of Captain Edward Fogarty Fegan Royal Navy. In the late afternoon of 5 November 1940, Jervis Bay was escorting a convoy of thirty-seven merchant ships in the mid-Atlantic. Suddenly, over the horizon, appeared the German pocket battleship *Admiral Scheer*. Captain Fegan immediately turned towards the *Scheer*, knowing his ship would be sunk and that he would most likely die, out-ranged and out-gunned as he was. *Jervis Bay* fought for half an hour before she was sunk and later, when a ship returned to pick up survivors, the Captain was not among them. Edward Fegan was awarded a posthumous VC. But that half hour bought vital minutes of the convoy to scatter and make the *Scheer*'s job of catching and sinking more than a few of them too difficult. His was the moment we all know we may have to face ourselves.

We are indoctrinated from earliest days in the Navy with stories of great bravery such as this and many others like it, from Sir Richard Grenville of the *Resolution* to Lieutenant-Commander Roope VC of the *Glowworm* who, in desperation, turned and rammed the big German cruiser *Hipper* with his dying destroyer sinking beneath him. We had all been taught the

same – each and every one of the captains who sailed with me down the Atlantic towards the Falklands in the late April of 1982 – that we will fight, if necessary to the last death, just as our predecessors have traditionally done. And if our luck should run out, and we should be required to face a superior enemy, we will still go forward, fighting until our ship is lost.
>   Admiral Sir J.F. 'Sandy' Woodward, *One Hundred Days: The Memoirs of the Falklands Battle Group Commander*, 1992, p.113.

The Traditions of the Foreign Legion – four main principles:
  the will to serve well
  a sense of discipline and honour
  pride in a task well done
  the cult of remembrance
>   Quoted in John Robert Young, *The French Foreign Legion*, 1984.

The end of the short-lived Cold War means that the Royal Navy can now get back to its proper business – fighting the French!
>   Rear-Admiral Guy F. Liardet, after-dinner speech, 'Trend and Change', quoted in James Goldrick and John B. Hattendorf, ed., *Mahan is Not Enough*, 1993.

TRAINING
General

The Lacedaemonians, both in their warsongs and in the words which a man spoke to his comrade, did but remind one another of what their brave spirits knew already. For they had learned that true safety was to be found in long previous training, and not in eloquent exhortations uttered when they were going into action.
>   Thucydides, of the Spartans at the Battle of Mantinea, summer 418 BC, *The Peloponnesian War*, V, c. 404 BC, tr. Benjamin Jowett, 1881.

Training is light, and lack of training is darkness. The problem fears the expert. If a peasant doesn't know how to plough, he can't grow bread. A trained man is worth three untrained: that's too little – say six – six is too little – say ten to one. We will beat them all, roll them up, take them prisoner! In the last campaign the enemy lost 75,000 counted, but more like 100,000 in fact. He fought with skill and desperation, but we didn't lose 500. You see, lads! Military training! Gentlemen, what a marvellous thing it is!
>   Field Marshal Prince Aleksandr V. Suvorov, *The Science of Victory*, 1796.

Do I want to see you eaten by vultures and hyaenas after the next battle, merely because you were too stupid or lazy to understand that what I am trying to teach you today will save you tomorrow?
>   Mgobozi, c. 1818, to his Zulu recruits, quoted in E.A. Ritter, *Shaka Zulu*, 1955.

Every officer and soldier who is able to do duty ought to be busily engaged in military preparation, by hard drilling, etc., in order that, through the blessing of God, we may be victorious in the battles which, in His all-wise providence, may await us. If the war is carried on with vigour, I think, under the blessing of God, it will not last long.
>   Lieutenant-General Thomas J. 'Stonewall' Jackson, quoted in R.L. Dabney, *Life of Jackson*, I, 1864.

Military sway lies not merely in warships and arms, but also in the immaterial power that wields them. If we consider that a gun whose every shot tells can hold its own against a hundred guns which can hit only one shot in a hundred, we seamen must seek military power spiritually. The cause of the recent victory of our navy [Tsushima], though it was in a great degree due to the Imperial virtue, must also be attributed to our training in

TRAINING

peaceful times which produced its fruit in war.
> Admiral Marquis Heihachiro Togo, 21 December 1905, his farewell to the Japanese Fleet, quoted in Ogasawara, *Life of Admiral Togo*, 1934.

More important even than military education is practical military training.
> Major-General John A. Lejeune, *The Reminiscences of a Marine*, 1929.

In war the issues are decided not by isolated acts of heroism but by the general training and spirit of an army.
> Field Marshal Albert Kesselring, *Memoirs of Field Marshal Kesselring*, 1953.

Nothing comes off – except what you have practised.
> German dive-bomber pilots' maxim, quoted by Colonel Hans Ulrich Rudel, *Stuka Pilot*, 1958.

The definition of military training is success in battle. In my opinion that is the only objective of military training. It wouldn't make any sense to have a military organization on the backs of the American taxpayers with any other definition. I've believed that ever since I've been a Marine.
> Lieutenant-General Lewis 'Chesty' Puller, 2 August 1956, quoted in Burke Davis, *Marine*, 1962.

Perhaps somewhere in the primal reaches of our Army's memory, left over from the days ten thousand years ago when armies first began, there's a simple and fundamental formula: SKILL + WILL = KILL.
> Colonel Dandridge M. Malone, *Army Magazine*, 9/1979.

The Conduct of Training

People die because they are incapable and are defeated because they are unsuitable. Therefore the most important thing in handling troops is to train them. If one man is trained for war he can teach ten, if ten are trained they can instruct a hundred, a hundred can train a thousand, a thousand ten thousand and ten thousand an army corps (37,500).
> Wu Ch'i (430–381 BC), *Art of War*, 3.5 (*Three Military Classics of China*, tr. A.L. Sadler, 1944).

We are informed by the writings of the ancients that, among their other exercises was that of the post. They gave their recruits round bucklers woven with willows, twice as heavy as those used on real service, and wooden swords double the weight of the common ones. They exercised them with these at the post both morning and afternoon...

Every soldier, therefore, fixed a post firmly in the ground, about the height of six feet. Against this, as against a real enemy, the recruit was exercised with the above mentioned arms, as with the common shield and sword, sometimes aiming at the head or face, sometimes at the sides, at others endeavoring to strike at the thighs or legs. He was instructed in what manner to advance and retire, and in short how to take every advantage of his adversary; but was thus above all particularly cautioned not to lay himself open to his antagonist while aiming his stroke at him.

...This was the method of fighting principally used by the Romans, and their reason for exercising recruits with arms of such a weight at first was, that when they came to carry the common ones, which were so much lighter, the difference might enable them to act with greater security and alacrity in time of action.
> Flavius Vegetius Renatus, *Military Institutions of the Romans*, c. AD 378, tr. Clark, 1776.

The man who spends more sleepless nights with his army and who works

TRAINING

harder in drilling his troops runs the fewest risks in fighting the foe.
  The Emperor Maurice, *The Strategikon*, c. AD 600 (*Maurice's Strategikon*, tr. George Dennis, 1984).

There is no other possible way, as far as strategy and experience are concerned, for you to prepare for warfare except by first exercising and training the army under your command. You must accustom them to and train them in the handling of weapons and get them to endure bitter and wearisome tasks and labors. They should not be allowed to become slack or lazy or to give themselves completely to drunkenness, luxury, or other kinds of debauchery. They certainly ought to receive their salaries and money for provisions regularly, as well as gifts and bonuses, more than are customary or stipulated.
  The Emperor Nikephoros II Phokas, *Skirmishing*, c. AD 969 (*Three Byzantine Military Treatises*, tr. George Dennis, 1985).

When emperor Manuel took over the imperial office, he became concerned as to how the Romans might improve their armament for the future. It had previously been customary for them to be armed with round shields and for the most part to carry quivers and decide battles by bows, but he taught them to hold ones [shields] reaching to their feet, and trained them to wield long lances and skillfully practice horsemanship. Desiring to make respites from war preparation for war, he was frequently accustomed to practice riding; making a pretence of battle, he placed formations opposite one another. Thus charging with blunted lances, they practiced maneuvering in arms. So in brief time the Roman [Byzantine] excelled the French and Italians.
  General John Kinnamos, *Deeds of John and Manuel Comnenus*, 1176.

Again, all wise and sufficient generals and colonels have always had special regard, when the enemy hath not been near at hand, that their sergeants major, captains, and other officers should oftentimes in the field reduce their bands and regiments into divers forms, and to teach their soldiers all orders military, with the use of their weapons in every degree, time, and place; as also how to lodge in their quarters orderly, and therewithal to understand the orders of watches, bodies of watches, sentinels, rounds, and counterrounds, with many other matters military whereby they might be made prompt and ready to encounter the enemy.
  John Smythe (c. 1580–1631), *Certain Discourses Military*, 1590.

The troops should be exercised frequently, cavalry as well as infantry, and the general should often be present to praise some, to criticize others, and to see with his own eyes that the orders …are observed exactly.
  Frederick the Great, *Instructions to His Generals*, 1747, tr. Phillips, 1940.

The soldiers like training provided it is carried out sensibly.
  Field Marshal Prince Aleksandr V. Suvorov (1729–1800), *The Suzdal Manual*, 1768, quoted in K. Ossipov, *Strategy*, 1945.

Training is hard work, but it does not have to be dull. Much military training is presented in boring fashion. The troops lose interest and do not absorb the instruction, the training program fails, and the morale of the troops drops. Good training requires a lot of mental effort; the commander must devise ways to make training intellectually and physically challenging to the troops. The unfortunate thing is that so many commanders don't recognize dull training. But their troops do.
  Lieutenant-General Arthur S. Collins, Jr., *Common Sense Training*, 1978.

TRAINING

Individual Training

We must remember that one man is much the same as another, and that he is best who is trained in the severest school.
> Thucydides, *The History of the Peloponnesian War*, 1, c. 404 BC.

Few men are born brave; many become so through training and force of discipline.
> Flavius Vegetius Renatus, *Military Institutions of the Romans*, c. AD 378, tr. Clark, 1776.

Every soldier must be brought to the state where it can be said of him: 'You have learned all there is to be learned, take care not to forget it.'
> Field Marshal Aleksandr V. Suvorov (1729–1800), favourite maxim on training, quoted in K. Ossipov, *Suvorov*, 1945.

Body and spirit I surrendered whole
To harsh insructors – and received a soul.
> Rudyard Kipling, 'The Wonder', *Epitaphs*, 1919.

The first day I was at camp I was afraid I was going to die. The next two weeks my sole fear was that I wasn't going to die. And after that I knew I'd never die because I'd become so hard nothing could kill me.
> Anonymous American soldier, quoted in Cowing, *Dear Folks at Home*, 1919.

Every normal man needs to have some sense of a contest, some feeling of resistance overcome, before he can make the best use of his facilities. Whatever experience serves to give him confidence that he can compete with other men helps to increase his solidarity with other men.
> Brigadier-General S.L.A. Marshall, *The Armed Forces Officer*, 1950.

One great difficulty of training the individual soldier in peace is to instil discipline and yet to preserve the initiative and independence needed in war. The best soldier in peace (officer or man) is not necessarily the best soldier in war – though he is so more often than not – and it is not always easy in peace conditions to recognize the man who will make good in war. The soldier who is a thorough nuisance in barracks is occasionally a treasure in the field, though not nearly as often as Hollywood and the sentimental novelists would have us believe.
> Field Marshal Viscount Wavell of Cyrenaica, *Soldiers and Soldiering*, 1953.

The whole training of an officer seeks to accomplish one purpose – to instill in him the ability to take over in battle in a time of crisis.
> General Matthew B. Ridgway, *Soldier*, 1956.

Lack of Training

That government is a murderer of its citizens which sends them to the field uninformed and untaught, where they are to meet men of the same age and strength mechanized by education and discipline for battle.
> Major-General Henry 'Light Horse Harry' Lee (1756–1818).

With a raw army it is possible to carry a formidable position, but not to carry out a plan or design.
> Napoleon, quoted in Liddell Hart, *Reputations Ten Years After*, 1928.

My fleet looked well at Toulon, but when the storm came on, things changed at once. The sailors were not used to storms; they were lost among the mass of soldiers; these from seasickness lay in heaps about the decks; it was impossible to work the ships; hence yardarms were broken and sails were carried away. Our losses resulted as much from clumsiness and inexperience as from defects in the materials supplied

TRAINING

by the dockyard.
> Admiral Pierre Villeneuve, early 1805, to Admiral Decrés on the effects of the short voyage between Toulon and Corsica on his poorly trained crews, quoted in David Howarth, *Nelson: The Immortal Memory*, 1988.

A collection of untrained men is neither more nor less than a mob, in which individual courage goes for nothing. In movement each person finds his liberty of action merged in a crowd, ignorant and incapable of direction. Every obstacle creates confusion, speedily converted into panic by opposition.
> Lieutenant-General Richard Taylor, *Destruction and Reconstruction*, 1879.

Untutored courage is useless in the face of educated bullets.
> General George S. Patton, Jr, in the *Cavalry Journal*, 4/1922.

In no other profession are the penalties for employing untrained personnel so appalling or so irrevocable as in the military.
> General of the Army Douglas MacArthur, *Annual Report of the Chief of Staff, US Army*, 1933.

Physical Training

The young recruits in particular must be exercised in running, in order to charge the enemy with vigor, to occupy on occasion an advantageous post with greater expedition, and balk the enemy in their designs upon the same; and that they may, when sent to reconnoiter, advance with speed, and return with celerity and more easily overtake the enemy in pursuit.
> Flavius Vegetius Renatus, *Military Institutions of the Romans*, c. AD 378, tr. Clark, 1776.

The foundation of training depends on the legs and not the arms. All the mystery of maneuvers and combats is in the legs, and it is to the legs that we should apply ourselves. Whoever claims otherwise is but a fool and not only in the elements of what is called the profession of arms.
> Field Marshal Maurice Comte de Saxe, *My Reveries, 1732*, 1732, tr. Phillips, 1940.

Is it really true that a seven-mile cross-country run is enforced upon all in this division, from generals to privates? ...It looks to me rather excessive. A colonel or a general ought not to exhaust himself in trying to compete with young boys running across country seven miles at a time. The duty of officers is no doubt to keep themselves fit, but still more to think of their men, and to take decisions affecting their safety or comfort. Who is the general of this division, and does he run the seven miles himself? If so, he may be more useful for football than for war. Could Napoleon have run seven miles across country at Austerlitz? Perhaps it was the other fellow he made run. In my experience, based on many years' observation, officers with high athletic qualifications are not usually successful in the higher ranks.
> Sir Winston S. Churchill, 4 February 1941, note to the Secretary of State for War.

A man who takes a lot of exercise rarely exercises his mind adequately.
> Captain Sir Basil Liddell Hart, *Thoughts on War*, 1944.

More emphasis will be placed on the hardening of men and officers. All soldiers and officers should be able to run a mile with combat pack in ten minutes and march eight miles in two hours. When soldiers are in actual contact with the enemy, it is almost impossible to maintain physical condition, but if the physical condition is high before they gain contact, it will not fall off sufficiently during contact

TRAINING

to be detrimental.
General George S. Patton, Jr, *War As I Knew It*, 1947.

Truly then, it is killing men with kindness not to insist upon physical standards during training which will give them a maximum fitness for the extraordinary stresses of campaigning in war.
Brigadier-General S.L.A. Marshall, *Men Against Fire*, 1947.

Small Unit Training

The training of an infantry company for war, considered by the uninitiated as one of the simplest things in the world, is in reality the most complex; it is one constant struggle against human nature, and incessant variations of the tactical situation and of the ground, to say nothing of the frequent changes in the company as regards junior officers and noncommissioned officers.
Captain R.C.B. Haking, *Company Training*, 1917.

Battles are fought by platoons and squads. Place emphasis on small unit combat instruction so that it is conducted with the same precision as close-order drill.
General George S. Patton, Jr, *War As I Knew It*, 1947.

Train as You Fight

If any one does but attend to the other parts of their [Roman] military discipline, he will be forced to confess that their obtaining so large a dominion, hath been the acquisition of their valor, and not the bare gift of fortune; for they do not begin to use their weapons first in time of war, nor do they the then put their hands first into motion, while they avoided so to do in times of peace; but as if their weapons did always cling to them, they have never any truce from warlike exercises; nor do they stay till times of war admonish them to use them; for their military exercises differ not at all from the real use of their arms, but every soldier is every day exercised, and that with great diligence, as if it were in time of war which is the reason why they bear the fatigue of battles so easily; for neither can any disorder remove them from their usual regularity, nor can fear affright them out of it, nor can labor tire them; which firmness of conduct makes them always to overcome those that have not the same firmness; nor would he be mistaken that should call those their exercises unbloody battles, and their battles bloody exercises.
Flavius Josephus (AD 37–c. 100), *The Wars of the Jews*, tr. William Whiston, 1736.

Make peace a time of training for war, and battle an exhibition of bravery.
The Emperor Maurice, *The Strategikon*, c. AD 600 (*Maurice's Strategikon*, tr. George Dennis, 1984).

My troops are good and well-disciplined, and the most important thing of all is that I have thoroughly habituated them to perform everything that they are required to execute. You will do something more easily, to a higher standard, and more bravely when you know that you will do it well.
Frederick the Great, *Principes Généraux*, 1748.

Buonaparte has often made his boast that our fleet would be worn out by keeping the sea, and that his was kept in order and increasing by staying in port; but now he finds, I fancy, if Emperors hear the truth, that his fleet suffers more in a night than ours in a one year.
Admiral Viscount Nelson, 1805, letter to Captain Collingwood, quoted in David Howarth, *Nelson: The Immortal Memory*, 1988.

All instructions must contemplate the assumption of vigorous offensive. This purpose will be emphasized in every

THE TROOPS

phase of training until it becomes settled habit in thought.
> General of the Armies John J. Pershing, Final Report of Gen. John J. Pershing, quoted in J.F.C. Fuller, *Decisive Battles of the U.S.A.*, 1953.

The commander must be at constant pains to keep his troops abreast of all the latest tactical experience and developments, and must insist on their practical application. He must see to it that his subordinates are trained in accordance with the latest requirements. The best form of welfare for the troops is first-class training, for this saves unnecessary casualties.
> Field Marshal Erwin Rommel, *The Rommel Papers*, 1953.

TREACHERY/TREASON

Once when the ephors said to him: 'Take the young man and march against this man's country: he will personally conduct you to the acropolis,' Agis son of Archidamus replied, 'And how is it proper, ephors, for so many young men to trust this one man who is betraying his own country?'
> Agis II, Eurypontid King of Sparta (427–400 BC), quoted in Plutarch, *The Lives*, c. AD 100 (*Plutarch on Sparta*, tr. Richard Talbert, 1988).

Such was the crop of traitors then sprouting everywhere in Greece that, when Philip wished to take a certain city protected by its inaccessibility, he asked if gold could not climb over the wall.
> Plutarch, *Moralia*, of King Philip II of Macedon, quoted in Alfred A. Bradford, *Philip II of Macedon: A Life from the Ancient Sources*, 1992.

It is not the fat, sleek-headed men I am afraid of, but the pale, lean ones.
> Julius Caesar, c. 44 BC; an appraisal of the type of men he feared most, having Cassius and Brutus in mind, quoted in Plutarch, *The Lives*, c. AD 100 (*Makers of Rome*, tr. Ian Scott-Kilvert, 1965).

The treason pleases, but the traitors are odious.
> Miguel de Cervantes, *Don Quixote*, 1605–1615, tr. Peter Motteux and John Ozell.

This [Major-General Benedict Arnold's treason] is an event that occasions me equal regret and mortification; but traitors are an outgrowth of every country and in a revolution of the present nature, it is more to be wondered at, that the catalogue is so small than there have been found a few.
> General George Washington, to the Comte de Rochambeau, 27 September 1780, *The Writings of George Washington*, XX, p. 97, John c. Fitzgerald, ed., 1937.

THE TROOPS
Take Care of the Troops

Pay heed to nourishing the troops; do not unnecessarily fatigue them. Unite them in spirit; conserve their strength. Make unfathomable plans for the movement of the army.

Thus, such troops need no encouragement to be vigilant. Without extorting their support the general obtains it; without inviting their affection he gains it; without demanding their trust he wins it.
> Sun Tzu, *The Art of War*, c. 500 BC, tr. Samuel Griffith.

Philip, when he heard that Pythias, a certain first rate soldier, was estranged from him, because he had three daughters whom in his poverty he could hardly feed and he received no help from the king – and some of Philip's Companions warned him that he should beware of Pythias – Philip said, 'What? If part of my body were sick, would I cut it off and not try to cure it?'

And, as though Pythias were a member of his family, he called in to himself, learned the truth, and arranged for some money. And he had a more

devoted and faithful man than he had had before he offended him.
> King Philip II of Macedon (382–336 BC), from Frontinus, *Strategematicon*, IV, vii, 37, quoted in Alfred S. Bradford, *Philip II of Macedon*, 1992.

Next day, despite a sword wound in his thigh, Alexander went round to see the wounded; he gathered together the dead and gave them a splendid military funeral, the whole army marshalled in their finest battle array. His speech contained citations of all whom he knew, from his own eyes or from the agreed report of others, to have distinguished themselves in the battle; he honoured each of them by a donation suitable to their worth.
> Arrian, of Alexander the Great at the Battle of Issus, October 333 BC, *The Anabasis of Alexander*, 2.12, c. AD 150, tr. P.A. Brunt, 1976.

You can be of no service to me, go to the soldiers, to whom you can be useful.
> Lieutenant-General Sir John Moore, 16 January 1809, to two surgeons who rushed to him as he lay mortally wounded on the field of Corunna, quoted in Carola Oman, *Sir John Moore*, 1953.

The officers of the army are much mistaken if they suppose that their duty is done when they have attended to the drill of their men, and to the parade duties of the regiment: the order and regularity of the troops in camp and quarters, the subsistence and comfort of the soldiers, the general subordination and obedience of the corps, afford constant subjects for the attention of the field officers in particular, in which, by their conduct in the assistance they will give the Commanding officer, they can manifest their zeal for the service, their ability and their fitness for promotion to the higher ranks, at least equally as by an attention to the drill and parade discipline of the corps.
> The Duke of Wellington, 16 September 1809, at Badajoz, *Selections from the General Orders of Field Marshal the Duke of Wellington During His Various Commands 1799–1818*, 1847.

The attention of commanding officers has been frequently called to the expediency of supplying the soldier with breakfast.
> The Duke of Wellington (1769–1852), during the Peninsular War, quoted in Arthur Bryant, *The Great Duke*, 1972.

I stand at the altar of murdered men and while I live I fight their cause.
> Florence Nightingale, letter of August 1856, quoted in Cecil Woodham-Smith, *Florence Nightingale, 1820–1910*, 1951.

I gained the hearts of my soldiers (who would do anything for me) not by my justice, etc., but by looking after them when sick and continually visiting the Hospitals.
> General Charles 'Chinese' Gordon, letter to Florence Nightingale, 22 April 1888, quoted in Cecil Woodham-Smith, *Florence Nightingale, 1820–1910*, 1951.

They hurried forward, often at a double-quick, waded the Shenandoah River, which was waist-deep to the men, ascended the Blue Ridge at Ashby's Gap, and, two hours after midnight, paused for a few hours' rest at the little village of Paris, upon the eastern slope of the mountain. Here General Jackson turned his brigade into an enclosure occupied by a beautiful grove, and the wearied men fell prostrate upon the earth, without food. In a little time an officer came to Jackson, reminded him that there were no sentries posted around his bivouac, while the men were all wrapped in sleep, and asked if some should be aroused, and a guard set. 'No,' replied Jackson, 'let the poor fellows sleep; I will guard the camp myself.' All the remainder of the night he

THE TROOPS

paced around it, or sat upon the fence watching the slumbers of his men...
An account of Lieutenant-General Thomas J. 'Stonewall' Jackson's solicitude towards his men after a gruelling march on 21 July 1861, in Dabney, *Life of Jackson*, I, 1866.

Be kindly and just in your dealing with your men. Never play favorites. Make them feel that justice tempered with mercy may always be counted on. This does not mean a slackening of discipline. Obedience to orders and regulations must always be insisted upon, and good conduct on the part of the men exacted. Especially should this be done with reference to civilian inhabitants of foreign countries in which Marines are serving.
Major-General John A. Lejeune, 19 September 1922, Letter No. 1 to the Officers of the Marine Corps from the Commandant.

Talk and Listen to the Troops
The great impediment to action, is in our opinion, not discussion, but the want of that knowledge which is gained by discussion preparatory to action. For we have a peculiar power of thinking before we act and of acting too, whereas other men are courageous from ignorance but hesitate upon reflection. And they are surely to be esteemed the bravest spirits who, having the clearest sense of both the pains and pleasures of life, do not on that account shrink from danger.
Pericles, 431 BC, funeral oration for the first Athenians to fall in the Peloponnesian War, quoted in Thucydides, *The Peloponnesian War*, 2.40, c. 404 BC, tr. Benjamin Jowett, 1881.

It is impossible for the General to explain to his army the motive for the movement he directs. The Commander of the Forces can, however, assure the army that he has made none since he left Salamanca which he did not foresee, and was not prepared for, and as far as he is a judge, they have answered the purposes for which they were intended. When it is proper to fight a battle, he will do it; and he will choose the time and place he thinks most fit: in the meantime he begs the Officers and Soldiers of the army to attend diligently to discharge their parts, and to leave to him and to the General Officers, the decision of measures which belong to them alone.

The army may rest assured that there is nothing he has more at heart than their honour – and that of their country.
Lieutenant-General Sir John Moore, General Order of 27 December 1808, on the retreat from Spain, quoted in Carola Oman, *Sir John Moore*, 1953.

You English are aristocrats. You keep a great distance between yourselves and the popolo. Nature formed all men equal. It was always my custom to go amongst the soldiers and rabble, to converse with them, hear their little histories, and speak kindly to them. This I found to be the greatest benefit to me.
Napoleon, 3 April 1817, on St Helena, quoted in R.M. Johnston, ed., *The Corsican*, 1910.

When he reviewed the troops he asked the officers, and often the soldiers, in what battles they had been engaged, and to those who had received serious wounds he gave the cross. Here, I think, I may appropriately mention a singular piece of charlatanism to which the Emperor had recourse, and which powerfully contributed to augment the enthusiasm of his troops. He would say to one of his aides de camp, 'Ascertain from the colonel of such a regiment whether he has in his corps a man who has served in the campaigns of Italy or the campaigns of Egypt. Ascertain his name, where he was born, the particulars of his family, and what he has done. Learn his number in the ranks, and to what company he belongs, and furnish me with the information.'

On the day of the review Bonaparte, at a single glance, could perceive the man who had been described to him. He would go up to him as if he recognised him, address him by his name, and say, 'Oh! so you are here! You are a brave fellow – I saw you at Aboukir – how is your old father? What! Have you not got the cross? Stay, I will give it you.' Then the delighted soldiers would say to each other, 'You see the Emperor knows us all; he knows our families; he knows where we have served.' What a stimulus was this to soldiers, whom he succeeded in persuading that they would all some time or other become Marshals of the Empire!

Louis Antoine Fauvelet de Bourrienne, Napoleon's private secretary (1797–1802), *The Memoirs of Napoleon Bonaparte*, II, 1855.

I have many a time crept forward to the skirmish-line to avail myself of the cover of the pickets' 'little fort', to observe more closely some expected result; and always talked familiarly with the men, and was astonished to see how well they comprehended the general object, and how accurately they were informed of the state of facts existing miles away from their particular corps. Soldiers are very quick to catch the general drift and purpose of a campaign, and are always sensible when they are well commanded or well cared for. Once impressed with this fact, and that they are making progress, they bear cheerfully any amount of labor and privation.

General of the Army William T. Sherman, *Memoirs*, II, 1875.

The power to command has never meant the power to remain mysterious, but rather to communicate, at least to those who immediately execute our orders, the idea which animates our plans. If any one ever had the chance of playing the mysterious role in war it was Napoleon. For his authority was beyond question, and he had taken upon himself to think out everything and decide everything for his army. Yet in his correspondence he always put his views and his programme for several days to come before the Commanders of his army corps. And if we call to mind a number of his proclamations we shall see that his very troops were made aware of the manoeuvre he intended. Souvarov said exactly the same thing. Every soldier should understand the manoeuvre in which he is engaged. He was convinced that one can get anything out of a force to which one speaks frankly, because such a force will understand what is asked of it and will then itself ask no better than to do what is required of it.

Marshal of France Ferdinand Foch, *Precepts and Judgments*, 1919.

...The led put the worst construction on the silence of the leaders, that they assume no news to be bad news, despite all the proverbs.

Captain Sir Basil Liddell Hart, *A Greater Than Napoleon: Scipio Africanus*, 1926.

In the Corps, we ask our young combat officers to share our faith in the superior intelligence of American fighting men by taking every man of their command, down to the newest Private, into their confidence during battle. The men must have the feeling of team effort, and the confidence that their officer is treating them all with fairness.

This can be created by letting them know the immediate plan of action, and why it has been chosen. A man fights much better when he has a sense of the common objective, rather than merely a knowledge of how things look from his own foxhole.

Time after time, I have seen men carry on when all their leaders have been knocked out, using the knowledge given them in talks, and conferences, in advance of the action.

This individual ability gives us an important margin over our enemies. It is generally acknowledged that we have better equipment, better weapons, and greater productive powers, than our enemies, but these advantages would lose their significance if our forces, man for man, were not superior in skill and initiative.
> General Alexander A. Vandegrift, 28 January 1944, address at the Pennsylvania Military College, Navy Department press release.

There is among the mass of individuals who carry rifles in a war an amount of ingenuity and efficiency. If men can talk naturally to their officers, the product of their resourcefulness becomes available to all.
> General of the Army Dwight D. Eisenhower (1890–1969), quoted in S.L.A. Marshall, *The Armed Forces Officer*, 1950.

The habit of talking down to troops is one of the worst vices that can afflict an officer.
> Brigadier-General S.L.A. Marshall, *The Armed Forces Officer*, 1950.

But it is not enough to have a worthy object; you have got to convince everyone in the party that it is a worthy object. What might have seemed obvious to me, sitting in Army HQs surrounded by maps, reports and returns, might not be so self-evident to the orderly at my door who hadn't seen his wife for four years, or to the wet, hungry soldier up there in the jungle who was being shot at. It may not be so plain to a lot of people now.
> Field Marshal Viscount Slim of Burma, *Courage and Other Broadcasts*, 1957.

I have always remembered Cromwell's saying that a soldier fights better when he knows what he is fighting for. Ever since I commanded twenty men in the first Desert Patrol in 1931, I have constantly collected all those under my command and explained to them every detail of the situation in which we have found ourselves. I have then invited questions and discussion.
> Lieutenant-General Sir John Glubb, *A Soldier With the Arabs*, 1957.

The need for truth is not always realized. A leader must speak the truth to those under him; if he does not they will find it out and then their confidence in him will decline. I did not always tell all the truth to the soldiers in the war; it would have compromised secrecy and it was not necessary.

I told them all they must know for the efficient carrying out of their tasks. But what I did tell them was always true and they knew it; that produced a mutual confidence between us.
> Field Marshal Viscount Montgomery of Alamein, *The Memoirs of Field Marshal Montgomery*, 1958.

I can answer questions and hold a detailed dialogue with a group of soldiers or even with a lone guard in a lookout post where there is a mutual interest in the subjects: What is happening on the enemy side: What are our defects? How can they be rectified: I am also interested in the personal lives of the men. Where do they come from? How do their families live? Do they rest or work when they are on leave? Do they support their parents?

I always found that the troops answered freely and were never afraid to criticize or complain. But these were more in the nature of question and answer than true dialogue. In this respect, I suppose there was no basic difference between my talk with an anonymous private in a Golan Heights

outpost and my discussions with the top officials of the Ministry of Defense.
  General Moshe Dayan, *The Story of My Life*, 1976.

Squint with your ears. The most important skill for leaders is listening. Introverts have great edge, since they tend to listen quietly and usually don't suffer from being an 'interruptaholic.' Leaders should 'squint with their ears.' Too many bosses are thinking about what they will say next, rather than hearing what is being said now.
  Major-General Perry M. Smith, 'Learning to Lead', *Marine Corps Gazette*, January 1997.

TRUCE

We broke our sworn truce. We fight as outlaws.
True, and what profit for us in the long run?
Nothing –
  Homer, *The Iliad*, 7.404–406, c. 800 BC, tr. Robert Fagles, 1990; Antenor speaking to the Trojan assembly after one of the gods had tricked a Trojan into breaking the truce with the Achaeans.

When without a previous understanding the enemy asks for a truce, he is plotting.
  Sun Tzu, *The Art of War*, 9, c. 500 BC, tr. Samuel Griffith, 1963.

I have issued the strictest orders that on no account is intercourse to be allowed between the opposing troops. To finish this war quickly we must keep up the fighting spirit... I am calling for particulars as to names of officers and units who took part in this Christmas gathering with a view to disciplinary action... War to the knife is the only way to carry out a campaign of this sort.
  General Sir Horace Smith-Dorrien, 26 December 1914, general order forbidding further instances of informal truces, as occurred on Christmas Day 1914 when British and German troops stopped fighting to celebrate the holiday together.

TRUST

King Wei asked, 'How can I get my people to follow orders as an ordinary matter of course?'
Master Sun said, 'Be trustworthy as a matter of course.
  Sun Bin, *The Lost Art of War*, c. 350 BC, tr. Thomas Cleary.

Trust is a distinguished reward for warriors.
  Sun Bin, *The Lost Art of War*, c. 350 BC, tr. Thomas Cleary.

I command a great army. It is founded on trust. The minds of those who have been given permission to depart are already set on home like arrows. Their wives and children are counting the hours that they must wait for them. Although the battle before us may be fraught with danger, that trust cannot be broken.
  Zhuge Liang (AD 180–234), The Way of the General, when urged by his officers, at the approach of an enemy army, to revoke the leave promised to a number of his troops. He personally urged the troops to go, but they refused, and the enemy suffered a decisive defeat, quoted in Bloodworth, *The Chinese Machiavelli*, 1976.

Mutual trust is the surest basis of discipline in necessity and danger.
  Generals Werner von Fritsch and Ludwig Beck, German Army, *Troop Leadership*, 1933, quoted in US War Department, *German Military Training*, 17 September 1942.

When in doubt abut honesty, trustfulness, and other moral qualities [of a candidate for a leadership position] a good test is to say to oneself, 'would I go in the jungle with that man?'
  Field Marshal Viscount Montgomery of Alamein, *The Path to Leadership*, 15, 1961.

TRUTH

The commander stakes his reputation, his career, all that he is and may be in his profession, on the issue of a fight in which his instinct must derive so much that is unknown and his science weigh all that is imponderable: above all, he stakes the coin he values most – the trust that his officers and men have in him.
  Lieutenant-General Sir Francis Tuker, quoted in D.M. Horner, *The Commanders*, 1984.

I built trust among my components because I trusted them...
  General H. Norman Schwarzkopf, quoted in *Joint Pub 1, Joint Warfare of the US Armed Forces*, 1991.

The notion of trust may convey even more than teamwork. It's critically important that you have trust, especially at the commander level. Issues are raised from time to time, but you can ask the questions that will defuse matters, because you're certain your fellow component commander wouldn't do or say that.
  Lieutenant-General Walter E. Boomer, quoted in *Joint Pub 1, Joint Warfare of the US Armed Forces*, 1991.

TRUTH

Truth gains strength by notoriety and time, falsehood by precipitancy and vagueness.
  Cornelius Tacitus, *Annals*, 4.35, c. AD 116 (*The Complete Works of Tacitus*, tr. Alfred Church and William Brodribb, 1942).

The path of nature and of truth is narrow but it is simple and direct; the devious paths are numerous and spacious; but they all lead to error and destruction.
  Robert Jackson, A *Systematic View of the Formation, Discipline and Economy of Armies*, 1804.

Nothing should be concealed from the Emperor, either good or bad: to deceive him, even about things that are likely to be disagreeable to him is a crime.
  Marshal of France Louis Berthier, Prince de Neuchatel, Prince de Wagram, February 1807, letter to Marshal Lannes.

Haig was an honourable man according to his lights – but his lights were dim. The consequences which have made 'Passchendaele' a name of ill omen may be traced to the combined effect of his tendency to deceive himself; his tendency, therefore, to encourage his subordinates to deceive him; and their 'loyal' tendency to tell a superior what was likely to conclude with his desires. Passchendaele is an object-lesson in this kind of well-meaning, if not disinterested truthfulness.
  Captain Sir Basil Liddell Hart, *Through the Fog of War*, 1938, referring to Field Marshal Douglas Haig, the commander of the British Expeditionary Force (BEF) during WWI.

The truth is uncontrovertible. Panic may resent it; ignorance may deride it; malice may destroy it, but there it is.
  Sir Winston S. Churchill (1874–1965).

Of course it's the same old story. Truth usually is the same old story.
  Prime Minister Margaret Thatcher, *Time Magazine*, 1981.

U–V

UNCERTAINTY/THE UNEXPECTED/THE UNKNOWN

War involves in its progress such a train of unforeseen and unsupposed circumstances that no human wisdom can calculate the end. It has but one thing certain, and that is to raise taxes.
Thomas Paine (1737–1809), *Prospects on the Rubicon*, 1787.

War is the realm of uncertainty; three quarters of the factors on which action in war is based are wrapped in a fog of greater or lesser uncertainty. a sensitive and discriminating judgment is called for; a skilled intelligence to scent out the truth.
Major-General Carl von Clausewitz, *On War*, 1.3, 1832, tr. Michael Howard and Peter Paret, 1976.

He who wars walks in a mist through which the keenest eye cannot always discern the right path.
General Sir William Napier, *History of the War in the Peninsula*, 1828–1840.

You will usually find that the enemy has three courses of action open to him, and of these he will adopt the fourth.
Field Marshal Helmuth Graf von Moltke (1800–1891).

The *unknown* is the governing condition of war. Everybody is familiar with the principle (so you might think), and being familiar with it will distrust the unknown and master it; the unknown will no longer exist.
This is not true in the least. all armies have lived and marched amidst the unknown.
Marshal of France Ferdinand Foch, *Precepts and Judgments*, 1919.

UNIFORMS

On May 31, 1740, Frederick William I died, and when those around him sang the hymn, 'Naked came I into the world and naked I shall go,' he had just sufficient strength to mutter, 'No, not quite naked; I shall have my uniform on.'
Frederick William I of Prussia, quoted by Major-General J.F.C. Fuller, *A Military History of the Western World*, II, 1955.

A well-dressed soldier has more respect for himself. He also appears more redoubtable to the enemy and dominates him; for a good appearance is itself a force.
Joseph Joubert (1754–1824).

I hear that measures are in contemplation to alter clothing, caps &c. of the army.
There is no subject of which I understand so little; and, abstractly speaking, I think it indifferent how a soldier is clothed, provided it is in a uniform manner; and that he is forced to keep himself clean and smart, as a soldier ought to be. But there is one thing I deprecate, and that is any imitation of the French in any manner.
It is impossible to form an idea of the inconveniences and injury which result from having anything like them… I only beg that *we* may be as different from the French in everything.
The Duke of Wellington, 1811, letter to the Horse Guards, *The Dispatches of Field*

UNITY OF COMMAND

Marshal the Duke of Wellington, During His Various Campaigns in India, Denmark, Portugal, Spain, the Low Countries, and France, VIII, 1834–1838, pp. 378–379.

Dress them in red, blue, or green – they'll run away just the same.
> Fernando I, King of the Two Sicilies, c. 1800, on the futility of trying to make good soldiers of his subjects by appealing to martial dress.

A soldier must learn to love his profession, must look to it to satisfy all his tastes and his sense of honor. That is why handsome uniforms are useful.
> Napoleon (1769–1821) to General Gaspard Gourgaud, 1815, on St Helena.

I am ashamed for strangers to see my barefoot, ragged boys in camp, but I would be glad for all the world to see them on the field of battle.
> General Robert E. Lee, to an English visitor to the Army of Northern Virginia as his ragged army passed on parade, quoted in Lieutenant-General D.H. Hill, 'Address before the Mecklenburg (N.C.) Historical Society', *Southern Historical Society Papers*, I, January–June 1876, p. 397.

The better you dress a soldier, the more highly will he be thought of by women.
> Field Marshal Lord Wolseley, *The Soldier's Pocketbook*, 1869.

The secret of uniform was to make a crowd solid, dignified, impersonal: to give it the singleness and tautness of an upstanding man.
> Colonel T.E. Lawrence, *Revolt in the Desert*, 1927.

Self-respect and *esprit de corps* demand some form of outward expression. Hence the justification of the uniform as the mark of a special class. The uniform indicates the soldier's responsibility; it is the outward sign of inward comradeship; it supports and confirms discipline.
> Colonel-General Hans von Seekt, *Thoughts of a Soldier*, 1930.

UNITY OF COMMAND
Principle of War

Rage – Goddess, sing the rage of Peleus'
    son Achilles,
murderous, doomed, that cost the
    Achaeans countless losses,
hurling down to the house of Death so
    many sturdy souls,
great fighters' souls, but made their
    bodies carrion,
feasts for dogs and birds,
and the will of Zeus was moving toward
    its end.
Begin, Muse, when the two first broke
    and clashed,
Agamemnon lord of men and brilliant
    Achilles.
> Homer, *The Iliad*, 1.1–8, c. 800 BC, tr. Robert Fagles, 1990; the invocation of the Muse, the opening lines that announce an underlying theme of *The Iliad*, the breakdown in Greek leadership.

Too many kings can ruin an army –
    mob rule!
Let there be one commander, one
    master only,
endowed by the son of crooked-minded
    Cronus
with kingly scepter and royal rights of
    custom:
whatever one man needs to lead his
    people well.
> Homer, *The Iliad*, 2.225–229, c. 800 BC, tr. Robert Fagles, 1990; Odysseus restoring order to the Achaean ranks.

For what should be done seek the advice of many, for what you will actually do take council with only a few trustworthy people; then off by yourself alone decide on the best and most helpful plan to follow, and stick to it.
> The Emperor Maurice, *The Strategikon*, c. AD 600 (*Maurice's Strategikon*, tr. George Dennis, 1984).

UNITY OF COMMAND

Now in war there may be one hundred changes in each step. When one sees he can, he advances; when he sees that things are difficult, he retires. To say that a general must await commands of the sovereign in such circumstances is like informing a superior that you wish to put out a fire. Before the order to do so arrives the ashes are cold. And it is said one must consult the Army Supervisor in these matters! This is as if in building a house beside the road one took advice from those who pass by. Of course the work would never be completed!

To put a rein on an able general while at the same time asking him to suppress a cunning enemy is like tying up the Black Hound of Han and then ordering him to catch elusive hares. What is the difference?

Ho Yen-hsi, Chinese military theorist of the Sung Dynasty, c. AD 1000, commentary on Sun Tzu, *The Art of War*, 3, c. 500 BC, tr Samuel Griffith, 1963.

Never will I accept a divided command – no, not even were God Himself to be my colleague in office. I must command alone, or not at all.

Albert von Wallenstein, quoted in B.H. Liddell Hart, *Great Captains Unveiled*, 1927.

The first security for success is to confer the command on one individual. When the authority is divided, opinions are divided likewise, and the operations are deprived of that *ensemble* which is the first essential of victory. Besides, when an enterprise is common to many, and not confined to a single person, it is conducted without vigour, and less interest is attached to the result.

Field Marshal Count Raimond Montecuccoli (1609–1680), quoted in David Chandler, ed., *The Military Maxims of Napoleon*, 1987.

Kellermann would command the army quite as well as I do; for I am certain that our victories are due to the bravery and daring of the men. But I am convinced that to combine Kellermann and myself in Italy would be to court disaster. I cannot willingly serve alongside a man who considers himself to be the best general in Europe. In any case, I am certain that one bad general is better than two good ones. Fighting is like governing; it needs tact.

Napoleon, 14 May 1796, letter to Citizen Carnot, after the Battle of Lodi on 10 May, complaining of the Directory's intention to appoint General Kellermann to share his command, *Correspondance de Napoléon Ier, publiée par ordre de l'Empereur Napoléon III*, I, 421, 1858–1870.

It has been said that one bad general is better than two good ones; and the saying is true, if taken to mean no more than that an army is better directed by a single mind, though inferior, than by two superior ones, at variance, and cross-purposes with each other.

And the same is true, in all joint operations wherein those engaged, can have none but a common end in view, and can differ only as to the choice of means. In a storm at sea, no one on board can wish the ship to sink; and yet, not unfrequently, all go down together, because too many will direct, and no single mind can be allowed to control.

President Abraham Lincoln, Annual Message to Congress, 31 December 1861, quoted in *Lincoln: Speeches, etc.*, 1989, p. 295.

An army is a collection of armed men obliged to obey one man. Every change in the rules which impairs the principle weakens the army.

General of the Army William T. Sherman (1820–1891).

No; I will not direct any one to do what I would not do myself under similar circumstances. I never held what might

VETERANS

be called formal councils of war, and I do not believe in them. They create a divided responsibility, and at times prevent that unity of action so necessary in the field. Some officers will in all likelihood oppose any plan that may be adopted; and when it is put into execution, such officers may, by their arguments in opposition, have so far convinced themselves that the movement will fail that they cannot enter upon it with enthusiasm, and might possibly be influenced in their actions by the feeling that a victory would be a reflection upon their judgment. I believe it is better for a commander charged with the responsibility of all the operations of his army to consult his generals freely but informally, get their views and opinions, and then make up his mind what action to take, and act accordingly. There is too much truth in the old adage, 'Councils of war do not fight.'
> General Ulysses S. Grant, comments on the suggestion that Sherman hold a council of war to determine whether he should march to the sea from Atlanta, early October 1864, quoted in Horace Porter, *Campaigning With Grant*, 1897.

You can't run a military operation with a committee of staff officers in command. It will be nonsense.
> Field Marshal Viscount Montgomery of Alamein, quoted in Francis de Guingand, *Operation Victory*, 1947.

When our front line infantry man or young sailor looked to his officer or NCO in the South Atlantic he was looking for guidance on how to improve his contribution. And this can be and was projected upwards. The ability to operate central joint command of our national force was war winning. Much was left unsaid because we knew our people and could rely on their flexibility, commonsense and sense of purpose. We were thus able to be truly joint. We won because we were unified, the enemy were not.
> Major-General Sir Jeremy Moore, 'The Falklands Experience', *RUSI*, 1 March 1983.

VALOUR

Now those who discuss generals always think of their valour. But valour is only one of the many qualities of a general, and the valorous type of general is apt to join battle lightly and carelessly, without knowing how to profit by it, so that he cannot win.
> Wu C'hi (c. 430–381 BC), *The Art of War*, 1.4 (*Three Military Classics of China*, tr. A.L. Sadler, 1944).

In valor there is hope.
> Cornelius Tacitus, *Annals*, c. AD 116.

Valor is superior to numbers.
> Flavius Vegetius Renatus, *Military Institutions of the Romans*, c. aD 378, tr. Clark, 1776.

Valor lies halfway between rashness and cowheartedness.
> Miguel de Cervantes, *Don Quixote*, 1615.

If valour can make amends for the want of numbers, we shall probably succeed.
> Major-General Sir James Wolfe, 1759, letter from Halifax to the Elder Pitt before the attack on Quebec.

Among the men who fought on Iwo Jima, uncommon valor was a common virtue.
> Admiral Chester W. Nimitz, March 1945.

Hidden valor is as bad as cowardice.
> Roman proverb.

VETERANS

As for the cavalry, it should never be touched; old troopers and old horses are good, and recruits of either are absolutely useless. It is a burden, it is an expense, but it is indispensible.

In regard to the infantry, as long as

VICTORY

there are a few old heads you can do what you want with the tails; they are the greatest number, and the return of these men in peace is a noticeable benefit to the nation, without a serious diminution of the military forces.
Field Marshal Maurice Comte de Saxe, *My Reveries*, 1732, tr. Phillips, 1940.

Call a meeting of the War Committee, and draft a measure to provide for discharged and wounded soldiers, by giving them preference in appointments to such administrative posts in the Forestry, Post Office, Tobacco, Taxation, and in fact all departments, as can profitably be held by discharged officers and men. It is unfair, and contrary to my intention, to give such posts to people who have done nothing.
Napoleon, 27 January 1811, to General Count Andréossy, President of the War Committee of the State Council, *Correspondance de Napoléon Ier, publiée par ordre de l'Empereur Napoléon III*, XXI, 17301, 1858–1870.

The citizen soldiers were associated with so many disciplined men and professionally educated officers, that when they went into engagements it was with a confidence they would not have felt otherwise. They became soldiers themselves almost at once.
General of the Army Ulysses S. Grant, *Personal Memoirs of U.S. Grant*, I, 1885.

What is our task? To make Britain a fit country for heroes to live in.
Prime Minister David Lloyd George, speech at Wolverhampton, England, 24 November 1918.

It takes very little yeast to leaven a lump of dough.... It takes a very few veterans to leaven a division of doughboys.
General George S. Patton, Jr, *War As I Knew It*, 1947.

VICTORY
General

Victory shifts, you know, now one man, now another.
Homer, *The Iliad*, 6.401, c. 800 BC, tr. Robert Fagles, 1990.

Thus it is that in war the victorious strategist only seeks battle after the victory has been won, whereas he who is destined to defeat first fights and afterwards looks for victory.
Sun Tzu, *The Art of War*, c. 500 BC, tr. Giles, 1910.

It is when you are proceeding against an enemy you expect to defeat that you should draw on all available resources. After all, an easy victory never gave anyone cause for regret.
Xenophon, *How To Be A Good Cavalry Commander*, c. 360 BC (*Hiero the Tyrant and Other Treatises*, tr. Robin Waterfield, 1997).

Alexander and Victory!
The battle cry of Alexander the Great (356–323 BC), quoted in Plutarch, *The Lives*, c. AD 100.

The greatest happiness is to vanquish your enemies, to chase them before you, to rob them of their wealth, to see those dear to them bathed in tears, to clasp to your bosom their wives and daughters.
Genghis Khan (1162–1227).

Man does not enter battle to fight, but for victory. He does everything he can to avoid the first and obtain the second.
Colonel Charles Ardant du Picq, *Battle Studies*, 1880, tr. Greely, 1957.

There is nothing certain about war except that one side won't win.
General Sir Ian Hamilton, *Gallipoli Diary*, 1920.

I'm violently opposed to losing wars.
Lieutenant-General Lewis B. 'Chesty' Puller (1898–1971), comment after his

VICTORY

The Consequences of Victory

The end and perfection of our victories is to avoid the vices and infirmities of those whom we subdue.
> Alexander the Great (356–323 BC), quoted in Plutarch, The Lives, c. AD 100, tr. John Dryden.

War will of itself discover and lay open the hidden and rankling wounds of the victorious party.
> Cornelius Tacitus, Histories, AD 104–109, 2.77, tr. Alfred Church and William Brodribb, 1942.

The problems of victory are more agreeable than those of defeat, but they are no less difficult.
> Sir Winston S. Churchill, 11 November 1942, in the House of Commons.

The shadow of victory is disillusion: the reaction from extreme effort and prostrations.
> Sir Winston S. Churchill (1874–1965).

The vengeful passions are uppermost in the hour of victory.
> Captain Sir Basil Liddell Hart, Defence of the West, 1950.

It is very important that if we commit again to any kind of battle that we are sure to understand the ramifications of what happens if we do accomplish our objectives.
> General H. Norman Schwarzkopf, 1991, interview with C.D.B. Bryan, quoted in Peter David, Triumph in the Desert, 1991.

The Elements of Victory

Thus we may know that there are five essentials for victory: (1) He will win who knows when to fight and when not to fight. (2) He will win who knows how to handle both superior and inferior forces. (3) He will win whose army is animated by the same spirit throughout all the ranks. (4) He will win who, prepared himself, waits to take the enemy unprepared. (5) He will win who has military capacity and is not interfered with by his sovereign. Victory lies in the knowledge of these five points.

Hence the saying: If you know the enemy and know yourself, you need not fear the result of a hundred battles. If you know yourself but not the enemy, for every victory gained you will also suffer a defeat. If you know neither the enemy nor yourself, you will succumb in every battle.
> Sun Tzu, The Art of War, 3, c. 500 BC, tr. Giles, 1910.

Victory in war does not depend entirely upon numbers or mere courage; only skill and discipline will insure it.
> Flavius Vegetius Renatus, The Military Institutions of the Romans, aD 378, tr. Clark, 1776.

Victory is achieved only through the combination of courage and military art.
> Field Marshal Prince Aleksandr V. Suvorov (1729–1800), quoted in M. Gareyev, Frunze, Military Theorist, 1985.

If in conclusion we consider the total concept of a victory, we find that it consists of three elements:
1. The enemy's greater loss of material strength
2. His loss of morale
3. His open admission of the above by giving up his intentions.
> Major-General Carl von Clausewitz, On War, 4.4, 1832, tr. Michael Howard and Peter Paret, 1976.

Victory is an inclined plain. On condition that you do not check your movement the moving mass perpetually increases its speed.
> Marshal of France Ferdinand Foch, Precepts and Judgments, 1919.

VICTORY

The Army had a policy that you don't just beat him [the enemy] a little bit.
> Lieutenant-General John Yeosock, quoted in Donnelly, 'The Generals Plan', *Army Times*, 2 March 1992.

Generalship and Victory

The expert at battle seeks his victory from strategic advantage (*shih*) and does not demand it from his men. He is thus able to select the right men and exploit the strategic advantage (*shih*). He who exploits the strategic advantage (*shih*) sends his men into battle like rolling logs and boulders. It is the nature of logs and boulders that on flat ground, they are stationary, but on steep ground, they roll; the square in shape tends to stop but the round tends to roll. Thus, that the strategic advantage (*shih*) of the expert commander in exploiting his men in battle can be likened to rolling round boulders down a steep ravine thousands of feet high says something about his strategic advantage (*shih*).
> Sun Tzu, *The Art of Warfare*, 5, c. 500 BC, tr. Roger Ames, 1993.

A good general not only sees the way to victory; he also knows when victory is impossible.
> Polybius, *Histories*, 1, c. 125 BC.

I have never in my life taken a command into battle, and had the slightest desire to come out alive unless I won.
> General Philip H. Sheridan, when told by Brigadier-General Porter that he had exposed himself too much at the Battle of Five Forks, 1 April 1865, quoted in Horace Porter, *Campaigning With Grant*, 1897.

Final victory has never been held the acid test of generalship, even where two professional armies have met. Ultimate defeat does not blind us to the skill of Hannibal, Napoleon, and Lee. (Jan. 1924.)
> Captain Sir Basil Liddell Hart, *Thoughts on War*, 1944.

The Necessity of Victory

If victorious, we have everything to live for. If defeated, there will be nothing left to live for.
> General Robert E. Lee, 3 May 1864, comment before the Battle of the Wilderness.

Victory at all costs, victory in spite of terror, victory however long and hard the road may be; for without victory there is no survival.
> Sir Winston S. Churchill, 13 May 1940, to the House of Commons.

If we win, nobody will care. If we lose, there will be nobody to care.
> Sir Winston S. Churchill, secret session of the House of Commons, 25 June 1941, *Winston S. Churchill: His Complete Speeches, 1897–1963*, VI, 1974, p. 6438.

In war there is no second prize for the runner-up.
> General of the Army Omar N. Bradley, *Military Review*, 2/1950.

But once war is forced upon us, there is no other alternative than to apply every available means to bring it to a swift end. War's very object is victory, not prolonged indecision.

In war, there is no substitute for victory.
> General of the Army Douglas MacArthur, 19 April 1951, address to Congress.

Pyrrhic Victory

The armies separated; and, it is said, Pyrrhus replied to one that gave him joy of his victory that one other such would utterly undo him. For he had lost a great part of the forces he brought with him, and almost all his particular friends and principal commanders; there were no others there to make recruits, and he found the confederates in Italy backward. On the other hand, as from a fountain continually flowing out of the city, the Roman camp was

VICTORY

quickly and plentifully filled up with fresh men, not all abating in courage from the loss they sustained, but even from their very anger gaining new force and resolution to go on with the war.
> Plutarch, *The Lives*, c. AD 100, tr. John Dryden, of the coining of the term 'Pyrrhic Victory' from the heavy, self-defeating losses taken in battle with the Romans, at Asculum in 279 BC, by Pyrrhus, King of Epirus.

The pursuit of victory without slaughter is likely to lead to slaughter without victory.
> The Duke of Marlborough (1650–1722).

If God give us another defeat like this, your Majesty's enemies will be destroyed.
> Marshal Claude-Louis-Hector de Villars, 11 September 1711, of the devastating losses he inflicted on Marlborough's infantry at Malplaquet, despite having to withdraw in good order from the field, a technical defeat.

It was a dear bought victory; another such would have ruined us.
> General Sir Henry Clinton speaking of the losses of 40 per cent which the British suffered on 17 June 1775 in taking Breed's Hill in attempting break Washington's siege of Boston.

If we have learned anything from our Vietnam experience, we should understand that the application of military power unsupported by adequate political consensus and resolve is insufficient to insure victory. Our military forces cannot operate in a political vacuum, for we are a consensus-oriented society, served rather than dominated by its defense establishment. The violation of this principle can only lead to the worst form of pyrrhic victory – when an army wins virtually every battle only to lose the highly political act that we call war. Thus, when reminded that his army had not defeated the U.S. Army on the battlefield, a North Vietnamese officer aptly replied, 'That may be so, but it is also irrelevant.'
> Colonel Stuart A. Herrington, *Peace With Honor*, 1983; quoting Colonel Harry G. Summers, Jr. *On Strategy: A Critical Analysis of the Vietnam War*, 1982.

Reports of Victory

I came, I saw, I conquered.
> Julius Caesar, 47 BC, the slogan carried in his Pontic Triumph, signifying how quickly he had dispatched this war, quoted in Suetonius, *Lives of the Twelve Caesars*, AD 121, tr. Philemon Holland, 1606.

The enemy came to us. They are beaten. God be praised! I have been a little tired all day. I bid you good-night and am going to bed.
> General Marshal Vicomte de Turenne, 14 June 1658, report of his victory that day in the Battle of the Dunes, quoted in Maxime Weygand, *Turenne, Marshal of France*, 1930.

I have not time to say more, but to beg you will give my duty to the queen, and let her know her army has had a glorious victory, Mons. Tallard and two other generals are in my coach and I am following the rest. The bearer, my aide-de-camp, Colonel Parke, will give her an account of what has pass'd. I shall doe it in a day or two by another more at large.
> The Duke of Marlborough, 13 August 1704, report of his victory that day at the Battle of Blenheim, quoted in Sir John Fortescue, *History of the British Army*, I, 1899.

We have met the enemy, and they are ours: Two ships, two brigs, one schooner and one sloop.
> Captain Oliver Hazard Perry, 10 September 1813, report to General William Henry Harrison of victory over the British at the Battle of Lake Eire on that date.

VICTORY

I yesterday, after a most severe and bloody contest, gained a complete victory, and pursued the French till dark. They are in complete confusion; and I have I believe, 150 pieces of cannon; and Blücher, who continued the pursuit all night, my soldiers being tired to death, sent me word this morning that he had got 60 more... The finger of providence was upon me, and I escaped unhurt.

> The Duke of Wellington, 19 June 1815, letter about Waterloo to Lady Frances Webster, *Supplementary Despatches, Correspondence, and Memoranda of Field Marshal Arthur Duke of Wellington*, KG, 1794–1818, X, 1863, p. 531.

It has been a fine battle. There is no doubt of the result... But we must not think that the party is over... We must keep up the pressure. We intend to hit this chap for six out of North Africa.

> Field Marshal Montgomery Viscount of Alamein, 5 November 1942, announcement of victory in the Battle of Alamein, quoted in Nigel Hamilton, *Monty: Master of the Battlefield*, 1942–1944, 1983.

Casualties many; percentage dead unknown; combat efficiency: We are winning.

> General David Shoup, 21 November 1943, situation report from Betio Island in Tarawa Atoll, quoted in James R. Stockman, *The Battle for Tarawa*, 1947.

The mission of this Allied force was fulfilled at 0300, local time, May 7th, 1945.

> General of the Army Dwight D. Eisenhower, message to the Combined Chiefs upon the surrender of Germany, quoted in Albert D. Chandler, *The Papers of Dwight D. Eisenhower*, 1970, IV, p. 2696. The time was corrected to 0241 by his clerk before sending it out.

Be pleased to inform Her Majesty that the White Ensign flies besides the Union Jack in South Georgia. God Save the Queen.

> Signal from HMS *Antrim*, 24 April 1982, on the recapture of South Georgia Island from the Argentines, the first British victory in the Falklands War, quoted in Sandy Woodward, *One Hundred Days: The Memoirs of the Falklands Battle Group Commander*, 1982, p. 105.

The Uses of Victory

For as we have often observed, it is no doubt a great thing to be successful in our undertakings, and to defeat our enemy in the field of battle; but it is a proof of greater wisdom, and requires more skill, to make a good use of victory. For many know how to conquer; few are able to use their conquest aright.

> Polybius, *Histories*, 10.36, c. 125 BC (*Familiar Quotations from Greek Authors*, tr. Crauford Ramage, 1895).

A victory on the battlefield is of little account if it has not resulted either in breakthrough or encirclement. Though pushed back, the enemy will appear again on different ground to renew the resistance he momentarily gave up. The campaign will go on.

> Field Marshal Alfred Graf von Schlieffen (1833–1913).

The true national object in war, as in peace, is a more perfect peace. The experience of history enables us to deduce that gaining military victory is not in itself equivalent to gaining the object of war. (Oct. 1925.)

> Captain Sir Basil Liddell Hart, *Thoughts on War*, 1944.

VIOLENCE

Those who despise violence are warriors fit to work for kings.

> Sun Bin, *The Lost Art of War*, c. 350 BC, tr. Thomas Cleary, 1996.

War therefore is an act of violence intended to compel our opponent to fulfill our will.
  Violence arms itself with inventions

VOLUNTEERS

of art and science in order to contend against violence. Self-imposed restrictions, almost imperceptible and hardly worth mentioning, termed usages of international law, accompany it without essentially impairing its power. Violence, that is to say, physical force (for there is no moral force without the conception of states and law), is therefore the *means*; the compulsory submission of the enemy to our will is the ultimate *object*.
    Major-General Carl von Clausewitz, *On War*, 1.1.2, 1832, tr. J.J. Graham, 1908.

I've got to get the maximum violence out of this campaign – now!
    General Wesley Clark, on escalating the bombing campaign against Serbia, quoted in 'The Commander's War', *The Washington Post*, 21 September 1999, A16.

Some men thrive on violence and killing. Some few enjoy it ecstatically. It takes but a small band of such men, armed and determined, to destroy a fragile society. If the worst thugs are not removed – or killed – the victimized society has no chance of reconstruction. Mass murderers do not reform, they just rearm. We love the idea of redemption, but the rest of the world wants revenge... those who thrive by the gun generally only understand a larger gun pointed at them. Violence cannot be halted without the will to respond in kind. Those who shake the hands of tyrants become their accomplices.
    Lieutenant-Colonel Ralph Peters, 'Sierra Leone's Blood Is On America's Hands', *Wall Street Journal*, 11 May 2000.

VOLUNTEERS

The patriot volunteer, fighting for his country and his rights, makes the most reliable soldier on earth.
    Lieutenant-General Thomas J. 'Stonewall' Jackson, quoted in Henderson, *Stonewall Jackson and the American Civil War*, 1898.

This conflict is one thing I've been waiting for. I'm well and strong and young – young enough to go to the front. If I can't be a soldier I'll help soldiers... when there is no longer a soldier's arm to raise the Stars and Stripes above our Capitol, may God give strength to mine.
    Clara Barton, 1861, quoted in National Park Service, *Clara Barton* (Handbook 110), 1981.

It is a striking feature, so far as my observation goes, of the present volunteer army of the United States, that there is nothing which men are called upon to do, mechanical or professional, that accomplished adepts cannot be found for the duty required in almost every regiment.
    General of the Army Ulysses S. Grant, 6 July 1863, report on the Vicksburg campaign.

If you are only ready to go when you are fetched, where is the merit of that? Are you only going to do your duty when the law says you must? Does the call to duty find no response in you until reinforced, let us rather say, superseded, by the call of compulsion.
    Field Marshal Earl Kitchener, 1915, speech at the Guildhall, London.

Our voluntary service regulars are the last descendents of those rulers of the ancient world, the Roman legionnaires.
    Major-General Sir Ian Hamilton, *Gallipoli Diary*, 1920.

The volunteer principle is all very well, and those who volunteer deserve honour; but it skims the cream off units.
    Brigadier Sir Bernard Fergusson, *The Trumpet in the Hall*, 1970.

W–Y

WAR
The Causes of War

... So they waited,
the old chiefs of Troy, as they sat aloft
  the tower.
And catching sight of Helen moving
  along the ramparts,
they murmured one to another, gentle,
  winged words:
'who on earth could blame them? ah,
  no wonder,
the men of Troy and Argives under
  arms have suffered
years of agony all for her, for such a
  woman.
Beauty, terrible beauty!'
A deathless goddess – so she strikes our
  eyes!
But still,
ravishing as she is, let her go home in
  the long ships
and not be left behind... for us and our
  children
down the years an irresistible sorrow.'
  Homer, *The Iliad*, 3.183–194, c. 800 BC, tr.
  Robert Fagles, 1990; Helen, the cause of
  the Trojan War, summoned to the walls of
  Troy to witness the duel between
  Menelaus and Paris.

The real though unavowed cause [of the Peloponnesian War] I believe to have been the growth of the Athenian power, which terrified the Lacedaemonians and forced them into war.
  Thucydides, *The Peloponnesian War*, 1.24, c. 404 BC, tr. Benjamin Jowett, 1881.

The possession of battle-ready troops, a well-filled state treasury and a lively disposition, these were the real things which moved me to war.
  Frederick the Great, on the invasion of Silesia, *a History of My Times*, 1741.

Two different motives make men fight one another: *hostile feelings* and *hostile intentions*. Our definition is based on the latter, since it is the universal element. Even the most savage, almost instinctive, passion of hatred cannot be conceived as existing without hostile intent; but hostile intentions are often unaccompanied by any sort of hostile feelings – at least by none that predominate. Savage peoples are ruled by passion, civilized peoples by the mind. The difference, however, lies not in the respective nature of savagery and civilization, but in their attendant circumstances, institutions, and so forth. The difference, therefore, does not operate in every case, but it does in most of them. Even the most civilized of peoples, in short, can be fired with passionate hatred of each other.
  Major-General Carl von Clausewitz, *On War*, 1.1.3, 1832, tr. Michael Howard and Peter Paret, 1976.

If ever there is another war in Europe, it will come out of some damned silly thing in the Balkans.
  Chancellor Otto von Bismarck (1815–1898).

Where evil is mighty and defiant, the obligation to use force – that is war – arises.
  Rear-Admiral Alfred Thayer Mahan, *Naval Strategy*, 1911.

WAR

I ask the House from the point of view of British interests to consider what may be at stake. If France is beaten to her knees... if Belgium fell under the same dominating influence and then Holland and then Denmark... if, in a crisis like this, we run away from these obligations of honor and interest as regards the Belgian Treaty... I do not believe for a moment that, at the end of this war, even if we stood aside, we should be able to undo what had happened, in the course of the war, to prevent the whole of the West of Europe opposite us from falling under the domination of a single power... and we should, I believe, sacrifice our respect and good name and reputation before the world and should not escape the most serious and grave economic consequences.

Sir Edward Grey, British Foreign Secretary, 3 August 1914, speech to the House of Commons on the grounds for war with Germany, quoted in Barbara Tuchman, *The Guns of August*, 1962.

Once in a generation a mysterious wish for war passes through the people. Their instinct tells them that *there is no other way* of progress and of escape from habits that no longer fit them. Whole generations of statesmen will fumble over reforms for a lifetime which are put into full-blooded execution within a week of a declaration of war. There is *no other way*. Only by intense sufferings can the nations grow, just as a snake once a year must with anguish slough off the once beautiful coat which has now become a strait jacket.

General Sir Ian Hamilton, *Gallipoli Diary*, 1918.

National insecurity is one of the fundamental causes of war, especially if the nation concerned is militarily powerful. All nations are impelled by the instinct of national preservation to seek secure frontiers, and, if secure frontiers cannot be gained by peaceful methods, powerful nations will seek to secure them by war. a strong frontier is nothing else than a natural fortress, which, when garrisoned, secures the nation against attack. The object is the security of the nation, consequently... the breaking down of the national will is the surest means of forcing the fortress to collapse.

Major-General J.F.C. Fuller, *The Foundation of the Science of War*, 1926.

The more I study war, the more I come to feel that the cause of war is fundamentally psychological rather than political or economic. When I come in contact with a militarist, his stupidity depresses me and makes me realize the amount of human obtuseness that has to be overcome before we can make progress towards peace. But contact with pacifists too often has the effect of making me almost despair of the elimination of war, because in their very pacifism the element of pugnacity is so perceptible. Moreover, with this active dislike, which in itself hinders understanding, there is mingled an attitude toward it rather like the Arabs' attitude to disease – that it is an affliction of Providence, against which 'charms' may avail, but not scientific study. Until we understand war in the fullest sense, which involves an understanding of men in war, among other elements, it seems to me that we can have no more prospect of preventing war than the savage has of preventing plague. (May 1934.)

Captain Sir Basil Liddell Hart, *Thoughts on War*, 1944.

There is no merit in putting off a war for a year, if when it comes, it is a far worse war or one much harder to win.

Sir Winston S. Churchill, *The Gathering Storm*, 1948.

WAR

From the beginning of man's recorded history physical force, or the threat of it, has been freely and incessantly applied to the resolution of social problems. It persists as an essential element in the social pattern. History suggests that as a society of men grows more orderly the application of forces tends to become better ordered. The requirement for it has shown no sign of disappearing.
General Sir John Hackett, *The Profession of Arms*, 1963.

One clear picture emerges: that the wars... none of them were inevitable. They were not the product of forces beyond human control, nor were they the inescapable result of the development of the armed forces themselves and the trade in arms which supplied them, although both influence the policies which governments pursued in support of what they believe to be the interests of their nations. It was those policies which were themselves the causes of war.
Field Marshal Lord Carver, *Twentieth Century Warriors*, 1987.

Civil War

Lost to the clan,
lost to the hearth, lost to the old ways, that one
who lusts for all the horrors of war with his own people.
Homer, *The Iliad*, 9.73–76, c. 800 BC, tr. Robert Fagles, 1990; speech by King Nestor of Pylos to reconcile Agamemnon and Achilles.

Alas for Greece, how many brave men have you killed with your own hands! If these men were still alive, they could have conquered all the barbarians in the world.
Agesilaus (444–360 BC), Eurypontid King of Sparta, his distress at the large number of Greeks killed in battle against other Greeks, quoted in Plutarch, *The Lives*, c. AD 100 (*The Age of Alexander*, tr. Ian Scott-Kilvert, 1973).

Unhappy it is though to reflect, that a Brother's Sword has been sheathed in a Brother's breast, and that, the once happy and peaceful plains of America are either to be drenched with Blood, or Inhabited by Slaves. Sad alternative! But can a virtuous Man hesitate in his choice?
General George Washington, 31 May 1775, to George William Fairfax, *The Writings of George Washington*, III, p. 292, John C. Fitzgerald, ed., 1931.

I am one of those who have probably passed a longer period of my life engaged in war than most men, and principally in civil war; and I must say this, that if I could avoid, by any sacrifice whatever, even one month of civil war in the country to which I was attached, I would sacrifice my own life in order to do it.
The Duke of Wellington, March 1829, to the House of Lords.

The Economy of War

War is a matter not so much of arms as of money, which makes arms of use.
Archidamus II, King of Sparta, 432 BC, speech on the prospect of war with Athens, quoted in Thucydides, *History of the Peloponnesian War*, 1.83, c. 404 BC, tr. Richard Crawley, 1910.

Fight thou with shafts of silver and thou shalt conquer in all things.
Response of the Delphic Oracle to Philip of Macedon asking how he might be victorious in war, quoted from Plutarch, *Apothegms*.

I would have sold London itself if I could have found a buyer.
Richard I (attributed), 1189, reflecting on his desperate search for money to finance his participation in the Third Crusade, quoted in M.J. Cohen, *History in*

WAR

Quotations, 2004.

Money, more money, always money.
> Marshal Gian Giacomo de Trivulce (1441–1518), when François I, King of France, asked him what he needed to make war.

Men, arms, money, and provisions are the sinews of war, but of these four, the first two are the most necessary; for men and arms will always find money and provisions, but money and provisions cannot always raise men and arms.
> Niccolò Machiavelli, *The Art of War*, 1521.

The army is a school in which the miser becomes generous, and the generous prodigal; miserly soldiers are like monsters, very rarely seen.
> Miguel de Cervantes, *Don Quixote*, 1615.

The army is a sack with no bottom.
> General Simon Bolivar (1783–1830), quoted in *Military History*, 8/1988.

He who has his thumb on the purse has the power.
> Chancellor Otto von Bismarck, 21 May 1869, speech.

For weeks and weeks now I have been trying to make bricks without straw, which in itself is bad enough, but which is made much worse when others believe you have the straw.
> Major-General George Alan Vasey (1895–1945), quoted in Horner, *The Commanders*, 1984.

Guerrilla/Partisan War

It is not to be expected that the rebel Americans will risk a general combat or a pitched battle, or even stand at all, except behind intrenchments as at Boston. Accustomed to felling of timber and to grubbing up trees, they are very ready at earthworks and palisading, and will cover and intrench themselves wherever they are for a short time left unmolested with surprising alacrity... Composed as the American Army is, together with the strength of the country, full of woods, swamps, stone walls, and other inclosures and hiding places, it may be said of it that every private man will in action be his own general, who will turn every tree and bush into a kind of temporary fortress, and from whence, when he hath fired his shot with all deliberation, coolness, and uncertainty which hidden safety inspires, he will skip as it were to the next, and so on for a long time till dislodged either by cannon or by a resolute attack of light infantry.
> Lieutenant-General John 'Gentleman Johnny' Burgoyne (1722–1792), quoted in Lloyd, *A Review of the History of Infantry*, 1908.

The military value of a partisan's work is not measured by the amount of property destroyed or the number of men killed or captured, but by the number he keeps watching.
> Colonel John S. Mosby, *Mosby's War Reminiscences*, 1887.

To oppose successfully such bodies of men as our burghers had to meet during this war demanded *rapidity of action* more than anything else. We had to be quick at fighting, quick at reconnoitring, quick (if it became necessary) at flying!
> Christiaan R. de Wet, *Three Years War*, 1902.

Guerrilla war is far more intelligent than a bayonet charge.
> Colonel T.E. Lawrence, 'The Science of Guerrilla Warfare', *Encyclopaedia Britannica*, 1929.

[Guerrilla War] must have a friendly population, not actively friendly, but sympathetic to the point of not betraying rebel movements to the enemy. Rebellions can be made by two percent active in a striking force, and 98 passively sympathetic.

Colonel T.E. Lawrence, 'The Science of Guerrilla Warfare', *Encyclopaedia Britannica*, 1929.

The advantages are nearly all on the side of the guerrilla in that he is bound by no rules, tied by no transport, hampered by no drill-books, while the soldier is bound by many things, not the least by his expectation of a full meal every so many hours. The soldier usually wins in the long run, but very expensively.

Field Marshal Viscount Wavell of Cyrenaica, 30 August 1932, critique on a counter-guerrilla exercise, Blackdown.

When the enemy advances, we retreat.
When he escapes we harass.
When he retreats we pursue.
When he is tired we attack.
When he burns we put out the fire.
When he loots we attack.
When he pursues we hide.
When he retreats we return.

Mao Tse-tung (1893–1976).

Guerrilla war, too, inverts one of the main principles of orthodox war, the principle of 'concentration' – and on both sides. Dispersion is an essential condition of survival and success on the guerrilla side, which must never present a target and thus can operate only in minute particles, though these may momentarily coagulate like globules of quicksilver to overwhelm some weakly guarded objective. For guerrillas as the principle of 'concentration' has to be replaced by that of 'fluidity of force'...

Captain Sir Basil Liddell Hart, *Strategy*, 1954.

In many ways this war [against Sinn Fein] was far worse than the Great War which had ended in 1918. It developed into a murder campaign in which, in the end, the soldiers became very skilful and more than held their own. But such a war is thoroughly bad for officers and men; it tends to lower their standards of decency and chivalry, and I was glad it was over.

Field Marshal Viscount Montgomery of Alamein, *The Memoirs of Field Marshal Montgomery*, 1958.

Concentration of troops to realize an overwhelming superiority over the enemy where he is sufficiently exposed in order to destroy his manpower; initiative, suppleness, rapidity, surprise, suddenness in attack and retreat. As long as the strategic balance of forces remains disadvantageous, resolutely to muster troops to obtain absolute superiority in combat in a given place, and at a given time. To exhaust little by little by small victories the enemy forces and at the same time to maintain and increase ours. In these concrete conditions it proves absolutely necessary not to lose sight of the main objective of the fighting that is the destruction of the enemy manpower. Therefore losses must be avoided even at the cost of losing ground.

General Vo Nguyen Giap, *People's War, People's Army*, 1961.

Advance like foxes, fight like lions, and fly like birds.

Northeastern American Indian tactical maxim.

This war is very long, and always think of this as the beginning, and always make the enemy think that yesterday was better than today.

Abu Musab al-Zarqawi, to the insurgents in Iraq just before the American attack on Fallujah on 8 November 2004, quoted in 'Victory in Fallujah,' *WSJ*, 17 November 2004.

Just War

You yourself acknowledge that you

WAR

were the aggressors and of this the gods are witnesses. They granted to us a righteous victory in that war, and shall grant us victory now.
 Scipio Africanus, 202 BC, his reply to Hannibal the day before the Battle of Zama, which ended the Second Punic War with the defeat of Carthage, quoted in Livy, Annals of the Roman People, c. 25 BC–AD 17 (Livy: A History of Rome, 30, tr. Moses Hadas and Joe Poe, 1962).

The cause of war must be just.
 The Emperor Maurice, The Strategikon, c. AD 600 (Maurice's Strategikon, tr. George Dennis, 1984).

The war is just which is necessary.
 Niccolò Machiavelli, The Prince, 22, 1513.

It is most necessary for a general in the first place to approve his cause, and settle an opinion of right in the minds of his officers and soldiers.
 General George Monck (1608–1670).

War justifies everything.
 Napoleon, 1808, quoted in Christopher Herold, ed., The Mind of Napoleon, 1955.

The most just war is one which is founded upon undoubted rights and which, in addition, promises to the state advantages commensurate with the sacrifices required and the hazards incurred.
 Lieutenant-General Antoine-Henri Baron de Jomini, Summary of the Art of War, 1838, tr. Mendell and Craighill, 1862.

War is a dreadful thing, and unjust war is a crime against humanity. But it is such a crime because it is unjust, not because it is war.
 President Theodore Roosevelt, 23 April 1910, speech at the Sorbonne.

The Love of War
One can achieve his fill of good things, even of sleep, even of making love... rapturous song and the beat and sway of dancing
A man will yearn for his fill of all these joys
before his fill of war.
 Homer, The Iliad, 13.743–746, c. 800 BC, tr. Robert Fagles, 1990; Menelaus appealing to Zeus to slake the Trojan blood lust.

...A man whose passion was war. When he could have kept at peace without shame or damage, he chose war; when he could have been idle, he wished for hard work that he might have war; when he could have kept wealth without danger, he chose to make it less by making war; there was a man who spent upon war as if it were a darling lover or some other pleasure.
 Xenophon, Anabasis, c. 360 BC (The March Up Country, tr. W.H.D. Rouse, 1959), on the Spartan mercenary general Clearchus.

A shot through the mainmast knocked a few splinters about us. He observed to me, with a smile, 'It is warm work, and this day may be the last to any of us at a moment;' and then, stopping short at the gangway, he used an expression never to be erased from my memory, and said with emotion, 'but mark you, I would not be elsewhere for thousands.'
 Admiral Viscount Nelson, 2 April 1801, during the Battle of Copenhagen, as related by Colonel William Parker, quoted in Alfred Thayer Mahan, The Life of Nelson, II, 1897.

We do not, generally speaking, like the thoughts of peace. I expect I shall remain abroad for three or four years, which, individually, I would sooner spend in war than in peace. There is something indescribably exciting in the former.
 General Charles 'Chinese' Gordon, 1856, quoted in Paul Charrier, Gordon of Khartoum, 1965.

It is well that war is so terrible – we would grow too fond of it.

WAR

General Robert E. Lee to Lieutenant-General James Longstreet, *Battle of Fredericksburg*, 13 December 1862.

The young bloods of the South: sons of planters, lawyers about towns, good billiard-players and sportsmen, men who never did work and never will. War suits them, and the rascals are brave, fine riders, bold to rashness, and dangerous subjects in every sense... The men must all be killed or employed by us before we can hope for peace.

General of the Army William T. Sherman, 17 September 1863, letter to Major-General Henry Halleck.

There are poets and writers who see naught in war but carrion, filth, savagery, and horror... They refuse war the credit of being the only exercise in devotion on the large scale existing in this world. The superb moral victory over death leaves them cold. Each one to his taste. to me this is no valley of death – it is a valley brim full of life at its highest power.

General Sir Ian Hamilton, 30 May 1915, diary entry at Gallipoli.

I'm afraid the war will end very soon now, but I suppose all good things come to an end sooner or later, so we mustn't grumble.

Field Marshal Earl Alexander of Tunis, 1917, letter to his mother while a subaltern during WWI.

Through a Glass, Darkly
Through all the travail of ages,
Midst the pomp and toil of war
Have I fought and strove and perished
Countless times upon this star.
In the forms of many peoples
In all panoplies of time
Have I seen the luring vision
Of the victory Maid, sublime.

General George S. Patton, Jr, 27 May 1922, quoted in Roger N. Nye, *The Patton Mind*, 1993.

War is conflict, fighting an elemental exposition of the age-old effort to survive. It is the cold glitter of the attacker's eye, not the point of the questing bayonet, that breaks the line. It is the fierce determination of the driver to close with the enemy, not the mechanical perfection of a Mark VIII tank that conquers the trench. It is the cataclysmic ecstasy of conflict in the flier not the perfection of his machine gun which drops the enemy in flaming ruin...

In war tomorrow we shall be dealing with men subject to the same emotions as were the soldiers of Alexander; with men but little changed... from the starving shoeless Frenchmen of 1796. With men similar save in their arms to those who the inspiring powers of a Greek or a Corsican changed at a breath to bands of heroes all enduring all capable...

General George S. Patton, Jr, from his 1926 lecture, 'The Secret of Victory', from the Patton Collection, United States Military Academy Library.

I love war and responsibility and excitement. Peace is going to be hell on me.

General George S. Patton, Jr (1885–1945).

I was satisfied by my years with a battalion in France that there is a breed of men who do not seem to fit into the structure of society; these men are vaguely discontented with the vast inhuman life of cities, its prudent, punctual existence, with its money saving, its daily stress on the need of security; they find in the army at least an alternative to the prison life of great towns. 'All warlike people are a little idle, and love danger better than travail.'

Lord Moran (quoting Bacon), *The Anatomy of Courage*, 1945.

If the Germans are a warrior race, they are certainly militarist also. I think they love the military pageant and the panoply of war; and the feeling of

WAR

strength and power that a well-organized and disciplined unit gives to each and every individual member of that unit. I am quite willing to admit that I myself share this curious attraction for the strength and elegence of beautifully trained and equipped formations, with all the art and subtlety of their movements in action against an enemy. I can well understand the enthusiasm which the soldiers – from marshals to the private soldier – showed for Napoleon; and why they followed their leader without doubt or question in his victorious campaigns. Feeling thus, they shared the glory of his conquests.
    Field Marshal Earl Alexander of Tunis, *The Alexander Memoirs*, 1961.

War is the most exciting and dramatic thing in life. In fighting to the death you feel terribly relaxed when you manage to come through.
    General Moshe Dayan, *The Observer* (London), 13 February 1972.

The Nature of War

The LORD is a man of war: The LORD is his name.
    Exodus, 15:3.

The art of war is of vital importance to the state. It is a matter of life and death, a road either to safety or to ruin. Hence it is a subject of inquiry which can on no account be neglected.
    Sun Tzu, *The Art of War*, 1, c. 500 BC, tr. Giles, 1910.

War is the last of all things to go according to programme.
    Thucydides, *The Peloponnesian War*, c. 404 BC.

It is not the object of war to annihilate those who have given provocation for it, but to cause them to mend their ways; not to ruin the innocent and guilty alike, but to save both.
    Polybius, *Histories*, 5, c. 125 BC.

...Not a river, nor a lake, but a whole ocean of all sorts of evil.
    Gustavus II Adolphus (1594–1632), quoted in Nils Ahnlund, *Gustavus Adolphus the Great*, 1940.

War is more than a true chameleon that slightly adapts its characteristics to the given case. As a total phenomenon its dominant tendencies always make war a remarkable trinity – composed of primordial violence, hatred, and enmity, which are to be regarded as a blind force; of the play of chance and probability within which the creative spirit is free to roam; and of its element of subordination, as an instrument of policy, which makes it subject to reason alone.
    Major-General Carl von Clausewitz, *On War*, 1.1.28, 1832, tr. Michael Howard and Peter Paret, 1976.

It is painful enough to discover with what unconcern they speak of war and threaten it. I have seen enough of it to make me look upon it as the sum of all evils.
    Lieutenant-General Thomas J. 'Stonewall' Jackson, April 1861, letter.

You cannot qualify war in harsher terms than I will. War is cruelty, and you cannot refine it, and those who brought war into our country deserve all the curse and maledictions a people can pour out.
    General of the Army William T. Sherman, 12 September 1864, letter to James C. Calhoun, the Mayor of Atlanta.

I shall not essay to enlighten you upon the subject of war. Were I to attempt it, I should doubtless miserably fail, for it has long been said, as to amount to an adage, that 'women don't know anything about war.'
    I wish men didn't either. They have always known a great deal too much about it for the good of their kind. They have worshipped at Valkyria's

WAR

shrine, and followed her siren lead, till it has cost a million times more than the whole world is worth, poured out the best blood and crushed the fairest forms the good God has ever created.
Clara Barton, quoted in Percy Epler, *The Life of Clara Barton*, 1915.

War is an integral part of God's ordering of the universe. In war, man's noblest virtues come into play. Courage and renunciation, fidelity to duty and a readiness for sacrifice that does not stop short of offering up life itself. Without war the world would become swamped in materialism.
Field Marshal Helmuth Graf von Moltke (1800–1891), quoted in Toynbee, *War and Civilization*, 1950.

War is a game to be played with a smiling face.

O, horrible war! Amazing medley of the glorious and the squalid, the pitiful and the sublime! If modern men of light and leading saw your face closer, simple folk would see it hardly ever.
Sir Winston S. Churchill, *London to Ladysmith*, 1900.

Let us learn our lessons. Never, never, never believe any war will be smooth and easy, or that anyone who embarks on the strange voyage can measure the tides and hurricanes he will encounter... Antiquated War Offices, weak, incompetent or arrogant Commanders, untrustworthy allies, hostile neutrals, malignant Fortune, ugly surprises, awful miscalculations – all take their seats at the Council Board on the morrow of a declaration of war. Always remember, however, sure you are that you can easily win, that there would not be a war if the other man did not think he also had a chance.
Sir Winston S. Churchill, *A Roving Commission*.

Four things greater than all things are, –
Women and Horses and Power and War.
Rudyard Kipling, 'The Ballad of the King's Jest', stanza 5, *The Collected Works of Rudyard Kipling*, XXV, 1941, p. 234.

In war there are nothing but particular cases; everything has there an individual character; nothing ever repeats itself.
Marshal of France Ferdinand Foch, *Principles of War*, 1913.

The story of the human race is war.
Sir Winston S. Churchill, *Marlborough*, 1933.

The sword is the axis of the world and its power is absolute.
General Charles de Gaulle, *Army of the Future*, 1934.

War is a series of local emergency measures.
Major-General J.F.C. Fuller, *Memoirs of an Unconventional Soldier*, 1936.

War is a contest of strength, but the original pattern of strength changes in the course of the war. Here the decisive factor is subjective effort – winning more victories and committing fewer errors. The objective factors provide the possibility for such change, but in order to turn this possibility into actuality both correct policy and subjective effort are essential. It is then that the subjective plays the decisive role.
Mao Tse-tung, *On Protracted War*, 1938.

War and truth have a fundamental incompatibility. The devotion to secrecy in the interests of the military machine largely explains why, throughout history, its operations commonly appear in retrospect the most uncertain and least efficient of human activities.
Captain Sir Basil Liddell Hart, *Through the Fog of War*, 1938.

This war... is one of those elemental conflicts which usher in a new

WAR

millenium and which shake the world.
> Adolf Hitler, 1942, speech in the Reichstag.

War creates such a strain that all the pettiness, jealousy, ambition and greed, and selfishness begin to leak out of the seams of the average character. On top of this are the problems created by the enemy.
> General of the Army Dwight D. Eisenhower, 16 December 1942, *Letters to Mamie*, 1978.

War is very simple, direct, and ruthless. It takes a simple, direct, and ruthless man to wage war.
> General George S. Patton, Jr, 15 April 1943, diary entry.

War, like politics, is a series of compromises.
> Captain Sir Basil Liddell Hart, *Thoughts on War*, 1944.

The main ethical objection to war for intelligent people is that it is so deplorably dull and usually so inefficiently run.... Most people seeing the muddle of war forget the muddles of peace and the general inefficiency of the human race in ordering its affairs.
> Field Marshal Viscount Wavell of Cyrenaica, unpublished 'Recollections', 1947.

War is always a matter of doing evil in the hope that some good may come of it.
> Captain Sir Basil Liddell Hart, *Defense of the West*, 1950.

War is not to be likened to a task which can be completed by installments, bit by bit, part of which when done is done forever. War is rather a race of an extraordinary character which, once started, has to be run through to the end.
> Sir Winston S. Churchill (1874–1965).

War is very cruel. It goes for so long.
> Sir Winston S. Churchill (1874–1965).

Some people asked me when we attacked retreating Iraqi forces near Basrah, wasn't that 'extreme violence.' They missed the point. War is extreme violence. And the way to halt the suffering is to get the war over as quickly and decisively as you possibly can.
> General Charles Horner, 'Extreme Violence', Al Santoli, ed., *Leading the Way: How Vietnam Veterans Rebuilt the US Military*, 1993.

...About the general conduct of war. There are, in the end, no humanitarian wars. War is serious and it is deadly. Casualties, including civilian casualties, are to be expected. Trying to fight a war with one hand tied behind your back is the way to lose it. We always regret the loss of lives. But we should have no doubt that it is the men of evil, not our troops or pilots, who bear the guilt.
> Margaret Thatcher, 20 April 1999, speech.

The Objectives of War

The legitimate object of war is a more perfect peace.
> General of the Army William T. Sherman, 20 July 1865, speech in St. Louis.

The object of war is to attain a better peace – even if only from your own point of view. Hence it is essential to conduct war with constant regard to the peace you desire. This is the truth underlying Clausewitz's definition of war as a 'continuation of policy by other means' – the prolongation of that policy through the war into the subsequent peace must always be borne in mind. A State which expends its strength to the point of exhaustion bankrupts is own policy, and future.
> Captain Sir Basil Liddell Hart (1895–1970).

If there should be a war between Japan

WAR

and America, then our aim, of course, ought not to be Guam or the Philippines, nor Hawaii or Hong Kong, but a capitulation at the White House, in Washington itself. I wonder whether the politicians of the day really have the willingness to make sacrifices, and the confidence, that this would entail?
  Admiral Isoroku Yamamoto, January 1941, letter to Sasakawa Ryoichi, quoted in Agawa, *The Reluctant Admiral*, 1979. Note: Misrepresented during WWII as a boastful reference to complete victory, in fact this statement was a cold-blooded, cautionary appraisal of the only possible objective of a war against an aroused America.

Politics/Policy and War

A great country can have no such thing as a little war.
  The Duke of Wellington, 1815.

It is clear... that war is not a mere act of policy but a true political instrument, a continuation of political activity by other means. What remains peculiar to war is simply the peculiar nature of its means. War in general, and the commander in any specific instance, is entitled to require that the trend and designs of policy shall not be inconsistent with these means. That, of course, is no small demand; but however much it may affect political aims in a given case, it will never do more than modify them. the political object is the goal, war is the means of reaching it, and means can never be considered in isolation from their purpose.
  Major-General Carl von Clausewitz, *On War*, 1.1.24, 1832, tr. Michael Howard and Peter Paret, 1976.

Politics uses war for the attainment of its ends; it operates decisively at the beginning and at the end, of course in such manner that it refrains from increasing its demand during the war's duration or from being satisfied with an inadequate success.... Strategy can only direct its efforts toward the highest goal which the means available make attainable. In this way, it aids politics best, working only for its objectives, but in its operations independent of it.... There is uncertainty in war, but the aims of policy remain. Policy must go hand in hand with strategy.
  Field Marshal Helmuth Graf von Moltke, *Ueber Strategie*, 1871.

Place in the hands of the King of Prussia the strongest possible military power, then he will be able to carry out the policy you wish; this policy cannot succeed through speeches, and shooting matches and songs; it can only be carried through blood and iron.
  Chancellor Prince Otto von Bismarck, 28 January 1886, speech in the Reichstag.

That the soldier is but the servant of the statesman, as war is but the instrument of diplomacy, no educated soldier will deny. Politics must always exercise an extreme influence on strategy; but it cannot be gainsaid that interference with the commanders in the field is fraught with the gravest danger.
  Colonel George F.R. Henderson, *Stonewall Jackson*, 1898.

Policy is a thought process, while war is merely a tool, and not the reverse.
  V.I. Lenin (1870–1924), quoted in Gorshkov, *Sea Power of the State*, 1979.

War is not the continuation of policy. It is the breakdown of policy.
  Colonel-General Hans von Seekt, *Thoughts of a Soldier*, 1930.

There are some militarists who say: 'We are not interested in politics but only in the profession of arms.' It is vital that these simple-minded militarists be made to realize the relationship that exists between politics and military affairs. Military action is a method used to

attain a political goal. While military affairs and political affairs are not identical, it is impossible to isolate one from the other.
  Mao Tse-tung, On Guerrilla Warfare, 1937.

History shows that gaining military victory is not in itself equivalent to gaining the object of policy. But as most of the thinking about war has been done by men of the military profession there has been a very natural tendency to lose sight of the basic national object, and identify it with the military aim. In consequence, whenever war has broken out, policy has too often been governed by the military aim – and this has been regarded as an end in itself, instead of as merely a means to the end.
  Captain Sir Basil Liddell Hart, Strategy, 1954.

Protracted War

Aren't you sick of being caged inside those walls?
Time was when the world would talk of Priam's Troy
as the city rich in gold and rich in bronze – but now
our houses are stripped of all their sumptuous treasures,
troves sold off and shipped to Phrygia, lovely Maeonia,
once great Zeus grew angry...
  Homer, The Iliad, 18.332–338, c. 800 BC, tr. Fagles, 1990; Hector's account of the cost to Troy of the Trojan War, then in its tenth year.

When you engage in actual fighting, if victory is long in coming, the men's weapons will grow dull and their ardour will be damped. If you lay siege to a town, you will exhaust your strength. Again, if the campaign is protracted, the resources of the State will not be equal to the strain.
  Now, when your weapons are dulled, your ardour damped, your strength exhausted and your treasure spent, other chieftains will spring up to take advantage of your extremity. Then no man, however wise, will be able to avert the consequences that must ensue.
  Thus, though we have heard of stupid haste in war, cleverness has never been associated with long delays. There is no instance of a country having been benefited by a prolonged war.
  Sun Tzu, The Art of War, 2, c. 500 BC, tr. Giles, 1910.

A friend came to see me on one of the evenings of the last week – he thinks it was on Monday August 3. We were standing at a window of my room in the Foreign office. It was getting dusk, and the lamps were being lit in the space below on which we were looking. My friend recalls that I remarked on this with the words: 'The lamps are going out all over Europe; we shall not see them lit again in our lifetime.'
  Sir Edward Grey, British Foreign Secretary, 3 August 1914, the day Germany declared war on France.

The German people, both at home and at the front, have suffered and endured inconceivable hardships in the four long years of war. The war has undermined and disintegrated patriotic feeling and the whole national *moral* [morale].
  Poisonous weeds grew in this soil. All German sentiment, all patriotism, died in many breasts. Self came first. War profiteers of every kind, not excluding the political variety, who took advantage of the country's danger and the Government's weakness to snatch political and personal advantages, became more and more numerous. Our resolution suffered untold harm. We lost confidence in ourselves.
  General Erich Ludendorff, My War Memories, 1914–1918, 1919.

The experience of history brings ample evidence that the downfall of civilized states tends to come not from the direct assaults of foes but from internal decay, combined with the consequences of exhaustion in war. A State of suspense is trying – it has often led nations as well as individuals to commit suicide because they were unable to bear it. But suspense is better than to reach exhaustion in pursuit of the mirage of victory. Moreover, a truce to actual hostilities enables a recovery and development of strength, while the need for vigilance helps keep a nation on 'its toes'.

Captain Sir Basil Liddell Hart, *Strategy*, 1954.

The Results of War

Surely it would have been best if the gods had granted our fathers forebearance – if yours had been content to be masters of Italy, ours, to rule over Africa. Even for you Sicily and Sardinia are not sufficient return for the loss of so many fleets, so many armies, so many great generals. But the past is more readily lamented than corrected. We coveted the wealth of others and we have had to fight to keep our own. It has meant war for us in Africa, for you in Italy: You have seen the standards of an enemy almost within your gates and on your walls; in Carthage we can hear the armies of the Roman camp.

Hannibal, 202 BC, his unsuccessful plea to Scipio Africanus for a negotiated peace on the day before the Battle of Zama, which ended the Second Punic War with the defeat of Carthage, quoted in Livy, *Annals of the Roman People*, c. 25 BC–ad 17 (*Livy: A History of Rome*, 30, tr. Moses Hadas and Joe Poe, 1962).

Weapons are instruments of ill omen, war is immoral.

Really they are only to be resorted to when there is no other choice. It is not right to pursue aggressive warfare because one's country is large and prosperous, for this ultimately ends in defeat and destruction. Then it is too late to have regrets. Military action is like a fire – if not stopped it will burn itself out. Military expansion and adventurism soon lead to disaster.

The rule is 'Even if a country is large, if it is militaristic it will eventually perish.'

Liu Ji (1310–1375), *Lessons of War* (*Mastering the Art of War*, tr. Thomas Cleary, 1989).

What a cruel thing is war: to separate and destroy families and friends, and mar the purest joys and happiness God has granted us in this world; to fill our hearts with hatred instead of love for our neighbors, and to devastate the fair face of this beautiful world.

General Robert E. Lee, 25 December 1862, letter to his wife.

I begin to regard the death and mangling of a couple thousand men as a small affair, a kind of morning dash – and it may well be that we become so hardened.

General of the Army William T. Sherman, July 1864, letter to his wife.

A war, even the most victorious, is a national misfortune.

Field Marshal Helmuth Graf von Moltke, 1880, letter.

I am not worried about the war; it will be difficult but we shall win it; it is after the war that worries me. Mark you, it will take years and years of patience, courage, and faith.

Field Marshal Jan Christian Smuts (1870–1950).

War stirs in men's hearts the mud of their worst instincts. It puts a premium on violence, nourishes hatred, and gives free rein to cupidity. It crushes the weak, exalts the unworthy, and bolsters tyranny. Because of its blind fury many of the noblest schemes have come to

nothing and the most generous instincts have more than once been checked. Time and time again it has destroyed all ordered living, devastated hope, and put the prophets to death. But, though Lucifer has used it for his purposes so, sometimes, has the Archangel. With what virtues has it not enriched the moral capital of mankind! Because of it, courage, devotion, and nobility have scaled the peaks. It has conferred greatness of spirit on the poor, brought pardon to the guilty, revealed the possibilities of self-sacrifice to the commonplace, restored honor to the rogue, and given dignity to the slave. It has carried ideas in the baggage wagons of its armies, and reforms in the knapsacks of its soldiers. It has blazed a trail for religion and spread across the world influences which have brought renewal to mankind, consoled it, and made it better. Had not innumerable soldiers shed their blood there would have been no Hellenism, no Roman civilization, no Christianity, no Rights of Man, and no modern developments.
General Charles de Gaulle, *The Edge of the Sword*, 1932.

To some the game of war brings prizes, honor, advancement or experience; to some the consciousness of duty well discharged... But here were those who had drawn the evil numbers – who had lost their all, to gain only a soldier's grave. Looking at those shapeless forms confined in a regulation blanket, the pride of race, the pomp of Empire, the glory of war appeared but the faint unsubstantial fabric of a dream.
Sir Winston S. Churchill (1874–1965).

The cost of war is constantly spread before me, written neatly in many ledgers whose columns are gravestones.
General of the Army George C. Marshall, quoted in *MHQ*, Spring 1995.

Mussolini said in the early 1930s: 'War alone brings all human energies to their highest tension, and sets a seal of nobility on the people who have the virtue to face it.' This is rubbish, and dangerous rubbish at that. War does not ennoble. Kant's view that war has made more bad people than it has destroyed is probably nearer the mark. But the interesting thing is that although war almost certainly does not ennoble, the preparation of men to fight in it almost certainly can and very often does.
General Sir John Hackett, *The Profession of Arms*, 1963.

A professional soldier understands that war means killing people, war means maiming people, war means families left without fathers. All you have to do is hold your first dying soldier in your arms, and have that terribly futile feeling that his life is flowing out and you can't do anything about it. Then you understand the horror of war. Any soldier worth his salt should be antiwar. And still there are things worth fighting for.
General H. Norman Schwarzkopf, *Reader's Digest*, 8/1995.

We have lost far too many people in this century – to war, revolution, persecution. A total of 75 million Russians have died in this way. We have enough space, enough resources, to ensure a life with human dignity in our territory. We have simply never been allowed to do so, because we have had to go to war all the time and kill.
Colonel-General Aleksandr I. Lebed, *Det fri Aktuelt* (Copenhagen), 2 July 1997.

A great war leaves the country with three armies: an army of cripples, an army of mourners, and an army of thieves.
American proverb.

WAR CORRESPONDENTS

I have made arrangements for the

WAR CORRESPONDENTS

correspondents to take the field... and I have suggested to them that they wear a white uniform to indicate the purity of their profession.
>Major-General Irvin McDowell, quoted in William Russell, My *Diary North and South*, 1861.

Good! Now we'll have news from hell before breakfast.
>General William Tecumseh Sherman, 1863, upon being told three journalists had been killed by artillery during the siege of Vicksburg, quoted in Dixton Wecter, *The Hero in America; A Chronicle of Hero-Worship*, 12.3, 1941.

Newspaper correspondents with an army, as a rule, are mischievous. They are the world's gossips, pick up and retail the camp scandal, and gradually drift to the headquarters of some general, who finds it easier to make reputation at home than with his own corps or division. They are also tempted to prophesy events and state facts which, to an enemy, reveal a purpose in time to guard against it. Moreover, they are always bound to see facts coloured by the partisan or politcal character of their own patrons, and thus bring army officers into the political controversies of the day, which are always mischievous and wrong. Yet, so greedy are the people at large for war news, that it is doubtful whether an army commander can exclude all reporters, without bringing down on himself a clamor that may imperil his own safety. Time and moderation must bring a just solution to this modern difficulty.
>General William T. Sherman, *Memoirs of General W.T. Sherman*, II, 1875.

Those newly invented curse to armies... that race of drones who are an encumbrance to an army; they eat the rations of the fighting man and do not work at all.
>Field Marshal Viscount Wolseley, *The Soldier's Pocket-book*, 1869.

The British public likes to read sensational news, and the best war correspondent is he who can tell the most thrilling lies.
>Field Marshal Earl Haig, 1898, of the Sudan campaign, letter to his sister.

So the presence of a correspondent on the firing-line, or his absence from it, does not prove that he is not doing his full duty to his paper. The best correspondent is probably the man who by his energy and resource sees more of the war, both afloat and ashore, than do his rivals, and who is able to make the public see what he saw.
>Richard Harding Davis, 'Our War Correspondents in Cuba and Puerto Rico', *Harper's Monthly*, May 1899.

'All the danger of war and one-half per cent of the glory': such is our motto, and that is the reason why we [the press] expect large salaries.
>Sir Winston S. Churchill, *Ian Hamilton's March*, 1900.

There is no parallel... between the work of a war correspondent and the obligation of the serving soldier. The latter has to stick it and has no choice about the risks he will run. With very few notable exceptions, our war correspondents will not stay with combat danger long enough to begin to understand the ordeal of troops. They are another variety of sightseer. They flit in and out of the scene, hear a few shells explode, take a quick look at frontal living conditions, ask a few trivial questions of the hometown boys, and then beat it back to secure billets to pound out tear-jerking pieces about the horrors of war. With few exceptions, they will not stay on the job.
>Brigadier-General S.L.A. Marshall, *Bringing Up the Rear*, 1979.

WAR OFFICE/PENTAGON

To attack journalists is as pointless as fighting against women. If you lose to them, you just stand there foolishly. If you win, you just stand there twice as foolishly.
   General Aleksandr I. Lebed, *Nezavisimaya Gazeta* (Moscow), 29 August 1997.

[O]ver the years, on battlefields all over the world, I have enjoyed encounters and made friendships with soldiers and sailors that I cherish, and which I think some of them, too, have gained pleasure from. My experiences among warriors have been among the most important of my life. On their side, it is compulsory to express disdain for publicity. Yet it is only natural for men who are risking their lives for their country to appreciate recognition. However much they dislike the media, commanders of Western armies have been obliged to acknowledge that soldiers in modern war want to know that their efforts and sacrifices are being reported home. They need journalists to tell their story. I am one of those who has often been happy to do so, although this book also plenty of occasions of deceiving military authority in various parts of the world. That has been my job as a journalist and, to be honest, great fun too. Even when we admire soldiers, it is not our business to record their doings on their terms. Of course, I have often been frightened and run away. I have seen soldiers do some terrible things. I have also, however, seen them do fine and even great ones.
   Max Hastings, *Going to the Wars*, 2000, pp. xvii-xviii.

It may not be the journalist's fault: it is a reporter's job simply to report what he or she finds. But without being framed in a broader understanding of strategy, instant pictures can mislead. So while viewers may be 'seeing'' more than ever before, they may actually be 'learning' less, albeit in a more spectacular fashion.
   Secretary of State for Defence Geoff Hoon, 29 March 2003, *Times of London*.

WAR OFFICE/PENTAGON

It is a good thing to see the inside of the War Office for a short time, as it prevents one from having any respect for an official letter, but it is a mistake to remain there too long.
   Field Marshal Earl Haig, 1909, upon leaving the War Office, quoted in J.H. Marshall Cornwall, *Haig as Military Commander*, 1973.

To Hell with the War Department,
   General John J. Pershing, Summer 1917, a favourite expression as he struggled to establish the structure American Expeditionary Force (AEF) in France, quoted in Lawrence Stallings, *The Doughboys*, 1963, p. 45.

Training was, in fact, largely a case of trying to make bricks without straw, and there was much truth in what a distinguished General once said to me: 'Never forget, Robertson, that we have two armies – the War Office army and the Aldershot army. The first is always up to strength, and is organized, reorganized, and disorganized almost daily. The second is never up to strength, knows nothing whatever about the first, and remains unaffected by any of these organizing activities. It just cleans its rifle and falls in on parade.'
   Field Marshal Sir William Robertson, *From Private to Field-Marshal*, 1921.

Sir Henry Wilson was right, for though it is true that, whatever the circumstances might have been, he could not have reorganised the army, with such an instrument as the War Office headed by the army Council it was impossible for any man, even had he possessed the wisdom of Athena and the driving

485

force of Mars, to organize anything. He was in the position of a painter of miniatures equipped with a boot brush. Seeing how foolish it all was, he daubed out caricatures to make people laugh, and in this he was eminently successful.

Before we can blame him or any soldier in his position, it is necessary to understand what a Government Department is like; yet in all probability, in 1919, the War Office was more alive than most others. I began to realise this soon after the war ended, when a naval officer came over from the Admiralty to see me about persuading us to introduce intelligence tests in our recruiting. I said to him: "Intelligence,' that is not exactly the word for us, surely?' To which he replied: 'Oh! You are far more alive on this side of Whitehall. Do you realise that on our side, there are old, old men lost in the cellars of the admiralty still counting the round shot we fired at Sebastopol?'

Major-General J.F.C. Fuller, *Memoirs of an Unconventional Soldier*, 1936.

It was a dull January morning when an elderly Major-General strolled into the War Office to pass the time of day. Finding his friend engaged, he waited in one of the passages on the second floor. a young Staff Captain with a large bundle of branch memoranda approached him, and the Major-General, no longer able to contain himself, turned to him and said:

'Isn't it ridiculous? Winston has been made Secretary for War.'

The Staff Captain looked at him wearily and answered:

'It is ridiculous. The only man who could put this business straight is Trotsky!'

Major-General J.F.C. Fuller, *Memoirs of an Unconventional Soldier*, 1936.

A different illustration of the dangers of taking things too seriously was provided by an episode which casts a startling light on how 'official machinery' can go wrong, and served to show that no staff system should be regarded as infallible. One day we received an order for the immediate demobilization of an officer whom we in the General Staff branch had specially asked to retain. Feeling that it must be a mistake I asked the Adjutant-General's branch to check the order with the War Office, but was told that it would be improper to question such a clear and precise order. Not willing to be put off, I set out in person to trace back the course of the order through the various channels, and eventually tracked it down to a typist's mistake in extending a bracket one line too far when making a combined list of officers who were to be respectively retained or demobilised.

Captain Sir Basil Liddell Hart, 1920, *The Memoirs of Captain Liddell Hart*, I, 1965.

The Pentagon was a sorry place to light after having commanded a theater of war.

General Dwight D. Eisenhower, 15 December 1945, letter to his son, John, quoted in Stephen Ambrose, *Eisenhower*, I, 1983.

WAR ON TERRORISM

This is a battle with only one outcome: our victory not theirs.

Prime Minister Tony Blair, speech to the Labour Party Conference, 2 October 2001

If you harbor a terrorist, if you support a terrorist, if you feed a terrorist, you're just as guilty as the terrorists. And, the Taliban found out what we meant.

President George W. Bush, 15 September 2003, address to the 3rd Infantry Division, Fort Stewart, Georgia, upon its return from Iraq, *AUSA News Release*.

[W]ill be won on the offensive.

President George W. Bush, 15 September 2003, address to the 3rd Infantry Division, Ft. Stewart, Georgia, upon its return from Iraq, *AUSA News Release*.

WEAPONS

This is the struggle which engages us. It is a new type of war. It will rest on intelligence to a greater degree than ever before. It demands a different attitude to our own interests. It forces us to act even when so many comforts seem unaffected, and the threat so far off, if not illusory. In the end, believe your political leaders or not, as you will. But do so, at least having understood our minds.
> Prime Minister Tony Blair, speech, 5 March 2004.

If the terrorists are able to get us to change our behavior dramatically so that we are no longer functioning as a free people, then they've won.
And we simply can't let that happen.
> Secretary of Defense Donald Rumsfeld, 8 September 2004, quoted in 'Rumsfeld: Ending Terrorism Could Take a Long Time,' American Forces Press Service.

WEAPONS

After God, we should place our hopes of safety in our weapons, not in our fortifications alone.
> The Emperor Maurice, *The Strategikon*, c. AD 600 (*Maurice's Strategikon*, tr. George Dennis, 1984).

The sword is the soul of the warrior. If any forget or lose it he will not be excused.
> Tokugawa Ieyasu (1543–1616), quoted in A.L. Sadler, *The Maker of Modern Japan: The Life of Tokugawa Ieyasu*.

The means of destruction are approaching perfection with frightful rapidity.
> Lieutenant-General Antoine-Henri Baron de Jomini, *Summary of the Art of War*, 1838, tr. Mendell and Craighill, 1862.

New weapons would seem to be regarded merely as an additional tap through which the bath of blood can be filled all the sooner.
> Captain Sir Basil Liddell Hart, *Paris, Or the Future of War*, 1925.

As long ago as January 1919 I pointed out in *Weekly Tank Notes* that 'Tools or weapons, if only the right ones can be discovered, form 99 per cent. of victory,' and that all other things are no more than the '1 per cent. which makes the whole possible. Indeed, "Savage animalism is nothing, inventive spiritualism is all."...' I pointed out that though Napoleon was an infinitely greater General than Lord Raglan, the Minié rifle of 1854 would have enabled Lord Raglan to have annihilated him in 1815, and that had Napoleon, in 1805, 'placed down as a challenge to the mechanical intellect of France 25,000,000 francs to produce a weapon 100 per cent. more efficient than the 'Brown Bess,' it is almost a certainty that, in 1815, he would have got it; that he would have won Waterloo, and that the whole course of history would have been changed....'
> Major-General J.F.C. Fuller, *Memoirs of an Unconventional Soldier*, 1936.

There are two universal and important weapons of the soldier which are often overlooked – the boot and the spade. Speed and length of marching has won many victories; the spade saved many defeats and gained time for victory.
> Field Marshal Earl Wavell of Cyrenaica, *The Good Soldier*, 1948.

I spy my service pistol. automatically I pick it up, remove the clip, and check the mechanism. It works with buttered smoothness. I weigh the weapon in my hand and admire the cold, blue glint of its steel. It is more beautiful than a flower; more faithful than most friends.
> First-Lieutenant Audie Murphy, *To Hell and Back*, 1949. Murphy was referring to the US Army's Colt .45-calibre semi-automatic pistol, but using the older term 'service pistol'.

WEATHER

England were but a fling, save for the crooked stick and the greygoose wing.
English proverb, referring to the longbow, the clothyard arrow and the deadly skill of the English archer.

WEATHER

Climate is what you expect but weather is what you get.
American military saying.

It is always necessary to shape operation plans... on estimates of the weather, and as this is always changing, one cannot imitate in one season what has turned out well in another.
Frederick the Great, *Instructions to His Generals*, 1747, tr. Phillips, 1940.

Complaints have been brought to my attention that the infantry have got their feet wet. That is the fault of the weather. The march was made in the service of the most mighty monarch. Only women, dandies, and lazy-bones need good weather.
Field Marshal Prince Aleksandr V. Suvorov, 1799, to the commander of the Austrian allied force, who complained of having to march in bad weather, quoted in Philip Longworth, *The Science of Victory*, 1966.

I cannot command winds and weather.
Admiral Viscount Nelson, April 1796.

Russia has two generals in whom she can confide – Generals Janvier and Février.
Emperor Nicholas I, *Punch*, 10 March 1853.

Weather is not only to a great extent a controller of the condition of ground, but also of movement. It is scarcely necessary to point out the influence of heat and cold on the human body, or the effect of rain, fog, and frost on tactical and administrative mobility; but it is necessary to appreciate the moral effect of weather and climate, for in the past stupendous mistakes have resulted through deficiency in this appreciation.

Unless the headquarters staff have intimate experience of the conditions surrounding the fighters, two types of battle are likely to be waged – the first between the brains of the army and the enemy, in which case this action will be rendered impotent on account of the muscles being unable to execute the commands of the brains; and the second between the muscles and the enemy, which battle will be disorganized, not so much through the enemy's opposition as through the receipt of orders which are impossible to carry out.
Major-General J.F.C. Fuller, *The Foundation of the Science of War*, 1926.

We must begin our next offensive early, in order to make the most of the fine weather. It is no good waiting until August and then gambling with the Almighty. '*Gott mit uns*' is all very well; but the Deity expects man to consult the almanac before He obliges him by tampering with the seasons.
Major-General J.F.C. Fuller, *Memoirs of an Unconventional Soldier*, 1936.

The Admiral cannot take up a position that only in ideal conditions of tide and moon can the operation be begun. It has got to be begun as soon as possible, as long as conditions are practicable, even though they are not the best. People have to fight in war on all sorts of days, and under all sorts of conditions.
Sir Winston S. Churchill, note to General Ismay regarding the Dakar operation, 19 August 1940.

This is sheer torture for the troops, and for our cause it is a tragedy, for the enemy is gaining time, and in spite of all our plans we are being carried deeper into winter. It really makes me sad. The best of intentions are wrecked by the

WILL

weather. The unique opportunity to launch a really great offensive recedes further and further, and I doubt if it will ever recur. God alone knows how things will turn out. One must just hope and keep one's spirits up, but at the moment it is a great test.
    Colonel-General Heinz Guderian, November 1941, quoted in von Mellenthin, *German Generals of World War II*, 1977.

Held a final meeting at 0415. This time the prophets [weather forecasters] came in smiling, conditions having shown a considerable improvement. It was therefore decided to let things be and proceed... I am under no delusions as the risks involved in this most difficult of all operations... We shall require all the help that God can give us & I cannot believe that this will not be forthcoming.
    Admiral Sir Bertram Ramsay, 5 June 1944, diary entry referring to the next day's Normandy invasion in which he commanded the Allied naval forces, *1944: The Year of D-Day*, 1994.

WHITEHALL

In case you should run away with any false ideas from what I say it will be as well to explain that I am a lunatic. I have just been told I am a lunatic by several people in Whitehall. That is a street of offices where all the wise men in England sit at desks and write papers to each other. Of course they don't read each other's papers; they are carefully docketed by a large staff of clerks and then go into pigeon-holes and stay there. I belong to the Office which deals with soldiers. Perhaps you didn't know that we have any soldiers – but we have. You can see two of them in Whitehall any day. On big occasions you may see a couple of thousand on the Horse Guards parade.
    Field Marshal Sir Henry Wilson, before 1914, quoted in C.R. Ballard, *Kitchener*, 1930, p. 204.

WILL

As each man's strength gives out, as it no longer responds to his will, the inertia of the whole gradually comes to rest on the commander's will alone. The ardor of his spirit must rekindle the flame of purpose in all others; his inward fire must revive their hope. Only to the extent that he can do this will he retain his hold on his men and keep control.
    Major-General Carl von Clausewitz, *On War*, 1832, 1, tr. Michael Howard and Peter Paret, 1976.

In battle, two moral forces, even more than two material forces, are in conflict. The stronger conquers. The victor has often lost... more men than the vanquished... With equal or even inferior power of destruction, he will win who is determined to advance.
    Colonel Charles Ardant du Picq, *Battle Studies*, 1880, tr. Greely, 1957.

The will to conquer: such is the first condition of victory, consequently the first duty of every soldier; and it is also the supreme resolution with which the commander must fill the soul of his subordinates.
    Marshal of France Ferdinand Foch, *Principles of War*, 1913.

In any war victory is determined in the final analysis by the state of mind of the masses which shed their blood on the field of battle.
    Mikhail V. Frunze (1885–1925).

I have been awarded the Iron Cross for my modest share in the battle of Tannenberg. I had never thought that this finest of all military decorations would be won by sitting at the end of a telephone line. However, I realize now that there must be someone there who keeps his nerve, and by brute determination and will to victory overcomes difficulties, panics and suchlike nonsense.
    General Max Hoffmann, 9 September

WIVES

1914, *War Diaries and Other Papers*, I, 1929.

In war the chief incalculable is the human will.
    Captain Sir Basil Liddell Hart,
    Encyclopaedia Britannica, 1929.

Hindenburg was right when he said that the man who wins a war is the one whose nerves are the strongest. Our nerves proved particularly feeble, because we had so long been forced by the inadequacy of our technical equipment to make up the deficit by an excessive wastage of life. One cannot fight successfully with bare hands against an enemy provided with all the resources of modern warfare and inspired by patriotism.
    General Aleksei A. Brusilov, *A Soldier's Notebook*, 1931.

If we clear the air of the fog of catchwords which surround the conduct of war, and grasp that in the human will lies the source and mainspring of all conflict, as of all other activities of man's life, it becomes clear that our object in war can only be attained by the subjugation of the opposing will. All acts, such as defeat in the field, propaganda, blockade, diplomacy, or attack on the centres of government and population, are seen to be but means to that end.
    Captain Sir Basil Liddell Hart, *Thoughts on War*, 1944.

The will does not operate in a vacuum. It cannot be imposed successfully if it runs counter to reason. Things are not done in war primarily because a man wills it; they are done because they are do-able. The limits for the commander in battle are defined by the general circumstances. What he asks of his men must be consistent with the possibilities of the situation.
    Brigadier-General S.L.A. Marshall, *Men Against Fire*, 1947.

Your job as an officer is to make decisions and to see them carried out; to force them through against the opposition not only of the enemy, that is fair enough, but against that of your own men. Against colleagues who want it done another way, of allies – there is only one thing worse than having allies and that is not to have them – and of all opposition of man and nature that will bar your way. You cannot be a leader at all without this strength of will, this determination.
    Field Marshal Viscount Slim of Burma, *Courage and Other Broadcasts*, 1957.

The ultimate test of willpower surely is the ability to dominate events rather than be dominated by them. I refer to the leader who can stand his ground cooly and imperturbably, when chaos surrounds him. A strong will is the function of a sound conscience. And judging from my own limited experience, the prime flaw in those who have cracked under pressure has usually been a lack of willpower to stand up to the pressures of people and events – or possibly an inability to relax.
    Lieutenant-General Sir James Glover, 'A Soldier and His Conscience', *Parameters*, 9/1983.

We are imposing our will. At times resistance has been venomous, but not now
    Major-General Robin Brims, Commander, British 1st Armoured Division, quoted by Paul Martin, *The Washington Times*, 3 April 2003.

WIVES

That cry – that was Hector's honored
    mother I heard!
My heart's pounding, leaping up in my
    throat,
the knees beneath me paralyzed – Oh I
    know it...
something terrible's coming down on
    Priam's children.

WIVES

Pray god the news will never reach my ears!
Yes but I dread it so – what if great Achilles
has cut my Hector off from the city, daring Hector,
and driven him out across the plain, and all alone? –
He may have put an end to that fatal headstrong pride
that always seized my Hector – never hanging back
with the main force of men, always charging ahead,
giving ground to no man in his fury!'

So she cried,
dashing out of the royal halls like a madwoman,
her heart racing hard, her women close behind her.
But once she reached the tower where soldiers massed
she stopped on the rampart, looked down and saw it all –
saw him dragged before the city, stallions galloping,
dragging Hector back to Achaea's beaked warships –
ruthless work. The world went black as night
before her eyes, she fainted, falling backward,
gasping away her life breath…
she flung to the winds her glittering headdress,
the cap and the coronet, braided band and veil,
all the regalia golden Aphrodite gave her once,
the day that Hector, helmet aflash in sunlight,
led her home to Troy from her father's house
with countless wedding gifts to win her heart.

Homer, The Iliad, 22.530–555, c. 800 BC, tr. Robert Fagles, 1990; of Andromache, Hector's wife, on his death at the hands of brilliant Achilles.

If I must say anything on the subject of female excellence to those of you who will now be in widowhood, it will be all comprised in this brief exhortation. Great will be your glory in not falling short of your natural character, and greatest will be hers who is least talked of among the men whether for good or for bad.

Pericles, 431/430 BC, funeral oration for the Athenian dead, quoted in Thucydides, The History of the Peloponnesian War, c. 404 BC, tr. Richard Crawley, 1910.

O Great Lord of All Things, remember your servant
Who has gone to exalt your honour and the greatness of your name.
He will offer blood in that sacrifice which is war.
Behold, Lord, that he did not go out to work for me
Or for his children! He did not abandon us to obtain things
to support his home, with his tump line on his head,
Or with his digging stick in his hand. He went for your sake.
In your name, to obtain glory for you. Therefore, O Lord,
Let your pious heart have pity on him, who with great labor
And affliction now goes through the mountains and valleys,
Hills and precipices, offering you the moisture from his brow,
His sweat. Give him victory in this war so that he may return
To rest in his home and so that my children and I may see
His countenance again and feel his presence.

The prayer of Aztec womanhood, recited every day at dawn until their husbands, sons, brothers and other relatives returned from the wars (15th/early 16th centuries), quoted in Durán, The Aztecs: The History of the Indies of New Spain, 1581, tr. Doris Heyden, 1994.

WOMEN IN WAR

Marriage is good for nothing in the military profession.
  Napoleon, *Political Aphorisms*, 1848.

Marriage for young warriors is a folly. Their first and last duty is to protect the nation from its enemies. This they cannot do efficiently if they have family ties... I tell you all, in future a man will have to prove his worth to be a father, before he receives permission to marry.
  Shaka Zulu, c. 1818, quoted in E.A. Ritter, *Shaka Zulu*, 1955.

A woman who hampers a man at the beginning of his career is a hateful abomination, and he always thinks so sooner or later. So you had best not consider me at all in making your decision [to stay in the Service]. The family might not thank me for telling you this but I think it.... A girl might just ruin a man's life by upsetting it at the beginning. You can decide better if you consider your self as one instead of as two. You must decide alone and then I will go with you any where.
  Beatrice Ayer, the future wife of George S. Patton, 17 January 1909, *The Patton Papers*, I, 1972.

A soldier has no business to be married.
  General Sir Ian Hamilton, *Soul and Body of an Army*, 1921.

Scott was the last man to paint an unduly dark picture. I knew his men were almost at the end of their strength and in a desperate position. I could not help wishing that he had not been so close a friend. I thought of his wife and of his boys. There were lots of other wives, too, in England, India, and Burma whose hearts would be under that black cloud a couple of miles away. Stupid to remember that now! Better get it out of my head.
  Field Marshal Viscount Slim of Burma, *Defeat Into Victory*, 1956.

I can't tell you how I've valued your letters. It's them, and the close touch with you, that have kept my spirits up when I was under most pressure.
  General Sir Peter de la Billière, to his wife, Bridget, at the end of the Gulf War, *Storm Command*, 1992.

WOMEN IN WAR

Lord Raglan, in his last visit to me, asked me 'if my father liked my coming out.' I said with pride my father is not as other men are, he thinks that daughters should serve their country as well as sons – he brought me up to think so – he has no sons – & therefore he has sacrificed me to my country – & told me to come home with my shield on [sic 'or'] upon it. He does not think, (as I once heard a father & a very good & clever father say,) 'The girls are all I could wish – very happy, very attentive to me, & very amusing.' He thinks that God sent women, as well as men, into the world to be something more than 'happy', 'attentive' & 'amusing'. Happy & *dull*, religion is said to make us – 'happy & *amusing*' social life is supposed to make us – but my father's religious & social ethics make us strive to be the pioneers of the human race & let 'happiness' & amusement' take care of themselves.
  Florence Nightingale, 14 November 1855, Balaklava, letter to her father, quoted in Sue M. Goldie, ed., *'I have done my duty' Florence Nightingale in the Crimean War 1854–56*, 1987.

Miss Nightingale was in an adventurous mood, and proposed to go still farther into the trenches up to the Three-Mortar Battery. Her Friends Mr. Bracebridge, Dr. Anderson, and M. Soyer were favourable to her wish, but the sentry was in a great state of consternation.

'Madam,' said he, 'if anything happens I call on these gentlemen to witness that I did not fail to warn you of

WOMEN IN WAR

the danger.'

'My good young man,' replied Miss Nightingale, 'more dead and wounded have passed through my hands than I hope you will ever see in the battlefield during the whole of your military career; believe me, I have no fear of death.'

> Florence Nightingale, c. 1855, quoted in Sarah A. Tooley, *The Life of Florence Nightingale*, 1905.

We were in a slight hollow and all shell which did not break our guns in front, came directly among or over us, bursting above our heads or burying themselves in the hills beyond.

A man lying upon the ground asked for a drink, I stopped to give it, and having raised him with my right hand, was holding him.

Just at this moment a bullet sped its free and easy way between us, tearing a hole in my sleeve and found its way into his body. He fell back dead. There was no more to be done for him and I left him to his rest. I have never mended that hole in my sleeve. I wonder if a soldier ever does mend a bullet hole in his coat?

> Clara Barton, diary notes at Antietam 1862, quoted in Percy Epler, *The Life of Clara Barton*, 1915.

You have just made the change from peacetime pursuits to wartime tasks – from the individualism of civilian life to the anonymity of mass military life. You have given up comfortable homes, highly paid positions, leisure. You have taken off silk and put on khaki. And all for essentially the same reason – you have a debt and a date. A debt to democracy, a date with destiny.

> Colonel Oveta Culp Hobby to the first Officer Candidate Class of the Women's Army Auxiliary Corps (WAAC), 23 July 1942, quoted in Mattie Treadwell, *The Women's Army Corps*, 1954.

I moved my WACs forward early after occupation of occupied territory because they were needed and they were soldiers in the same manner that my men were soldiers. Furthermore, if I had not moved my WACs when I did, I would have had mutiny... as they were eager to carry on where needed.

> General of the Army Douglas MacArthur, statement to Colonel Boyce in Tokyo, 14 October 1945, quoted in Mattie Treadwell, *The Women's Army Corps*, 1954.

I remember we received our first consignment of Women's Army Corps personnel, then known as Women's Auxiliary Army Corps. Until my experience in London I had been opposed to the use of women in uniform. But in Great Britain I had seen them perform so magnificently in various positions, including serving in active anti-aircraft batteries, that I had been converted. In Africa many officers were still doubtful of women's usefulness in uniform – the older commanders in particular were filled with misgivings and open skepticism. What these men had failed to note was the changing requirements of war. The simple headquarters of a Grant or Lee were gone forever. An army of filing clerks, stenographers, office managers, telephone operators, and chauffeurs had become essential, and it was scarcely less than criminal to recruit these from needed manpower when great numbers of highly qualified women were available. From the day they first reached us their reputation as an efficient, effective corps continued to grow. Toward the end of the war the most stubborn die-hards had become convinced – and demanded them in increasing numbers. At first the women were kept carefully back at GHQ and secure bases, but as helpfulness grew, so did the scope of their duties in positions progressively nearer the front. Nurses had, of course, long been

WORK

accepted as a necessary contingent of a fighting force. From the outset of this war our nurses lived up to traditions tracing back to Florence Nightingale; consequently it was difficult to understand the initial resistance to the employment of women in other activities. They became hospital assistants, dieticians, personal assistants, and even junior staff officers in many headquarters. George Patton, later in the war, was to insist that one of his most valuable assistants was his WAC office manager.
   General of the Army Dwight D. Eisenhower, *Crusade in Europe*, 1948.

How wise you were to bring your women into your military and into your labor force. Had we done that initially, as you did, it could well have affected the whole course of the war. We would have found out, as you did, that women were equally effective, and for some skills, superior to males.
   Albert Speer (1905–1981), master of the German war industries, conversation with Lieutenant-General Ira C. Eaker, quoted in *Air Force Magazine*, 6 December 1976.

WORK

And call each man by his name and
   father's line,
show them all respect. Not too proud
   now.
We are the ones who ought to do the
   work.
On our backs, from the day that we
   were born,
it seems that Zeus has piled his pack of
   hardships.
   Homer, *The Iliad*, 10.78–82, c. 800 BC, tr. Robert Fagles, 1990; Agamemnon urging the Greek leaders to set an example by doing their utmost to resist Hector's assault on their camp.

For a man who is a man, work, in my belief, if it is directed to noble ends, has no object beyond itself.
   Alexander the Great, July 326 BC, in India, to his men when they refused to go on, quoted in Arrian, *The Campaigns of Alexander*, 5.26, c. AD 150, tr. Aubrey de Sélincourt, 1971.

Nothing more than effort and work.
   Frederick William I (1688–1740), motto of the father of Frederick the Great.

THE WOUNDED

After the battle the general should give prompt attention to the wounded and see to burying the dead. Not only is this a religious duty, but it greatly helps the morale of the living.
   The Emperor Maurice, *The Strategikon*, c. AD 600 (*Maurice's Strategikon*, tr. George Dennis, 1984).

On this day the Turks made four assaults… They made all these assaults hoping to wear out our small force… During these engagements, many convalescents, although not yet fully recovered, used to come to the works and helped as best they could, because, like men of spirit, they preferred to die fighting rather than be cruelly butchered in their quarters if it were our misfortune that the Turks should win.
   Francesco Balbi de Correggio, 1565, at the siege of Malta.

Think also of the poor wounded of both armies. Especially have paternal care for your own and do not be inhuman to those of the enemy.
   Frederick the Great, *Instructions to His Generals*, 1747, tr. Phillips, 1940.

There are no greater patriots than those good men who have been maimed in the service of their country.
   Napoleon, *Political Aphorisms*, 1848.

My wounded are behind me, and I will never pass them alive.
   Major-General Zachary Taylor, 22 February 1847, when advised to retreat at Buena Vista.

WOUNDS

The wounded had been taken off the battle-field by their general, who ordered his medical director, Dr McGuire, to send them to the rear. As the army was retreating, the surgeon said: 'But that requires time. Can you stay to protect us?' 'Make yourself easy about that,' replied he; 'this army stays here until the last wounded man is removed.' And then with deep feeling he said: 'Before I will leave them to the enemy I will lose many more men.'

> Mary Anne Jackson, of General Jackson's care for the wounded at the Battle of Kernstown, 22 March 1862, *The Life and Letters of General Thomas A. Jackson*, 1892.

Passing through Douclon woods, we heard the moans of wounded men all around us. It was a gruesome sound. A low voice from a nearby bush called '*Kamerad, Kamerad!*' A youngster from the 127th lay with a breast wound on the cold stony ground. The poor lad sobbed as we stooped over him – he did not want to die. We wrapped him in his coat and shelter half, gave him some water, and made him as comfortable as possible. We heard the voices of wounded men on all sides now. One called in a heart-breaking way for his mother. Another prayed. Others were crying with pain and mingled with the voices we heard the sound of French: '*Des blessés, camarade!*' It was terrible to listen to suffering and dying men. We helped friend and foe without distinction.

> Field Marshal Erwin Rommel, of his experiences during 1914, *Infantry Attacks*, 1937.

It is more fruitful to wound than to kill. While the dead man lies still, counting only one man less, the wounded man is a progressive drain upon his side. Comrades are often called upon to bandage him, sometimes even to accompany him back; stretcher-bearers and ambulance drivers to carry him back; doctors and orderlies to tend him in hospital. And on his passage thither the sight of him tends to spread depression among the beholders, acting on morale like the drops of cold water which imperceptibly wear away the stone. (April 1930.)

> Captain Sir Basil Liddell Hart, *Thoughts on War*, 1944.

Evacuate me, hell! Take that tag and label a bottle with it. I will remain in command.

> Lieutenant-General Lewis B. 'Chesty' Puller, 8 November 1942, on Guadalcanal, to a doctor who had just pinned an evacuation tag on him after he had received seven wounds that day, Puller Collection, Marine Corps Historical Division.

Men, all I can say is, if I had been a better general, most of you would not be here.

> General George S. Patton, Jr, 1945, to wounded soldiers at Walter Reed Army Hospital in Washington, DC.

I realized vividly now that the real horrors of war were to be seen in hospitals, not on the battlefield.

> Lieutenant-General Sir John Glubb, *Into Battle: A Soldier's Diary of the Great War*, 1978.

WOUNDS

But does any man among you honestly feel that he has suffered more for me than I have suffered for him? Come now – if you are wounded, strip and show your wounds, and I will show mine. There is no part of my body but my back which has not a scar; not a weapon a man may grasp or fling the mark of which I do not carry upon me. I have sword-cuts from close fight; arrows have pierced me, missiles from catapults bruised my flesh; again and again I have been struck by stones or clubs – and all for your sakes: for your glory and gain.

YOUTH AND AGE

Alexander the Great, July 323 BC, quelling the mutiny of the Macedonians at Opis, quoted in Arrian, *The Campaigns of Alexander*, 7.10, c. AD 150, tr. Aubrey de Sélincourt, 1971.

In the naval battle of Lepanto he lost his left hand as a result of a harquebus shot, a wound which, however unsightly it may appear, he looks upon as beautiful, for the reason that it was received on the most memorable and sublime occasion that past ages have known or those to come may hope to know; for he was fighting beneath the victorious banner of the sons of that thunderbolt of war, Charles V of blessed memory.
> Miguel de Cervantes, of himself, quoted in Daniel Boorstin, *The Creators*, 1992.

Let me alone: I have yet my legs left, and one arm. Tell the surgeon to make haste, and get his instruments. I know I must lose my right arm; so the sooner it is off the better.
> Admiral Viscount Nelson, 24 July 1797, after the assault on Santa Cruz de Tenerife, as he refused help in climbing up the side of his ship with his right arm almost shot away, quoted in Robert Southey, *Life of Nelson*, 1813.

When he [Tsar Alexander I] reviewed the French Army and at Napoleon's side watched the Old Guard march past he was struck by the scars and wounds which many of these veterans bore. 'And where are the soldiers who have given these wounds?' he exclaimed to Ney. 'Sire, they are dead.'
> Sir Winston S. Churchill, *The Age of Revolution*, 1957, of the meeting of Napoleon and Alexander I of Russia at Tilsit, 7–9 July 1807.

I consider these wounds a blessing; they were given me for some good and wise purpose, and I would not part with them if I could.
> Lieutenant-General Thomas J. 'Stonewall' Jackson, as he lay on his deathbed after the Battle of Chancellorsville, quoted in J.M. Daniells, *The Life of Stonewall Jackson*, 1863.

YOUTH AND AGE

Thou dost know
The faults to which the young are ever
  prone;
The will is quick to act, the judgment
  weak;
> Homer, *The Iliad*, Vol. 4, tr. William Cullen Bryant, 1905, p. 139.

I never should if I had not been very young.
> The Duke of Wellington, 3 July 1803, to Lieutenant-General Stuart, describing how he was able to cope with the responsibilities of command in India, quoted in George Chad, *The Conversations of the First Duke of Wellington with George William Chad*, 1956.

One has only a certain time for war. I will be good for six years more; after that even I must cry halt.
> Napoleon, after 1805, comment to his valet, Baron W. Constant, *Mémoires*, V, 1896, quoted in David Chandler, *The Campaigns of Napoleon*, 1966.

I have to perform the labours of Hercules at an age when strength forsakes me, debility increases, in one word when hope, the comforter of the distressed, begins to fail me.
> Napoleon, 1817. Major-General J.F.C. Fuller, *Generalship: its Diseases and Their Cure*, 1933.

I consider it a great advantage to obtain command young, having observed as a general thing that persons who come into authority late in life shrink from responsibility, and often break down under its weight.
> Admiral David G. Farragut, 1819, journal entry.

YOUTH AND AGE

Boldness grows less common in the higher ranks... Nearly every general known to us from history as mediocre, even vacillating, was noted for dash and determination as a junior officer.
   Major-General Carl von Clausewitz, *On War*, 3.6, 1832, tr. Michael Howard and Peter Paret, 1976.

Routine is the death of any institution, but it is not the peculiar attribute of age – you will find young-fogies as well as old fogies; and more of them – some men's ideas never grow old, those of others are never young.
   Rear Admiral John Dahlgren, letter, 26 January 1856, John A. Dahlgren Papers, Library of Congress, box 2, quoted in Robert J. Schneller, Jr., *A Quest for Glory: A Biography of Rear Admiral John A. Dahlgren*, 1996, pp.151–2.

Through great good fortune, in our youth our hearts were touched with fire. It was given to us to learn at the outset that life is a profound and passionate thing.
   Oliver Wendell Holmes, associate justice, Supreme Court of Massachusetts, address to the John Sedgwick Post No. 4, Grand Army of the Republic, Keene, New Hampshire, 30 May 1884, *Speeches of Oliver Wendell Holmes*, 1934.

I told him that our real difficulty was not man-power, but brain-power, and that so long as our Generals would think in terms of 1870, there could be no progress.
   '1870?' he queried after a long pause, '1870?' – they must be very old men.'
   'Very old,' I answered; 'from their knowledge of warfare, most of them might have fought at the battle of Hastings.'
   Major-General J.F.C. Fuller, 1918, to Field Marshal Sir Henry Wilson.

As regards myself, you wonder how it is that I have served for over twenty years without getting blunted. I think the answer is, that it is in no way necessary for a doctor whose work compels him to look after lunatics to go mad himself. I quite agree that one's position is frequently very trying, and for long I have been convinced that the two secrets of continuing mentally young are: (a) never get obsessed by detail, and (b) never be contented with anything.
   Major-General J.F.C. Fuller, April 1922, letter to a friend, quoted in *Memoirs of an Unconventional Soldier*, 1936.

I am doubtful whether the fact that a man has gained the Victoria Cross for bravery as a young officer fits him to command an army twenty or thirty years later. I have noticed more than one serious misfortune which arose from such assumptions....
   Sir Winston S. Churchill, on Sir Redvers Buller, *My Early Life*, 1930.

Few youths of spirit are content at eighteen with comforts or even caresses. They seek physical fitness, movement, and the comradeship of their equals under hard conditions. They seek distinction, not favour, and exult in their manly independence.
   Sir Winston S. Churchill, *Marlborough*, 1933.

Long training tends to make a man ever more expert in execution, but that skill may be gained at the expense of fertility of ideas, originality of conception, and elasticity of views. It is too hopeful to expect generals whose minds have become set to make effective use of new weapons or tactics. Youth is surprise and surprise is war. (March 1923.)
   Captain Sir Basil Liddell Hart, *Thoughts on War*, 1944.

The present general officers of the line are for the most part too old to command troops in battle under the

terrific pressures of modern war.... I do not propose to send our young citizen-soldier into action, if they must go into action, under commanders whose minds are no longer adaptable to the making of split-second decisions in the fast-moving war of today.... They'll have their chance to prove what they can do, but I doubt that many of them will come through satisfactorily. Those that don't will be eliminated.
> General of the Army George C. Marshall (1880–1959), quoted in Eric Larrabee, *Commander in Chief: Franklin Delano Roosevelt, His Lieutenants and Their War*, 1987.

A leader must have the physical resources for the part. A man is as old as his arteries – the old tag is most true in war – and he who makes these calls upon his will power is spendthrift of his days. I have the impression that many senior soldiers and sailors who held high command in the first German war died before their time; they were worn out.
> Lord Wilson Moran, *The Anatomy of Courage*, 1945.

It is sad to remember that, when anyone has fairly mastered the art of command, the necessity for that art usually expires – either through the termination of the war or through the advanced age of the commander.
> General George S. Patton, Jr, *War As I Knew It*, 1947.

Next comes the vexed question of age. One of the ancient Roman poets has pointed out the scandal of old men at war and old men in love. But at exactly what age a general ceases to be dangerous to the enemy and a Don Juan to the other sex is not easy to determine. It is impossible really to give exact values to the fire and boldness of youth as against the judgment and experience of riper years; if the mature mind still has the capacity to conceive and to absorb new ideas, to withstand unexpected shocks, and to put into execution bold and unorthodox designs, its superior knowledge and judgment will give the advantage over youth. At the same time there is no doubt that a good young general will usually beat a good old one.
> Field Marshal Viscount Wavell of Cyrenaica, *Soldiers and Soldiering*, 1953.

Acknowledgements

This book is based on my book *Warriors' Words: A Quotation Book* (Arms & Armour Press, 1992) and on the revised *Greenhill Dictionary of Military Quotations* (Greenhill Books, 2000). This current edition is result of a quarter-century of effort borne out of a persistent frustration – that of being unable to call up at will the distilled wisdom of the profession of arms in the clear, penetrating, and often elegant or pungent words of the great captains. This dilemma found a solution in thousands of file cards which eventually found their way into a computer database as the technology developed. Through these three editions, this collection has grown, evolved in organisation and substance, and sharpened in focus.

My appreciation for the help I received in this journey though military history still stands, perhaps even more so now that almost fifteen years have passed since *Warriors' Words* was published. Without the cheerful and efficient assistance of military librarians, colleagues in and out of uniform, and fellow authors of military history in both the United States and the United Kingdom this book would have been a poor and far more limited thing. In particular, I would like to thank the officers of the British Army with whom I had the high honour to serve. They were the embodiment of the living tradition of the 'Thin Red Line.' I can still hear one of them humming, 'Over the hills and far away'. Not least, I would like to acknowledge the professionalism, enthusiasm, and friendship of my publisher, Lionel Leventhal, and his dedicated staff at Greenhill Books, the premier military publishing house in the UK.

The Art of War by Sun Tzu, translated by Lionel Giles, part of a collection within *Roots of Strategy*, edited by Thomas R. Phillips; used with permission of Stackpole Books.

The Art of War by Sun Tzu, translated by Samuel Griffith, translation copyright © 1963 by Oxford University Press; used by permission of Oxford University Press.

Battle Studies by Charles Ardant du Picq, translated by John L. Greely and Robert C. Cotton; used with permission of Stackpole Books.

Bringing up the Rear by S.L.A. Marshall, 1979, with permission of the S.L.A. Marshall Collection, University of Texas in El Paso.

The Broken Spears by Miguel Leon-Portilla, copyright © 1962, 1990 by Miguel Leon-Portilla; expanded and updated edition copyright © 1992 by Miguel Leon-Portilla; reprinted by permission of Beacon Press, Boston.

The Campaigns of Alexander by Arrian, translated by Aubrey de Sélincourt (published as *Arrian: The Life of Alexander the Great*, 1958, by Penguin Classics; revised edition, 1971); copyright © the Estate of Aubrey de Sélincourt, 1958.

The Edge of the Sword by Charles de Gaulle; copyright © 1960 by Criterion Books, Inc., and Faber and Faber Ltd; reprinted by permission of HarperCollins Publishers, Inc.

Fighting for the Future by Ralph Peters; copyright © 1999 by Ralph Peters; used with permission of Stackpole Books.

First to Fight by Victor Krulak; copyright © 1984, 1991 by the United States Naval Institute; used with permission of the Naval Institute Press.

The History of the Indies of New Spain by Fray Diego Duran, translated by Doris Heyden; copyright © 1994 by the University of Oklahoma Press; reprinted by permission of the publisher.

ACKNOWLEDGEMENTS

The Iliad by Homer, translated by Robert Fagles; translation copyright © 1990 by Robert Fagles; Introduction and Notes copyright © 1990 by Bernard Knox; used by permission of Viking Penguin, a division of Penguin Putnam, Inc.

Ideals of the Samurai by William Scott Wilson; copyright © 1982 by William Scott Wilson; reprinted by permission of the publisher, Ohara Publications, Inc.

Instructions of Frederick the Great to His Generals, Frederick the Great, 1747, translated by Thomas R. Phillips; part of a collection within *Roots of Strategy*, edited by Thomas R. Phillips; used with permission of Stackpole Books.

It Doesn't Take a Hero by General H. Norman Schwarzkopf and Peter Petre; copyright © 1992 by H. Norman Schwarzkopf; used by permission of Bantam Books, a division of Random House, Inc.

Jomini's Art of War by Antoine Henri Jomini, edited by J.D. Hittle; used with permission of Stackpole Books.

Leading the Way, edited by Al Santoli; copyright © 1993 by Al Santoli; reprinted by permission of Ballantine Books, a division of Random House, Inc.

Marine! The Life of Lt. Gen. Lewis B. (Chesty) Puller, USMC by Burke Davis; copyright © 1990 by Burke Davis; used with permission of Little, Brown and Company, Inc.

Mastering the Art of War (the works of Zhuge Liang and Liu Ju), translated by Thomas Cleary, copyright © 1989; reprinted by arrangement with Shambhala Publications, Inc., Boston.

Maurice's Strategikon: Handbook of Byzantine Military Strategy, translated by George T. Dennis; copyright © 1984 University of Pennsylvania Press; reprinted by permission of the publisher.

Memoirs of Field Marshal Montgomery by Bernard Law Montgomery, reprinted with permission of A.P. Watt Ltd, on behalf of Viscount Montgomery of Alamein, CBE.

Men Against Fire by S.L.A. Marshall, 1947, with permission of the S.L.A. Marshall Collection, University of Texas in El Paso.

The Military Institutions of the Romans by Vegetius, translated by John Clarke, part of a collection within *Roots of Strategy*, edited by Thomas R. Phillips; used with permission of Stackpole Books.

The Mind of Napoleon: A Selection of His Written and Spoken Words, translated by J. Christopher Herold; copyright © 1995 Columbia University Press; reprinted with permission of the publisher.

My Reveries Upon the Art of War, Maurice de Saxe, 1832, translated by Thomas R. Phillips; part of a collection within *Roots of Strategy*, edited by Thomas R. Phillips; used with permission of Stackpole Books.

The Military Maxims of Napoleon by Napoleon, translated by General Burnod, 1827; part of a collection within *Roots of Strategy*, edited by Thomas R. Phillips; used with permission of Stackpole Books.

On War by Carl von Clausewitz, translated by Michael Howard and Peter Paret; copyright © 1976 by Princeton University Press; reprinted by permission of Princeton University Press.

The Patton Papers (two volumes), edited by Martin Blumenson, published by Houghton Mifflin Co., 1972–1974; reprinted with permission of Blanche C. Gregory, Inc., on behalf of the author.

Pre-Columbian Literatures of Mexico, by Miguel Leon-Portilla, translated by Grace Lobanov and the author, University of Oklahoma Press, 1969.

Plutarch on Sparta, translated by Richard J.A. Talbert, Penguin Classics, 1988; translation copyright © J.A. Talbert, 1988.

The Rise of the Roman Empire by Polybius, translated by Ian Scott-Kilvert, Penguin Classics, 1979; copyright © Ian Scott-Kilvert, 1979.

The Rommel Papers by Field Marshal Erwin Rommel, 1953; used with permission of HarperCollins Publishers Ltd.

Strategy by B.H. Liddell Hart; published by Faber and Faber Ltd, 1954; reprinted with permission of David Higham Associates, on behalf of B.H. Liddell Hart.

Thoughts on War by B.H. Liddell Hart; published by Faber and Faber Ltd, 1944; reprinted with permission of David Higham Associates, on behalf of B.H. Liddell Hart.

War As I Knew It by George S. Patton; copyright © 1947 by Beatrice Patton Walter, Ruth Patton Totten and George Smith Totten; copyright © renewed 1975 by Major General George Patton, Ruth Patton Totten, John K. Waters, Jr. and George P. Waters; reprinted by permission of Houghton Mifflin Co.; all rights reserved.

Index

Note: some contributors appear more than once on the cited page.

Abrams, Creighton general, US Army 121, 186, 375, 406, 414
Abu Hafs al-Qaida leader 438
Abu Musab al-Zarqawi al-Qaida leader 474
Agesilaus II King of Sparta 157, 191, 195, 240, 247, 263, 294, 312, 376, 472
Agis II King of Sparta 148, 453
Ahmose son of Abana, Egyptian general 352, 372
Ahmose-Pen-Nekhbet Egyptian general 381
Ahuitzotl 8th Aztec emperor 325
Alexander The Great King of Macedonia 92, 100, 134, 142, 151, 159, 173, 176, 194, 204, 242, 376, 464, 465, 494, 496
Alexander, Harold R. Viscount of Tunis field marshal, British Army 141, 276, 384, 476, 477
Alfred 'The Great' King of Wessex 242
Allen, Ethan colonel, Continental Army 244
Allenby, Edmund H. Viscount of Megiddo field marshal, British Army 113, 190, 259, 360, 420
Allon, Yigal general, Israeli Defence Force 49, 84, 103, 257, 275, 307
Alvarez de Toledo, Fernando Duke of Alva Spanish general 180, 376
Ammenhotep III Pharaoh of Egypt 78
Anonymous 161, 168, 194, 211, 228, 248, 265, 276, 317, 336, 361, 367, 386, 393, 403, 408, 414, 445, 447, 450, 491
Inscriptions: 111, 157, 175, 338
Instructions, etc.: 59, 166, 217, 282, 341, 344, 351, 364, 402
Mottoes: 39, 45, 110, 155, 364, 409, 415, 435, 443
Proverbs: 42, 225, 265, 333, 385, 463, 483, 488
Sayings: 182, 248, 315, 348, 406, 474
Songs/poems: 50, 143, 227, 410
Antigonus II Gonatas King of Macedonia 376
Antony, Mark Roman general 420
Archidamus II King of Sparta 148, 166, 338, 492
Archidamus III King of Sparta 236, 237
Archilochus Greek lyric poet 123

Ardant du Picq, Charles colonel, French Army 32, 48, 64, 78, 125, 129, 158, 211, 262, 285, 287, 316, 326, 428, 435, 464, 489
Armistead, Lewis A. brigadier-general, Confederate States Army 245
Arrian, Flavius Arrianus Xenophon Roman general 73, 88, 164, 178, 215, 319, 454
Asakura Norikage Japanese general 117
Asakura Toshikage Japanese general 266, 358
Asquith, H.H. British prime minister 218, 333
Atahualpa last independent Inca emperor 243
Atatürk, Kemal Mustapha Turkish general and statesman 153
Auchinleck, Sir Claude field marshal, British Army 74, 308
Augereau, Pierre F.C. marshal of France 382
Augustus Roman emperor 60
Aurelius Antoninus, Marcus Roman emperor 34, 69

Bach, Christian A. major, US Army 59, 102, 306
Bacon, Francis English philosopher 395
Baden-Powell, Lord R.S.S. lieutenant-general, British Army 74, 145, 203
Balbi di Correggio, Francesco Italian soldier 494
Ballou, Sullivan major, US Volunteers 441
Barret, Robert English soldier 67, 385
Barton, Clara founder of American Red Cross 111, 246, 269, 280, 289, 309, 469, 478, 493
Baumer, William H. general, US Army 327
Beatty, Sir David admiral, British Navy 393
Beck, Ludwig colonel-general, German Army 70, 114, 249, 307, 458
Belisarius Byzantine general 32, 109, 192, 237, 238, 240, 290, 297, 321, 323, 398, 410, 414
Bennett, H. Gordon major-general, Australian Army 324
Beowulf hero of English 175, 194, 215, 308
Berthier, Louis marshal of France 459

INDEX

Bible 134, 170, 203, 217, 419, 477
Bickerdyke, Mary mother, US Civil War, 45
Billière, Sir Peter de la general, British Army 62, 72, 262, 263, 492
Binyon, Laurence poet and soldier, British Army 157
Bismarck-Schönhausen, Otto E.L. von chancellor, German Empire, 189, 341, 470, 473, 480
Blair, Tony British Prime Minister 22, 31, 35, 115, 116, 127, 210, 241, 320, 438, 486, 487
Blount, B. 'Buff' III major-general, US Army 434
Blücher, Gebhard L. von field marshal, Prussian Army 71, 126, 245, 273
Bolivar, Simon South American general 245, 385, 410, 473
Boomer, Walter K. lieutenant-general, US Marine Corps 308, 459
Botha, Louis South African general 324
Bourrienne, L.A. Fauvelet de secretary to Napoleon 456
Boyington, G. 'Pappy' colonel, US Marine Corps 445
Bradley, Omar N. general, US Army 62, 82, 94, 167, 188, 228, 259, 299, 310, 335, 369, 401, 466
Brasidas Spartan general 178, 312, 403
Brims, Robin major-general, British Army 226, 318, 490
Brunne, Guillaume M.A. marshal of France 245
Brusilov, Alexei A. general, Russian Imperial and Red Army 153, 363, 372, 383, 490
Brutus, M. Junius Roman general 242, 420
Buckner, Simon Bolivar Jr lieutenant-general, US Army 238
Bugeaud de la Piconnerie, T.-R. marshal of France 80
Bullard, Robert Lee lieutenant-general, US Army 339
Burgoyne, John major-general, British Army 51, 339, 473
Burke, A. '31-knot' admiral, US Navy 25, 27, 236
Bush, George W. President of the USA 62, 98, 121, 126, 193, 336, 380, 386, 486
Butler, Sir William lieutenant-general, British Army 324
Byrtwold huscarl 137

Caesar, G. Julius Roman general 33, 66, 79, 81, 113, 169, 192, 195, 242, 329, 363, 453, 467
Cambronne, Pierre J. general, French Army 120
Campbell, Sir Archibald lieutenant-general, British Army 337
Carnot, Comte Lazare general, French Army 75, 276
Carver, Lord Michael field marshal, British Army 62, 351, 472
Cathcart, Sir George general, British Army 51
Cato, M. Porcius Roman politician 124
Cavell, Edith L. British nurse 246
Cervantes, Miguel de Spanish novelist 39, 382, 453, 463, 473, 496
Chabrias Athenian mercenary general 247
Chamberlain, A. Neville British prime minister 35, 115
Chamberlain, Joshua L. major-general, US Army 51, 70, 131, 152, 371, 384
Charles II King of England 265
Charles XII King of Sweden 184, 311
Charteris, J. brigadier, British Army 134
Chelmsford, Lord lieutenant general, British Army 118, 174
Chia Lin (Jia Lin) commentator on Sun Tzu 76
Churchill, Sir Winston L.S. British statesman 22, 25, 30, 31, 34, 35, 38, 44, 47, 50, 52, 54, 58, 64, 67, 74, 76, 82, 84, 94, 95, 99, 101, 102, 104, 108, 111, 113, 123, 124, 126, 131, 139, 145, 150, 152, 158, 165, 182, 184, 186, 190, 198, 205, 207, 214, 217, 218, 229, 233, 234, 249, 253, 262, 263, 268, 269, 281, 282, 283, 285, 288, 291, 296, 299, 321, 334, 335, 340, 347, 353, 356, 360, 361, 367, 368, 374, 378, 380, 381, 387, 391, 392, 397, 399, 401, 404, 410, 411, 416, 419, 424, 425, 433, 435, 442, 443, 445, 451, 459, 465, 466, 471, 478, 479, 483, 484, 488, 496, 497
El Cid, Rodrigo Diaz de Vivar Spanish national hero 180, 242
Clark, Wesley K. general, US Army 401, 469
Clarke, Bruce C. lieutenant-general, US Army 282
Clausewitz, Carl von major-general, Prussian Army 22, 23, 35, 37, 41, 42, 44, 47, 53, 54, 64, 65, 66, 69, 77, 80, 91, 100, 101, 109, 110, 117, 119, 125, 146, 154, 160, 167, 169, 173, 177, 181, 186, 189, 193, 199, 201, 211, 213, 216, 217, 224, 234, 236, 238, 247, 253, 269, 270, 276, 285, 286, 297, 301, 331, 343, 347, 349, 360, 377, 383, 394, 399, 400, 409, 415, 419, 420, 423, 428, 430, 436, 437, 438, 439, 444, 460, 465, 469, 470, 477, 480, 489, 497
Clearchus Spartan mercenary general 252
Clemenceau, George French politician 77, 170

INDEX

Clinton, Sir Henry general, British Army 467
Clive, Lord Robert founder of British India 235, 420
Collins, Arthur S. Jr lieutenant-general, US Army 449
Collins, Tim lieutenant colonel, British Army 154
Constantine XI Palaeologus Byzantine emperor 143, 243, 389
Cornwallis, Marquis, Charles major-general, British Army 73, 172, 194, 424
Cortes, Hernan Spanish adventurer 190, 244, 369, 382, 386
Craig, Malin general, US Army, chief of staff 258
Crawford, Earl of British Army 382
Crazy Horse Oglala Lakota war chief 165
Croesus King of Lydia 171
Cromwell, Oliver English statesman 76, 134, 235, 265, 291, 311, 323, 361, 369, 379, 396
Cuauhtemoc Aztec emperor 143, 243
Cuchulain Irish hero 171
Cunningham, Viscount Andrew Browne admiral, Royal Navy 143, 445
Curtius Rufus, Q. Roman author 304
Custer, George A. major-general, US Army 123, 174
Cyrus the Great founder of the Persian Empire 156, 242

Dahlgren, John A.B. rear-admiral, US Navy 172, 409, 497
Daly, Dan sergeant-major, US Marine Corps 165
Danjou, Jean captain, French Foreign Legion 120
David King of Israel 88, 170, 171, 336
Davidson, Philip B. lieutenant-general, US Army 188, 318
Davis, Richard Harding American journalist 265, 484
Day, J.L. major-general, US Marine Corps 334
Dayan, Moshe Israeli politician 111, 175, 232, 458, 477
Decatur, Stephen commodore, US Navy 111, 321
Delphic Oracle 395, 472
Demosthenes Athenian orator 75, 80
Dewey, George admiral, US Navy 246
Dietrich, J. 'Sepp' colonel-general, Waffen SS 333
Dill, Sir John Greer field marshal, British Army 363, 368
Douhet, Guilio general, Royal Italian Air Force 28, 67

Dragomirov, Mikhail I. general, Russian Imperial Army 439
Drake, Sir Francis English admiral 92, 120, 204, 325, 326
Dupuy, Trevor N. colonel, US Army 200, 232

Edward 'The Black Prince' 369
Edward I King of England 243
Edward III King of England 183
Eichelberger, Robert L. general, US Army 424
Eisenhower, Dwight D. 'Ike' President of the US 21, 28, 31, 86, 94, 97, 125, 190, 193, 199, 210, 235, 240, 246, 270, 281, 333, 335, 354, 457, 468, 479, 486, 494
Elizabeth I Queen of England 120, 165
Elles, Hugh J. major-general, British Army 158, 287
Emerson, Ralph Waldo American essayist 361
Epaminondas Theban 50, 242
Eugene of Savoy Prince of Saxony, Austrian general 255, 286
Ewell, Richard Stoddert lieutenant-general, Confederate States Army 222

Fabius Maximus, Q. 'The Delayer' Roman general 60, 107
Farragut, David G. admiral, US Navy 118, 144, 161, 183, 261, 313, 349, 353, 375, 417, 496
Ferguson, Sir Bernard brigadier, British Army 469
Fernando I King of the Two Sicilies 461
Ferry, Abel J.E. lieutenant, French Army 285
Fisher, John Arbuthnot admiral, Royal Navy 270, 291, 302, 314, 396, 423
Fletcher, Frank J., 'Black Jack' admiral, US Navy 114
Foch, Ferdinand marshal of France 22, 24, 30, 36, 41, 42, 49, 52, 53, 82, 84, 118, 131, 135, 146, 183, 201, 212, 266, 285, 296, 298, 301, 302, 308, 367, 377, 423, 425, 436, 456, 460, 465, 478, 489
Folard, Jean-Charles de French soldier 348
Forrest, Nathan Bedford lieutenant-general, Confederate States Army 91, 161, 181, 229, 365, 394
Fortescue, Sir John W. British military historian 80
Franks, Fred Jr general, US Army 98, 127, 400
Fraser, Sir David general, British Army 28, 155
Frederick III German Emperor 230
Frederick Augustus Duke of York, field

503

INDEX

marshal of the British Army 263
Frederick William, Elector of Brandenburg and founder of Prussian state 29
Frederick William I King of Prussia 38, 244, 494
Frederick William II, 'The Great' King of Prussia 23, 26, 32, 39, 40, 42, 49, 53, 63, 69, 76, 90, 99, 119, 123, 126, 128, 146, 149, 156, 158, 165, 166, 182, 184, 186, 187, 191, 192, 197, 209, 211, 219, 225, 228, 236, 244, 252, 253, 259, 272, 273, 275, 279, 286, 297, 300, 302, 303, 335, 342, 352, 354, 365, 382, 386, 388, 397, 398, 415, 422, 437, 438, 449, 452, 460, 470, 488, 494
French, Sir John Denton field marshal, British Army 168, 378
Fritsch, Werner von general, German Army 70, 114, 249, 307, 458
Frontinus, S. Julius Roman general 92, 155, 206, 218, 222
Frunze, Mikhail V. military theorist 261, 283, 489
Fuller, J.F.C. major-general, British Army 27, 34, 35, 36, 38, 42, 43, 44, 57, 61, 67, 70, 87, 91, 96, 97, 108, 109, 122, 130, 135, 136, 137, 148, 156, 162, 163, 179, 182, 198, 200, 202, 207, 209, 212, 221, 224, 225, 231, 232, 233, 237, 249, 266, 274, 276, 283, 284, 293, 298, 308, 310, 314, 316, 327, 331, 344, 346, 356, 367, 373, 378, 386, 392, 398, 399, 402, 409, 412, 416, 423, 424, 429, 430, 432, 471, 478, 486, 487, 488, 497

Gallery, Daniel V. rear-admiral, US Navy 288
Galway, Lord 154
Gaulle, Charles A.M.J. de general, French Army, statesman 22, 36, 38, 70, 85, 113, 123, 126, 132, 135, 154, 192, 207, 210, 220, 231, 234, 236, 237, 247, 314, 321, 332, 339, 350, 354, 388, 394, 408, 433, 478, 483
Gavin, J.M. 'Jumping Jim' lieutenant-general, US Army 112
Gehlen, Reinhardt general, German Army 77
Genghis Khan Mongol khan 157, 325, 464
Getman, Andrey L. general, Soviet Army 299
George II King of England 352
Giap, Vo Nguyen general, People's Army of Vietnam 133, 226, 325, 385, 474
Gibbon, Edward British historian 21, 163
Gibbon, John major-general, US Army 362
Glover, Sir James M. general, British Army 71, 95, 105, 235, 490
Glubb, John Bagot British soldier 457, 495
Gneisenau, August W. von field marshal, Prussian Army 32, 42, 297, 384, 415

Goering, Hermann Reichsmarshal, German Third Reich 175
Goltz, Wilhelm L.C. von der, field marshal, German Imperial Army 119, 208
Gonzalo de Cordoba, 'El Gran Capitan' Spanish general 363
Gordon, Charles G., 'Chinese' general, British Army 111, 246, 331, 454, 475
Gordon, John Brown lieutenant-general, Confederate States Army 80, 153
Gorgo wife of King Leonidas of Sparta 289
Gorshkov, Sergei G. admiral of the fleet, Soviet Navy 397
Grandmaison, Louis Louzeau de brigadier-general, French Army 48
Grant, Ulysses S. general, President of the USA 24, 32, 41, 43, 75, 76, 95, 96, 98, 117, 121, 122, 140, 144, 151, 177, 181, 246, 256, 273, 285, 297, 311, 318, 325, 326, 333, 353, 362, 368, 400, 425, 428, 436, 463, 464, 469
Graves, Robert R. British soldier poet 153
Gray, Alfred M. general, US Marine Corps 310, 436
Greene, Nathaniel major-general, Continental Army 237, 251, 325, 376, 382
Greflinger, Georg German mercenary 272
Grey, Sir Edward British Foreign Secretary 471, 481
Guderian, Heinz W. colonel-general, German Army 43, 52, 77, 200, 373, 407, 424, 433, 434, 446, 489
Guibert, Jacques A.H. de marshal of France 39, 97, 158, 253, 262, 409
Guingand, Sir Francis de major-general, British Army 68, 75, 198, 413
Gustavus II Adolphus King of Sweden 121, 134, 160, 165, 174, 175, 244, 255, 277, 292, 369, 389, 477
Gylippus, Spartan general 27, 122

Hackett, Sir John general, British Army 37, 163, 180, 217, 278, 325, 351, 367, 472, 483
Haig, Earl Douglas field marshal, British Army 44, 139, 337, 484, 485
Haile Selassie Emperor of Ethiopia 282
Haking, R.C.B. captain, British Army 452
Halleck, Henry W. major-general, US Army 406
Halsey, W.F., Jr 'Bull' admiral of the fleet, US Navy 354, 401, 416
Hamilton, Sir Ian M. general, British Army 64, 105, 108, 170, 202, 222, 231, 238, 285, 303, 314, 316, 320, 321, 355, 374, 388, 404, 427, 433, 444, 445, 464, 469, 471, 476, 492
Hammerstein Equord, K. von general, German Army 303, 412

INDEX

Hancock, Winfield S. major-general, US Army 160, 390
Hannibal Barca Carthaginian general 76, 125, 137, 148, 191, 237, 242, 304, 482
Hanrahan, Brian British journalist 65
Harald III 'Hardraade' King of Norway 242
Harbord, James lieutenant-general, US Army 45
Hari, Mata Dutch spy 303
Harold Godwinson King of England 119, 264
Harris, Sir Arthur 'Bomber' air marshal, Royal Air Force 28, 62, 65
Harrison, William English antiquary 402
Hastings, Max British journalist 485
Hattusili I Hittite King 106
Hawke, Lord Edward admiral, British Navy 138
Hawkwood, Sir John de English mercenary general 219, 348
Hemingway, Ernest M. American writer 107, 118
Henderson, George F. colonel, British Army 236, 261, 480
Henry II King of England 242
Henry IV 'of Navarre' King of France 309, 387
Henry V King of England 88, 137, 254
Herodotus Greek historian 188, 402
Herrington, Stuart colonel, US Army 467
Hershey, Lewis B. general, US Army 96
Hill, Sir Jacob Parliamentarian in the English Civil War 336
Hindenburg, Paul von field marshal, Imperial German Army 64, 72, 200, 206, 258, 402
Hippocrates of Cos Greek physician 268
Hitler, Adolf German dictator 30, 74, 126, 189, 420, 426, 479
Hobby, Oveta C. colonel, US Army 493
Hoffman, Max major-general, Imperial German Army 23, 150, 210, 224, 489
Hollingsworth, J.F. lieutenant-general, US Army 333
Holmes, Oliver Wendell captain, US Army 82, 334, 497
Homer, Greek epic bard 21, 46, 53, 55, 59, 66, 73, 78, 88, 96, 106, 110, 127, 136, 141, 143, 147, 163, 171, 174, 175, 176, 180, 183, 204, 218, 229, 263, 264, 278, 289, 295, 308, 319, 335, 375, 376, 380, 382, 389, 403, 416, 418, 435, 440, 458, 461, 464, 470, 472, 475, 481, 491, 494, 496
Hooker, J. 'Fighting Joe' major-general, US Army 174, 314
Hoon, Geoff British politician 31, 68, 399, 485

Hopton, Sir Ralph English lieutenant-general 196
Horner, Charles A. 'Chuck' general, US Air Force 479
Horrocks, Sir Brian G. lieutenant-general, British Army 51, 190, 308
Howze, Hamilton H. general, US Army 106, 280, 365

Jackson, Andrew 'Old Hickory' President of the United States 81, 101, 120, 124, 135, 144, 219, 370, 495
Jackson, Michael lieutenant-general, British Army 110, 182, 315, 342, 352, 356, 436
Jackson, Robert British military doctor 93, 260, 459
Jackson, Thomas J. 'Stonewall' 23, 41, 51, 56, 69, 91, 112, 144, 160, 166, 170, 178, 181, 201, 216, 245, 264, 290, 313, 321, 329, 341, 347, 349, 360, 368, 371, 377, 399, 411, 447, 455, 469, 477, 496
Jan III Sobieski of Poland 165, 244
Joan of Arc French soldier 79, 98, 119, 134, 160, 164, 206, 243, 310, 329, 361
Joffre, Joseph J.C. marshal of France 30, 139, 381
Johnston, Joseph E. brigadier-general, US Army, and general, Confederate States Army 93
Jomini, Baron Antoine-Henri de, military theorist 26, 41, 44, 64, 71, 73, 77, 80, 86, 91, 100, 101, 198, 199, 211, 273, 276, 279, 301, 320, 344, 352, 383, 397, 411, 430, 439, 475, 487
Jones, John Paul admiral, US Navy 54, 215, 273, 387, 396, 418, 422
Joseph chief of the Nez Perce 425
Josephus, Flavius Jewish historian general 452
Joubert, Joseph French novelist 460

Katukov, Mikhail Y. marshal of tank troops, Soviet Army 178
Kennedy, John F. President of the United States 65, 209, 408, 416
Kesselring, Albert field marshal, German Air Force 381, 447
Khayer ad-Din, 'Barbarossa' Ottoman general 395
Khrushchev, Nikita S. Soviet Premier 388
King, Ernest J. admiral of the fleet, US Navy 27, 114, 214, 287, 308
Kinnamos, John general, Byzantine Empire 185, 449
Kipling, J. Rudyard British writer and poet 34, 139, 153, 184, 192, 241, 265, 292, 293, 309, 334, 337, 365, 443, 444, 450, 478

INDEX

Kitchener, Herbert H. field marshal, British Army 145, 366, 469

Kluck, Alexander von general, Imperial German Army 177

Kolokotrones, Theodoros Greek general 324

Koniev, Ivan S. marshal of the Soviet Union 332

Krulak, Charles C. general, US Marine Corps 323

Krulak, Viktor H., 'The Brute' lieutenant-general, US Marine Corps 78, 170, 258, 259, 316, 363, 382

Kublai Khan Mongol khan 76

Kutuzov, Mikhail I. field marshal, Imperial Russian Army 176, 297

Lafayette, Marie de French statesman 270

Lamachus Athenian general 280

Lannes, Jean marshal of France 160, 245

Lawrence, James captain, US Navy 138

Lawrence, T.E. 'Lawrence of Arabia' British soldier and writer 100, 102, 132, 141, 146, 150, 153, 212, 214, 234, 274, 309, 341, 355, 366, 375, 378, 388, 402, 429, 431, 436, 443, 445, 461, 473, 474

Leahy, William D. admiral, US Navy 175

Lebed, Aleksandr I. lieutenant-general, Russian Army 115, 124, 175, 229, 263, 334, 385, 414, 432, 483, 485

Lee, H. 'Light Horse Harry' major-general, Continental Army 149, 260

Lee, Robert E. 'The Incomparable' colonel, US Army, general, Confederate States Army 22, 42, 69, 75, 81, 93, 118, 121, 129, 130, 142, 144, 150, 151, 160, 203, 205, 210, 221, 246, 256, 278, 280, 281, 295, 333, 341, 362, 368, 371, 380, 406, 425, 461, 466, 476, 482

Lefebvre, Pierre-François marshal of France 317

Lejeune, John A. major-general, US Marine Corps 131, 161, 233, 247, 334, 354, 410, 435, 448, 455

LeMay, Curtis E. general, US Air Force 38, 83, 114, 135, 185, 198, 218, 362

Lenin, Vladimir I. Russian leader 439, 480

Leonidas King of Sparta 119, 214, 294, 389

Leopold I, 'The Old Dessauer' field-marshal, Prussian Army 337

Lepic, Comte Louis general, French Army 120

Letterman, Jonathan major, US Army, military doctor and medical reformer 268

Lettow-Vorbeck, Paul von field marshal, German Imperial Army 147, 229, 426

Liardet, Guy F. rear-admiral, Royal Navy 447

Li Ch'üan (Li Quan) commentator on Sun Tzu 241

Liddell Hart, Sir Basil H. captain, British Army, military historian 22, 24, 31, 33, 34, 36, 41, 44, 48, 51, 52, 58, 68, 74, 82, 84, 87, 92, 93, 96, 100, 108, 119, 131, 132, 133, 147, 150, 162, 167, 169, 173, 179, 183, 186, 187, 188, 189, 193, 202, 206, 207, 210, 213, 214, 217, 219, 220, 223, 224, 225, 226, 233, 234, 237, 247, 248, 251, 253, 258, 261, 267, 272, 279, 281, 283, 284, 286, 290, 294, 298, 299, 300, 302, 306, 308, 309, 311, 314, 319, 324, 327, 331, 332, 345, 350, 353, 362, 366, 379, 381, 387, 394, 398, 402, 409, 411, 412, 416, 419, 421, 423, 427, 429, 431, 433, 438, 443, 445, 451, 456, 459, 465, 466, 468, 471, 472, 474, 478, 479, 481, 482, 486, 487, 490, 495, 497

Liggett, Hunter lieutenant-general, US Army 168, 412

Lincoln, Abraham President of the United States 43, 76, 97, 136, 144, 149, 177, 181, 201, 205, 210, 289, 352, 363, 385, 396, 397, 436, 462

Lin Piao senior commander, Chinese People's Liberation Army 114, 388, 437

Liu Ji Chinese general 45, 58, 67, 154, 160, 284, 305, 338, 358, 482

Lloyd, Henry British mercenary general 97, 201, 220, 393

Lloyd George, David British Prime Minister 324, 464

Loeffke-Arjona, Bernard major-general, US Army, Special Forces and Airborne 69

Longstreet, J. 'Old Pete' major, US Army, and lieutenant-general, Confederate States Army 314

Louis XIV 'The Sun King' King of France 244

Lovat, Simon Fraser brigadier-general, British officer 194

Ludendorff, Erich von field marshal, Imperial German Army 32, 61, 185, 198, 199, 217, 227, 231, 301, 324, 355, 412, 432, 481

Lysander Spartan commander 126

MacArthur, Arthur major-general, US Army 235

MacArthur, Douglas 25, 29, 68, 99, 103, 118, 121, 123, 125, 140, 213, 222, 251, 259, 266, 287, 304, 309, 311, 316, 339, 345, 350, 368, 373, 379, 380, 391, 410, 411, 413, 443, 466, 493

McAuliffe, Anthony C. general, US Army 121

McCaffrey, William R. lieutenant-general, US Army 356

INDEX

McCain, John S. admiral, US Navy 291
McClellan, George B. major-general, US Army 35, 406
McDowell, Irvin major-general, US Army 484
Machiavelli, Niccolò 27, 44, 53, 63, 67, 81, 90, 112, 123, 126, 138, 180, 192, 226, 267, 272, 275, 292, 311, 316, 322, 376, 397, 398, 403, 418, 437, 473, 475
Magee, John flying-officer, Royal Canadian Air Force 337
Mahan, Alfred Thayer rear-admiral, US Navy 25, 29, 57, 61, 67, 91, 96, 118, 125, 136, 154, 189, 211, 238, 251, 261, 283, 291, 296, 298, 301, 302, 311, 344, 349, 356, 360, 380, 394, 396, 409, 415, 419, 420, 430, 436, 445, 451, 470
Maharabal Barca Carthagian general 169, 359
Makarov, Stepan O. admiral, Russian Imperial Navy 27, 400
Malone, Dandridge M. colonel, US Army 448
Manstein, Fritz E. von field marshal, German Army 70, 296, 364
Manteuffel, Hasso von general of Panzer troops, German Army 328
Mao Tse-tung Chinese leader 43, 92, 112, 133, 167, 186, 229, 261, 299, 306, 319, 324, 333, 334, 340, 385, 474, 478, 481
Marcellus, Claudius Roman general 157
Marius, G. Roman general 90
Marlborough, John Churchill Duke of Marlborough, English statesman 160, 244, 252, 259, 275, 305, 320, 372, 467, 468
Marshall, George C. general, US Army 48, 61, 178, 186, 227, 231, 239, 247, 271, 281, 283, 287, 324, 330, 368, 390, 405, 483, 498
Marshall, Samuel L.A. brigadier-general, US Army Reserve 33, 45, 51, 52, 57, 78, 102, 105, 109, 114, 121, 133, 169, 178, 180, 203, 214, 218, 221, 229, 239, 249, 257, 258, 261, 280, 286, 294, 296, 307, 315, 321, 373, 388, 393, 401, 408, 409, 410, 411, 418, 450, 452, 457, 484, 490
Martel, Gifford Le Quesne lieutenant-general, British Army 44, 238
Mattis, James lieutenant-general, US Army 182, 319
Mauldin, W.H. 'Bill' American military cartoonist 228
Maurice Byzantine emperor and general 92, 117, 123, 126, 149, 159, 166, 185, 190, 196, 206, 219, 253, 259, 268, 276, 280, 305, 310, 335, 357, 363, 365, 369, 386, 398, 400, 407, 410, 431, 443, 449, 452, 461, 475, 487, 494
Meade, George G. major-general 96

Mehmed II 'The Conquerer' Ottoman Sultan 176, 255
Meinertzhagen, Richard colonel, British Army 241, 250, 432
Meir, Golda Israeli prime minister 140
Menzies, John author 365
Merneptah Pharaoh of Egypt 355
Mgobozi Msane Zulu general 111, 447
Mitchell, W. 'Billy' brigadier-general, US Army 28, 378
Mohammed founder of Islam 112
Moltke, Helmut J.L. von 'The Younger' general, Imperial German Army 282
Moltke, Helmut K.B. von 'The Elder' field marshal, Imperial German Army 26, 67, 91, 98, 100, 113, 126, 150, 199, 211, 218, 230, 234, 259, 273, 279, 301, 309, 311, 314, 331, 387, 412, 415, 419, 430, 460, 478, 480, 482
Monash, Sir John general, Australian Army 254, 331, 349
Monck, George English general 348, 475
Montcalm, Marquis Louis-Joseph de major-general, French Army 244
Montecuccoli, Raimondo field marshal in Austrian service 85, 462
Montgomery, Bernard L. Viscount of Alamein field marshal, British Army 25, 28, 29, 36, 41, 72, 83, 86, 87, 89, 91, 94, 99, 100, 104, 114, 140, 186, 188, 190, 198, 228, 246, 248, 250, 254, 258, 262, 269, 271, 275, 287, 288, 290, 300, 307, 315, 332, 336, 354, 363, 371, 377, 400, 408, 413, 414, 418, 424, 426, 439, 457, 458, 463, 467, 474
Montrose, Marquis of Scottish general 443
Moore, Harold G. lieutenant-general, US Army 44, 80, 90, 441
Moore, Sir John general, British Army 76, 78, 138, 152, 166, 236, 245, 284, 305, 387, 454, 455, 463
Moran, Lord Charles M.W. British military physician 70, 102, 103, 107, 121, 178, 179, 237, 239, 249, 280, 367, 476, 498
Morillon, Phillipe general, French Army 62
Morrow, Lance American journalist 342
Morshead, Leslie lieutenant-general, Australian Army 229, 335
Mortimore, H.D. captain, British Army 432
Mosby, John S. colonel, Confederate States Army 24, 57, 64, 177, 184, 251, 419, 473
Motecuhzoma II Xocoyotl 'The Younger' Aztec emperor 142
Mountbatten, Earl Louis of Burma British admiral and statesman 189, 139, 426
Mowat, Farley captain, Canadian Army 78, 90

507

INDEX

Murat, Joachim marshal of France 245
Murphy, Audie first lieutenant, US Army 68, 371, 487
Musashi Miyamoto samurai 151, 348, 414, 427, 442
Myers, Richard B. general, US Army 320

Nabeshima Naoshige Japanese general 417
Napier, Sir William lieutenant-general, British Army 201, 221, 227, 295, 383, 419, 444, 460
Napoleon I emperor of the French 21, 24, 25, 27, 29, 34, 35, 39, 40, 41, 42, 44, 46, 47, 49, 50, 51, 54, 55, 56, 58, 59, 60, 63, 66, 69, 71,76, 78, 79, 80, 81, 83, 86, 87, 91, 95, 96, 98, 99, 101, 103, 108, 110, 113, 117, 120, 123, 125, 126, 129, 135, 142, 149, 151, 152, 154, 155, 156, 165, 167, 168, 169, 173, 174, 176, 178, 180, 183, 187, 189, 191, 192, 197, 201, 205, 206, 208, 209, 210, 211, 215, 219, 220, 221, 224, 225, 227, 233, 240, 245, 247, 251, 253, 255, 259, 260, 262, 264, 265, 266, 267, 268, 272, 276, 279, 281, 282, 283, 284, 285, 286, 290, 291, 293, 295, 297, 300, 302, 305, 311, 312, 313, 317, 322, 323, 325, 326, 329, 331, 334, 335, 340, 342, 343, 347, 348, 349, 352, 355, 358, 360, 362, 363, 364, 366, 367, 370, 372, 374, 376, 377, 379, 383, 385, 387, 390, 392, 393, 394, 396, 397, 402, 403, 404, 406, 407, 408, 409, 417, 420, 422, 424, 425, 432, 437, 442, 444, 450, 455, 461, 462, 464, 475, 492, 494, 496
Nelson, Viscount Horatio admiral, Royal Navy, master tactician 27, 34, 47, 54, 66, 69, 88, 98, 109, 124, 138, 143, 152, 169, 203, 205, 215, 230, 239, 244, 267, 270, 281, 292, 313, 325, 337, 348, 354, 369, 370, 392, 442, 452, 475, 488, 496
Neville, W.C. 'Buck' major-general, US Marine Corps 383
Ney, Michel marshal of France 205, 227, 245, 383
Nezahualcoyotl King of Tezcoco 110, 134, 243
Nicholas I Russian tsar 488
Nightingale, Florence English nurse 22, 144, 208, 268, 375, 404, 454, 492, 493
Nikephoros II Phokas Byzantine emperor 166, 422, 449
Nimitz, Chester A. admiral of the fleet, US Navy 125, 463
Nivelle, Robert general, French Army 120
Nogi, Maresuki general, Imperial Japanese Army 50, 61, 390

Ogarkov, Nikolai V. marshal of the Soviet Union 68, 302
Ohnishi, Takijiro vice-admiral, Imperial Japanese Navy 111, 391
Omar I ibn al-Khattab Muslim caliph 310, 358
O'Neil, James colonel, US Army 338
Orwell, George British essayist 340
Osama bin Ladin al-Qaida leader 116, 127, 190, 241, 288, 320, 348, 438
Osman I, 'Osman Gazi' founder of the Ottoman Empire 243
Owen, Wilfred poet and soldier, British Army 142

Pachacutec Inca Yupanqui 'The Cataclysm' founder of the Inca Empire 25
Pagonis, William G. lieutenant-general, US Army 37, 253
Paine, Thomas American political philosopher 460
Palmer, Dave R. lieutenant-general, US Army 44
Patton, George S. Jr 'Old Blood and Guts' general, US Army 36, 41, 42, 43, 44, 48, 52, 56, 61, 78, 81, 82, 84, 86, 102, 107, 114, 121, 130, 133, 142, 145, 150, 153, 156, 162, 163, 178, 179, 183, 185, 187, 207, 209, 212, 226, 229, 232, 241, 246, 249, 258, 261, 268, 270, 277, 288, 296, 299, 302, 304, 306, 315, 319, 327, 329, 332, 335, 344, 347, 350, 364, 371, 374, 375, 377, 381, 387, 391, 392, 394, 400, 405, 410, 417, 419, 429, 434, 435, 443, 448, 451, 452, 464, 476, 479, 492, 495, 498
Paullus, L. Aemilius Roman general 26, 191
Pelopidas Theban general 269
Pericles Athenian statesman 59, 142, 156, 214, 279, 280, 291, 310, 321, 323, 335, 392, 395, 455, 491
Perry, Oliver H. captain, US Navy 467
Pershing, J.J. 'Black Jack' general, US Army 54, 88, 221, 248, 306, 394, 453, 485
Pétain, Henri-Phillipe marshal of France 183, 252
Peters, Ralph lieutenant-colonel, US Army 27, 152, 210, 241, 427, 469
Petraeus, David H. major-general, US Army 124, 318
Philip II King of Macedonia 247, 252, 382, 454
Phormio Athenian general 116
Pickett, George E. major-general, Confederate States Army 165
Picton, Sir Thomas lieutenant-general, British Army 51
Pike, Hugh W.R., lieutenant-colonel,

508

INDEX

British Army 28, 52
Pilsudski, Joseph K. field marshal, Polish Army 184, 249
Pitt, William 'The Younger' Prime Minister of Britain 291
Pizarro, Francisco Spanish conqueror of Peru 244
Plutarch Greek historian 23, 96, 107, 112, 289, 453, 466
Polybius Greek historian 23, 75, 109, 115, 149, 169, 191, 195, 262, 267, 271, 317, 323, 357, 359, 466, 468, 477
Porter, David Dixon vice-admiral, US Navy 56, 106, 256, 265, 359
Pound, Sir David admiral, British Navy 397
Powell, Colin general, US Army 122, 251, 258, 312, 419
Puller, Lewis B. 'Chesty' lieutenant-general, US Marine Corps 25, 53, 94, 97, 162, 263, 266, 268, 278, 330, 354, 373, 384, 408, 413, 464, 495
Pyle, Ernest Taylor, 'Ernie' American journalist and war correspondent 227
Pyrrhus King of Epirus 128

Quarles, Francis English poet 444

Raleigh, Sir Walter English courtier 210, 396
Rameses II, Pharaoh of Egypt 158, 421
Rameses III Pharaoh of Egypt 116, 208, 338, 388
Ramsay, Sir Bertram admiral 489
Ramsay, Francis M. rear-admiral, US Navy 427
Raus, Erhard colonel, Austrian Army, and colonel-general, German Army 328
Remarque, Erich Maria German novelist 89
Reynolds, John F. major-general, US Army 245
Rhigas, Konstantinos Greek patriotic hero 272
Richard I 'The Lion Heart' King of England 191, 243, 323, 336, 346, 377, 472
Richthofen, M. von, 'The Red Baron' captain, German Imperial Army 246
Rickover, Hyman G. admiral, US Navy, founder of the nuclear Navy 282
Riddell-Webster, Mike commander, British Army 400
Ridgway, Matthew B. general, US Army 71, 72, 75, 103, 104, 162, 191, 250, 328, 332, 350, 361, 414
Roberts, Earl Frederick S. field marshal, British Army 263
Robertson, Sir William R. field marshal, British Army 77, 306, 327, 485
Rommel, Erwin J. field marshal, German Army 24, 29, 45, 55, 92, 102, 136, 154, 162, 163, 167, 184, 224, 233, 246, 252, 269, 275, 281, 307, 315, 328, 345, 387, 394, 409, 434, 436, 446, 453, 495
Roosevelt, Franklin D. President of the United States 25, 116, 189, 283
Roosevelt, Theodore President of the United States 172, 475
Root, Elihu US Secretary of War 274
Rudel, Hans U. colonel, German Air Force 405, 448
Rumsfeld, Donald US Secretary of Defense 31, 191, 342, 360, 363, 380, 487
Rundstedt, Gerd von field marshal, German Army 175, 200
Russell, William H. British journalist 341

Saddam Hussein Iraqi dictator 175
Sahagún, Fray Bernadino de 21, 55, 196
St Vincent, Lord Sir J.J. admiral, British Navy 98, 138, 230, 265, 380, 396, 418, 422
Sakai, Saburo lieutenant, Japanese Army 153, 290
Saladin (Salah ad-Din) Sultan of Egypt 53, 98, 310, 369
San Martin, Jose F. de Argentinian general 53
Sassoon, Siegfried L. British writer 278, 290
Saxe, Maurice Comte de marshal of France 42, 49, 55, 66, 67, 79, 85, 101, 128, 140, 187, 213, 253, 260, 277, 279, 286, 311, 322, 335, 348, 352, 359, 361, 376, 431, 444, 445, 451, 464
Scharnhorst, Gerhard J. von general, Prussian Army 199, 266, 379
Schlieffen, Alfred von field marshal, German Imperial Army 156, 212, 246, 468
Schmitt, John major, US Marine corps 436
Schurz, Carl major-general, US Army 321
Schwarzkopf, H. 'Stormin'' Norman general, US Army 33, 34, 62, 65, 108, 125, 150, 154, 156, 174, 180, 194, 248, 348, 382, 401, 441, 459, 465, 483
Scipio Africanus 'The Elder' Roman general 98, 137, 157, 195, 208, 242, 475
Scott, Winfield lieutenant-general, US Army 53, 93, 163, 292, 406
Sedgwick, John major-general, US Army 174
Seekt, Hans von colonel-general, German Army 72, 89, 93, 99, 132, 145, 168, 187, 200, 212, 319, 350, 361, 374, 391, 439, 461, 480
Sesostris III Pharaoh of Egypt 42
Severus, L. Septimius Roman emperor 242, 321

INDEX

Seydlitz, Freiherr F.W. von general, Prussian Army 229
Shaka Zulu creator of the Zulu nation 124, 165, 245, 360, 440, 492
Sharon, Ariel general, Israeli Defence Force 122, 234, 333, 347, 434
Shazly, Saad el lieutenant-general, Egyptian Army 155, 191
Shelton, Hugh general, US Army 342
Sheridan, Phillip H. general, US Army 61, 181, 292, 329, 415, 466
Sherman, William T. general, US Army 35, 59, 88, 93, 101, 109, 121, 129, 161, 179, 194, 222, 228, 254, 256, 257, 261, 262, 292, 293, 322, 326, 359, 360, 372, 374, 392, 399, 412, 456, 462, 476, 477, 479, 482, 484
Shiba Yoshimasa Japanese warrior 236, 260, 400
Shinseki, Erick K. general, US Army 68
Shoup, David A. general, US Marine Corps 182, 468
Sickles, Daniel E. major-general, US Army 371
Sidney, Sir Philip English soldier 59
Simonides of Keos Greek lyric poet 156
Simonov, Konstantin M. Russian playwright 441
Singlaub, John K. major-general, US Army 347, 351
Sitting Bull (Tatanka Iyotake) leader of the Lakota (Sioux) 118
Slessor, Sir John C. air marshal, Royal Air Force 29, 31, 319, 340
Slim, Viscount Sir William J. field marshal, British Army 26, 48, 50, 56, 83, 97, 102, 104, 106, 126, 128, 129, 133, 134, 198, 214, 232, 249, 250, 270, 275, 282, 287, 288, 312, 328, 340, 346, 350, 354, 366, 375, 393, 405, 407, 408, 427, 446, 457, 490, 492
Smith, H.M. 'Howlin' Mad' general, US Marine Corps 303
Smith, Perry M. major-general, US Air Force 458
Smith, Walter B. lieutenant-general, US Army 126
Smith-Dorrien, Sir Horace L. general, British Army 458
Smuts, Jan C. South African Boer statesman 324, 482
Smythe, Sir J. English soldier 60, 81, 211, 277, 323, 449
Socrates Athenian philosopher 173, 195, 204, 330, 407, 445
Sokolovskiy, V.D. marshal of the Soviet Union 386, 416
Solon Athenian lawgiver 134

Speer, Albert German war minister 494
Speidel, Hans general, German Army 31
Spruance, Raymond A. admiral, US Navy 192
Ssu Jang Chu Chinese general 317
Stalin, Joseph V. Soviet dictator and war leader 30, 68, 125, 394, 416
Stanton, Charles E. colonel, US Army 30
Stark, John brigadier-general, Continental Army 138
Stilwell, J.W. 'Vinegar Joe' general, US Army 168
Stockdale, James B. admiral, US Navy 71, 209, 213, 235, 248, 251, 275
Stuart, J.E.B. major-general, Confederate States Army 54, 64, 74, 142, 144, 230, 245, 364
Suleyman I, 'The Magnificent' Ottoman sultan 325
Sullivan, Gordon R. general, US Army 361
Sun Bin Chinese strategist 234, 240, 248, 272, 417, 421, 458, 468
Sun Tzu Chinese master strategist 23, 32, 52, 58, 83, 90, 100, 112, 137, 143, 150, 159, 172, 195, 206, 224, 225, 228, 238, 252, 262, 266, 294, 296, 300, 304, 310, 315, 357, 363, 364, 365, 397, 398, 403, 408, 414, 417, 420, 428, 430, 437, 453, 458, 464, 465, 466, 477, 481
Suvorov, Aleksandr V. field marshal, Imperial Russian Army 21, 23, 40, 51, 55, 60, 87, 92, 98, 99, 101, 119, 121, 124, 129, 138, 149, 151, 160, 163, 166, 169, 178, 183, 192, 209, 219, 220, 244, 263, 277, 279, 284, 292, 294, 303, 305, 312, 330, 342, 348, 359, 375, 382, 389, 393, 408, 411, 422, 431, 435, 437, 442, 444, 447, 449, 450, 465, 488
Svechin, Aleksandr A. general, Imperial Russian Army and Red Army 135, 298

Tacitus, Cornelius Roman orator, politician, historian 173, 188, 209, 317, 391, 403, 407, 422, 459, 463, 465
T'ai Kung general and originator of strategic studies in ancient China 39, 159, 357
Tamerlane Turkic conqueror of Central Asia, the Middle East and India 441
Taylor, Maxwell D. general, US Army 28, 90, 95, 304 362
Taylor, R. 'Dick' lieutenant-general, Confederate States Army 451
Taylor, Zachary President of the United States 39, 120, 383, 494
Tecumseh Shawnee chief 339
Tedder, Baron Arthur W. air chief marshal, Royal Air Force 29, 270
Terrail, Pierre de Bayard French hero 243
Thatcher, Margaret British Prime Minister 27,

INDEX

65, 127, 265, 380, 388, 416, 419, 459, 479
Themistocles Athenian General 26, 173, 395, 418
Thucydides Athenian historian 21, 55, 106, 213, 223, 278, 323, 330, 391, 395, 400, 421, 447, 450, 470, 477
Thutmose III Pharaoh of Egypt 254
Tilly, Count J.T. 'The Monk in Armour' Flemish general in Bavarian service 255
Tlacaelel generalissimo of the Aztecs 60, 358
Togo, Heihachiro admiral, Imperial Japanese Navy 42, 96, 145, 190, 363, 417, 448
Tokugawa Ieyasu Japanese shogun 257, 334, 487
Travis, William B. lieutenant-colonel, Army of the Republic of Texas 120
Trivulce, Gian G. Italian soldier in French service 473
Trotksy, Leon revolutionary 189, 262, 274, 279, 287, 385, 439
Ts'ao Ts'ao 'The Martial Emperor' founder of the Wei Dynasty 266, 365, 417
Tuchman, Barbara W. American historian 361
Tucker, Jedediah British sailor, 290
Tuker, Sir Francis I.S. lieutenant-general, British Army 459
Tukhachevskiy, Mikhail N. marshal of the Soviet Union 48, 84, 91, 231, 246, 430, 433
Turenne, Henri de la Tour d'Auvergne marshal of France 174, 280, 295, 305, 467
Tvardovskiy, Aleksandr T. Russian poet 39
Twining, Merrill B. general, US Marine Corps 278, 414

Valette, Jean P. de French soldier 119, 138, 164, 295, 389
Vandegrift, Alexander A. general, US Marine 54, 118, 239, 437, 457
Vasey, George A. major-general, Australian Army 327, 473
Vauban, Sebastien le P. de military engineer 155
Vegetius, Flavius Renatus Roman author 55, 85, 92, 100, 123, 137, 143, 148, 166, 264, 272, 275, 294, 310, 330, 338, 357, 364, 398, 422, 437, 448, 450, 451, 463, 465
Victoria of Battenberg, Princess Beatrice Mary mother of Louis Mountbatten 290
Vigny, Alfred V. Comte de French writer 216
Villars, Claude L.H. de marshal of France 467
Villeneuve, Pierre C.J.B.S. de vice-admiral, Imperial French Navy 451

Wainwright, Jonathan M. general, US Army 426
Walker, Walton H. major-general, US Army 140
Wallenstein, Albert E. von Duke of Friedland and Mecklenburg 462
Walters, Vernon A. lieutenant-general, US Army 276
Walton, Clifford colonel, British Army 445
Ward, Orlando major-general, US Army 368, 412
Washington, George President of the United States 38, 69, 75, 76, 88, 94, 127, 135, 174, 184, 203, 215, 244, 255, 266, 272, 291, 292, 303, 322, 339, 348, 358, 362, 364, 367, 374, 381, 396, 397, 453, 472
Wavell, Earl Archibald P. field marshal, British Army 25, 38, 49, 50, 54, 61, 70, 82, 83, 84, 94, 141, 148, 163, 174, 198, 203, 207, 208, 213, 234, 240, 254, 261, 274, 296, 306, 315, 327, 355, 360, 366, 379, 392, 401, 405, 413, 427, 429, 431, 446, 450, 474, 479, 487, 498
Wellington, Arthur Wellesley, Duke of field marshal, British Army 24, 38, 41, 45, 46, 50, 57, 59, 60, 63, 81, 85, 87, 88, 93, 108, 112, 123, 124, 129, 130, 131, 135, 138, 143, 144, 146, 151, 155, 166, 168, 170, 178, 181, 184, 185, 187, 189, 192, 213, 216, 219, 223, 226, 227, 228, 235, 238, 240, 251, 252, 253, 255, 265, 281, 284, 293, 295, 305, 313, 326, 330, 331, 334, 340, 346, 355, 359, 363, 367, 370, 371, 377, 383, 384, 385, 387, 389, 403, 411, 418, 419, 430, 442, 454, 460, 468, 472, 480, 496
Westmoreland, William C. general, US Army 65, 342
Wet, Christiaan R. de Boer soldier 473
Wheeler, Earle G. general, US Army 187
Wilhelm II, Kaiser of Germany 174
William I of Orange Dutch soldier and statesman 120
Williams-Ellis, Clough major, British Army 433
Wilson, Sir Henry field marshal, British Army 339, 489
Wilson, T. W. President of the United States 115
Wingate, Orde major-general, British Army 337, 423
Wolfe, Sir James major-general, British Army 66, 107, 113, 123, 168, 211, 239, 244, 278, 302, 365, 396, 422, 427, 463
Wolseley, Viscount Garnet J. field marshal, British Army 205, 222, 258, 274, 366, 388, 407, 427, 461, 484

INDEX

Wood, Leonard major-general, US Army 339

Woodward, Sir J.F. 'Sandy' admiral, British Navy 97, 122, 145, 283, 336, 342, 447, 468

Woolls-Sampson, Sir Aubrey colonel, British Army 444

Wu Ch'i Chinese general 85, 92, 137, 443, 448, 463

Xenophon Athenian historian 40, 55, 62, 107, 112, 127, 159, 172, 178, 217, 222, 228, 247, 286, 295, 304, 359, 369, 417, 421, 464, 475

Yadin, Yigael general, Israeli Defence Force 345

Yagyu Munenori 40

Yakir, Iona E. army commander, first rank, Red Army 246

Yamamoto Isoroku admiral, Imperial Japanese Navy 152, 175, 226, 356, 379, 381, 423, 480

Yeosock, John general, US Army 466

York, The Duke of (*see* Frederick Augustus)

Zhuge Liang Chinese writer 63, 196, 257, 267, 310, 312, 357, 458

Zhukov, Georgi K. marshal of the Soviet Union 266, 294, 329, 373

Ziska, Jan Bohemian Hussite general 243